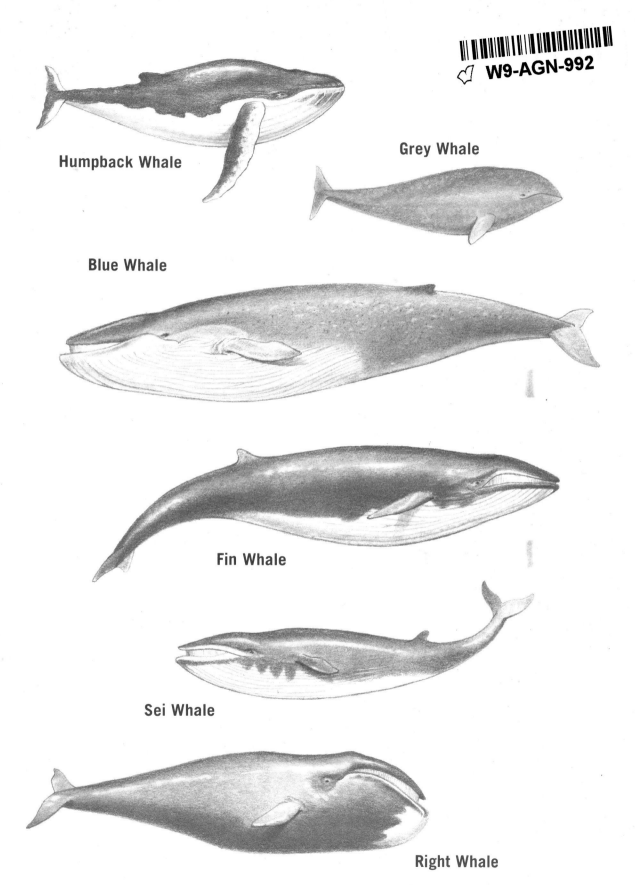

Humpback Whale

Grey Whale

Blue Whale

Fin Whale

Sei Whale

Right Whale

SAUNDERS GOLDEN SUNBURST SERIES IN ENVIRONMENTAL STUDIES

ENVIRONMENTAL SCIENCE

Second Edition

AMOS TURK, Ph.D.
The City College of the City University of New York

JONATHAN TURK, Ph.D.
Naturalist

JANET T. WITTES, Ph.D.
Hunter College of the City University of New York

ROBERT E. WITTES, M.D.
Department of Medicine,
Memorial Sloan-Kettering Cancer Center, New York

W. B. SAUNDERS COMPANY
Philadelphia, London, Toronto

W. B. Saunders Company: West Washington Square
Philadelphia, PA 19105

1 St. Anne's Road
Eastbourne, East Sussex BN21 3UN, England

1 Goldthorne Avenue
Toronto, Ontario M8Z 5T9, Canada

Library of Congress Cataloging in Publication Data

Main entry under title:

Environmental science.

(Saunders golden sunburst series in environmental studies)

Includes index.

1. Ecology. 2. Pollution. 3. Environmental protection.
   I. Turk, Amos.

QH541.E57   1978   301.31   77–24005

ISBN 0–7216–8927–2

Cover photograph of caribou passing through the Gates of the Arctic, by Steven C. Wilson. Courtesy of Entheos Communications, Bainbridge Island, Washington.

Environmental Science   ISBN 0-7216-8927-2

01234 147 987654

PREFACE

Patterns of instruction in environmental science continue to evolve in various directions, particularly toward greater integration of technical and social issues. To this end, the present edition devotes considerably more attention to legal, economic, and social aspects of environmental problems and to their interactions with technical developments. In fact, Unit I of the book, which comprises the first two chapters, introduces these important matters and sets the stage for their specific applications in subsequent chapters.

In addition, the following new features have been added:

(a) Organization into Study Units

The book has been organized into instructional units, each comprising several chapters. They are:

Unit   I: Introduction—The Social Background
Unit II: The Biological Background
Unit III: Human Population
Unit IV: Resources and Energy
Unit V: Rural Land Use
Unit VI: Pollution
Epilogue

This arrangement should offer the instructor considerable flexibility in organizing the course, since the units are somewhat self-contained and need not necessarily be presented in the sequence in which they appear in the text.

(b) Case Histories

A case history at the end of each chapter after Unit I dramatizes the formal material.

(c) Take-Home Experiments

At the end of most chapters, the reader will find a miniature "laboratory manual," which provides directions for a few simple experiments. These experiments are designed to illustrate various topics in the chapter and to require no more, or very little more, equipment and supplies than are commonly available in the household. Some may be used as classroom projects. They should be fun to do.

(d) Glossary

An extensive glossary of terms used in the text is provided at the end of the book.

(e) Use of the Metric System

Since metrication is both inevitable and (we believe) desirable, this edition has been extensively converted to metric units in the text, tables, and figures. In most instances where British units are more familiar to American readers, both units are given. Measurements of heat are shown in calories, with the conversion factor to joules, the official SI energy unit. All the needed conversions are also provided in the Appendix (see below).

(f) Appendices and Supplements to Chapters

Various items of a technical nature not essential for pedagogical development will nevertheless be wanted by some readers as reference material. Examples are chemical formulas of pesticides and calculations of decibel levels. Such material, as well as other helpful information, is given in the several Appendices.

(g) Student Guide

A separate student guide is available to summarize the chapter highlights and to provide sample examination questions and answers.

Acknowledgments

The various reviewers of the first edition had offered valuable suggestions and comments, which helped the present text, even if indirectly, and we thank them all again. In addition, attorney Edward Gallant of Stony Creek, Connecticut, reviewed the legal portions of this edition.

We also owe special thanks to Mrs. Pearl Turk, who gathered and classified source material and photographs. Finally, it was the many able people at W. B. Saunders Company who put it all together.

CONTENTS

Unit 4   Resources and Energy

8

9

10

NUCLEAR ENERGY AND THE ENVIRONMENT

Unit 5   Rural Land Use

11

AGRICULTURAL SYSTEMS

Unit I

Introduction — The Social Background

1

THE ENVIRONMENT – ECONOMICS, POLICY, AND THE LAW

1.1 THE NATURE OF ENVIRONMENTAL DEGRADATION

The pollution of Lake Erie in the United States (Fig. 1.1), the transformation of the sparkling Rhine into a sewer flowing through central Europe, and the despoiling of the deepest lake in the world, Lake Baikal in Siberia, all attest to the polluting potential of industrial development under various economic, social, and political systems. Many segments of society contribute to the increasing environmental insult. Mining and timber operations ravage the landscape. Municipalities dump raw or partially treated sewage into waterways, and use the air as a sink for wastes. The by-products of manufacturing are found everywhere.

National governments, too, have abetted environmental degradation. In the United States, for example, laws which permit and encourage the transfer of public property into the private domain and which aid industrialization have had, until recently, few provisions designed to protect the environment. The Homestead Acts of the nineteenth century, current laws governing mineral rights, such tax incentives as oil, mineral, and timber depletion allowances, and contracts for bombers and space programs, as well as for housing projects and urban renewal, have all neglected consideration of environmental side effects. Oil spills, chemical toxins, radioactive wastes, improperly insulated homes, overpackaging of goods, and the overheating of dwellings in winter and overcooling them in summer all add to the environmental burden.

These various disruptions are not perceived by individuals or by governments as being all of a similar nature,

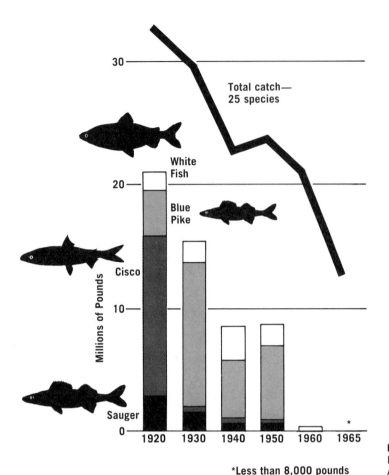

Figure 1.1 Commercial fish catch in Lake Erie. (Adapted from the *New York Times,* April 20, 1970.)

*Less than 8,000 pounds

and consequently no single public policy can apply to them all. It will therefore be helpful to group them into categories, as follows:

(a) Pollution. As later chapters will explain in more detail, pollution is the deterioration of the quality of the environment by the introduction of impurities. Smoke pollutes the air; sewage pollutes the waters; junk cars pollute the land. Such contaminations are usually incontestable; often our direct sensations provide the evidence. The effects of pollution on human welfare or on the economy, however, may be a matter of considerable disagreement. Nonetheless, when issues related to pollution control are being considered, there is at least a reasonable common agreement that everyone is talking about the same contaminants.

(b) Depletion of Resources. When you turn on your air conditioner or electric heater, you are cooling or warming the air around you, but you are not polluting it. However, if the energy you use is generated at a power plant that burns oil or coal, the fuel must first be extracted from the earth. As will be shown in later chapters, the energy of a fuel cannot be recycled. When the useful life of your air conditioner is over, it may be taken to some garbage dump.

If it remains there, the factory that makes your new air conditioner must get its metals, ultimately, from a mine, and the resources of the earth will again have been depleted. Of course, the old unit in the dump could be recovered and recycled, but energy must be utilized to carry it back and reprocess it, and as we have just seen, this expenditure may deplete a source of fuel.

Even if one assumes that all such actions could be carried out without causing pollution, the depletion of fuel and mineral resources is an environmental loss. It would seem reasonable to minimize such losses by policies of conservation and recycling. However, there is less common ground here for making decisions. The first difficulty is that the effect of depletion is more delayed than that of pollution. This difference is not always sharp, but it may be. so perceived, as the following example will illustrate:

Picture yourself walking along a road, and seeing five automobiles pass you. The first is a small, old car, carrying four passengers. It seems to be in poor mechanical conditions, and you note that a haze of blue smoke comes out of its exhaust pipe. The next four are late-model luxury cars, each carrying one passenger. The first car pollutes, but in terms of miles per gallon of gasoline *per passenger,* it conserves fuel. The other four, let us assume, do not pollute nearly so much, but they are certainly wasteful of fuel. The pollution from the first car makes you cough, which means that it affects you as soon as it is produced. The much greater depletion of the Earth's petroleum reserves caused by the other four, however, is easily forgotten, because there will still be gasoline for sale at the service station tomorrow.

Conservation is therefore often seen as a measure whose benefit will be realized later, perhaps only by our children or grandchildren, and not all makers of policy are equally concerned for future generations.

The second difficulty in making decisions about conservation is that there is generally little consensus on the amount of resources available, on future rates of consumption, or on alternative sources of energy and materials. It would be unrealistic, for example, to expect agreement on energy policy between one group who expects oil to run out by the end of the century, coal to last for only a few hundred years, and nuclear energy to be unavailable, and another group who believes that most fossil fuel reserves have not yet been discovered, and that nuclear energy will be plentiful for thousands of years.

(c) Disturbance of the Natural Condition. If you are planning a trip next winter to Africa, or next summer to Scandinavia, and need to know what clothes to pack, you may ask your travel agent, who will consult a book on world climate and tell you what temperatures and how much rain to expect. Of course, the predictions will not pretend to be precise, but the fact that such references exist

"I ask you, what's wrong with the environment?"
(Drawing by Alan Dunn; © 1972 The New Yorker Magazine Inc.)

attests to the relative constancy of global conditions from year to year. You will also be aware that there is some chance of an extreme departure from expectations—there may be a drought when the season should be rainy, or there may be an unusual hot spell or cold snap. Nonetheless, such aberrations have been known in all past history, and they are attributed to natural causes, such as sunspots or unusual geological processes.

Only recently have we begun to confront the question of whether human activities can affect the natural global condition. Can the exhaust from high-flying aircraft or a stable gaseous pollutant that drifts to the stratosphere weaken the thin shield of ozone which filters out harmful solar radiation? Will contamination of the lower atmosphere affect global temperatures, rainfall, or agricultural productivity? Will the pollution of the oceans destroy its living organisms, which account for a large measure of the Earth's photosynthesis? Could such deprivation reduce the oxygen content of the atmosphere? All of these questions have been explored by scientists, and they will be considered in more detail in appropriate sections of this book. The question here is whether such problems have entered the arena of public policy. To politicians, the environmental effects involved seem much more remote and conjectural than the categories of pollution and depletion discussed above. However, many environmental disruptions have a combined effect—they pollute now and they pose a future threat to global systems. Examples are oil spills in the ocean and industrial dust in the atmosphere. In such instances public policy is focused on the immediate pollution problem, but the overhanging uncertainties add a measure of anxiety, and perhaps urgency, to the public response.

1.2 THE ECONOMIC COSTS OF ENVIRONMENTAL POLLUTION

Imagine yourself driving your own new automobile, and consider how this experience affects the quality of your life. Obviously there are both benefits and drawbacks. Suppose an economist wishes to measure these effects in dollars. It is easy to assess the benefit—it is measured by the amount of money you were willing to pay to own and operate the automobile. If its price had been higher than what you thought it was worth, you would not have bought it. If the price had been too low, classical economic theory tells us that a generally increased demand would have pushed the price up until supply and demand were in balance.

The market value of your automobile, together with that of all other goods and services produced by the economy in a given year, is called the **gross national product (GNP).** The GNP is the generally accepted measure of economic activity and, by implication, of economic health. But as a measure of improvements to the quality of life, it is a misleading index, because it neglects many of the negative environmental effects. The cost of some of these effects, such as the decrease in your feeling of well-being caused by traffic noise, cannot be expressed in dollars. Other negative costs, such as damage to vegetation from automobile exhaust, can perhaps be measured, but they are not borne by the manufacturer of the automobile and therefore are not reflected in its price. Under current economic structures, the price of a product reflects mostly the cost of raw materials and labor, the amount of capital investment, and the demand.

The costs that are not accounted for by the manufacturer but that are borne by some other sector of the society, are external costs, or **externalities.** They involve social, including environmental, costs. If the smoke from manufacturing (or from disposal) darkens nearby houses and dirties the clothes of local residents, the costs of more frequent repainting and laundering are part of the total environmental cost of the product. Hidden costs of products whose manufacture involves occupational health hazards

An example of an economic externality. A home in Appalachia collapses as the overburden on which it stands subsides into an abandoned, unsupported deep mine shaft. The price of coal is not high enough to allow the coal industry to pay for the effort necessary to prevent such destruction. (Courtesy of Bureau of Mines, U.S. Department of the Interior.)

include medical bills, loss of work because of illness, decreased feeling of well-being, and increasing deafness.

Whenever a discharge from a factory pollutes a river that must be purified by a town downstream, the cost of water is reflected elsewhere than in the price of the product. Loss of tourist trade because of pollution is another economic burden shouldered by parties other than the manufacturer or the purchaser. Somehow, costs of all these factors should be assessed, for all are part of the total cost to society for the manufacture, use, and eventual disposal of the product. The total cost of all externalities is formidable. As a very rough approximation, the costs of environmental damages to health, of depression of property values, and of losses of crops and materials are *each* somewhere between five and ten billion dollars per year. Of course, such estimates (especially those of damages to health) are subject to large errors, for they represent only known dangers and do not include possible long-term effects.

Many environmental externalities, such as the traffic noise mentioned earlier, are not readily quantifiable, for they involve the deterioration of the quality of human life. As such, they require subjective judgments and the willingness to assign numerical values to qualitative feelings. An extreme example is the human suffering and misery caused by illness, or the value of a human life that ends too soon. Many people feel, on religious grounds, that such costs are beyond our right to judge.

Recreational opportunities, too, are changing as an increasing number of natural areas across the globe are being transformed into farmland and cities, and as others are being destroyed by pollution. Fifty years ago, for instance, on the south shore of Lake Erie, people fished, boated, swam, and explored the beach and marshes. Children chased turtles and lovers chased each other along the shore. Today, swimming there is unsafe, the number of tasty game fish has declined dramatically, the wild secluded beaches have disappeared, and boats ply polluted waters. Alternative recreational opportunities are now available. Swimmers can find artificial swimming pools, children can play in approved parks, people who want to see turtles can drive to the Cleveland Zoo, and lovers can still find places to play. For many, these recreational changes involve poignant losses, but their dollar value is difficult to assess.

One might argue that the industrialization of the past fifty years has brought unprecedented leisure, comfort and health, affording the modern residents of the Lake Erie shore an easier and more relaxing life than their grandparents had. Moreover, the same industrialization that has polluted this area has increased wealth, thereby opening travel opportunities for residents. However, areas like Yellowstone Park, formerly relatively untouched, are now so popular that the natural fauna and flora are threatened by the large crowds of tourists.

On the other hand, many believe that the comforts of modern society are a poor substitute for the joys of nature, and that by losing certain pleasures we are losing an important part of our humanity. The anxieties, tensions, and social problems of the day, this argument runs, might be alleviated if people were able to enjoy natural things. Moreover, the poor do not have travel opportunities; they never get to see Yellowstone or the Olympic National Park.

1.3 THE ECONOMIC IMPACT OF POLLUTION CONTROL

A concerted effort to clean up the environment would benefit all of us in a great many ways, but it would be expensive. The development, installation, and operation of control equipment all cost money. Of course, this cost must be weighed against the external costs of pollution itself, but there is little doubt that pollution control is cheaper than pollution.

Various studies of the probable effect of pollution control on American business have led to the following conclusions: (1) Only a portion of the additional cost of manufacturing need be ultimately reflected in consumer costs and (2) companies that have been on a strong economic footing will continue to prosper even under strict pollution guide-

Swimming in a pool is different from swimming in a natural environment.

lines, but (3) a large proportion of companies which are currently in economic jeopardy will fail. The consequences of economic failure will of course depend on the nature of the company. The failure of a small company in a large industrial area will cause relatively little unemployment; the failure of a company in a single-industry town will create severe local economic problems. (4) As a result of pollution controls and increased consumer prices, U.S. goods will be relatively more expensive than foreign imports. Unless foreign countries impose the same types of pollution control, protective tariffs may be fought for by industrial interests.

All these predictions are based on cost estimates that are very difficult to assess adequately. There are vast discrepancies in the estimates cited by those who seek to control pollution and those who are subject to control. In fact, the prediction that a company will fail if environmental controls are required often turns out to be false. In some cases, pollution abatement systems become part of an overall improvement in the manufacturing process. In others, the facilities are converted to making a new model, or even a new product, whose manufacture produces less pollution.

Some factors tend to make pollution abatement appear more costly than it is. Capital expenditures for pollution control can often be spread over 20 years, but, for income tax advantages, the depreciation can be compressed to as little as five years. When the control devices are long-lasting, industry benefits from such a tax structure.

It is often stated as an "obvious" conclusion that the imposition of pollution abatement will lead to higher costs of production, which, in turn, engender higher consumer prices, and reduced demand for goods, both domestically and abroad. By contrast, it is implied, foreign goods will be comparatively less expensive, and the balance of payments deficit will grow. However, careful estimates* indicate that the costs needed to satisfy Federal environmental legislation were only about 1.1 percent of the gross national product (GNP) in 1973, that they will rise to about 1.5 percent in 1979, level off through 1983, and then decrease.

Environmental programs are frequently accused of being a major cause of unemployment. However, all the analyses seem to indicate that they have probably increased, not decreased, the number of available jobs. This is so because of expenditures stimulated by the air and water pollution control deadlines and by the municipal grants program. Construction of sewers and sewage treatment plants, for instance, is relatively labor-intensive. Each billion dollars spent generates 20,000 to 25,000 jobs at the construction site, a similar number of jobs in manufacturing the equipment and materials required for the facilities, and still more jobs to produce goods and services for the workers, for a grand total of up to 85,000 jobs. Not only are workers needed to manufacture the pollution control equipment and construct the facilities, but also over the longer term many will be needed to operate and maintain the equipment. In the same vein, a National Academy of Sciences study has pointed out the need for more trained mechanics to maintain automobile emissions control systems.

Of course, the overall picture does not predict changes in individual industries. Some, especially those involved in pollution abatement, will prosper. Others will be subject to local recessions. Suppose, for example, that you work for a

"Which would you rather see built on this site? (A) An intercontinental jetport; (B) an atomic powerplant; (C) a mall-type shopping center; or (D) a 3,000-unit middle-income housing development."

(Reprinted from *Audubon,* the magazine of the National Audubon Society.)

PRICE ELASTICITY AND INELASTICITY

If you ride a bus to your job, and there is no other means of getting there, you will not ride less frequently because of a fare increase. Such a utilization, which is insensitive to changes in price, is said to be **price inelastic.** On the other hand, even if you like the taste of caviar, you don't have to buy any more than you can afford, and if the price is too high, you can do without it altogether. Such consumption, therefore, does depend on price, and is said to be **price elastic.**

These economic concepts are important to environmental issues because attempts to manipulate prices or impose taxes to discourage the utilization of environmentally damaging products operate only under conditions of price elasticity.

---

*The Sixth Annual Report of the Council on Environmental Quality, Dec., 1975.

chemical company whose effluents pollute the air and which cannot afford the cost of control equipment. The plant closes down. Everyone now enjoys the clean air, but you are out of a job. If enough others like you feel that they are being unfairly treated, there will be vigorous opposition to programs of environmental control.

Industrial decision-making procedures often affect policies regarding pollution control. For example, in large corporations that operate many plants each, the plant manager is rewarded with an annual bonus, the amount of which increases with the profitability of the plant during a specific year. From a corporate point of view, such a system of incentives encourages each plant to be highly profitable without requiring constant surveillance by the corporate management. By encouraging yearly computation of profits, however, efficient and frugal long-term approaches to pollution control are discouraged, for these may involve very high initial costs. If costs are high for a year or two, the plant's profitability, the manager's bonus, and even his job may all be jeopardized. Thus, the bonus system encourages the manager to opt for some piecemeal devices and to install one this year, another the next year, and so on. Moreover, the stockholders are more likely to be favorably impressed by high profits than by large expenses. Finally, the seemingly obvious solution—assigning to corporate management the detailed tasks of pollution control—is fraught with difficulties, for the different operating procedures of each plant and the various legal standards of each location are best known by the individual plant managers and not by the corporate leaders.

1.4 POLLUTION CONTROL: ALTRUISM, LEGISLATION, AND ENFORCEMENT

ALTRUISM

Altruism is devotion to the interests of others. An altruistic approach to environmental control is the action of an individual person, firm, or government to help preserve the environment voluntarily. If taken by large numbers of individuals, such actions would benefit the environment. For example, if no woman bore more than two children, population size would decrease. From the point of view of limiting population growth, therefore, an altruistic act would be performed by a woman who decided to have only two children, though she wanted more. How have altruistic approaches succeeded, and what is their prognosis? Consumers recycle paper, cans, and bottles, carefully read detergent boxes and study books written to aid concerned individuals in keeping the environment healthy. In spite of

---

WHAT SHOULD COVER THE BABY'S BOTTOM? (A Problem in Household Economics)

The urban baby, or the baby growing up in a temperate or cool climate, cannot run around naked. What kind of diaper is least environmentally disruptive? The choice is between disposable paper diapers and reusable cloth ones. Conventional wisdom ridicules the paper diaper as wasteful, an environmentally noxious product for the lazy parent. A paper diaper is made in a factory, which uses energy and produces pollution. Every paper diaper represents a small portion of a felled tree. Its disposal overloads the sewers or adds to solid wastes.

But what about the environmental effects of the cloth diaper? Diapers must be soaked in soapy water, washed in hot water, and rinsed several times in warm water. If a baby has diaper rash, diapers should be boiled or rinsed even more thoroughly. Moreover, diaper rash precludes the use of rubber pants, so that each time a baby urinates, its clothing and its sheets become wet and all must be washed in hot, soapy water. The costs of the energy and water needed, and the environmental damage caused by soap or synthetic detergents, are difficult to estimate. The environmentally optimal choice depends on the age of the baby, the sensitivity of its skin, its frequency of urination, the availability of water and energy in the locality where the baby lives, and the type of waste treatment facility. The choice is not clear.

all these efforts, however, only a very small proportion of domestic waste is recycled.

The problem is that incentives operate against altruistic approaches. The homemaker who brings empty glass bottles and flattened aluminum cans to the recycling center, dries clothes on the line, and composts household garbage is paying a personal price for environmental improvements that all others share, whether they make the same sacrifices or not. This example points out why it is unrealistic to rely on the good will of individuals to clean the environment.* For altruism to work, the great majority of people must help. Perhaps public education to increase each person's notion of his own self-interest may encourage many individuals to act with care toward the environment.

LEGAL APPROACHES TO POLLUTION CONTROL

Imagine that a factory in your neighborhood is polluting the air and water around you. What can you do? One recourse is to take its managers to court. The Anglo-American legal system has developed judicial rules and procedures governing controversies in which one party, the plaintiff, alleges that he has been harmed by another party, the defendant. If the court judges that harm has indeed occurred, the defendant is often ordered to pay the plaintiff money to offset the damage. Perhaps, if you were the plaintiff you would want not money but an injunction, a legal order to the defendant requiring him to stop doing the wrongful act. Classic English examples were cases in which a plaintiff sought an injunction against a defendant to prevent the latter from hunting on the plaintiff's land.

A plaintiff who fulfills all necessary requirements to pursue a case in court is said to have **legal standing**. Standing for damage suits is granted only if an individual's harm is distinguishable from that of the public in general, or, as the United States Supreme Court† has stated, one is required to have a "personal stake in the outcome of the controversy." The question of standing has rendered the courts inaccessible to many environmental suits. For example, an individual man in Chicago could not have sued Union Oil Company for the oil spill in Santa Barbara simply because of his own personal anguish over the killing of birds. However, if he had owned waterfront land nearby, he would have been able to sue for damage to his own property. Nor can an individual woman sue the Department of Interior for destroying a wilderness area by commercializing it, even though part of her taxes are being used, and she is thus being economically "hurt" by the development. She lacks standing, for her harm cannot be distinguished from that of the public in general.

Sometimes a group is granted standing where an individual is not; some judges are particularly likely to grant standing in environmental cases. Recent statements by

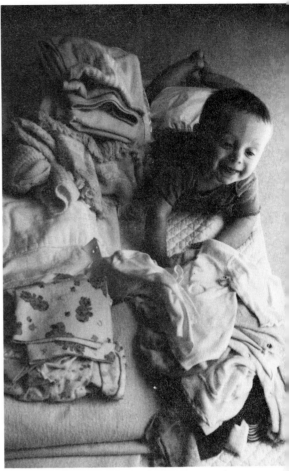

Urban baby with one day's laundry.

*Of course, these arguments are not to be interpreted as recommendations for selfish behavior and therefore should not be used as an excuse to curtail your own altruistic practices. Garrett Hardin's essay, "The Tragedy of the Commons" (found in his book *Exploring New Ethics for Survival*, N.Y., Viking Press, 1972), deals with the problems of relying on altruism for environmental protection.

†Flast *v.* Cohen. 392 U.S. 83 (1968).

LOUIE

(From the *New York Daily News*, April 14, 1972.)

SIERRA CLUB v. MORTON. SUPREME COURT OF THE UNITED STATES 405 U.S. 727 (1972). DISSENT BY JUSTICE DOUGLAS

Inanimate objects are sometimes parties in litigation. A ship has a legal personality, a fiction found useful for maritime purpose. The corporation sole—a creature of ecclesiastical law—is an acceptable adversary and large fortunes ride on its cases. The ordinary corporation is a "person" for purposes of the adjudicatory processes, whether it represents proprietary, spiritual, aesthetic, or charitable causes.

So it should be as respects valleys, alpine meadows, rivers, lakes, estuaries, beaches, ridges, groves of trees, swampland, or even air that feels the destructive pressures of modern technology and modern life. The river, for example, is the living symbol of all the life it sustains or nourishes—fish, aquatic insects, water ouzels, otter, fisher, deer, elk, bear, and all other animals, including man, who are dependent on it or who enjoy it for its sight, its sound, or its life. The river as plaintiff speaks for the ecological unit of life that is part of it. Those people who have a meaningful relation to that body of water—whether it be a fisherman, a canoeist, a zoologist, or a logger—must be able to speak for the values which the river represents and which are threatened with destruction.

---

*Summarized from Boomer *v.* Atlantic Cement Company. Court of Appeals of New York. 26 N.Y. 2d 219 Supp. 2d 312 (1970).

some Supreme Court judges suggest that perhaps valleys and mountains, like corporations, should be granted standing. Some cases are heard because they are considered "public interest" cases. For instance, a factory spewing noxious gases into the atmosphere might be sued by a town and legally declared a "public nuisance." Certain legal scholars declare that a pollution-free and healthy environment is one of the unenumerated rights of the Ninth Amendment, which says, "The enumeration in the Constitution, of certain rights, shall not be construed to deny or disparage others retained by the people."

An adversary approach to environmental protection is fraught with problems. The issue of standing is still in flux. Moreover, litigation is very expensive and slow. Another disadvantage is that lawsuits may be addressed to a very small problem. In addition, a given decision is not generally binding on future offenders. Each violation requires a new case. Environmentalist groups may be pitted against an array of giant corporations and government agencies, and an individual case may cost as much as half a million dollars. Often, however, the threat of a suit and adverse publicity will discourage the defendant.

Finally, even when it is clear that a plaintiff is being damaged, a judge may deny an injunction on the basis that its cost would be unfair to the defendant (that is, the polluter). Consider the following example:*

A group of residents complain that a nearby cement company has created a nuisance in the form of dust and excessive vibrations from blasting and has thereby deprived them of the reasonable use of their property. The total permanent dollar loss to the plaintiffs, estimated from the decline in the value of their properties, is $185,000. The cement plant, however, was erected at a cost of $40,000,000. The judge therefore denies the injunction on the basis of the large economic difference between the damages to the plaintiff and the loss that the company (and the community) would suffer if its operations were shut down. The plaintiffs are therefore awarded money damages, but the cement company is allowed to continue operating.

Can valleys and mountains, like corporations, have legal standing? (Photo of Mount Robson, highest peak in the Canadian Rockies.)

1.5 INCENTIVES FOR POLLUTION CONTROL

Regulations which simply prohibit pollution ignore the resulting hardships that are imposed on individual companies or communities. An alternative approach is to offer economic incentives that will persuade polluters not to despoil the environment. Two major economic incentives have been proposed. First is a subsidy for cleaning up; second is a tax on pollutants. Subsidies would operate as follows. An industrial firm declares that its normal operating procedures would yield a certain amount of pollutants. However, by installing antipollution devices, the amount of pollution would be reduced. Then the government rewards the firm with a subsidy that depends directly on this reduction. Such a system is open to abuse. In many cases, but not all, it may be relatively easy to measure the total amount of

Red Clay Creek, a small stream in Delaware, has fish living in it for the first time in 50 years. NVF Company, founded as a flour and saw mill in 1763, is located on the Creek; it manufactures 45 per cent of the world's vulcanized fiber. The zinc used in the manufacturing process has in the past been dumped into the creek. In the late 1960's, Delaware declared that the NVF fiber plant failed to meet state requirements for maximum acceptable zinc concentrations. Today, NVF recycles 50,000 pounds of zinc chloride per month at roughly the same cost as fresh zinc chloride. (Reprinted with permission from *Environmental Science and Technology*, Oct., 1972. Copyrighted by the American Chemical Society.)

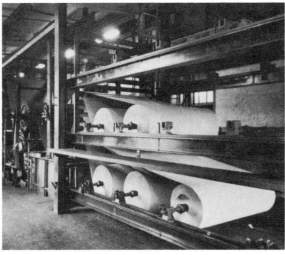

pollution, but very difficult to estimate the amount that would have occurred without controls. Opportunities for corruption and bribery are obvious.

Another approach is the so-called **pollution tax, or residual charge.** Each polluter is charged an amount proportional to the quantity of pollutant emitted. Such a scheme would encourage ecologically sound manufacturing, for it penalizes processes which pollute. Surveys of industries have generally found that this approach is favored by most firms, for it would simultaneously internalize pollution costs for all companies. Of course, standards for particularly hazardous or toxic materials would still need to be established and enforced. The President's Council on Environmental Quality, many environmental groups, and several members of Congress have endorsed this approach. Its main disadvantage is that, like sales taxes and value-added taxes, a pollution tax is passed on to the consumer in a regressive fashion. Thus, under this system, the poor would ultimately pay a higher proportion of their income for pollution control than would the rich.

Ideally, residual charges present the polluter with a choice—either dump waste and pay the fine, or deal with the waste in some other way—treat it, recycle it, store it, or minimize it in manufacturing. If the residual charge is too low, there will be little economic incentive to prevent waste, and pollutants will continue to be dumped untreated into the environment.

Pollution taxes have often been ridiculed as "licenses to pollute," but if such taxes are administered carefully and with advance notice, plants are likely to incorporate production changes designed to prevent pollution. Although the effective implementation of residual charges will involve great care and foresight, as well as adequate monitoring procedures, the idea is supported by many as a general approach to environmental control.

A variation on the pollution tax is the establishment of methods for reducing consumption, and hence, indirectly, production. This approach is much less regressive than the pollution tax itself. Individuals could be given economic incentives, or punishments, to encourage reduction of pollution, waste, and other environmental stresses. Then pollution charges could be levied against individuals. The prices of many consumer goods do not adequately reflect their effect on nature. Three goods which are grossly underpriced if we consider environmental effects are water, electricity, and automobiles. Many of our environmental problems would be alleviated if people refrained from profligate use of water, excessive reliance on electricity, and unnecessary dependence on the automobile. Pollution taxes for these three commodities could greatly decrease their use. For water and electricity, a steeply increasing tax might be levied; the tax should allow a reasonable amount of usage at very low rates, but should tax heavy users severely. Rationing is another approach. If electric costs reflected externalities, individual homemakers might be quite reluctant to use electric heat and air conditioning, and would rarely use an electric clothes dryer for a very small load. A decrease in per capita demand of electricity would obviate the need for so many new power plants.

1.6 ULTIMATE PROSPECTS FOR ENVIRONMENTAL IMPROVEMENT

The kinds of economic incentives and the legal paths discussed in the preceding sections would not lead to a sudden return of the environment to a former, unsullied state. Rather, they would probably lead to a lessening in the rate of deterioration of the environment in some areas, a gradual improvement in other regions, and perhaps an occasional dramatic improvement.

There is no guarantee, however, that an overall cleaning of the environment would be accomplished. Some pessimists hold that natural processes alone or, perhaps more accurately, natural calamities triggered by environmental abuse, will reverse the trend. Others take the position that the environment can be healed only by methods that sacrifice human rights and liberties. The most effective way to control population, for example, is to prevent the birth of babies by methods such as compulsory sterilization after the second child, or compulsory vasectomy at age 35. Even less punitive measures, such as limiting tax relief or free

Female students on Los Angeles campus make an appeal for cleaner breathing room. (From *All Clear,* January, 1970.)

education to the eldest two or three children in a family, though effective, would be inequitable, because the burden would fall on some children and not on others. A fourth child never asked to be born. Why should it be deprived of education because of a parental decision or mistake?

Other examples of harsh but effective policies come readily to mind. The skies would be cleaner if the managers of polluting factories would be jailed, and the streets would be unlittered if, as in Singapore, an individual caught throwing a cigarette on the ground would be fined an amount equivalent to half a year's pay.

Is the human prospect, then, a gloomy choice between environmental degradation, natural disaster, and the loss of individual liberty? Such an unreservedly pessimistic outlook ignores the deep sources of the human will to survive, the great plasticity of human behavior, and the human genius to devise alternatives for bettering its condition. Surely one positive step is the achievement of a better understanding of the relationships between humankind and its terrestrial environment. To this end, the remainder of the book is a beginning.

PROBLEMS

1. **Environmental disruptions.** What are the three categories of environmental degradation? Which one makes the most demands on public response? Which the least? Give reasons for your selections.

2. **Vocabulary.** Define economic externality.

3. **Cost of environmental control.** Suppose you were a reporter for your school newspaper and were assigned to compare the cost of stopping the pollution of the local river with the cost of the economic externalities that the pollution produces. What sources of information would you seek out?

Which category of costs would be more difficult to estimate? Why?

4. **Compensation.** If the burden of environmental control falls heavily on some particular segments of society, do you think it would be fair to provide compensation? Would you favor granting such compensation to workers who lose their jobs? To companies whose costs for environmental clean-up are high? To stockholders whose equities are reduced in value? For each of these segments, what kinds of abuses are possible? What benefits to the environment would such compensations provide?

5. **Economic externalities.** On February 26, 1972, heavy rains destroyed a dam across Buffalo Creek in Logan County, West Virginia. The resulting flood killed 75 people and rendered 5000 homeless. The dam was built from unstable coal-mine refuse or "spoil." A United States Geological Survey had warned of the instability of many dams in Logan County, especially the one at Buffalo Creek. Soil stabilization and reclamation programs directed at mine spoil dams would have added to the cost of producing coal. After the flood, the Bureau of Mines denied responsibility for corrective action, contending that although mine spoil is within the Bureau's jurisdiction, dams are not. (a) How would you relate the concept of economic externalities to this tragedy? (b) Suggest legislation designed to prevent future Appalachian dam disasters. (c) How will your legislation affect the price of coal and its competitive position in the fuel market?

6. **Economic externalities.** What information would you need to assess whether a cotton or a cotton-polyester shirt is more ecologically sound? In your answer, consider that pesticides are used to grow cotton, that the spinning of both natural and artificial fibers has environmental side effects, and that electricity is needed to iron the shirt. The usual blend in no-iron shirts is 35 percent cotton, 65 percent polyester. What do you think were the factors entering into this choice of blend? How would you establish the composition of a shirt that minimizes harm to the environment?

7. **Economics.** Explain why the control of pollution may result in a net increase in the number of jobs. Give examples.

8. **Pollution tax.** Why is it difficult for firms with small profits to borrow money for pollution control?

9. **Vocabulary.** Define price elasticity and inelasticity.

10. **Price elasticity.** The following consumer items involve environmental degradation of some kind: (a) electricity; (b) leopard coats; (c) luxury cars; (d) compact cars; (e) soda in non-returnable bottles; (f) paper napkins. How does the manufacture, use,

or disposal of each harm the environment? Are the prices of each elastic or inelastic? What kinds of legislation would be needed to prevent the harm?

11. **Price elasticity.** For each of the six items in Problem 10, suggest a product that may be substituted. How is ease of substitution related to elasticity of demand for these items? (Can you generalize?) How is price elasticity related to the likelihood that pollution control costs can be passed on to the consumer?

12. **Usage of utilities.** Most gas companies charge less per cubic foot as the amount used increases. A typical rate schedule for a Northeastern city is:

HUNDREDS OF CUBIC FEET PER MONTH	CENTS PER 100 CUBIC FEET
5.0 or less	minimum bill
Next 10	17.0
Next 15	15.0
Next 570	13.2
Next 5400	10.6
Next 94,000	9.5
Next 100,000	9.0
Over 200,000	8.5

How do you think this rate structure influences use of gas in homes? (Roughly 300 cubic feet of gas are sufficient for the monthly operation of a gas stove for a family of four.)

13. **Usage of utilities.** About 10 percent of the gas consumed by a household kitchen stove with a pilot light is used to maintain the pilot light. Water that is simmering is just as hot as water that is boiling vigorously. An egg covered by half a pint of boiling water will harden just as rapidly as in a quart. Outline specific guidelines for a family of four that might cut its consumption of natural gas in half.

14. **Altruistic approaches.** Suggest one action you personally might take to help preserve the environment. How do you assess the importance of your own action? How would you estimate the number of people who would have to join you before the action would have a significant environmental impact?

15. **Damage suits.** How can legal suits against polluters be used? What are the drawbacks of depending on litigation as the primary social tool for environmental protection?

16. **Legislation.** Discuss the following alternative proposals for reducing water consumption. (a) Tax all water usage at a constant rate per gallon. (b) Tax water usage at a progressive rate, that is, allow the tax to rise with increasing use. (c) Shut off all water from 2 to 4 P.M. (d) Shut off all water from 2 to 4 A.M. (e) Ration water at some reasonable level. What is the purpose of the legislation? What are the side effects? Can you propose better legislation?

17. **Public policy.** If you were the mayor of a small town, and a prosperous factory, the largest single employer in the town, were illegally dumping untreated wastes into a stream, what action would you recommend? What if the factory were barely profitable?

18. **Vocabulary.** Define residual charge.

19. **Residual charge.** Do you think a pollution tax on automobile mileage should be proportional to the miles driven, or progressive? Should miles driven on each car be used, or miles per family? Can you think of a progressive tax structure that would encourage more than one car?

20. **Environmental policy.** Rendering plants convert animal wastes into useful products. However, the conversion is very smelly, unless costly odor-control systems are installed. Would you recommend that air pollution ordinances grant special exemptions to rendering plants? Why, or why not?

21. **Housing.** Compare the environmental impact of single-family homes with other types of housing.

BIBLIOGRAPHY

A good introduction to environmental economics is given by:
Donald T. Savage, Melvin Burke, John D. Coupe, Thomas D. Duchesneau, David F. Wihry, and James A. Wilson: *The Economics of Environmental Improvement.* Boston, Houghton Mifflin Co., 1974. 210 pp.

For an emphasis on natural resources, refer to:
Ferdinand E. Banks: *The Economics of Natural Resources.* New York, Plenum Press, 1976. 267 pp.

Issues of public policy are covered by:
Daniel H. Henning: *Environmental Policy and Administration.* New York, American Elsevier Publishing Co., 1974, 205 pp.

Citations of legal cases appear in:
Jerome G. Rose (ed.): *Legal Foundations of Environmental Planning.* New Brunswick, N.J., Center for Urban Policy Research (Rutgers University), 1974. 318 pp.

A somewhat apocalyptic view is given by:
Robert L. Heilbroner: *An Inquiry into The Human Prospect.* New York, W. W. Norton & Co., 1974. 150 pp.

Two well-written and provocative books discussing economy and environment are:
Gerald Garvey: *Energy, Ecology, Economy.* New York, W. W. Norton & Co., 1972. 235 pp.
Edwin G. Dolan: *TANSTAAFL: The Economic Strategy for Environmental Crisis.* New York, Holt, Rinehart, & Winston, 1971. 113 pp.

For economic projections, the following volume is useful:
Environmental Quality, The Sixth Annual Report of the Council on Environmental Quality, Washington, D.C., 1975. 763 pp.

An engaging volume on social policies for the environment is:
Garrett Hardin: *Exploring New Ethics for Survival.* New York, Viking Press, 1972. 273 pp.

A guide to actions individuals may take to protect the environment is:
Paul Swatik: *A User's Guide to the Protection of the Environment.* New York, Ballantine Books, 1971. 312 pp.

A book which has had much impact in persuading others that nature cannot support mankind much longer unless patterns of growth and consumption change is:
Donella H. Meadows, Dennis L. Meadows, Jørgen Randers, and William W. Behrens, III: *The Limits to Growth.* New York, Universe Books, 1972. 205 pp.

A volume dealing with current social issues is:
Michael Micklin: *Population, Environment, and Social Organization: Current Issues in Human Ecology.* Hinsdale, Ill., Dryden Press, 1973.

For a survey of the ways in which state, federal, and international bodies have responded to the growth of environmental problems, see:
Thomas W. Wilson, Jr.: *International Environmental Action—A Global Survey.* Cambridge, Mass., Dunellen, 1971. 382 pp.

A handbook on the legal rights of citizens is provided by:
National Resources Defense Council (Elaine Moss, ed.): *Land Use Controls in the United States.* New York, Dial Press, 1977. 362 pp.

2

THE AUTOMOBILE: A CASE HISTORY IN ENVIRONMENTAL SCIENCE

2.1 THE AUTOMOBILE AND THE ENVIRONMENT

Remember the mythical magic lamps and their genies? One had only to rub the lamp and speak one's wish, and it was done. The "fully equipped" automobile seems like a modern version of those magic lamps. A touch of the finger, hardly even a push, brings warming, or cooling, or music; it opens or closes windows, locks or unlocks the car doors, or even the garage doors. A touch of the toe brings a surge of acceleration, or a swift stop.

The automobile, however, is no isolated fantasy, for it interacts both with the natural systems of the Earth and with human society. You shall see throughout your study of environmental science that small changes in one system may lead to large changes elsewhere. Furthermore, as noted in the previous chapter, environmental stresses do not bear equally on all sectors of society. Here, too, the social and economic problems centering around the automobile are typical of many other environmental issues.

In motion, the automobile and its occupants are not an isolated entity, independent of the world outside. If people are to use the automobile as the vehicle of choice to go anywhere, there must be roads everywhere. Furthermore, since the roads are not parked in some garage at night, they are a permanent fixture of the landscape—a feature of the environment.

Having an automobile to drive and roads to travel on,

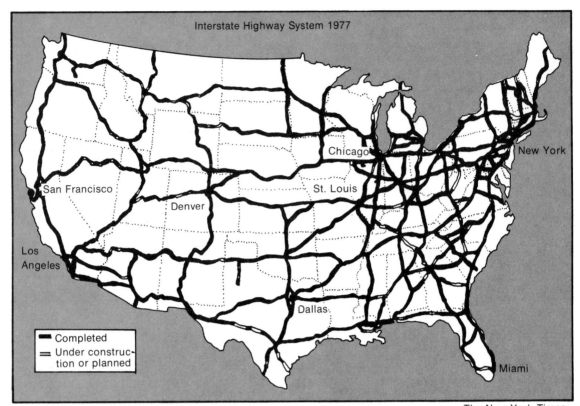

The New York Times

one has less need for other forms of transportation, such as railroads and streetcars. As an added bonus, one need not live near a railroad station or near any transit route established for clustered masses of population. One may choose an isolated farm or woodland, or a small village far from one's place of work. In fact, distance is no longer measured in units of length like miles or kilometers. Instead, it is counted in units of time, such as "three quarters of an hour from the factory," with the understanding that one refers to an interval of privacy and comfort.

The alternative means of transit are therefore used less, sometimes even abandoned, and their rights-of-way given over to ever-lengthening miles of highways. Such freedoms, however, exact their price. The "Sunday drivers," who use the automobile only for sport and recreation, are a privileged few who, like those who ride horses on bridle paths, remind us of our less hurried past. Most of us now have little choice; we have become dependent on the automobile, for there may be no other way to get to work, to shop for food, or to visit friends and relatives. Furthermore, the rural areas that once surrounded our cities have become crowded suburbs.

The interior of a modern car, with the windows closed, is relatively quiet, but it is not soundproof. Since other people, too, must drive to work, traffic is heavy. As tires roll, their treads establish little areas of suction on the road sur-

Automobile traffic.

face, and as the treads lift, air rushes in noisily to fill these tiny vacuoles. The rubber, too, squeaks and squeals. All these small noises combine to make a steady roar outside, which intrudes on the occupants of even the most lavishly insulated cars. Engines, too, are noisy, and so are the horns sounded by impatient motorists. In aggregate, the sounds of the highway are heard all along its length for a kilometer or so on either side and have therefore become another widespread feature of our environment.

An automobile runs because its engine converts the internal energy of gasoline into mechanical motion. This conversion is accompanied by a chemical transformation—combustion. Gasoline is produced in a refinery from petroleum that was drilled out of the earth. It is held in large storage tanks until it is shipped to local service stations; and finally it is pumped through a hose into the automobile's tank. None of these transfers is perfect, and some gasoline vapor escapes in every instance. Nor is the combustion in the engine all that it should be—some of the gasoline misses the flame completely, and some is only partially burned, yielding noxious exhaust vapors.

An automobile does not last forever. When its final mile has been run, it is either dismantled and salvaged, or it is abandoned, sometimes in a mound of other hulks, at other times on some lonely urban sidestreet, rural byway, or empty field. Not only do these discards despoil the landscape but also they deplete our resources, for as new cars are built, minerals must be extracted from the earth to supply the factories with raw materials.

Abandoned railroad tunnel and right-of-way in Colorado.

Chemical manufacturing, petroleum refinery, and storage areas, northern New Jersey. (From New Jersey Dept. of Environmental Protection.)

A Mustang II is shown going through the final assembly process at the Ford Motor Company assembly plant at Dearborn, Mich., in its Rouge manufacturing complex.

Figure 2.1 The automobile cycle.

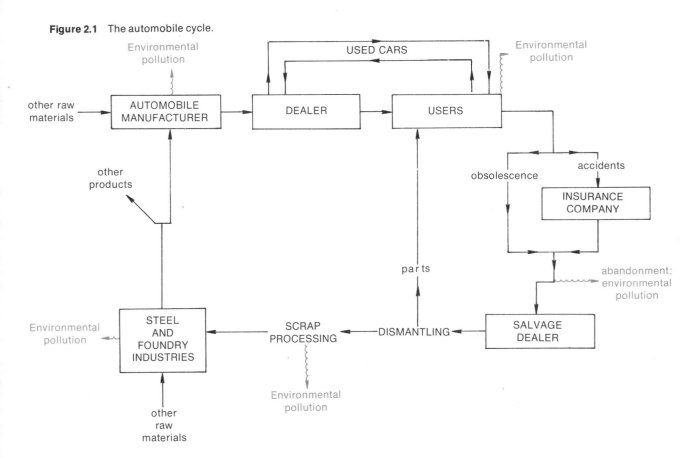

2.2 THE AUTOMOBILE CYCLE

To help identify the stages at which the use of the automobile engenders environmental pollution, let us follow its life cycle, starting at the factory where it is made. The factory ships the automobile to a dealer, who sells it to a customer. This sale is usually only the first of a series, during which the car goes from owner to owner, perhaps by a route that involves the dealer again. The last user eventually discards it, or sells it to a salvage dealer who converts it to scrap, some of which is used by a mill to manufacture steel from which new automobiles are made. Of course, the cycle is not nearly so simple, nor is it complete, for there are other paths that divert the flow of materials. Figure 2.1 shows some more details of the cycle and its interruptions, and makes note of the stages at which significant environmental pollution occurs. For the present, let us focus our attention on the pollution that is associated with the use of the automobile, rather than with its manufacture or disposal.

2.3 PREVENTING POLLUTION FROM AUTOMOBILES: TWO APPROACHES

How can pollution from automobiles be prevented? One approach can be stated as follows:
- The automobile is a product of science and engineering, and the environmental disruptions it causes are therefore a technical problem, for which a technical solution is needed.

The automobile, especially, is remarkably addictive. I have described it as a suit of armor with 200 horses inside, big enough to make love in. It is not surprising that it is popular. It turns its driver into a knight with the mobility of the aristocrat and perhaps some of his other vices. The pedestrian and the person who rides public transportation are, by comparison, peasants looking up with almost inevitable envy at the knights riding by in their mechanical steeds. Once having tasted the delights of a society in which almost everyone can be a knight, it is hard to go back to being peasants. I suspect, therefore, that there will be very strong technological pressures to preserve the automobile in some form, even if we have to go to nuclear fusion for the ultimate source of power and to liquid hydrogen for the gasoline substitute. The alternative would seem to be a society of contented peasants, each cultivating his own little garden and riding to work on the bus, or even on an electric streetcar. Somehow this outcome seems less plausible than a desperate attempt to find new sources of energy to sustain our knightly mobility.

From K. E. Boulding, "The Social System and the Energy Crisis." *Science*, **184**:255–257, April, 1974. Copyright 1974 by the American Association for the Advancement of Science.

Abandoned automobile, pre–World War II model.

Another view is expressed in the following terms:

• Automobiles are used by people and are involved in the social and economic structures of society. Therefore the pollution it causes is a problem requiring social remedies.

Such differences in approach recur in many environmental issues. The two categories of solutions are not necessarily mutually exclusive, for it is possible to combine them in attacking a specific problem.

The technical concepts involved in the control of air pollution from automobiles are not difficult. There are two categories of emissions to be dealt with, as illustrated in Figure 2.2: (a) leaks of gasoline vapor from the engine compartment or from the tank, and (b) pollutants from the engine exhaust. The first category requires, simply, that the leaks be stopped. This objective is met by mechanical means which, in effect, isolate the gasoline vapors in a closed system. The second category—the exhaust—requires a chemical transformation which will complete the oxidation of the gasoline that was incompletely burned in the engine.

It must be noted that the various complications which have arisen thus far have proved to be so difficult that the problem has not yet been fully solved. Furthermore, no one can yet be sure that the problem will ever be solved in a way that is economically practical. The difficulty seems to lie in the fact that each partial solution leads to a new snag. For example, the device that has been in use since 1976, the "catalytic muffler," is rendered ineffective by the lead compounds that are supposed to prevent engine knock. The obvious solution has been nonleaded gasoline. But this

Figure 2.2 Potential sources of air pollutants from an automobile.

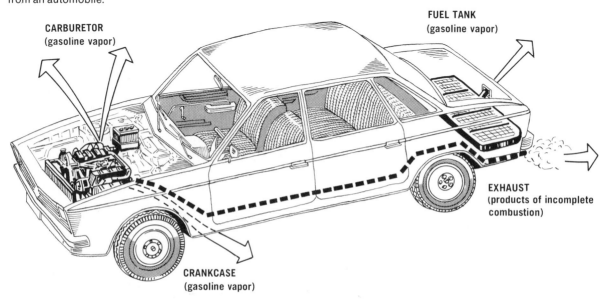

CARBURETOR
(gasoline vapor)

FUEL TANK
(gasoline vapor)

EXHAUST
(products of incomplete combustion)

CRANKCASE
(gasoline vapor)

type of gasoline performs poorly. This deficiency can be counteracted by changing the chemistry of the gasoline. But the new compounds thereby introduced may be carcinogenic (cancer-producing). Alternatively, one could accept the poorer gasoline and modify the engine. But such engines are less efficient and consume more gasoline. Furthermore, if all *these* problems are solved and complete combustion is attained, the trace of sulfur present in gasoline could be oxidized to sulfuric acid. But then a way might be found to remove the acid ... and so on and on.

These circumstances have led some environmentalists to the conclusion that the solution to the problem of air pollution from automobile exhaust does not lie in a technical approach at all but rather in some change of public policy, such as a shift to less polluting or more efficient means of transportation, or even to modes of life that require *less* transportation. Since questions of this kind recur frequently in environmental issues, and since for the most part they have not yet been resolved either by experience or by consensus among experts, it will be helpful to outline the conceptual foundations of the two sets of viewpoints.

Many environmentalists believe that technical solutions to environmental problems are inherently faulty because they address only one aspect of a complex and *interrelated* set of processes, which will never yield to a piecemeal sequence of remedies. Engineers do not agree with this argument. They point out that many complex mechanical systems that have been developed *over a long period of time* work very well indeed. To appreciate this position, you need only lift the hood of a modern automobile and look at the maze it covers. The engine compartment is in fact a

Under the hood of a modern car (1977 Cadillac).

complex system of interrelated, not independent, components that provides much more than the simple transportation of 50 or 60 years ago. No one invented—in fact, no one could have invented—a 1978 automobile in 1918. It is the product of many inventions, many trials and errors, by many scientists and engineers, in many countries, over many years, yielding many *thousands* of models, most of which have long since been abandoned. Better, almost, to say that the automobile has *evolved*, rather than been developed. Furthermore, it works (most of the time), with all its interrelated parts functioning in harmony. The reason that the control of air pollution from automobile exhaust has not yet been perfected, this argument continues, is that the time available for its development has been too short. Many different approaches must be tried as part of the total system, and the poor designs must be abandoned as new ones evolve, so the matter cannot be rushed. Eventually, technology will succeed, and there will be a complete, efficient, pollution-free system just as there is a self-starter, temperature control, and automatic transmission.

Such differences in outlook are hardly confined to the automobile. As we shall see, they pervade many aspects of environmental issues. In some cases, however, the technical argument has another edge. In the nuclear controversy, for example, it cannot be held that the design of nuclear plants has evolved, like the automobile, through thousands of models. The student should therefore be alert to the appropriateness of such arguments in their proper contexts.

2.4 OTHER ENGINES, OTHER FUELS

The piston-driven internal combustion engine powered by gasoline is not the only option available for the automobile. There are also gas-turbine engines, diesel engines, steam engines, and electric cars powered by batteries. Why are such alternatives so conspicuously absent from our highways? The answers lie in a complex of technical, social, and economic issues. Some insight into the complexity of the problem can be gained by considering one example among those cited above: the electric car. The electric car is truly nonpolluting as it drives along, but the accommodations it would require could have complex economic and social consequences. Some of them are described in the following paragraphs.

Need for More Central Power Plants. The energy for electric cars would necessarily result in increased output by central power plants, which would be the source of the power required for recharging batteries. Furthermore, there are many more steps in the power plant sequence, each of them suffering some loss of useful energy. Power

plants are experiencing difficulty now in keeping up with the demand, especially during the summer (air conditioning) season. The required increase in capacity would take some time to establish, so the changeover would have to be gradual. Furthermore, power plants create their own pollution. This fact would be strong persuasion in favor of nuclear power plants, which could be responsible for a different set of problems. In general, however, it is expected to be easier to control emissions from a small number of large power plants than from millions of automobiles. Therefore the overall result should be a decrease in air pollution.

Economic Depression in the Oil Industry. It is difficult to predict accurately how much loss would be suffered by the oil industry; this would depend in part on how many of the new central power plants used oil. A fair guess is that total sales would be cut in half. The result would necessarily be pockets of unemployment and depression, and the shift to new industries would involve dislocations.

Changes in Automobiles and the Automobile Industry. It is not feasible to manufacture electric cars that are powerful enough to resemble the heavy, luxury-type gasoline-driven automobile. It is therefore likely that they would resemble the new "mini" sub-compacts. There would be fewer "extras" sold because there would be less room for accessories in the smaller cars. This loss of profit for the manufacturer, added to the fact that battery components are expensive, might make the small electric cars cost about as much as or more than the larger gasoline-driven ones. The entire "parts" industry would have to change. Instead of furnishing spark plugs, mufflers, carburetors, and so forth, it would have to supply motor commutators, electrical controls, and the like. During the changeover, which would take years, both kinds of parts would be needed.

Changes in Service Stations. Mechanical repairs would partially give way to electrical repairs. Gasoline sales would have to be replaced by battery sale, rental, exchange, and recharging services.

Changes in Highway Design. The current high-speed interstate highway system is designed for the rapidly accelerating gasoline-driven automobiles. The requirement for fast pickup is especially critical when an automobile enters the stream of swiftly moving highway traffic. The existing acceleration lanes would be insufficient for safe entry of the slower electric cars. This difference might prove to be especially difficult during the transition period when both types of vehicles were on the road.

Taxation. The federal government and the states gain revenue from gasoline taxes. The money from these taxes is generally earmarked for the building and maintenance of roads; thus, the people who pay for the highway system are those who use it. With the use of electric cars and the consequent decrease in gasoline purchases, this income would have to be replaced, most probably by an increase in

Electric car. (United Press International photo.)

"This may be your last chance to acquire a superpowered, oversized, hyper-polluting gas guzzler. Don't blow it." (Drawing by Mirachi, © 1977 The New Yorker Magazine, Inc.)

tax on electricity. This increase, however, would be imposed on everyone, and the nondriver would be paying money that would primarily benefit the operators of automobiles. Moreover, the resulting increased cost of electricity would influence people to avoid its use where possible. This would result in the use of fewer electric stoves and electric hot water heaters, for example. Instead, gas and oil would be used, leading to the same kind of pollution that the electric car was designed to eliminate.

Are we to conclude from all these circumstances that the internal combustion automobile using gasoline is to be with us for the forseeable future? Not necessarily. First of all, there are circumstances where the problems cited above do not apply or are not significant and where the electric car offers real benefits. In enclosed spaces, such as warehouses or airport terminal buildings, for example, the *absolute* absence of exhaust pollutants is an overriding advantage. Vehicles used in urban routes at modest speeds and with frequent stops within a limited range, such as light delivery trucks, small commuter buses, or taxis, constitute another reasonable application. Such vehicles spend a significant portion of the time either stationary or coasting; these are conditions in which gasoline-powered engines waste energy but batteries do not. Such a combination of place and function is called a **niche,** a term much used by ecologists in biological contexts. One may say, for example, that a postal delivery route offers a niche for electric vehicles.

Finally, fuels derived from petroleum constitute a much more limited supply of energy than other sources which can be converted to electricity. The niche, therefore, may well grow larger. Such possibilities illustrate the fact that environmental science is concerned with a broad range of phenomena, including natural systems and technology, as well as social and economic processes. Furthermore, these phenomena are by no means independent of one another. Although their relationships are complex and often hidden from view, a change in one sector may produce a disturbance where it is least expected.

PROBLEMS

1. **Automobile cycle.** Trace the cycle of an automobile that the original owner trades in to a dealer. You then buy it as a used car, and later sell it to a friend who drives it until he has no further use for it. He strips it himself and sells the salvaged parts to other owners and the hulk to a junkyard. At the dump the combustible material is burned out, and the compressed hulk is shipped to a steel mill. Identify the points at which pollution occurs.

2. **Social and economic factors.** Suggest some of the social and economic problems that would accompany a transition from an automobile-based transportation system to a cheap and efficient system of public transportation.

3. **"Technological fix."** It has been pointed out that complex mechanical systems that have evolved over a long period of time and that have been tested in many models work very well. To which of the following types of systems is this argument most applicable? To which is it least applicable? Defend your answers. (a) Airplanes; (b) space vehicles; (c) refrigerators; (d) nuclear plants; (e) automobiles.

4. **The electric car.** In each of the following applications, suggest whether the preferred choice would be the electric car, the gasoline-driven car, or whether the two would be equally satisfactory: (a) A family of four drives from their home in Columbus, Ohio, to visit Yellowstone National Park. (b) A public-opinion polling service establishes a route to interview residents of a large suburb. (c) An employee of an urban police department cruises the side streets to issue summonses to illegally parked cars. (d) A sales agent for a manufacturer of citizens-band radios is assigned to cover the state of Texas.

BIBLIOGRAPHY

A nontechnical book that pleads the case against the automobile is:

John Burby: *The Great American Motion Sickness.* Boston, Little, Brown and Co., 1971. 408 pp. (Also available from Consumers Union, Mt. Vernon, N.Y.)

Two technical books that deal with the relationships between the automobile and the environment are:

Robert U. Ayres and Richard P. McKenna: *Alternatives to the Internal Combustion Engine: Impacts on Environmental Quality.* Baltimore, Johns Hopkins University Press, 1972. 324 pp.

D. J. Patterson and N. A. Henein: *Emissions from Internal Combustion Engines and Their Control.* Ann Arbor, Mich., Ann Arbor Science Publishers, 1972. 355 pp.

A report that considers the problem from the viewpoint of manufacturers of motor vehicles is:

Policies to Abate Pollution from Motor Vehicles: An Evaluation of Some Alternatives. Princeton, N.J., Mathematica, Inc., 1975. 245 pp.

A brief article that offers a good chemical review of the problem is:

Thomas R. Wilderman, "The Automobile and Air Pollution." *J. Chem. Educ.,* **51**:290–294. 1974.

Finally, many reports on the subject are available from the U.S. Environmental Protection Agency, Washington, D.C. One of them is:

The Automobile Cycle: An Environmental and Resource Reclamation Problem, 1972. 115 pp.

32

Unit II

The Biological Background

THE
ECOLOGY
OF
NATURAL
SYSTEMS

3.1 INTRODUCTION

Environmental science is a study of systems. A system is a collection of parts that are interrelated or that act together in some way. Think of the human body. The heart, liver, brain, and stomach are all separate organs, yet none of them can operate properly unless the entire system is functioning. Our modern world contains many·types of systems. Some, like a nuclear power plant, are designed, engineered, constructed, and operated by people; others, such as a wheat field or a commercial forest, are managed by people but are simultaneously dependent on the natural growth of living organisms. Still others, like the alpine tundra, or the deep sea ocean floor, operate almost independently of human intervention. The systems studied in environmental science vary widely in form and function. Yet they all share some similarities. Each one is far more than an independent collection of parts, for the individual components are interconnected in such a way that an event affecting one component often affects the whole. Thus, the bursting of a single pipe carrying cool water through a nuclear power plant triggers a series of events that shuts down the entire plant. A slight rise in the average temperature of the water in a mountain stream may alter the number of species living in the stream. The interactions among the components of a complex system are often so delicate and subtle that it is impossible to predict how a single event will affect the whole. If we disturb an environment, say by changing its temperature, sometimes it may regulate itself so as to restore normality. In other situa-

Alpine tundra in early spring in the Colorado Rockies.

tions, a seemingly small perturbation throws the system permanently out of balance. An understanding of these internal regulatory mechanisms is fundamental to environmental science.

Moreover, environmental science studies the complex manner in which one system interrelates with others. No single environmental system operates independently, for each uses energy from outside itself and each exchanges at least some raw materials with other systems.

Plants and animals occurring together, plus that part of their physical environment with which they interact, constitute an **ecosystem.** An ecosystem is defined to be nearly self-contained, so that the matter which flows into and out of it is small compared to the quantities which are internally recycled in a continuous exchange of the essentials of life. The dynamics of the flow of energy and materials in a given geological environment, as well as the adaptation made by the individual and the species to find a place within the environment, constitute the subject matter of the ecology of natural systems.

Ecosystems differ widely with respect to size, location, weather patterns, and the types of animals and plants that live in them. A watershed in New Hampshire, a Syrian desert, the Arctic ice cap, and Lake Michigan are all distinct ecosystems. Common to them all is a set of processes. In each there are plants which use energy from the sun to convert simple chemicals from the environment into complex, energy-rich tissues. Each houses various forms of plant-eaters, predators who eat the plant-eaters, predators who eat the predators, and organisms that cause decay.

Throughout this book we shall sometimes refer to ecosystems as geographically distinct areas with unique characteristics. But ecosystems are of course interconnected, so we shall also discuss the nature of their linkages.

3.2 ENERGY

Energy is the capacity to do work or to transfer heat. In ordinary speech we refer to "physical work" or "mental work" to describe a variety of activities that we think of as energetic. To the physicist, however, "work" has a very specific meaning: work is done on a body when a body is *forced*

There are many different forms of energy.

A moving freight train has kinetic energy.

This pile of coal can burn to produce heat. It has chemical potential energy.

This campfire is producing heat energy.

A stone perched high above the floor of the Grand Canyon has potential energy.

37

to move. Merely holding a heavy weight requires force but is not work, because the weight is not being moved. Lifting a weight, however, is work. Climbing a mountain or a flight of stairs is work. So is stretching a spring, or compressing a gas in a cylinder or in a balloon, because all of these activities move something.

When heat is transferred to a body, its temperature rises or its composition or state is changed, or both. For example, when heat is transferred to air, the air gets warmer; when ice is heated enough, it melts; when sugar is heated enough, it decomposes. All of these changes require energy.

Now imagine that you see an object lying on the ground. Does it have energy? If it can do work or transfer heat, the answer is yes. If it is a rock lying on a hillside, then you could nudge it with your toe and it would tumble down, doing work by hitting other objects during its descent and forcing them into motion. Therefore, it must have had energy by virtue of the potential to do work that was inherent in its hillside position. If it is a lump of coal, it could burn and be used to heat water, or cook food, and therefore it too has energy. If it is an apple, you could eat it and it would enable you to do work and to keep your body warm, and so an apple has energy (see Table 3.1).

Of course, if you roll an apple down a hill, it has energy as a falling body. It is proper to assign a quantity of energy to an apple as a food (one small apple contains 64 Calories) rather than as a falling body because people eat apples; they don't build power plants in orchards to generate electricity from the falling fruit. This means that the energy you assign to a body depends on the process you have in mind. A person who looks at a warm lake and says, "There is energy in all that water," may be thinking of its ability to spill over a dam and force turbines to generate electricity, or its ability to melt ice, or its potential for use in a fusion reactor (see Chapter 10), where some of its hydrogen would be converted to helium with accompanying release of energy. Therefore, we refer to the energy of a body only in relation to a specific process. The potential energy of the water to be used in the hydroelectric plant is the *difference* between its energy at the foot of the dam and its energy at the top of the dam. It doesn't matter what absolute value we assign to any one of the two states—it is the energy difference between them that counts.

Let us return to the apple. The apple has 64 Calories because that is the amount of energy released when it is converted by oxidation to carbon dioxide and water. It does not matter whether this oxidation is performed by metabolism in a human being, or in a worm, or by combustion in a fire. It is only the difference between the final and initial states that matters. But if the apple is left in a closed vessel where there is insufficient oxygen to convert it all to carbon dioxide, it may ferment to yield alcohol. It will thus have reached a different state and the amount of energy

TABLE 3.1 Energy: What It Can Do

	HEAT A BODY ABOVE THE SURROUNDING TEMPERATURE	WORK
Apple	Eating it helps you to maintain your body temperature at 37°C (98.6°F), even when you are surrounded by air at 20°C (68°F)	Eating it helps you to be able to pull a loaded wagon up a hill.
Coal	Burning it keeps the inside of the house warm in winter.	Burning it produces heat that boils water that makes steam that drives a piston that turns a wheel that pulls a freight train up a hill.

transferred will be different—only about two Calories. Of course, the alcohol thus produced may also be considered to have energy, because it could be burned completely to carbon dioxide and water, releasing another 62 Calories. So, either path yields 64 Calories (Fig. 3.1).

Energy is essential for all living organisms because many biochemical reactions require energy. In complex ecosystems such as those that exist today, all organisms are interrelated with other organisms by the food- (energy-) gathering process. Therefore, the study of the energy relationships within an ecosystem is of primary interest to the ecologist.

Refer to Appendix A and to the glossary for the definition of calorie and for an explanation of the difference between large Calorie (capital C) and small calorie. The energy released when an apple falls 4.5 meters is only about 1/1000 Calorie, Isaac Newton's head notwithstanding. The Appendix also explains other energy units.

3.3 ENERGY RELATIONSHIPS WITHIN AN ECOSYSTEM

Plants are able to trap the sun's energy and transform it into chemical energy to build molecular structures such as those of sugars, starches, proteins, fats, and vitamins.

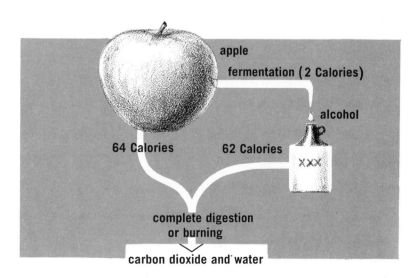

Figure 3.1 Energy from an apple.

TABLE 3.2 Ecological Classifications of Organisms Based on Energy Flow

ECOLOGICAL CLASSIFICATION	TROPHIC LEVEL	LEVEL OF CONSUMPTION	EXAMPLES
Autotroph	First	Nonconsumer	Trees, grass
Heterotroph	Second	Primary consumer	Grasshopper, field mouse, cow
	Third	Secondary consumer	Praying mantis, owl, wolf
	Fourth	Tertiary consumer	Shrew, owl

A *carnivore* is an organism that eats the flesh of animals (living or dead). Thus carnivores occupy the third or higher trophic levels. The definition of *predator* is ambiguous. Some authors consider predator to be synonymous with carnivore, although this definition would include carrion-eaters, and thus would eliminate the sense of "attack" or "pillaging" which is the essence of the older meaning of predation. This text, as well as many others, uses a much broader definition: A predator is an organism that eats other living organisms. By this definition, a cow is predator of grass.

For this reason plants are called **autotrophs**,* meaning self-nourishers. All other organisms obtain their nourishment (energy) from other sources and are called **heterotrophs** (other-nourishers). This broad classification includes such widely diverse species as cows, grasshoppers, mountain lions, sharks, maggots, and amoebae. The way to classify heterotrophs depends on the type of study one wishes to conduct. A comparative anatomist would make separate categories based on evolutionary and morphological similarities. Ecologists focus their attention on function, specifically on the position of the organism in the energy flow, and are therefore interested in levels of nourishment, or **trophic levels.** Autotrophs occupy the *first trophic level*. All heterotrophs that obtain their energy directly from autotrophs are known as **primary consumers** and are said to occupy the *second trophic level*. They may be as different from each other as a grasshopper is from a cow, but both have similar ecological functions: they are grazers. Praying mantises eat grasshoppers, and owls eat field mice. Therefore, both of these predators are **secondary consumers,** for they obtain energy from the plants only indirectly, in two steps, and are said to occupy the *third trophic level*. Let's take another step: shrews eat praying mantises, and martens eat owls; therefore, both are tertiary consumers. Owls who eat shrews are quaternary consumers, and martens who eat the owls who have eaten shrews are still another step removed from the original plant. Now just a minute—how can an owl be both a secondary and a quaternary consumer? The answer is that these categories are not mutually exclusive; a hungry owl doesn't care about human definitions, but dives at whatever is little and running about, regardless of how we classify him.

In very simple ecosystems the flow of food energy progresses through a food chain in which one step follows another. In most natural systems, such as the one partially outlined above, the term **food web** is a more accurate description of the observed interactions (Figs. 3.2 and 3.3). Food webs are further complicated by the presence of **om-**

*An auto- (self) troph (nourish) obtains energy by *itself* from the sun. A hetero- (other) troph obtains energy by eating *other* organisms.

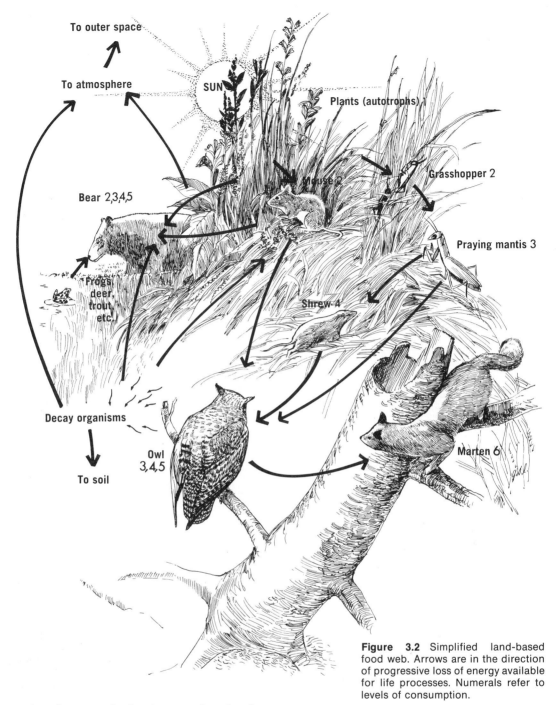

To outer space

To atmosphere

SUN

Plants (autotrophs) 1

Bear 2,3,4,5

Mouse 2

Grasshopper 2

Praying mantis 3

Frogs, deer, trout, etc.

Shrew 4

Decay organisms

Owl 3,4,5

Marten 6

To soil

Figure 3.2 Simplified land-based food web. Arrows are in the direction of progressive loss of energy available for life processes. Numerals refer to levels of consumption.

nivores, species that eat both plant and animal matter. Bears, rats, pigs, people, chickens, crows, and grouse are all omnivores. These species are often very difficult to place within any simple scheme. When a grizzly bear eats roots and berries, he is a primary consumer; when he eats a deer, he is a secondary consumer; eating a frog makes him a tertiary consumer; and eating a trout qualifies him as a quaternary consumer.

Of course, individuals often die without being immediately consumed, but instead remain to decay. The process

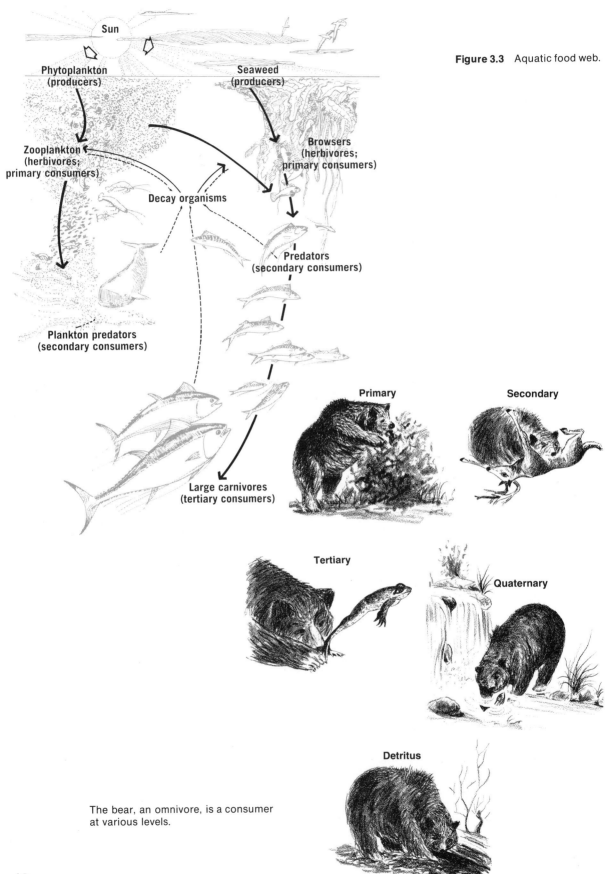

Sun

Phytoplankton
(producers)

Seaweed
(producers)

Zooplankton
(herbivores;
primary consumers)

Decay organisms

Browsers
(herbivores;
primary consumers)

Predators
(secondary consumers)

Plankton predators
(secondary consumers)

Large carnivores
(tertiary consumers)

Figure 3.3 Aquatic food web.

Primary

Secondary

Tertiary

Quaternary

Detritus

The bear, an omnivore, is a consumer
at various levels.

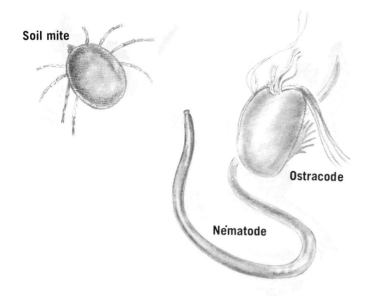

Soil mite

Ostracode

Nematode

Land snail

Figure 3.4 Saprophytes.

of decay is carried out in many widely varying types of species, such as fungi, bacteria, soil mites, nematodes, ostracodes, and snails. These organisms, both plants and animals, are also classified by function and are called **saprophytes** (Fig. 3.4). Since saprophytes are also subject to predation, and some are omnivorous, the food web within a forest floor or on a lake bottom is quite complex. All the non-living organic matter in an ecosystem is known collectively as **detritus.** This debris is consumed by a group of organisms known as **detritus feeders.** Thousands of different species of plants and animals, and billions of individuals, feed on detritus. Although the complex interactions among the organisms are poorly understood, we do know that all stable ecosystems depend on the detritus food web to maintain stability.

One characteristic of ecosystems must be reemphasized: Subclassifications are helpful in focusing attention on a particular aspect of study, but the lines are not sharply drawn in nature. Consider the bear (clearly not a decay organism) who rips open a rotting log in search of termites. In consuming them, he derives energy. Some energy has now been removed from the detritus food web and has entered the food web of the larger animals.

Energy is continuously received from the Sun at a constant rate. Ultimately this heat is radiated off into space. Because we observe that the average annual temperature of the Earth remains relatively constant from year to year, we know that the energy gain and loss of the Earth must be in balance. The condition in which the inflow equals the outflow is called **steady state.** The flow and ultimate fate of energy from the Sun is represented schematically in Figure 3.5.

The chemical energy available for life processes con-

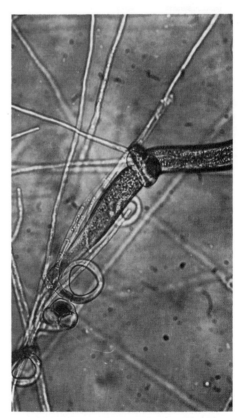

The thicker, wormlike creature in this photograph is a nematode that has been captured by the thin predaceous fungus *Dactylella dreschleri* (× approx. 560). (Courtesy of Dr. David Pramer.)

43

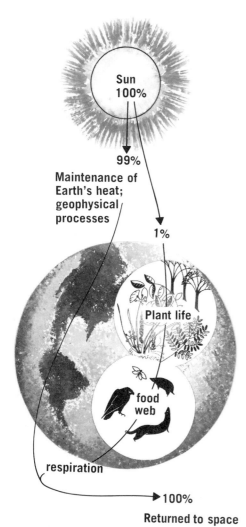

Figure 3.5 The flow of energy on Earth.

stantly decreases through the food chain.* To understand the workings of an ecosystem, one must unravel the energy flow through the food web. Therefore, the ecologist is interested in the total production of organic tissue and the total consumption, which can be measured as respiration. The ratio of total (gross) production to total respiration is an indicator of the energy balance of the system; it tells us whether the system is growing, aging, or maintaining a steady state. Another important variable that is related to both the present and past production/respiration ratios is the **biomass** — the total mass of organic matter present at one time in an ecosystem. Finally, the *biomass at each trophic level* is of fundamental importance. For terrestrial ecosystems, a large primary production is required to support a small weight of predators, so that a diagram of biomass of a self-contained ecosystem can usually be represented as a pyramid (Fig. 3.6).

We shall now follow the energy flow through a simple food chain. Consider a plant that receives 1000 Calories of light energy from the Sun in a given day (Fig. 3.7). The efficiency of conversion of the Sun's energy to chemical energy depends on the species of plant and the conditions of growth but, in any case, most of the energy is not absorbed; instead, it is reflected or transmitted through the tissue. Of the energy that is absorbed, most is stored as heat and used for evaporation of water from leaf surfaces and other types of physical processes. Of the energy that is used in the life processes of the plant, most is expended in respiration. The remainder, about five Calories, is stored in the plant tissue as energy-rich material, suitable as food for animals.

Now suppose a herbivore, say a doe, eats a plant containing five Calories of food energy. She would dissipate about 90 percent of the energy received to maintain her own metabolism as well as to retain some energy for muscle action for movement. As a result, the doe would store only about ½ Calorie in the form of body weight. A carnivore eating the doe is likewise inefficient in converting food to body weight, so the energy available to each succeeding trophic level is progressively diminished. The energy advantage of the herbivores is one important reason why there are so many more herbivores than carnivores. It is obvious that humans, who can occupy primary, secondary, or tertiary positions in the food chain, use the Sun's energy most efficiently when they are primary consumers, that is, when they eat plants.

3.4 NUTRIENT CYCLES

Energy alone is insufficient to support life. Imagine, for example, that some aquatic plants such as algae were sealed into a sterilized jar of pure water and exposed to ad-

---

*See Chapter 8 for a discussion of the First and Second Laws of Thermodynamics.

equate sunlight. The process of photosynthesis which utilizes the atmospheric carbon dioxide and released oxygen would not be balanced by the plants' own respiration, and soon the plants would starve for lack of carbon dioxide. If the experimenter maintained the proper atmosphere through some gas-supply and exhaust lines, the plants still would not survive. They would starve for lack of the chemicals necessary for life. Suppose, then, that the experimenter fertilized the jar with the proper inorganic chemicals, which had been sterilized to insure that no additional living organisms entered the system. The plants would grow, and the cells which died in the normal processes of life would accumulate. Soon there would be no more room to introduce new fertilizer, and unless the jar were enlarged, the fertilization would have to be stopped and the algae would die. The jar might be stabilized indefinitely, however, if some plant-consumer, a snail for example, were introduced, and the pure water were replaced by pond water to supply inorganic nutrients and saprophytes. Assuming an initially balanced ratio of snails to algae, the jar would now perhaps become a balanced, stable ecosystem in which the nutrients would be continuously recycled. The algae use water, carbon dioxide, sunlight, and the dissolved nutrients to support life and build tissue. Oxygen is one major waste product of algae and at the same time is an essential requirement of snails and other heterotrophs.

In general, plants produce more food than they need. This overproduction allows animals to eat parts of the plants, thus obtaining energy-rich sugars and other compounds. These compounds, in turn are broken down during respiration, a process that uses oxygen and releases carbon dioxide. Other animal waste products, such as urine, though relatively energy-poor, are consumed by microorganisms in the pond water. The waste products of the microorganisms are even simpler molecules which can be reused by the algae as a source of raw materials. The consumption of dead tissue by aquatic microorganisms is essential both as a means of recycling raw materials and for waste disposal. The interaction among the various living species permits the community of the jar to survive indefinitely. In fact, sealed aquariums in which life has survived for a decade or more can be seen in some biology laboratories. Such models, however, are greatly simplified pictures of a natural ecosystem. In fact, natural systems are so complex that they are not yet fully understood in all their details.

The biosphere itself can be considered to be a sealed jar, for although it receives a continuous supply of energy from outside, it exchanges very little matter with the rest of the universe. Thus, for all practical purposes, life started on this planet with a fixed supply of raw materials. There are finite quantities of each of the known elements. The chemical form and physical location of each element can be changed, but the quantity cannot.*

150 lbs. human

3,300 lbs. beef

27,000 lbs. alfalfa

Figure 3.6 Food pyramid. The mass shown in each box is the amount required to produce the mass of tissue in the box above it. The areas in the boxes are proportional to the masses.

*This statement does not apply to radioactive elements.

45

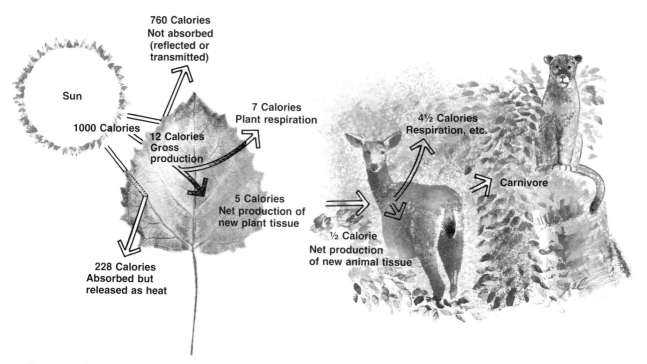

Figure 3.7 Energy flow through a simple food chain.

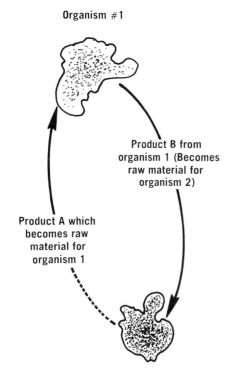

Figure 3.8 Generalized food cycle.

A generalized nutrient cycle can be visualized as follows (Fig. 3.8): Suppose substance A is a vital raw material for a certain species of organism 1, and substance B is one of the organism's products. Substance B must then be a vital raw material for a second organism, which then produces a new product. This alternation of products and the organisms that consume them must continue until the cycle is complete; that is, until some organism produces our original substance A. When averaged over a long period of time, each product must be produced and consumed at the same rate; consumption must balance production. If any substance in the cycle, for example, substance B, were produced faster than it was consumed, it would begin to accumulate. The population of organism 2 would then increase in response to its greater food supply, and it would begin to consume more of substance B. Eventually the consumption of B would be increased enough to match its production, and the cycle would thus have regulated itself. Conversely, if B were produced at a rate *less* than the demand for it, the population of organism 2 would decline because of this insufficiency. The resulting decreased demand would enable production to catch up with consumption, and balance would be reestablished. Any permanent interruption of the cycle, for example by extinction of one of the species of organisms and the failure of any other organism to fulfill its function, would necessarily lead to the death of the entire system.

The substances produced by living organisms are widely varied. First, there are wastes associated with the act of living. Plants produce more oxygen than their life processes require; animals exhale carbon dioxide. However, organisms themselves can be viewed as products. A tree produces leaves, twigs, a trunk; a lion produces bones, a mane, a tail. All these parts of an organism, indeed the entire organism, when it is either living or dead, is a product which can be consumed by other organisms.

When studying nutrient cycles, it is helpful to separate those elements that can be found in large quantities in the air from those which are dissolved in water or located in the soil or the Earth's crust. The continuous state of turbulence of the atmosphere insures a constant composition (omitting water vapor and air pollutants) all over the globe. Moreover, gases in the air are relatively quickly assimilated and released by living organisms. The net result is the existence of a large, constant, readily available pool of nutrients. This does not mean an infinite, unchanging pool, for atmospheric composition has, in fact, changed considerably in the course of time.

Of all the known elements, about 40 are needed by living systems to maintain life. Many of these 40 are needed in only trace amounts, while others such as carbon, oxygen, hydrogen, and nitrogen constitute a large proportion of the mass of a system.

OXYGEN CYCLE

Oxygen atoms are present on Earth in the following forms: (a) As molecular oxygen, O_2, in the atmosphere; (b) in water, H_2O; (c) in gaseous carbon dioxide, CO_2; (d) in many organic compounds, such as sugars, starches, and

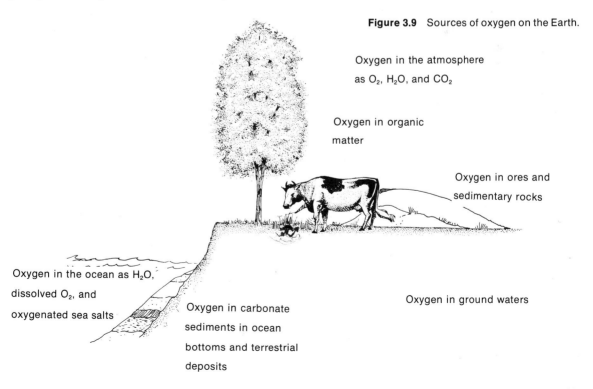

Figure 3.9 Sources of oxygen on the Earth.

Oxygen in the atmosphere as O_2, H_2O, and CO_2

Oxygen in organic matter

Oxygen in ores and sedimentary rocks

Oxygen in the ocean as H_2O, dissolved O_2, and oxygenated sea salts

Oxygen in carbonate sediments in ocean bottoms and terrestrial deposits

Oxygen in ground waters

47

proteins; (e) in ions such as nitrate, NO_3^-, and carbonate, CO_3^{2-}, that are dissolved in water, and (f) in many kinds of rocks, minerals, and other geological formations in the Earth's crust, such as limestone, $CaCO_3$, quartz, SiO_2, and metallic ores like bauxite (aluminum ore), Al_2O_3, and hematite (iron ore), Fe_2O_3 (Fig. 3.9). This last category accounts for most of the oxygen present in the biosphere. Most of the rapid *exchange of oxygen*, however, occurs by the action of living organisms (Fig. 3.10).

A plant synthesizes carbohydrates by combining carbon dioxide with water in the presence of sunlight and discharging oxygen as a byproduct:

$$6CO_2 + 6H_2O \xrightarrow{\text{sunlight}} \underset{\substack{\text{sugar, a} \\ \text{carbohydrate}}}{C_6H_{12}O_6} + 6O_2 \qquad (1)$$

This process is called **photosynthesis.** Most molecules in plant tissue contain oxygen atoms. Heterotrophs, which are unable to build sugars from carbon dioxide and water, consume plants and assimilate their complex chemicals, including the oxygen contained therein. As these chemicals move through the food web to consumers of other orders, they are all ultimately converted to carbon dioxide and

Figure 3.10 Oxygen-carbon cycle.

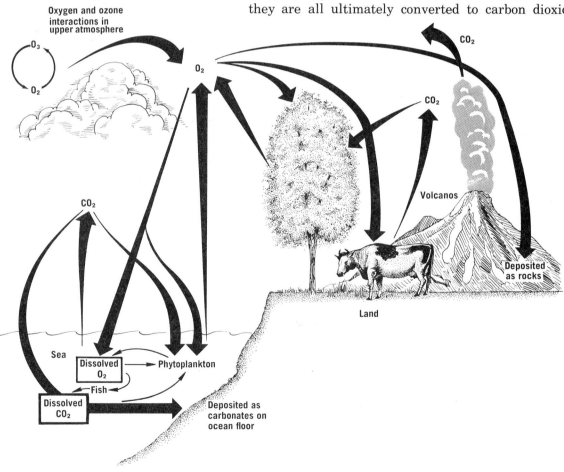

water by respiration or other oxidative processes. The organic oxygen participates in these conversions and so it, too, finds its way into CO_2 and H_2O molecules:

$$6O_2 + C_6H_{12}O_6 \xrightarrow{\text{heterotrophs}} 6\ CO_2 + 6H_2O \qquad (2)$$

This completes the cycle, for the carbon dioxide and water may then be reused as raw materials for new synthesis.

CARBON CYCLE

Carbon, like oxygen, is present in the Earth in many different forms. (See Table 3.3.*) It can be seen readily from Equations 1 and 2 that the *biological* carbon cycle is intimately related to the oxygen cycle, for whenever oxygen is cycled by life processes, carbon must accompany it. (See Figure 3.10.) Carbon is incorporated into organic tissue by photosynthesis and released by respiration and decay. However, less than half of the total cycling of carbon occurs through organic pathways. Atmospheric carbon dioxide dissolves readily in water, and some dissolved molecules escape from the sea to the air. This exchange occurs as a dynamic equilibrium; that is, atmospheric carbon dioxide molecules are constantly being dissolved into the ocean, while those that were previously dissolved are constantly escaping from the sea to the air. There is little net change in the ratio of dissolved carbon compounds to atmospheric carbon compounds. This geochemical cycling is independent of any living process; it is an inherent property of the chemistry of carbon dioxide and water.

Some of the dissolved carbon dioxide reacts with sea water to form carbonates, which settle to the ocean floor as calcium carbonate either in the form of inorganic precipitates (limestone) or as skeletons of various forms of sea organisms. This loss is partially balanced by the action of

---

*Note that less than 0.003 percent of all carbon on the Earth is contained in living plants and animals!

TABLE 3.3 Sources and Quantities of Carbon In the Biosphere

SOURCE	QUANTITY (BILLIONS OF METRIC TONS)	PERCENT OF TOTAL
Limestone, oil shale, and other sedimentary deposits	18,000,000	99.75
Oceans	32,000	.17
Coal and oil	9,100	.05
Dead organic matter	3,400	.018
Atmosphere	640	.003
Plants on land	410	.002
Plankton	9	.00005

A chalk cliff in Sussex, England. The chalk contains large quantities of carbon in the form of calcium carbonate. (Courtesy of British Tourist Authority.)

The burning of fossil fuels (coal, gas, and oil) is an oxidation process that converts the carbon in the fuel to carbon dioxide. The processes are represented by the following simplified equations:

$$\underset{\text{coal}}{C} + \underset{\text{oxygen}}{O_2} \rightarrow \underset{\substack{\text{carbon}\\\text{dioxide}}}{CO_2}$$

$$\underset{\substack{\text{natural}\\\text{gas}}}{CH_4} + 2O_2 \rightarrow CO_2 + 2H_2O$$

The same products (CO_2 and H_2O) are also obtained when liquid components of petroleum, such as octane, C_8H_{18}, are burned in air.

inland water which slowly dissolves limestone deposits on land and carries the carbonates to sea.

Currently there is about 15 times as much carbon locked into fossil fuel deposits as there is present in the atmosphere. These deposits were caused by past imbalances in the carbon cycle. The burning of coal and oil since the Industrial Revolution has released much carbon, permitting it to reenter biotic cycles. In playing an active role in the long-term recycling processes of the biosphere, industrial activity is measurably changing the human environment. Chapter 13 will consider the physical effects and possible harmful consequences of introducing large quantities of carbon dioxide into the air.

NITROGEN CYCLE

Nitrogen is a necessary constituent of proteins and therefore is essential to both plants and animals. Although nitrogen is roughly four times as plentiful in the atmosphere as oxygen, it is chemically less accessible to most organisms. Almost every plant and animal can utilize atmospheric oxygen, but relatively few organisms can utilize atmospheric nitrogen (N_2) directly. The nitrogen cycle (Fig. 3.11) must therefore provide various bridges between the atmospheric reservoir and the biological community. Lightning, photochemical reactions, modern fertilizer factories,

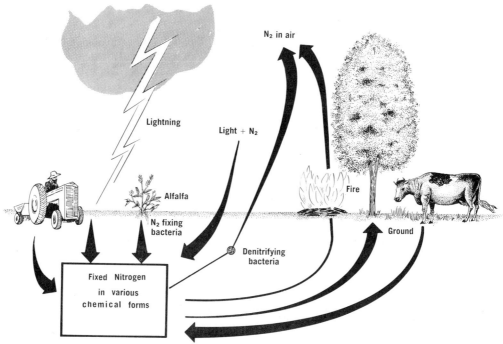

Figure 3.11 Pathways whereby nitrogen is removed from and returned to the atmosphere (nitrogen cycle).

and specialized bacteria and algae (called nitrogen-fixers) transform molecular nitrogen into forms usable by living organisms. The most common forms of soil nitrogen that are readily usable by plants (so-called "fixed" nitrogen) are nitrite ion (NO_2^-), nitrate ion (NO_3^-), and ammonia (NH_3) or its acidic form, ammonium ion (NH_4^+). Usually, fixed nitrogen is assimilated by plants and then travels into the heterotrophic chain. It may now cycle for a considerable time within the food web. Organic decay might return it to the soil, or it might be digested and returned to the soil through feces or urine. Once in the soil, it may reenter plant systems. Denitrifying bacteria and fire provide mechanisms for the return of nitrogen to the atmosphere. The rates of assimilation and denitrification are generally slow compared to the rates of nitrogen cycling. Therefore, unless some process removes nitrogen from an ecosystem, a given nitrogen atom may cycle many times before it is returned to the atmosphere. As a result, a large natural biomass can be supported by a small rate of nitrogen assimilation.

In agricultural systems, on the other hand, nitrogen-rich crops are harvested and generally trucked to distant cities. If the soil is to remain fertile, either the biological rate of nitrogen fixation must be high or farmers must return nitrogen to the soil by artificial means. Only a few organisms—for instance, the bacteria on the roots of legumes such as alfalfa, peas, and beans—are able to fix atmospheric nitrogen, that is, to convert N_2 to ammonium ion, NH_4^+. Most grain crops lack this ability, and therefore must be supplied with fixed nitrogen from some outside

To "fix" means to make firm or stable. In this sense, a gas (which is not firm) can be fixed by binding it in some form of solid or liquid.

source if the fields are to be harvested annually. Nitrogen fertilization can be effected by any of three techniques. The simplest method and the one used most commonly by the ancients is to fertilize the soil by recycling plant and animal waste. Another technique is crop rotation, that is, the planting of legumes and grain in alternate years, so as to maintain soil nitrogen. Finally, chemists have learned to convert atmospheric nitrogen to plant fertilizer by producing ammonia ($N_2 + 3H_2 \rightarrow 2NH_3$), which is then readily converted to ammonium ion. Today, enormous quantities of ammonia are manufactured in this manner; by some estimates, industrial fixation accounts for *one-third* of the total annual production of nitrogen compounds in the biosphere.

MINERAL CYCLES

The cycling of minerals such as phosphorus, calcium, sodium, potassium, magnesium, or iron is much more fragile than the cycling of oxygen, nitrogen, or carbon because of the absence of a large mineral reservoir in a readily available form.

Let us consider the mineral cycles in a particular watershed in the mountains of New Hampshire.* This area includes a ring of mountains and the enclosed valley. Rain is the area's only significant source of water; the only important exit for flowing water in this watershed is through a single stream, because the geology of the area is such that seepage is negligible. In such a system the total input of minerals is limited to two processes: the rain deposits some mineral matter, and weathering of rocks frees some minerals from the earth's crust. This gain is partially balanced by losses.

In the particular watershed that was studied, it appeared that the stream outflow removed fewer minerals than weathering and rainwater were bringing in. (See Figure 3.12.) However, the net gain for the ecosystem was very small compared to the requirement of all the life forms in the valley. This means that the valley life was operating on mineral capital, not mineral income. In fact, the entire biosphere depends for its supply of mineral matter upon the ability of each individual incoming inorganic ion to undergo countless transformations from soil to plant tissue, from plant tissue into the organic detritus or on to animal tissue, and then to the detritus, back to the soil, and around again and again before being washed out of the system. Therefore, input and output rates bear little relationship to the quantity of inorganic matter in the private pool of any given ecosystem.

How important is the presence of healthy vegetation in maintaining this delicate mineral balance? In order to study this problem, all the trees, saplings, and brush in one

*For the details of this example, refer to Chapter 4 of G. M. Van Dyne, *The Ecosystem Concept in Natural Resource Management.* (See references at the end of this chapter.)

+ 2.7 lbs./acre yearly from rain

+ 5 lbs./acre yearly from weathering of rocks and minerals

181 lbs./acre within living organisms

44 lbs./acre exchanged between soil and organisms

325 lbs./acre in soil

- 7.1 lbs./acre yearly in stream outflow

Figure 3.12 Calcium balance in a New Hampshire watershed.

watershed were cut or chemically killed. No organic matter was removed, so that outright soil erosion could be minimized. In the year following this clearcutting experiment, the stream flow was 40 percent greater than it would have been had the system remained undisturbed. The water carried away many mineral salts, including calcium, and the net nutrient drain was then greater than the net input.

In a healthy ecosystem, water and nutrients in the soil are absorbed by plants and carried up to leaf tissues. The nutrients are generally assimilated, but much of the water is lost through **transpiration,*** evaporation from leaf surfaces. In a dead system no water is lost through transpiration, and therefore most of the rainwater must ultimately flow into the stream bed. Since all ground water carries nutrients out of the system, a large increase in stream flow disrupts the nutrient balance.

3.5 MAJOR ECOSYSTEMS OF THE EARTH

The **biosphere** is the region including all the life-supporting portions of our planet and its atmosphere. This section briefly describes some of the major ecosystems of the biosphere.

*Transpire (*Latin,* breathe through) means to pass a vapor (usually moisture) through a porous barrier, such as skin or leaves. The word is also used in the sense of leaking out, as of information, or becoming known, or, very loosely, coming to pass, occurring, or happening.

The Atlantic Ocean. (Photo by Frank E. Karelsen.)

OCEAN SYSTEMS

The ocean, far from being homogeneous, contains a varied and intricately interwoven set of ecosystems. Despite the fact that the ocean is known to be more than 6000 meters (19,000 feet) deep in places, light does not penetrate more than about 200 meters (650 feet) in sufficient quantities for photosynthesis to occur. The depth of this illuminated section, or **euphotic zone,** varies considerably with the turbidity of the water, reaching its maximum limits in the central oceans and narrowing to 30 meters (100 feet) in coastal regions. Although the depth of the photosynthetic zone is greatest in the central ocean, the rate of photosynthesis is not greatest there, for the concentration of plant nutrients is low. Indeed, the central ocean has often been likened to a great desert.

The primary autotrophs in all marine ecosystems are **phytoplankton,** the one-celled chlorophyll-bearing organisms suspended in the water. In areas near shore, various species of algae are abundant enough to account for a significant portion of the net quantity of living matter.

A large portion of the primary consumption in the sea is carried out by myriads of species of small grazers called **zooplankton** (Fig. 3.13). These animals range in size from

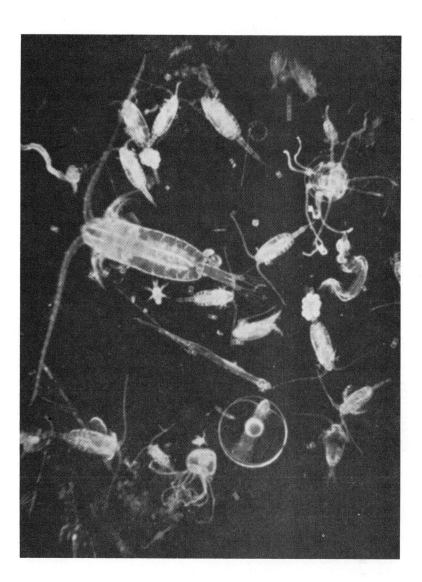

Figure 3.13 Living zooplankton. The shrimplike animals in the picture are various small crustaceans. Two tiny jellyfish with long tentacles are also seen (magnification 16×). (From A. Hardy: *The Open Sea.* London, William Collins Sons, 1966.)

roughly 0.2 mm (.008 in) to 20 mm (0.8 in) in diameter and consist both of permanent zooplankton and larvae of larger forms of life. The composition by species of a sample of plankton varies with the climate, and plankton occupies several trophic levels. There are few large grazers in the sea analogous to land-based bison, deer, or cattle.

Many omnivores and predators live in the sea. These species represent various biological classifications, including fish, mammals, invertebrates, and reptiles. Organic debris from the life processes on the surface and in the body of the sea are continuously raining down toward the ocean floor. In the deep ocean, some of this food is intercepted, some decomposes, and some falls to the bottom, where **benthic species** (organisms that live in and on the bottom sediments) receive their nourishment (Fig. 3.14). Some of these are plants. Others are semimobile or im-

A

mobile animals. Still others crawl or swim actively in search of food.

While gravity is constantly pulling nutrients to the bottom of the sea, the mechanisms for upward recycling of nutrients are comparatively feeble. Currents and water turbulence serve to move some nutrients back to the surface, but a net loss does occur to the deep sea floor. These chemicals may eventually be recycled by geological upheavals. One net result of poor recycling in the ocean is that the concentration of nutrients dissolved in the bulk of our ocean waters is low and imposes limits on the growth rate of organisms. By contrast, in the food-rich coastal areas, wave actions, nutrient inflow from rivers, and a relative shallowness all combine to support a large biomass.

ESTUARY SYSTEMS

Coastal bays, river mouths, and tidal marshes are all physically contiguous to the open ocean, yet their proximity to land and fresh water affects the salinity and nutrient composition to such a large extent that these areas, known as **estuaries,** are characteristic neither of fresh nor of salt water (Figs. 3.15 and 3.16). The combination of such factors as (a) easy access to the deep sea, (b) a high concentration and retention of nutrients originating from land and sea,

B

Figure 3.14 *A,* A sluglike nudibranch crawling over encrusting sponges, which are also inhabited by sea anemones. (Photo by Harold Wes Pratt.) *B,* Deep sea benthic scavenger fish attack a piece of bait. (Courtesy of Richard A. Schwartzlose, Scripps Institution of Oceanography.)

Figure 3.15 Schematic representation of shoreline features: coastal bay, estuary, tidal marsh, and coral lagoon behind coral reef.

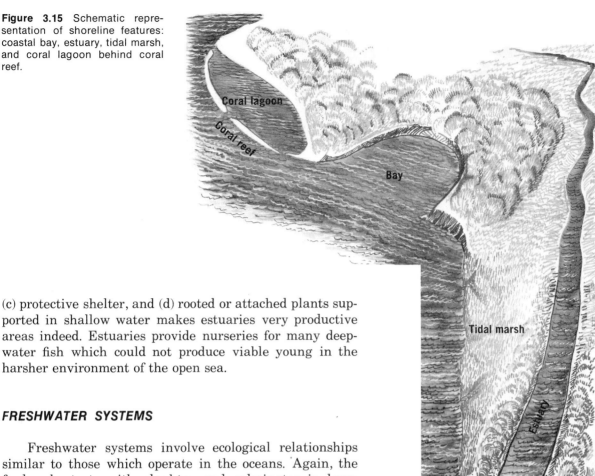

Coral lagoon

Coral reef

Bay

Tidal marsh

Estuary

(c) protective shelter, and (d) rooted or attached plants supported in shallow water makes estuaries very productive areas indeed. Estuaries provide nurseries for many deep-water fish which could not produce viable young in the harsher environment of the open sea.

FRESHWATER SYSTEMS

Freshwater systems involve ecological relationships similar to those which operate in the oceans. Again, the food web starts with plankton and culminates in large predators. A major difference between the two systems is that there are more trophic levels in salt water than in fresh. Freshwater systems, unlike other aquatic systems, are continuously fertilized by nutrients leached from the nearby soil. Because bodies of fresh water are shallower

Figure 3.16 Salt marsh estuary system.

than the oceans, rooted plants, marsh grasses, and lilies, as well as algae, are much more important in the food webs. Finally, as one would expect, ponds, lakes, mudpuddles, springs, creeks, and rivers all have unique and characteristic species.

TERRESTRIAL SYSTEMS

The characteristics of large, stable, terrestrial ecosystems, known as **biomes,** result from the interactions of many environmental, biological, and evolutionary factors. Rainfall, average and seasonal temperatures, altitude, and soil conditions all have profound influences. Specialized physical factors such as seasonal changes or ocean mists can help to create unique biomes.

A general characterization of the major terrestrial biomes of Canada and the United States is outlined in the following paragraphs. (See Figure 3.17.)

In the coldest regions, where summer temperatures

Figure 3.17. Biomes of Canada and the United States. 1, Tundra. 2, Taiga. 3, Mountain and conifer forest. 4, Conifer forest. 5, Broadleaf forest. 6, Prairie. 7, Desert.

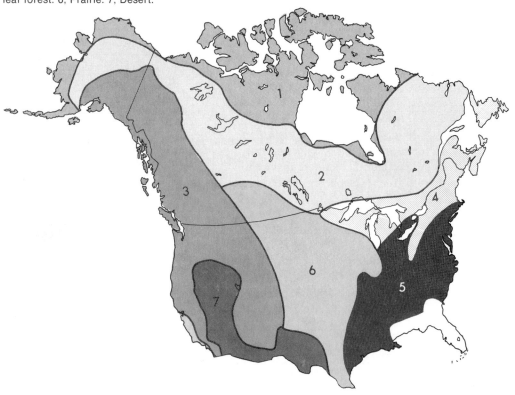

average under 10°C (50°F) and trees cannot survive, stable plant cover is known as **tundra** (Fig. 3.18). Some of the grazer-plant and predator-prey relationships in this area will be discussed later.

The region between the tundra and the northern forest is characterized by tundra grasses growing between dwarfed, gaunt, windblown conifers. (Such border ecosystems, known as **ecotones,** will be discussed further on page 98.) Animals from both the tundra and the forest systems live here.

Still farther south, where the weather is warmer, is the **taiga,** or northern evergreen forest, which supports deer, elk, moose, caribou, wood buffalo, and rodents as the primary grazers and wolves, lynx, mountain lions, and people as the largest predators.

South of the taiga are forests consisting largely of deciduous trees. Such forests need 75 to 150 cm (30 to 60 in) of rain per year and an average temperature during the growing season of about 15° to 18°C (60° to 65°F). Therefore, they do not exist farther north than central Maine or southwestern Quebec or farther south than central Georgia or southern Louisiana.

In the temperate areas in which the rainfall is too low to support forests, the stable ecosystems are **prairies** or **grasslands** (Fig. 3.19). In wet areas, the virgin grasses of North America were as tall as a person; elsewhere, they were short enough that the early scouts could "see the horizon under the bellies of millions of buffalo."

In very dry areas, where rainfall is less than 25 cm (10 in) per year, the stable ecosystem, called a **climax** (page 93), is a **desert,** either barren or able to support only scrub brush and cactus (Fig. 3.20). Deserts can be hot, as in Nevada, or relatively cold, as in eastern Washington.

The world contains other ecosystems in addition to those mentioned above. **Tropical rain forests** are stable systems of the Amazon basin, parts of Africa, Central America, and parts of Southeast Asia, Malaysia, and New Guinea. Here the seasonal temperature changes are less pronounced than daily ones. Vegetation is thick, and most of the available nutrients are found in the biomass. Decay and recycling are rapid.

Tropical **savannas** are fire-dependent systems occurring in areas with annual wet and dry seasons where the yearly rainfall is 100 to 150 cm (40 to 60 in). Fires are common during the dry season. The flora consists of a few fire-resistant trees and annual vegetation **blooms,** or periods of rapid growth. Many of the animals are migratory.

When one studies different biomes, it is of interest to compare similar climatic regions that have been separated for millennia. The classic comparison between North American plains and those of Australia shows similar trophic structure but dissimilarities among species. Thus,

Figure 3.18 Tundra in July on the coastal plain near the Arctic Research Laboratory, Point Barrow, Alaska. (Photos by the late Royal E. Shanks, E. E. Clebsch, and John Koranda. From Odum: *Fundamentals of Ecology,* 3rd ed. Philadelphia, W. B. Saunders Co., 1971.)

A

B

Figure 3.19 Natural temperate grassland (prairie) in central North America. *A,* Lightly grazed grassland in the Red Rock Lakes National Wildlife Refuge, Montana, with a small herd of pronghorns. *B,* Short-grass grassland, Wainwright National Park, Alberta, Canada, with herd of bison. (From Odum: *Fundamentals of Ecology,* 3rd ed. Philadelphia, W. B. Saunders Co., 1971.)

Figure 3.20 A desert region in Arizona.

kangaroos and bison, both important grazers, are morphologically quite different.

3.6 ECOSYSTEMS AND NATURAL BALANCE

We have seen that food webs perpetuate the flow of energy and the cycle of nutrients. All these processes are being carried on simultaneously. A diagram illustrating all the energy and nutrient transfers of even a simple system would be a blur of arrows showing sugars, mineral ions, nitrogens, oxygens, and carbons going this way and that. How is the ensemble regulated?

If we place a few square centimeters of grassy sod and a grasshopper together in a sealed aquarium, the grasshopper will eventually eat all the available grass. But in a large prairie ecosystem, the grasshoppers do not eat all the grass; studies have shown that under normal conditions 80 to 90 percent of the plant matter falls to the ground uneaten, and the 10 to 20 percent that is eaten is shared by insects, birds, reptiles, rodents, large mammals, and other herbivores. Only rarely does the population of herbivores increase enough to defoliate a region. Yet each individual animal strives to obtain as much food as it needs. What factors control the relative abundance of individuals in an ecosystem? How can a natural meadow, lake, or forest exist in an orderly and relatively unchanged state for years?

Many opposing forces operate within a natural ecosystem. Organisms eat and in turn are eaten; fertility rates vary; migration is common; weather varies and climates change; moisture and nutrients travel into and out of the soil. The net effect of all these events acting together is that in general, when an ecosystem is disrupted, the system tends to maintain its existence by regulatory mechanisms that oppose the disruption. Such a tendency is called a balance of nature, or, more formally, **ecosystem homeostasis.** For example, drought in a grassland inhibits the growth of plants. The meadow mice become malnourished, their fertility rate drops, and their birth rate decreases. In addition, they have a protective behavioral response to lack of food; they retreat to their burrows and sleep. Thus their death rate also decreases because they are less exposed to predation. Their behavior protects their own population balance as well as that of the grasses, which are not consumed by hibernating mice.

Dynamic balance in nature is controlled by the actions of the organisms themselves. Ecologists have observed that in natural, stable ecosystems the biomass is quite large compared with the biomass of less stable systems. Thus many individuals living together in a complex food web generally counteract perturbations more effectively than fewer individuals in simpler systems do.

A redwood forest in northern California.

Stable ecosystems are not necessarily in balance all the time, but if they are out of balance in one direction, they must become out of balance in the opposite direction at some time in the future if they are to survive with no essential change in character. In fact, all ecosystems naturally fluctuate. For example, climate varies from year to year. Other disruptions, such as migration, drought, flood, fire, or unseasonable frost can cause imbalance in an ecosystem. The ability of an ecosystem to survive depends on its ability to adjust to an imbalance.

Let us consider as an example the relative populations of tigers, grazing animals, and grass in a valley in Nepal. Assume that the mountains surrounding the valley are so high that no animals can enter or leave the valley. One year rainfall is limited and mild drought conditions exist. Because water is naturally stored in pools, the drinking water supply is adequate. However, the grasslands suffer from the lack of rainfall. Therefore, food is scarce for the grazing animals, and many are hungry and weak. Such a situation is actually beneficial to the tigers because hunting becomes easier, and the tiger population thrives.

The following spring, when rains finally arrive, the grazing population is low. This permits the grass to grow back strongly because not much of it will be eaten. However, because the food supply for the tigers is low (and the grazers that are left are the strongest of the original herd), the tigers suffer a difficult time the summer after the drought. The third year there are fewer tigers, so the herd resumes its full strength and balance is achieved again.

Stable ecosystems, if examined superficially, do not seem to go out of balance at all. Actually, their homeostatic mechanisms function so well that slight imbalances are corrected before they become severe. Such sensitive responses are, in fact, the essence of stability, because if severe imbalances are prevented, it is very unlikely that the system will be destroyed and, as a result, it lasts for a long time. Usually, when we refer to a stable ecosystem, we mean one which regenerates itself again and again with little overall change. Sometimes, however, an ecosystem is stable not because it regenerates often without change, but because the dominant plants are long-lived. A wonderful example is a redwood forest. To the hiker entering a redwood forest on a hot summer day, two characteristics of the environment are immediately apparent. It is cool, and there is a thick floor of spongy matter. The coolness, caused by the extensive shade of the tall trees, reduces water loss due to evaporation. The spongy floor is formed by a large bed of decomposed organic matter known as **humus,** as well as a liberal supply of fallen needles and other as yet undecomposed litter that catches and holds the rainwater, and thus serves as a water and nutrient reservoir, and as a home for the members of the detritus food chain (Fig. 3.21). Thus, the effect of a drought does not have to be balanced by the death of trees. In addition, the large amount of organic material available ensures a steady functioning of the recycling system. Thick mats of partially decomposed organic matter also retain the heat produced by saprophytes and help maintain a warm environment known as a **microclimate,** which is often more favorable for metabolism than the climate on the surface. Anyone who has ever operated a compost heap has observed this warming effect.

The coolness of the forest offers still another advantage. Clouds passing overhead are more likely to discharge their moisture where it is cool, for cold enhances the tendency of water to agglomerate into raindrops. Thus, the redwood forest tends to cause rain, to restrict evaporation, and to retain a substantial supply of water. Furthermore, decay organisms require moist environments to be effective, so the redwood forest insures the proper recycling of materials by environmental control.

This unusually stable ecosystem also exhibits other highly effective homeostatic abilities. One of these can be observed readily by walking through the forest on the west side of Route 1 as it runs through the town of Big Sur, California. Some time ago a tremendous fire swept through the forest. However, because the bark and the wood of the redwood tree are fire-resistant, the big trees were scarred but not killed. Because the forest environment was maintained, the forest community was quickly regenerated. If a fire of equal intensity had burned through a stand of pine, the trees would have died and many more years would have been needed for the forest to be rebuilt.

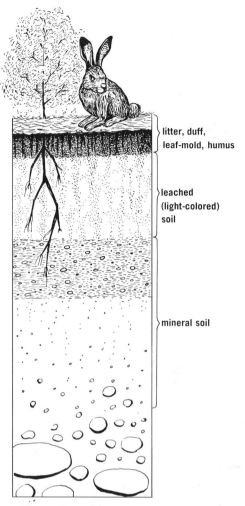

litter, duff, leaf-mold, humus

leached (light-colored) soil

mineral soil

Figure 3.21 Cross section of forest soil.

3.7 CASE HISTORY: *Westhampton Beach—A Coastal Ecosystem*

This chapter has emphasized the ways in which living organisms mold and regulate their own environment. The appearance of the land, the quality of the soil, the health of streams and lakes, and even some aspects of local weather and climate are partly defined by the plants and animals that live there. Biological environments exist within a framework defined by geological forces and are naturally affected by temperature, precipitation, wind, and sometimes tidal actions. Perhaps nowhere are geological forces more powerful and dynamic than along coastal ecosystems. Waves constantly strike the beach, tidal currents move incessantly back and forth, and storm systems travel unimpeded across flat stretches of ocean to attack the shore with periodic fury.

When waves strike a beach, sand particles and small pieces of rock are dislodged from the shoreline and carried along the coast, where they are later redeposited. This continuous movement of material creates many distinct land forms and ecological habitats. Along the southern coast of Long Island, movement of sediment has formed a series of thin, low-lying barrier islands (Fig. 3.22). Between the barrier islands and the mainland there lies a series of quiet, protected salt marshes. These marshes represent an estuary system that provides home and shelter for many forms of aquatic life, including the young of many species of deep water fish. The predominant movement of water along this coastline is from east to west, as shown in Figure 3.22. On the eastern end of Long Island, near Montauk Point, there is a region of tall cliffs composed of sand and gravel originally

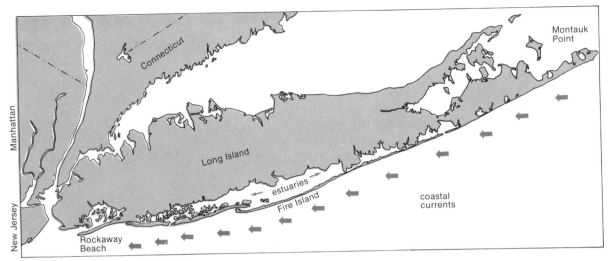

Figure 3.22 Map of Long Island and vicinity.

deposited by the glaciers of the great ice ages. At the present time, the ocean waters are eroding the sediment from these cliffs and carrying the material westward toward Rockaway Beach and New Jersey. If we studied a beach midway along the island, say Fire Island (shown on the map), we would find that the existing beach sand is constantly being eroded and carried westward, only to be replaced by sand from the eastern end of the island. The beach, in dynamic equilibrium, remains stable.

Superimposed on this east-west movement are variations caused by seasonal changes and severe storms. In the wintertime, series of steep, sharply undercut waves remove large quantities of beach sand, carry the sediment out to sea, and deposit it in a formation known as a sand bar, which lies parallel to the shore. Thus, in winter the beach erodes and grows smaller. This erosion is only temporary, however, for the gentler waves of summer carry sand from the bar inland and rebuild the beach. In this manner, long-term stability is maintained despite seasonal fluctuations.

In any coastal system, severe storms periodically strike the shore. In a natural barrier system such as the one in southern Long Island, storms completely overrun the low-lying outer island, but the system is not permanently damaged by these periodic inundations. The high waves that wash across the beach and roll inland over the dunes and salt marshes flatten some dunes, build others higher, and move sand and rock here and there. When the storm is over, there is little significant long-lasting erosion. Beaches remain intact, salt marshes rejuvenate quickly, and the dune grasses that had evolved in such a system grow back within a few months.

We do not mean to imply that geological forces have no effect on coastlines, for of course they do. But the change is generally slow. When the sand deposits near Montauk Point become depleted and no new inflow of sand can travel westward, the island will be reshaped and beaches will erode. But such changes are generally measured in centuries or millennia rather than in months or years.

The biological-geological balance that exists on Long Island has given rise to a fertile estuary system. However, the ebb and flow of beach sand and the rise and fall of storms are not always compatible with human activity. A shore–sand-dune system is dynamic—always changing. It is also prime waterfront real estate. So people build houses, resorts, and hotels on the shifting sands. Then, in winter, the beach starts to recede. Although this recession is entirely natural, many property owners have tried to save their personal stretch of beach from these cycles of the ocean. This can be done (temporarily) by building a **groin,** commonly called a **breakwater.** As mentioned previously, sand steadily moves from east to west along this particular stretch of coast. If a large stone barrier is built just west of a person's property (Fig. 3.23), it will trap sand moving from the east and keep that particular beach from receding in winter. But now the overall flow of sand has been impeded (Fig. 3.23*B*). The property just west of the breakwater receives no sand, but the currents that move the sand originate from farther out to sea and are not impeded. So behind the breakwater the beach is eroded as usual, but not replenished. The beach then recedes, as shown in the figure.

This unnatural recession aided by the up-

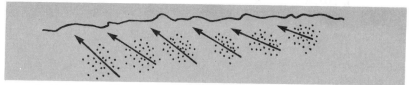

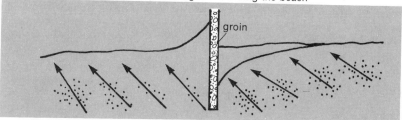

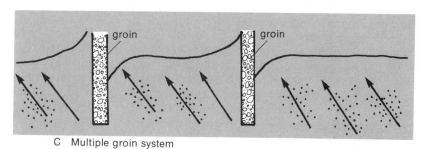

A Undeveloped beach; ocean currents (arrows) carry sand along the shore, simultaneously eroding and building the beach

B A single groin or breakwater. Sand accumulates on upstream side and is eroded downstream

Figure 3.23 Effect of breakwaters on a beach.

C Multiple groin system

1. A beach house on Atlantic Beach, Long Island. The photograph was taken in February; note that natural rocks are exposed and there is no sandy beach.

2. This photograph, taken from inside the house, shows a newly constructed breakwater.

3. The breakwater and sand fences trap the sand and build a beach.

4. The same house the following winter has a sandy beach. However, the sand had to come from somewhere, perhaps from elsewhere along the coast. (Photos courtesy of Frank E. Karelsen.)

65

stream breakwater is much more extensive than the natural ebb and flow of the beach, and erosion could remove the entire beach and any beach house behind it in a short period of time. So now, the person living behind the first breakwater may decide to build another one, west of his or her land, and pass the problem on downstream. The situation can perpetuate itself indefinitely, with the net result that millions of dollars must be spent to attempt to stabilize a system that was originally stable in its own dynamic manner.

Storms pose another dilemma. Periodic storms have flooded the dune lands every decade or so for millennia. A developer building a luxury home or an elite resort hotel cannot allow the buildings to be flooded or washed away every 10 or 15 years. Therefore he constructs large sea walls just inland of the beach itself. When a storm wave rolls over a low-lying dune, it dissipates its force gradually over the hills and pushes the beach sand inland. But if a sea wall interrupts this orderly flow, the waves will crash violently against the barrier. The steep breaking waves establish a circular turbulence that carries sand out to sea and erodes the beach.

An example of the futility of these human interferences is the case of Hurricane Ginger, which struck Long Island on September 30, 1971. Undeveloped islands were nearly completely flooded, and dunes were flattened, but within 10 months the beach had restabilized and was actually larger than it had been before the storm. By contrast, other beaches, which had been "protected" by sea walls, were severely eroded. In the spring of 1972 one community spent $500,000 to replenish the beach, but a storm later that year destroyed it again. In 1973, the National Park Service announced that "after spending $21,000,000 since the 1930's to erect and maintain artificial barriers against waves and storms, the agency has concluded that such work does more harm than good."

The social and political issues involved in beach erosion neatly illustrate some of the complications that develop when people try to occupy naturally delicate ecosystems. Perhaps a dynamic system such as a dune land should not be open to development at all. After all, the community as a whole benefits from unspoiled recreational areas, and the economic value of the estuaries as a breeding ground for commercial fish represents a valuable national asset. Furthermore, when an individual corporation owns a section of beach, poor people are generally excluded. Since breakwaters and sea walls traditionally have been built with government support, this type of activity is a means whereby the poor pay taxes for the benefit of the rich. Therefore, many feel that shore reconstruction should be abandoned.

The counter argument is that significant development along the beach has, in fact, already occurred, and that the tourist business is an economic boon to local communities. In any event, the next hurricane must not be allowed to wash innumerable houses, businesses, and roadways to the sea. Therefore, new and better sea walls and breakwaters should be built. Since breakwaters must be built along the entire stretch of coast to be effective, the government should support such projects.

This particular case was battled in the courts for a number of years. On December 20, 1974, the Suffolk County legislature voted to suspend all beach reconstruction projects and to allow the ocean to maintain its own coast.

TAKE-HOME EXPERIMENTS

1. **Saprophytes.** Obtain about a kg of fertile soil from a farm, woodland, or garden shop. Spread the soil carefully on a smooth piece of paper, and, using a magnifying glass, search for any living organisms. How many do you find? Draw pictures of them. If a microscope is available, place a small quantity of the soil on a slide and examine it. Can you see more organisms? What are they doing? How can the quantity and variety of life in your sample influence the quality of the soil?

2. **Ecosystem homeostasis.** Place about half a cupful of clean sand in one container and an approximately equal volume of fertile soil in another. Add enough tap water to each, with stirring, until the sample is thoroughly wet and a thin layer of sandy or muddy water remains on top. Carefully pour off and discard this layer. (It contains small particles that may obscure your test.) Now measure the acidity of each sample with indicator paper. (Litmus paper is not precise enough; you will need "universal" indicator paper, which

changes color over a range of acidities and can be matched to a pH scale. The scale is explained in Appendix C-3.) To carry out the test, touch the paper to the top of the sample; the paper will soak up the liquid so that you will be able to see the color above the level where sand or soil particles interfere. Now add about 5 ml of vinegar (which is a solution of acetic acid) to each sample.

For convenience, use a medicine dropper; there are about 20 drops to a milliliter, so 100 drops will do. Stir both samples well and let them sit for about 15 minutes. Measure the pH again. Did both samples change by the same amount? Why or why not? Interpret your results in terms of soil homeostasis.

PROBLEMS

1. **Systems.** The human body and an automobile factory can both be classified as systems. Show how in each case a malfunction of one component can affect the whole. Show how a malfunction of a different component will not necessarily affect the whole. Briefly outline how each of these systems exchanges energy and raw materials with the outside environment.

2. **Vocabulary.** Define ecology; ecosystem; biosphere.

3. **Energy.** Classify each of the following substances as energy-rich or energy-poor with specific reference to its ability to serve as a food or fuel: sand, butter, paper, fur, ice, marble, and paraffin wax.

4. **Vocabulary.** Define autotroph; heterotroph; trophic level.

5. **Trophic levels.** Name two organisms that occupy the first trophic level; two that occupy the second; two that occupy the third.

6. **Energy.** Organisms have evolved to obtain energy from foods or from the process of photosynthesis. Would it have been possible for life forms to evolve which obtain their energy by some other process, such as rolling down hills? Defend your answer.

7. **Food web.** What is a food web? Sketch a diagram of a food web that primarily involves life in the air, such as birds and insects. Will you need links to terrestrial or aquatic systems?

8. **Food web.** Consider the food web in the accompanying diagram. Arrow number 15 can be explained by the relationship: mountain lion eats deer. Write similar statements for each of the other 27 arrows.

9. **Food pyramid.** The discussion presented in Section 3.3 and illustrated in Figures 3.6 and 3.7 shows that more energy from the Sun is used to nourish a human being who eats meat than a human being who eats vegetables. Explain why this fact, by itself, is *not* an argument for or against vegetarianism. Take into account the following questions: Does the choice of human diet affect the total energy flow through the food web? Does it affect the total biomass of plant matter on Earth? Does the choice between the alternatives of a large human population living largely on vegetable food or a smaller human population living largely on meat and fish significantly affect the total biomass of animal matter on Earth? In your answer do not take into account the possible destruction of species; this topic is treated separately in Chapter 5.

10. **Food pyramid.** An experiment has shown that the total weight of the consumers in the English Channel is five times the total weight of the plants. Does this information agree with the food pyramid shown in

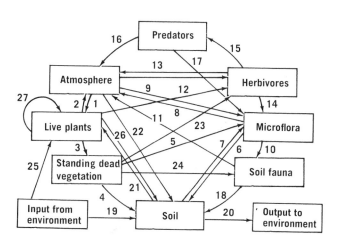

Figure 3.6? Can you offer some reasonable explanations for the findings?

11. **Food pyramid.** It is desired to establish a large but isolated area with an adequate supply of plant food, equal numbers of lion and antelope, and no other large animals. The antelope eat only plant matter, the lions, only antelope. Is it possible for the population of the two species to remain approximately equal if we start with equal numbers of each and then leave the system alone? Would you expect the final population ratio to be any different if we started with twice as many antelope? Twice as many lions? Explain your answers. (Assume that lions and antelope have the same body weight.)

12. **Nutrient cycles.** We speak about nutrient cycles and energy flow. Explain why the concepts of nutrient *flow* and energy *cycle* are not useful.

13. **Nutrient cycles.** Give three examples supporting the observation that nutrient cycling hasn't been 100 percent effective over geological time.

14. **Oxygen cycle.** Trace an oxygen atom through a cycle that takes (a) days, (b) weeks, (c) years.

15. **Nutrient cycles.** Why don't farmers need to buy carbon at the fertilizer store? Why do they need to buy nitrogen?

16. **Carbon cycle.** The carbon dioxide concentration in the air just above trees varies considerably between night and day. From your knowledge of the biochemistry of carbon, predict whether the atmospheric carbon dioxide concentration will be higher during the night or during the day.

17. **Nutrient cycles.** Certain essentials of life are abundant in some ecosystems but rare in others. Give examples of situations in which each of the following is abundant and in which each is rare: (a) water, (b) oxygen, (c) light, (d) space, and (e) nitrogen.

18. **Nitrogen cycle.** Describe three pathways whereby atmospheric nitrogen is converted to fixed forms that are usable by plants, and three pathways whereby fixed nitrogen is returned to the atmosphere.

19. **Mineral cycles.** Loggers can harvest timber either by clearcutting (removing all the trees from an area) or by selective cutting (removing only the most desirable trees). Explain why clearcutting is an ecologically unsound practice in most woodlands.

20. **Vocabulary.** Define euphotic zone; phytoplankton; benthic species.

21. **Estuary systems.** Copy Figure 3.15 and indicate which bodies of water are salty, fresh, and brackish.

22. **Terrestrial ecosystems.** List four terrestrial biomes and discuss the physical characteristics of each.

23. **Ecosystem homeostasis.** What is meant by the term ecosystem homeostasis? Cite two examples to show how homeostatic mechanisms operate.

24. **Ecosystem.** Could a large city be considered a balanced ecosystem? Defend your answer.

25. **Homeostasis.** Consider two outdoor swimming pools of the same size, each filled with water to the same level. The first pool has no drain and no supply of running water. The second pool is fed by a continuous supply of running water and has a drain from which water is flowing out at the same rate at which it is being supplied. Which pool is better protected against such disruptions of its water level as might be caused by rainfall or evaporation? What regulatory mechanisms supply such protection?

26. **Ecology of soil.** Some fixed nitrogen is returned to the atmosphere by the action of certain soil organisms known as **denitrifying bacteria.** Do you feel that it would be wise to poison bacteria in the soil to conserve the fixed nitrogen supply? Justify your answer.

27. **Ecosystem balance.** A sand dune ecosystem can survive inundation by salt water without suffering permanent damage, yet if a pine forest is similarly flooded it will not regain full productivity for many years. Is the pine forest less well balanced than the dune system? Discuss in terms of the homeostatic stability of ecosystems.

28. **Energy.** Assume that a plant converts 1 percent of the light energy it receives from the Sun into plant material, and that an animal stores 10 percent of the food energy that it eats in its own body. Starting with 10,000 Calories of light energy, how much energy is available to a man if he eats corn? If he eats beef? If he eats frogs that eat insects that eat leaves? Of the original 10,000 Calories, how much is eventually lost to space?

29. **Energy flow in food chain.** Referring to the values shown in Figure 3.7 for the energy content of organisms in a food chain, complete the following table:

Trophic level	Food value of organism for each 1000 Calories received by autotroph from the Sun
First	
Second	
Third	
Fourth	

(Assume that the efficiency of energy conversion remains constant after the second trophic level.)

BIBLIOGRAPHY

Three basic textbooks on ecology are:
Edward J. Kormondy: *Concepts of Ecology.* Englewood Cliffs, N.J., Prentice-Hall, Inc., 1969. 209 pp.
Charles J. Krebs: *Ecology.* New York, Harper and Row, 1972. 694 pp.
Eugene P. Odum: *Fundamentals of Ecology.* 3rd Ed. Philadelphia, W. B. Saunders Co., 1971. 574 pp.

A periodical issue devoted in its entirety to "The Biosphere" is:
Scientific American. September, 1970. 267 pp.

Three books dealing with specific areas of natural ecology are:
R. Platt: *The Great American Forest.* Englewood Cliffs, N.J., Prentice-Hall, 1965. 271 pp.

B. Stonehouse: *Animals of the Arctic: the Ecology of The Far North.* New York, Holt, Rinehart and Winston, 1971. 172 pp.
G. M. Van Dyne, ed.: *The Ecosystem Concept in Natural Resource Management.* New York, Academic Press, 1969. 383 pp.

A classic study of ecology and conservation as seen through the eyes of a naturalist is:
A. Leopold: *A Sand County Almanac.* New York, Sierra Club/Ballantine Books, 1966. 296 pp.

A book that specifically discusses shore erosion is:
Joseph M. Heikoff: *Politics of Shore Erosion: Westhampton Beach.* Ann Arbor, Michigan, Ann Arbor Science Publishers, 1976. 173 pp.

4

NATURAL GROWTH AND REGULATION OF POPULATIONS

4.1 POPULATION GROWTH AND CARRYING CAPACITY

Old ecosystems are typically well balanced because the interactions within such natural environments regulate the population levels of all native species. Of course, oscillations in population size do occur even in stable systems, but these oscillations are generally small, and the total population of each species as well as the *ratio* between populations fluctuate only slightly. In contrast to this observed stability, calculations show that the **biotic potential** of any species is extremely high. For example, a bacterium can split into two bacteria in approximately 20 minutes. If enough food is available and there is no predation, these two bacteria can grow into four after another 20 minutes; by the end of an hour the original bacterium will have become eight. By the end of a day and a half, a growing colony would have increased through 108 generations (36 hours at 3 generations per hour). Since each generation leads to a doubling of the number of individuals, the colony would consist of 2^{108}, or roughly 10^{33} individuals. This number of bacteria could cover the entire surface of the Earth to a uniform depth of 1 foot. Such a growth pattern is known as a geometric rate* of increase. Another example of a geometric growth rate would be tripling (or, in fact, multiplying by any constant number) in each generation, as shown in Table 4.1 and Figure 4.1*A*. Most other organisms have longer generation times than 20 minutes, and consequently, their biotic potential is smaller than that of bacteria. Nevertheless, the breeding capacity of *all* species is

> The biotic potential of a species is defined as the maximum rate of population growth which would result if all females bred as often as possible, and all individuals survived past their reproductive age. To grow at the biotic potential, a population must have ample food and living space and be free from disease and predation.

---

*See Appendix C-6 for a further discussion of growth rates.

TABLE 4.1 Geometric Growth

GENERATION	NUMBER OF ORGANISMS	
	Doubling	Tripling
1	1	1
2	2	3
3	4	9
4	8	27
5	16	81
6	32	243
7	64	729

large, and no ecosystem could support the geometric growth rate of any species for very long.

Some combination of environmental pressures must, therefore, act to inhibit the potential growth of every species. Examples of many of these environmental pressures are easy to observe: you swat a mosquito, a bird dies during a cold spell immediately following a winter rain, a puppy dies of diphtheria, you mow your grass before it goes to seed, a cat catches a rat, a bluejay chases a sparrow away from a crust of bread and the smaller bird loses a meal, or you eat an apple and throw the core (including the seeds) into a garbage disposal unit. Other pressures are

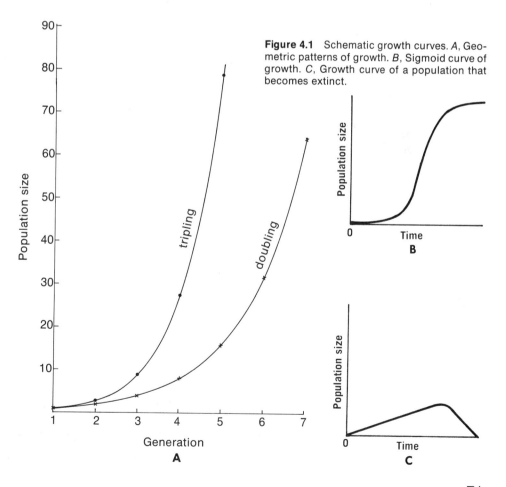

Figure 4.1 Schematic growth curves. *A*, Geometric patterns of growth. *B*, Sigmoid curve of growth. *C*, Growth curve of a population that becomes extinct.

TABLE 4.2 Four Patterns of Population Growth

PATTERN	EXAMPLE
Sigmoid	Yeast in experimental culture
Sharp peak followed by sharp decline	Reindeer on St. Paul Island
Relatively constant	Sheep in Tasmania
Oscillating sharply	Arctic lemmings

less easy to observe, and yet are occurring all around us: a paramecium eats a bacterium, a small crustacean eats a paramecium, a wild oat seed competes with a wild barley seed for an almost microscopic hole in the earth favorable for growth, a parasite infects a beetle, a seed fails to germinate during a drought, an acorn rots during a particularly wet spring, or a hailstone hits a caterpillar on the head. The sum of all the environmental interactions which collectively inhibit the growth of a population is known as the **environmental resistance.** Individual components of the environmental resistance will be discussed in Sections 4.3 through 4.7, but for the moment we will be concerned with the effect that the environmental resistance has on growth. The question we wish to discuss can be stated simply: We know that natural populations cannot grow at geometric rates; how do they actually change in size as time goes on?

Consider some population which is initially very small. The very fact that it is small places it in danger of extinction, because it may not be able to recover from such setbacks as epidemics, famine, or poor breeding. Even if such factors do not totally destroy the population, they will limit its growth rate, and therefore the population will increase only slowly at first. However, once the population is established, its size will rise more rapidly as long as there are adequate food sources, relatively few predators and favorable living conditions. When the population becomes very large with respect to its food supply, availability of shelter, and vulnerability to predators, and becomes so dense that disease spreads rapidly, the environmental resistance increases and the growth rate decreases. The entire curve of growth looks like an S and is said to be **sigmoid** or S-like, as shown in Figure 4.1*B*. The rate of growth shown by the upper right portion of the sigmoid curve is very nearly zero. A zero growth rate does not mean that there are no births and no deaths. It simply means that the total number of births plus immigrations equals the total number of deaths plus emigrations. When this equilibrium is reached, the biotic potential is balanced by the environmental resistance. The magnitude of this upper population level is characteristic for a given species in a given ecosystem; thus, we say that each ecosystem has a given **carrying capacity** for each species. When the carrying capacity has been reached, the system cannot continue to support any more individuals of that species. The carrying capacity is not constant from region to region; for ex-

---

The **carrying capacity** is the maximum number of individuals of a given species which can be supported by a particular environment.

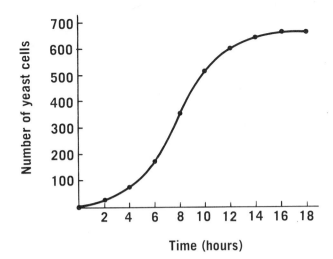

Figure 4.2 Growth function of yeast cells in the laboratory. (From Raymond Pearl: *The Biology of Population Growth,* Copyright 1925 by Alfred A. Knopf, Inc., and renewed 1953 by Maude de Witt Pearl. Reprinted by permission of Alfred A. Knopf, Inc.)

ample, a wheat field has an inherent ability to support more locusts than a short-grass prairie. Indeed, the carrying capacity may vary from time to time: the carrying capacity of a region can be altered by some calamity such as fire or by annual variations in temperature, rainfall, etc.

Do real populations in natural ecosystems grow in a smooth sigmoid fashion and eventually approach a constant size? Yes, sometimes. For example, yeast cells experimentally introduced into a culture do exhibit a sigmoid growth pattern (Fig. 4.2). On the other hand, when 4 male and 21 female European reindeer were experimentally introduced onto St. Paul Island near the coast of Alaska in 1912, a different result was observed. St. Paul Island was free of predators and environmentally favorable to the deer, and the population growth rate initially increased according to a sigmoid pattern, but the expected orderly approach to equilibrium was not observed. (See Figure 4.3.) Instead, a nearly geometric growth rate continued long after the carrying capacity of the island had been exceeded, but then the population declined abruptly. Why did this

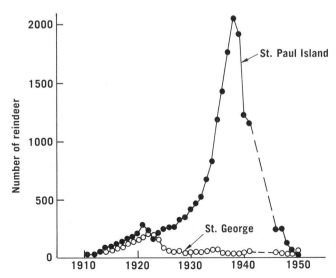

Figure 4.3 Growth function for reindeer on two similar islands. (From C. J. Krebs: *Ecology.* New York, Harper & Row, Publishers, 1972, p. 197.)

73

occur? Initially, the food supply on St. Paul was abundant, and since there was no predation, the animals were healthy and their fertility was high. As the population continued to increase, the reindeer were eating plant matter faster than the island could replace it by photosynthesis. The reindeer, however, were unaware of the instability of their situation; their food was still abundant and they continued to multiply. Then quite suddenly almost all the food was gone. The island was barren, and mass starvation and death occurred, until in 1950, only eight animals remained alive on the island.

This explanation for the observed behavior of reindeer on St. Paul Island may be valid, but if we use our hypothesis to predict that other newly introduced species will follow a similar growth pattern, we may be embarrassed by the facts. For example, at the same time that reindeer were introduced onto St. Paul Island, a similar herd was introduced onto nearby St. George Island, which resembled St. Paul in size and ecological properties. The size of the herd on St. George, however, grew in a more controlled fashion than the St. Paul herd. (See the lower plot in Figure 4.3.) Other introductions of reindeer into Alaska have exhibited population growth patterns intermediate between these two examples. No general theory has been able to explain the observed discrepancies.

Although the pattern exhibited by the reindeer on St. Paul Island is somewhat extreme, the tendency for any population to oscillate is quite common. Thus, when sheep were introduced into Tasmania the population grew to about two million individuals, then declined slightly, rose again, and continued to vary (Fig. 4.4). In this case the observed fluctuations were small, and the population of sheep was nearly constant over a long period of time.

Some populations oscillate sharply. North of the Canadian forest lies the arctic tundra, and one common rodent of the tundra is the lemming. Lemmings exhibit predictable three or four year population cycles. One summer the population is extremely high; the next year the population rapidly declines, or crashes. For another year or two, the population recovers slowly; then it skyrockets for a season and the cyclic pattern repeats itself.

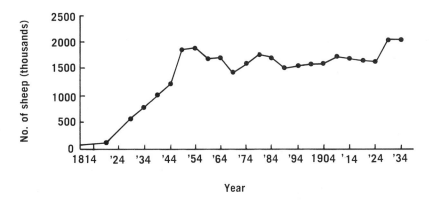

Figure 4.4 Growth of a sheep population introduced into a new environment on the island of Tasmania. (Redrawn from Odum: *Fundamentals of Ecology,* 3rd ed. Philadelphia, W. B. Saunders Co., 1971.)

74

Reindeer in the tundra. (Photograph by Eric Hosking. From National Audubon Society.)

Lemming abundance is associated with forage cycles (Fig. 4.5). During an abundant lemming year, the tundra plants are plentiful and healthy. Most of the available nutrients exist in plant tissue, and there is little stored in the soil reservoir. In the spring following the overabundance of lemmings, the heavily overgrazed plants become scarce, and most of the nutrients in the ecosystem are locked in the dead and dying lemmings. The process of decay and return to the soil requires another year or two, during which time the population and health of both plants and rodents increase.

Unfortunately, recognition of the relationship between lemming and forage cycles does not prove either that forage cycles cause rodent cycles or vice versa.* Some researchers argue that perhaps internal genetic factors

*Judith H. Myers and Charles J. Krebs: "Population Cycles in Rodents." *Scientific American,* June, 1974, p. 38.

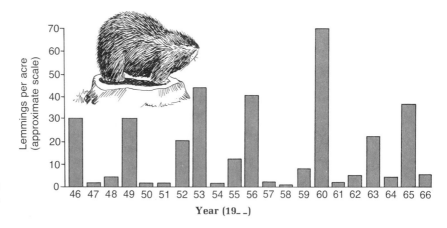

Figure 4.5 Lemming population cycles at Point Barrow, Alaska, from 1946 to 1966.

75

A bloom is a period of very rapid growth or sudden development. We use the word most frequently to describe the growth of a flowering plant in the spring, but other usages are also correct. For example, "The skinny awkward teenager bloomed into a robust adult." When a population of any species, plant or animal, expands *quickly*, it is said to bloom.

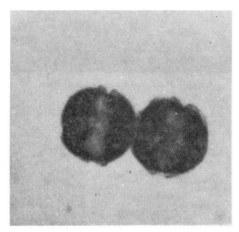

Red tide organism.

may be responsible for the cycles in some rodent populations.

Erratic **population blooms** occur even in old, seemingly stable systems. Since recorded history, a species of red phytoplankton has often been observed to enter a period of rapid growth in coastal areas and turn the ocean red. There seems to be no predictable cyclic pattern associated with this phenomenon, but instead, the plankton grow very rapidly at unexpected times—seemingly in response to a new influx of nutrients. Traditionally, these "red tides" were often observed to occur after flood waters washed large quantities of soil nutrients into the sea. This influx created a plentiful food source. The red-colored plants were best able to take advantage of it. Unfortunately, these organisms discharge toxic substances into the sea, killing fish and aquatic mammals, and are therefore considered a great hazard. Recently, a series of red tides apparently unrelated to natural phenomena have been tentatively attributed to the nutrient content of pollutants discharged into the sea.

The regulatory mechanisms of most ecosystems usually maintain balanced conditions. However, ecosystems that superficially seem "well behaved" do reveal perturbations when examined more intimately. For example, the arctic tundra, stretching to the horizon and colored with many flowers, looks nearly the same whether the lemmings are scarce or abundant. The thin layer of soil resting on icy permafrost is slippery and uneven, and in summer mosquitos are abundant. Herds of caribou migrate across the land. Wolves, bears, foxes, wolverines, lynx, and other animals all leave their tracks. Moreover, this landscape probably has been maintained for many years and will continue far into the future. Thus, over the long term, the system is stable. You have already learned, however, that lemming populations fluctuate from year to year. Thus, whether a system is considered stable depends in part on the length of time it is observed.

4.2 ENVIRONMENTAL RESISTANCE

A study of the population of phytoplankton near the surface of the central oceans reveals that there is a lot of available sunlight but relatively few plankton. Of course, there is sufficient water to support life; why, then, aren't there more plankton? The central ocean is an old ecosystem and one would expect that population equilibrium has been reached; that is, one would not expect average phytoplankton populations to rise appreciably in the next few years. Imagine that you tried to answer this question by studying the size of the grazing population in the deep sea, thinking that perhaps the low level of plant population is due to a high degree of predation. If, however, you determined the total biomass of the system (the weight of plants plus all the heterotrophic organisms) and compared it to the biomass of coastal aquatic ecosystems, or to the theoretical biomass that all the available sunlight could support, you would again find the phytoplankton population of the central oceans to be abnormally low. Why? The answer is that many nutrients such as nitrates and phosphates are scarce in the deep sea. The shortage of nutrients is due to the nature of the physical environment. On land, nutrients are generally retained within an ecosystem for a long time, but in the sea small pieces of debris settle below the euphotic zone and become unavailable. Moreover, nutrients moving into the sea from sources on land are diluted by the vastness of the oceans. Thus, the populations of phytoplankton are limited by a shortage of a few nutrients, even though water, light, and many mineral salts are available in quantity.

In general, the population level of any species in any ecosystem is regulated by those essentials of life which are available in the *minimum quantity*. This **Law of Limiting Factors** applies not only to nutrients but also to various physical conditions. Thus, an organism cannot survive without sufficient light, heat, moisture, and space. Additionally, many organisms are limited by the turbulence of their environments. For example, benthic organisms in streams cannot tolerate an excessive water flow; ocean dwellers are limited by the force of waves; and many plants cannot survive on exposed windswept areas.

Populations may be limited not only by physical deficiencies but also by excesses. Desert plants cannot survive where there is too much water, polar bears cannot survive in the tropics, and even saltwater fish would die in the Great Salt Lake, where salt concentrations are unusually high.

The growth of organisms is limited by the presence of other organisms in the ecosystem, as well as by the physical environment. Thus, a population is partially controlled by the abundance of its prey and by the pressures of its predators. In addition, if an essential nutrient is scarce, then different species will compete for it; if winds, water, currents, waves, or changes in temperature make survival difficult, plants and animals will compete for shelter. The biological component of environmental resistance consists largely of pressures from predation and competition. These two topics will be discussed in greater detail in Section 4.4.

Populations are regulated not only by external factors such as weather, predation, or competition, but sometimes also by internal factors. Internal regulation of populations may be genetically influenced, as in the case of the species of rodents cited earlier, or may be triggered by some other mechanism when an individual, a family, or a group spontaneously restricts its birth rate. A human female may choose to bear two children even though she has enough money to feed, clothe, and shelter many more, and is bio-

Recall that phytoplankton are the microscopic free-floating plants that are responsible for most of the primary food production in aquatic systems (page 54).

The growth of trees in alpine regions is limited by weather conditions.

logically capable of bearing a larger family. Patterns of voluntary birth control are difficult enough to analyze for human populations (see Chapter 6), but when we study animal populations, the patterns and motivations are even more obscure. For example, a wolf pack typically consists of a dominant male (the leader), his mate (the dominant female), and a number of subordinate males and females. The dominant pair mate every year, but other members of the pack generally do not copulate even though they are sexually mature. In one study of a wolf pack kept in a large fenced enclosure, the dominant male became aroused and tried to mate with a subordinate female, but she would not allow it, and with tail between her legs, she cowered and avoided his advances. It is easy to understand that this voluntary birth control is beneficial to a population of wolves in the wild, for otherwise the wolves, who are not themselves subject to predation, would overpopulate their range and face mass starvation. It isn't easy, however, to understand why the wolves act this way or what drives individuals to abstain. The female in the experimental enclosure was well fed all the time and was not acting in direct response to a food shortage. Self-regulation has been observed in many systems. The phenomenon is not limited to mammals, for some species of insects, birds, fish, and reptiles are also known to limit their own populations even if food is plentiful and predation is minimized.

4.3 THE ECOLOGICAL NICHE

There are roughly one and a half million different species of animals and one-half million species of plants on Earth. Each species performs unique functions and occupies specific habitats. The combination of function and habitat is called an **ecological niche.** To describe a niche fully, one would first have to describe all physical characteristics of a species' home. One might start with specifying the gross location (for example, the Rocky Mountains or the central floor of the Atlantic Ocean) and the type of living quarters (for example, a burrow under the roots of trees). For plants and the less mobile animals, one would describe the preferred microenvironment, such as the water salinity for species living at the interfaces of rivers and oceans, the soil acidity for plants, or the necessary turbulence for stream dwellers. An animal's trophic level, its exact diet within the trophic structure, and its major predators are also important in the description of its niche. Mobile animals generally have a more or less clearly defined food-gathering territory, or **home range,** which is another factor in establishing the physical niche.

A niche is not an inherent property of a species, for it is governed by factors other than genetic ones. Social and environmental factors contribute to the choice of niche. For

Animals live in many different types of micro-environments. These tent caterpillars congregate in their tent, where they keep warm and relatively dry in the early spring.

example, a certain population of tropical jellyfish, *Aurelia aurita,* swims fastest in water that is at a temperature of 29°C. Of course, it would be unreasonable to expect all the jellyfish of this population to be living in waters of exactly 29°C all the time. The weather changes, cold and warm spells occur, yet organisms survive. Imagine a situation in which a warm, sunny, sheltered bay provided an optimal physical environment for jellyfish but exposed them to abnormally high predatory pressures. Since the conditions in the bay would not be optimal, many individuals might migrate to a less favorable physical environment to find more favorable biotic surroundings. Thus, the observed niches of a bay-dwelling jellyfish and of a migrating jellyfish are different from each other and from the theoretical optimal niche. The niche of a given species in a given ecosystem is not a set of conditions that would be best suited to the genetic makeup of the organism, but rather it is the best accommodation that the organism can make to the realities of its environment.

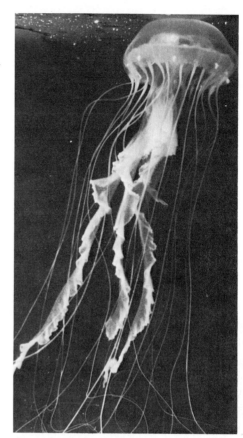

The sea nettle, *Chrysaora quinquecirrha,* a common jellyfish along the Atlantic coast. (Courtesy of William H. Amos.)

4.4 INTERACTIONS AMONG SPECIES

Food webs are complex. Populations overlap in place and function. Many homeostatic equilibria exist simultaneously in the same system. These complexities—the interplay of the checks and balances, the interactions between species—maintain population levels and thereby stabilize the system. Interactions among species are made up of separate events; an individual of one species interacts with one individual of another species at a given time. Of course, every isolated encounter has importance to the two interacting individuals. But when analyzing the system, we wish to focus attention upon the effects of the encounter on the total community. In a larger sense, we are ultimately concerned with the effects of the sum of encounters on the total community. For example, we need to know the relationship between deer and aspen saplings, but that bit of information is only one portion of a more comprehensive understanding of the relationship between the community of grazers and the community of plants.

When two individuals come into contact, the interactions are beneficial, harmful, or neutral to either one or both of them. Interactions between two species can be catalogued into eight major types: neutralism, competition, amensalism, predation, parasitism, commensalism, protocooperation, and mutualism. These are discussed in the ensuing paragraphs.

Neutralism is the inconsequential case of very little interaction. Wild rose bushes and lynx have practically no direct relations with each other.

Competition is an interaction in which two or more organisms try to gain control of a limited resource. When

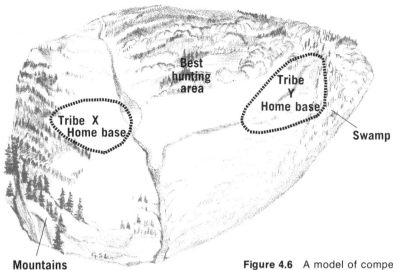

Mountains

Figure 4.6 A model of competition.

competition for a given commodity is severe, individuals often modify their demands. Thus, competition is a driving force toward diversity in natural ecosystems. To understand the interplay between competition and diversity, imagine two tribes, X and Y, of nomadic hunters living in the area of the fictitious map in Figure 4.6. Suppose that the best hunting grounds are found along the flood plain of the river. Naturally, both tribes would compete for this prime hunting area, perhaps by sporadic warfare. One probable result of this competition is that tribe X and tribe Y would maintain different secure territories on the edge of the flood plain. Between these two home bases, some of the hunting area would be visited by the hunters of both tribes. If tribe X had sole possession of the best hunting areas, the people could be fed easily from the population of lowland animals, but in the face of competition from tribe Y, the hunters from tribe X must obtain part of their game by hunting the mountainous area near their home base. Similarly, the hunters in tribe Y must search the swamp for a portion of their food. If some catastrophe, such as fire, destroys the lowland hunting areas, members of both tribes, experienced in hunting alternate sources of food, could probably feed their people.

Competition between animal species can be considered analogous to the competition between the two tribes. If the words "niche" and "species" are used in place of the words "home territory" and "tribe," the example just given illustrates a simplified case of competition between two animal groups.

These relationships can be further illustrated by several examples taken from natural systems. Consider two species of barnacles that live on the Scottish coast. The larger species, *Balanus,* dominates between the high and low tide marks, where the individuals can be assured of daily contact with the sea. The smaller species of barna-

80

Several species of barnacles and mussels compete for space in this intertidal zone in northern California.

Close-up of barnacles.

cles, *Chthamalus*, lives predominately higher up along the beach, where they must survive frequent and more prolonged periods of desiccation. Experiments have shown that in the absence of competition, the smaller barnacles can survive over a larger range (see Figure 4.7), while the larger species survives only in the region of plentiful moisture. The smaller species, however, is unable to compete with the stronger one in a wetter environment, and any larvae of the smaller species that attempt to grow outside of their observed range are competitively displaced or dislodged and eaten by predators. One species survives by virtue of its ability to adapt to extreme physical pressure, the other by virtue of its competitiveness and its ability to survive predation, and thus diversity is built into the system.

We can imagine that if the niches for two separate species were identical, competition for food and shelter

Desiccate, which means to deprive thoroughly of moisture, is from the Latin word meaning "to dry completely." An obvious cognate is the French word for dry, "sec." Sec is used in English to describe certain wines.

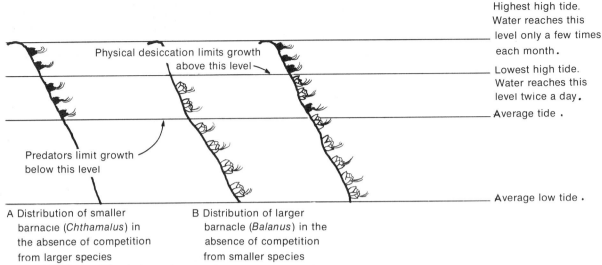

Highest high tide. Water reaches this level only a few times each month.

Lowest high tide. Water reaches this level twice a day.

Average tide.

Physical desiccation limits growth above this level

Predators limit growth below this level

Average low tide.

A Distribution of smaller barnacle (*Chthamalus*) in the absence of competition from larger species

B Distribution of larger barnacle (*Balanus*) in the absence of competition from smaller species

Figure 4.7 Competition between two species of barnacles.

would be particularly intense. In fact, both experiment and field observation have led to the generalization that two species in the same ecosystem cannot both survive in the same niche. If there are two species with the same preferred niche, competition will lead either to the elimination of one or to the adaptation of one of them to fit a new niche. A corollary to this rule is the generalization that in an ecosystem which houses many species, not only is the diversity of niches very large, but also species are coadapted to overlapping niches so that elimination through competition is reduced.

It is reasonable to ask whether or not two species of plants living side by side, with their roots in the same soil and their tops touching, are not, in fact, occupying the same niche. Naturally, not every set of plants living near each other has been studied, but those studies that have been conducted have shown that if there isn't a difference in nutrient or light requirements, there is usually some difference in root depth or in timing of life cycles. Of two species of clover that coexist, one was observed to grow faster and spread its leaves sooner than the other. The second clover species ultimately grows taller than the first. Thus, each has its period of peak sunlight, and they are both able to survive.

In almost all cases, competition between two species retards the growth of both competitors, and therefore it is tempting to say that one species would be "better off" somehow if its competitors were eliminated. Natural ecosystems, however, can be fascinatingly complex. In one section of the shoreline in the state of Washington, barnacles and sea anemones competed for suitable living space in the intertidal regions. When all the barnacles were removed during an experiment, the anemone population grew rapidly, as one might have expected, but this bloom was followed by a rapid decline, as shown in Figure 4.8. The decline did not

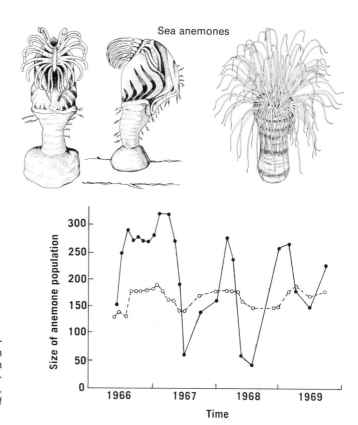

Sea anemones

Figure 4.8 Effect of removal of barnacles from intertidal region. Solid circles: Anemone population in an area where barnacles were removed. Open circles: Control; anemone population in an undisturbed area. (From P. K. Dayton: Ecolog. Mongr., *41*(4), 1972, p. 373. Reprinted with permission of author.)

result from a food shortage; rather, it was found that the anemones died of desiccation. In an unperturbed system anemones grow close to their competitors, the barnacles, thereby finding shelter from the hot summer wind.

We have shown some relationships among niches that occur when several species gather food in a single geographic area, or when several species of stationary organisms, such as barnacles, live along a gradient of climate. Sometimes niche interactions are defined by patterns of migrations. Of the numerous grazers in the Serengeti grasslands of Africa, the relationship between the zebra and the Thompson gazelle is illustrative. The former is a large, horse-like animal which does not chew its cud. Its physiology is therefore quite different from that of the cud-chewing gazelle. Since the zebra is able to eat large quantities of food and excrete the unnecessary carbohydrates, it can live on plants with a low protein content. The gazelle's digestive system is unable to handle such large quantities of food, and it must subsist on protein-rich grasses. During the abundant wet season, both animals graze together on the hillsides, but when the yearly drought arrives, the zebra cannot subsist on the sparse growth and moves down into the flood plain. There it eats the tall grasses, which are plentiful but nutritionally poor. The gazelle, meanwhile, remains to wander about the hillsides, picking out the small, rich surface plants which the zebra did not con-

Migrating herd in Africa. (From C. A. Spinage: *Animals of East Africa.* Boston, Houghton Mifflin Co., 1963.)

Top, Zebras; *Bottom,* Gazelle. (Courtesy of Dr. C. A. Spinage, College of African Wildlife Management, Mweka, Moshi, Tanzania.)

sume. Because the gazelle is small and needs less total food than the zebra, it can afford the time to search for food. The zebra would have starved had it stayed. After some time, even the gazelle must migrate. The tall grasses of the flood plain have been cut down by this time by the zebras, leaving the lower, smaller, richer surface vegetation for the gazelle. The zebra continues to migrate in a path that eventually brings it back to the hillsides at the start of the next rainy season.

Amensalism is an interaction in which the growth of one species is inhibited, while the growth of another species is unaffected. As an example, certain shrubs native to southern California secrete toxic chemicals which kill nearby grasses (Fig. 4.9).

Predation is an interaction in which certain individuals eat others. Since all heterotrophs must eat to survive, predation is an integral part of the function of any ecosystem. In stable ecosystems, growth and predation are balanced in such a way that all species maintain viable populations. In the example on page 62, we saw that the grazer population in our hypothetical valley in Nepal was regulated by both the availability of its supply of food (the grass) and the size and vitality of its predator population (the tigers). In turn, the tiger population was regulated mostly by the size and health of the herds of grazers.

Of course, not all ecosystems are so finely regulated, and predators have destroyed certain prey populations in some regions. The homeostasis of an ecosystem is depend-

84

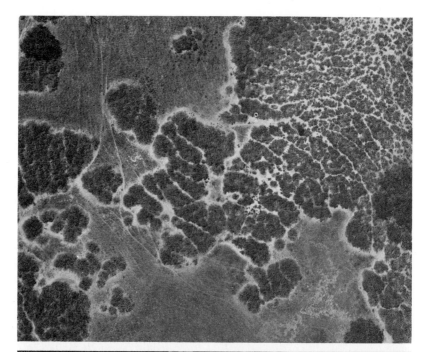

Figure 4.9 *Top,* Aerial view of aromatic shrubs *Salvia leucophylla* and *Artemisia californica* invading an annual grassland in the Santa Ynez Valley of California and exhibiting biochemical inhibition. *Bottom,* Close-up showing the zonation effect of volatile toxins produced by *Salvia* shrubs seen to the center-left of *A.* Between *A* and *B* is a zone two meters wide, bare of all herbs except for a few minute, inhibited seedlings (the root systems of the shrubs which extend under part of this zone are thus free from competition with other species). Between *B* and *C* is a zone of inhibited grassland consisting of smaller plants and fewer species than in the uninhibited grassland seen to the right of *C.* (Photos courtesy of Dr. C. H. Muller. University of California, Santa Barbara.) (From Odum: *Fundamentals of Ecology,* 3rd ed. Philadelphia, W. B. Saunders Co., 1971.)

ent not on individual restraint by a particular predator but on the statistical balance between offense and defense. Such natural controls do not always function smoothly. We saw in Section 4.1 that dramatic population fluctuations occur, and we will see in Chapter 5 that sometimes the predator population causes a species of prey to become extinct. In relatively stable ecosystems, however, consumption and growth are nearly balanced.

Predation is a delicate and fascinating process. One of us (Jon) was fortunate enough to watch a lone timber wolf stalk a moose by a river in the northern Canadian forest. The wolf followed the moose along a river bank, always maintaining a separation of several hundred yards. While the moose frequently looked toward her predator, she never broke into a run, but continued to feed and move slowly downstream. After about five minutes the wolf turned abruptly, trotted over the hill, and disappeared from view. Both animals knew that a wolf was no match for a young, healthy cow moose. Empirical analyses of caribou killed by wolves have shown that the old, the crippled, the sick, and the very young are killed in disproportionately high numbers. A healthy adult caribou or moose is attacked only very rarely by a wolf or wolf pack. By selecting the less fit animals as victims, predation is a force in the natural selection of the hunted species.

This story is only a small part of the total picture of predation. It has already been mentioned that two species cannot occupy exactly the same niche because one always dominates or displaces the other. It has been shown that in the absence of predation, two species with different but similar niches often are unable to coexist. Yellowstone Park has a food web that was once balanced and rich with diverse species. Human destruction of the predator population has led to a rapid rise in the number of elk. The abundance of elk has put so much competitive pressure on the deer that they are threatened with elimination from the area.

In other systems it has been shown that predatory pressures either *increase* the number of species of prey in a

Predators.

Left below, Archer fish knocks ladybug off a leaf with accurately aimed drops of water. (From Roy Pinney—Globe Photos, Inc.)

Right below, Wolves and moose on Isle Royale, Michigan. (Photo by L. David Mech.)

Opposite page, Artists' renditions of predacious fish, reptile, insect, and bird.

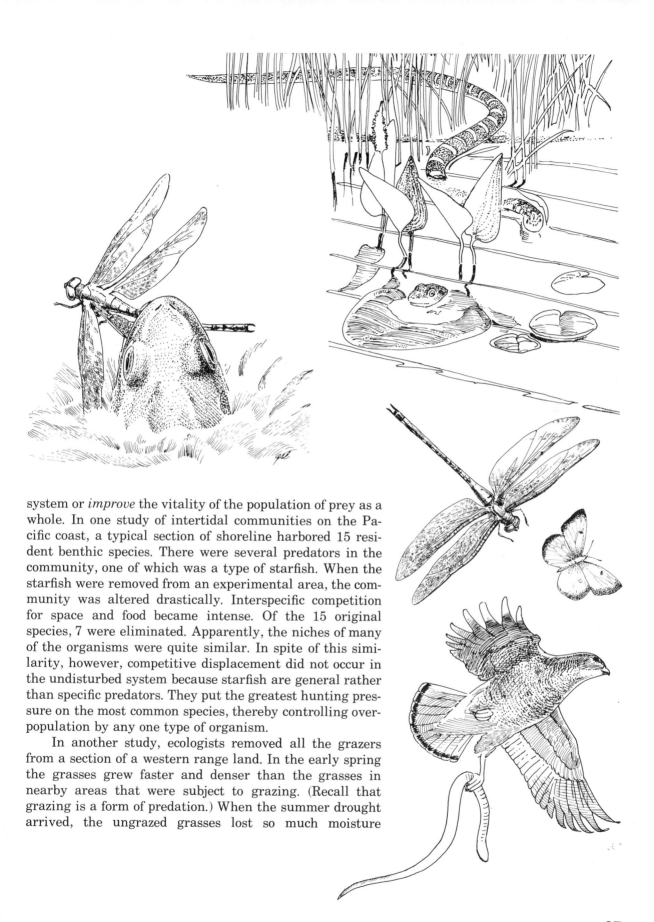

system or *improve* the vitality of the population of prey as a whole. In one study of intertidal communities on the Pacific coast, a typical section of shoreline harbored 15 resident benthic species. There were several predators in the community, one of which was a type of starfish. When the starfish were removed from an experimental area, the community was altered drastically. Interspecific competition for space and food became intense. Of the 15 original species, 7 were eliminated. Apparently, the niches of many of the organisms were quite similar. In spite of this similarity, however, competitive displacement did not occur in the undisturbed system because starfish are general rather than specific predators. They put the greatest hunting pressure on the most common species, thereby controlling overpopulation by any one type of organism.

In another study, ecologists removed all the grazers from a section of a western range land. In the early spring the grasses grew faster and denser than the grasses in nearby areas that were subject to grazing. (Recall that grazing is a form of predation.) When the summer drought arrived, the ungrazed grasses lost so much moisture through their extensive leaf systems that many plants yellowed, so that by the end of the season the ungrazed plot was less healthy than the natural one.

Parasitism is a special case of predation, in which the predator is much smaller than the victim and obtains nourishment by consuming the tissue or food supply of a living organism known as a **host.** Just as predator-prey interactions are balanced in healthy ecosystems, parasite-host relationships have also become part of the mechanism of homeostasis in nature. It must be stressed that this type of balance observed in old systems does not imply that a new parasite (or predator or grazer), artificially imported from another continent, will immediately establish itself as part of a stable system. On the contrary, a new species may find a new niche for itself and increase unchecked until, perhaps many years later, food supplies decline or another species migrates, is introduced, or evolves to control the rampant one.

Commensalism is a relationship in which one species benefits from an unaffected host. Several species of fish, clams, worms, and crabs live in the burrows of large sea worms and shrimp. They gain shelter and often eat their host's excess food or waste products, but do not seem to affect their benefactors.

A relationship favorable to both species is called **protocooperation.** Crabs often carry coelenterates on their backs, and move them from one rich feeding ground to another. In turn, the crabs benefit from the camouflage and protective stingers of their guests. (Not all crabs and coelenterates are mutually cooperative.)

Mutualism is an interaction beneficial and necessary to both parties. Lichen, which grows on bare rock, resembles an extremely thin layer of vegetation. Actually, the lichen is a mixture of a fungus and an alga. The fungus, which does not contain chlorophyll and thus cannot produce its own food by photosynthesis, obtains all of its food energy from the alga. In turn, the alga cannot retain water and, in some harsh environments, would dehydrate and die if it were not surrounded by fungus. Here the dependence is direct because the organisms must grow together in order to survive.

Another example of a mutualistic interaction can be found within our own bodies. Millions of bacteria live in the digestive tracts of every person. These organisms depend on their host for food but, in return, aid in the digestive process and are essential to our survival.

4.5 COMMUNITY INTERACTIONS

Interactions within an ecological community are composed of a multitude of two-species encounters. The simplest type of community interaction is a chain reaction. Cats eat rats, rats attack beehives, bees pollinate flowers and produce honey. Thus, the population of wild flowers and the price of honey are partly dependent on the population of cats. Because every animal has a place in a food web, the removal, addition, or exploitation of any species necessarily causes reverberations, large or small, beneficial or harmful, throughout the system. This rule holds for destruction of "pests" as well as for the introduction of new game species.

A previous example suggested that the plant life of an area is generally healthier if grazed than if left ungrazed. Upon closer examination, it is found that plant communities thrive best if they are grazed by a community of grazer species instead of by a single species. A single selective grazer is apt to put pressure on a few species of plants, allowing others to dominate.

Interactions between diverse species tend to promote stable ecosystems. A good example is afforded by the role of three species of grazers—moose, elk, and American bison—on the ecological balance of the Elk Island National Park in Canada. Moose eat saplings and small bushes. With brush growth held in check, grasses find room to grow. Bison eat grass. Elk eat either leaves or grasses. In this way, the community of grazers acts on the community of plants to ensure that an equilibrium is maintained. Of course, there are many other interactions, such as the relationships between the community of predators and the community of grazers. The net result is a balanced, continuous, self-perpetuating ecosystem.

Diversity is also important in the plant kingdom. A single species does not constitute a stable system capable of buffering itself against changing weather. For example, both annual and perennial grasses grow in the prairie. The perennial grasses and some low bushes have deep roots, while the annual plants depend on much shorter and less extensive root systems. During dry years there is so little water that many annuals die. However, the perennials, which use water deep underground, are able to live; in doing so, they hold the soil and protect it from blowing away with the dry summer winds. In years of high rainfall the annuals sprout quickly, fill in bare spots and, with their extensive surface root systems, prevent soil erosion from water runoff. Survival of both types of grasses is ensured by minimization of root competition because the different plants have root systems that reach different depths. In addition, not all species flower at the same time of the year, so that the seasons of maximum growth and consequent maximum water consumption differ.

In canyons in the Colorado Rockies, ponderosa pine, juniper, and small cacti predominate on the dry sunny side, and the blue spruce, Douglas fir, and flowering plants predominate on the wetter, shady side. Although both sides

Figure 4.10 Grazers in Elk Island National Park. *A*, Moose; *B*, elk; *C*, American bison.

89

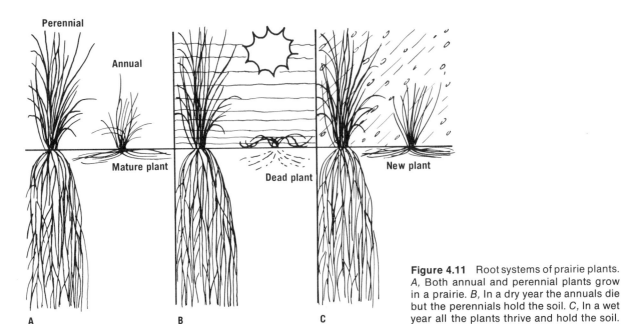

Figure 4.11 Root systems of prairie plants. *A,* Both annual and perennial plants grow in a prairie. *B,* In a dry year the annuals die but the perennials hold the soil. *C,* In a wet year all the plants thrive and hold the soil.

The south-facing slope of this ridge is covered with grasses, cacti, and a few pine trees, while the north slope is densely wooded. Photograph taken in Sunshine Canyon, Boulder, Colorado.

of the canyon contain representatives of all species, the niche of the pine is sufficiently different from that of the spruce to preserve the separation. If a prolonged drought were to strike the canyon, there would probably be some shift in the plant population in favor of those species that are more effective in conserving water. Thus, changing con-

ditions often change the order of dominance of species. The advantage of community diversity is that the changes that do occur are relatively mild and the ecosystem is not disrupted.

Numerous examples throughout this chapter have shown that most stable systems are diverse and that most diverse natural systems are stable. It might, therefore, seem logical to think that diversity *causes* stability. But this conclusion may not be correct. After all, if diversity were *inherently* stabilizing, why would natural systems not be even more diverse than they are? In other words, why are there not more species? In fact, some ecologists* suspect that extremely complex systems, like very simple ones, are unstable. Unfortunately, we cannot devise experiments to make natural systems more complex and then test their stability. Therefore, we don't know the answers.

4.6 GLOBAL INTERACTIONS AMONG ECOSYSTEMS

We have emphasized that the various species within an ecosystem are interdependent. This dependence can be observed within a radius of a few inches, as in the case of lichen, or within several square miles, as is the case for many predator-prey relationships. It is also apparent that entire ecosystems depend on other ecosystems which may be located thousands of miles away. The life around a large river depends on the yearly cycles of river flow. In turn, the river flow depends on the water balance at the tributaries. This water balance is controlled by the forest systems. Thus, the overall state of the forests on banks of a small creek has a direct effect on the life cycles of organisms at the mouth of the river.

In a larger sense, the life forms on the continental and oceanic masses of the Earth are linked together into a single interdependent system. Southerly winds and warm ocean currents bring heat to the northern tundra, an area that does not receive enough direct solar energy to support much life. The global balance of carbon dioxide and oxygen in the atmosphere is the classic example of biospheric homeostasis. Before life evolved, the primary constituents of Earth's atmosphere were nitrogen, ammonia, hydrogen, carbon monoxide, methane, and water vapor (Fig. 4.12). Oxygen was present only in trace quantities. Although some scientists believe that geological processes altered atmospheric composition, most feel that the excess oxygen released by the first autotrophs built up slowly over the millennia until its concentration reached about 0.6 percent

Nitrogen, $N \equiv N$

Ammonia,

Hydrogen, $H{-}H$

Carbon monoxide, $C \equiv O$

Methane,

Water,

Figure 4.12 Chemicals present in the Earth's primitive atmosphere.

*See, for example, Robert M. May: *Stability and Complexity in Model Ecosystems.* Princeton, N.J., Princeton University Press, 1973. 235 pp.

The Earth consists of a delicately interconnected set of ecosystems. (Courtesy of NASA.)

James E. Lovelock is an English chemist who has done important work on the analysis of trace gases, including atmospheric pollutants.

of the atmosphere. Multicellular organisms could have evolved only at this point, because aerobic respiration is a prerequisite for their development. The emergence of various organisms about 600 million years ago triggered an accelerated biological production of oxygen. The present oxygen level of about 20 percent of the atmosphere was reached some 450 million years ago. While there have been several more or less severe oscillations since that time, an overall oxygen balance has always been maintained.

If the oxygen concentration in the atmosphere were to increase even by a few percentage points, fires would burn uncontrollably across the planet; if the carbon dioxide concentration were to rise by a small amount, plant production would increase drastically. Since these apocalyptic events have not occurred, the atmospheric oxygen must have been balanced to the needs of the biosphere during the long span of life on Earth. By what mechanism has this gaseous atmospheric balance been maintained? The answer appears to be that it is maintained by the living systems themselves. The existence of an effective homeostatic mechanism of the whole biosphere has led J. E. Lovelock to liken the biosphere to a living creature. He calls that creature by the Greek name for Earth, *Gaia*. He believes that not only is the delicate balance of oxygen and carbon dioxide biologically maintained but also that the very presence of oxygen in large quantities in our atmosphere can be explained only by biological maintenance. If all life on Earth were to cease and the chemistry of our planet were to rely on abiological laws alone, oxygen would once again become a trace gas.

The concept of biological control over the physical environment warrants careful consideration. It says that, just as an organism is more than an independent collection of its organs and an ecosystem is more than an independent collection of its organisms, the biosphere is more than an independent collection of its ecosystems. The body chemistry of a human being can function only at or within a few degrees of $37°$ C ($98.6°$ F), but normally we are not in mortal danger of very high or low body temperatures, because there are mechanisms by which our bodies maintain the proper temperature. Similarly, Lovelock believes that modern life can exist only in an atmosphere at or very close to 20 percent oxygen, but we are not normally in mortal danger of conflagration or starvation because there are mechanisms by which we, the species of Earth, maintain the proper oxygen concentration.

An alternate theory claims that our physical environment has evolved through a series of inorganic reactions, and that biological and physical evolution were independent. The difference between these two beliefs is not trivial. If Lovelock is correct, then a large biological catastrophe such as the death of the oceans or the destruction of the rain forest in the Amazon Basin could cause reverberations throughout our physical world that might create an inhospitable environment for the rest of the biosphere. Alternatively, if the physical world did evolve independently of the biological and is currently controlled by inorganic processes, such a doomsday prediction concerning oxygen balance might be considered unnecessarily alarming.

4.7 NATURAL SUCCESSION

We have discussed the mechanisms by which natural ecosystems maintain balance, but we must remember that these mechanisms do not prevent change; rather, they provide compensating processes that tend to drive the system back to its original state. These compensations do not always work.

Natural succession is defined as the sequence of changes through which an ecosystem passes as time goes on. The **climax** is the final stage, the stage that is "unchanging." Of course, the word "final" is used with reservation, because the slow process of evolution changes everything. The composition of the climax depends on temperature, altitude, seasonal changes, and patterns of rainfall and sunlight.

As succession continues, changes occur not only in the

terrain and the types of species present but also in the trophic structure of the system. The food web grows more complex, diversity generally increases, and organisms tend to develop highly specialized niches. In the final climax, photosynthesis and respiration have reached a balance. Much of the total respiration that does occur is initiated by decay organisms. Although a climax system supports great quantities of organic matter, there is little net growth. Such a system can be compared to a human adult who does not gain or lose weight appreciably, but is alive nevertheless. A climax would be a poor agricultural system, for agriculture must produce excess food for harvest. But the climax structure is the type of system that has produced things which we traditionally consider beautiful — the redwood forests, the Great Plains, and the majestic hardwood stands that used to grow along the eastern seaboard.

To a large extent, our perception of change in a system depends on how coarse or fine a time scale we choose for our measurements. A small lake is often used as an example of a stable ecological system. Plants, large animals, and microorganisms all exist in balanced relationships. Temporary imbalances occur and are adjusted, as in all natural systems. However, for most lakes the incoming streams and rivers bring more mud into the environment than is removed by the outgoing streams. The effect is small; it may not be noticed in one human lifespan. A short-term study of the ecology of a lake would conclude that the ecosystem was in balance. But mass balance is disrupted by the steady addition of solid matter from incoming streams. In time, the lake begins to fill up with mud. The vegetable and animal life changes. New plants appear that can root in the bottom mud and extend to the surface where light is available. The trout give way to carp and catfish. If an ecologist studies the lake at this stage he might again say that it was a stable system. Again, minor imbalances and adjustments could be observed, but the overall system appears stable. Yet mud continues to flow into the lake. In addition, since it is common for the plants in the lake at this stage to produce more food than is consumed by the herbivores, the bottom fills up with humus. Eventually the lake may become so shallow that marsh grass can grow.

The marsh system is characterized by a net overproduction of organic matter, as indicated by imbalance in the following equation:

$$6CO_2 + 6H_2O \underset{\text{respiration}}{\overset{\text{production}}{\rightleftharpoons}} C_6H_{12}O_6 + 6O_2$$

This formulation tells us that photosynthesis and respiration (shown by Equations 1 and 2 on pp. 48–49) are reversible. The unequal lengths of the two arrows denote that

A marsh evolving into a meadow. (From *North American Reference Encyclopedia*.)

more organic matter is produced than is oxidized by respiration. In biological terms, this means that there are not enough grazers to consume the large bloom of vegetation, resulting in a net accumulation of litter. Thus, the energy of photosynthesis is stored in plant tissue. This tissue remains largely uneaten and fills the marsh.

A marsh is considered a stable system over a short range of observation, but in most cases it is gradually evolving into a meadow. Litter accumulates because decay in the marsh system is slow. Slow decay thus implies poor nutrient recycling. Such a system is inherently unstable, as is typical of any early successional stage. The accumulated litter gradually fills in the marsh, which slowly becomes a meadow.

The meadow, too, may change. If the climate is right, trees can start to grow. First, shrubs appear, then quick-growing soft woods like birch, poplar, and aspen. The soft woods are replaced by pine; and finally, in what is called the climax, the pine is replaced by hardwoods.

During the succession of plant species, each species prepares the way for the next while contributing to its own extinction. Marsh grass could never grow without the rich soil of the partially decayed algae. However, by producing the rich soil, the algae helps to fill the lake and thereby destroy its own environment.

We cannot estimate the amount of time typically required to fill a lake, because that depends on many factors, such as the original volume of the lake and the net rate of accumulation of solids. For example, Lake Tahoe in California is so deep and clear that succession, if it occurs at all, would take geological time. The ecologist is generally not concerned with processes that are so slow. For all practical purposes, many lakes may be considered to be in a climax condition.

A meadow evolving into a forest.

The time for progression from grassland to a climax forest has been measured. In the southwestern United States, grasslands give way to shrub thickets in 1 to 10 years, the shrubs become pine forests in 10 to 25 years, and the pine forests give way to hardwoods after about 100 years. However, for an ecosystem to be indefinitely stable there must be a complete balance. This is never the case on Earth. Daily, seasonal, yearly, and long-term fluctuations occur in all natural systems, even in the most stable ones. The ultimate consideration for the continuation of life on this planet is that the sum total of all the changes results in worldwide balance.

The preceding discussion does not explain how a grassland, a marsh, or the tundra can be a climax ecosystem. We have presented situations that sometimes, but not always, represent reality. The great plains that once stretched unfenced from the eastern slope of the Rocky Mountains almost to the Mississippi River never became forests because the rainfall wasn't sufficient. Instead, the grasslands evolved into a climax system, with all the trophic structure, energy balance, and stabilizing mechanisms characteristic of the most stable forest ecosystems.

Climax systems are often determined in large part by seasonal variations and geological cycles. A prairie plentiful with buffalo grass grew in Kansas because there

The Everglades. This river of grass, or Pay-hay-Okee as the Seminoles call it, is a strange mix of temperate and tropical zones. Forest and jungle, fresh and salt water ecosystems survive together in the 2000 square miles that comprise the national park. The glades are home to rare species of wildlife, such as the crocodile, manatee, and wood stork, which are found nowhere else in the United States. (National Park Service.)

wasn't enough rain to support the forests, while the Everglades marsh has existed for a long time because the rain comes in definite seasons. During the flood times, the Everglades behaves like an early successional marsh and is characterized by an overproduction of plant matter and a general silting of the streams and pools. If allowed to persist at this stage, the Everglades would soon follow the successional path toward a forest system. During the periodic droughts, the pools dry up, and the litter decomposes and sometimes burns rapidly. When the rainy season returns, the fallen leaves and logs have been recycled back to fertilizer, the pools have been cleared, and conditions are once again ripe for a bloom of marsh life. The total cycle in the Everglades can't be classified neatly either as an early successional stage or as a climax system. During periods of bloom, when net production greatly exceeds net consumption, we are reminded of a transient marsh, yet the overall stability and continuity of the system resembles that of a climax system.

The role played by fire in succession is also important. When the pioneers settled near the edge of the prairie in northern Wisconsin, plowed it, planted it, and controlled it, they discovered that any area left uncultivated soon became thick with new tree seedlings which sprouted and grew rapidly. A study of the area showed that the prairie

The Everglades, the dominant biological community south of Lake Okeechobee in Florida, includes extensive regions of sawgrass, cypress swamp, and mangrove. Its brackish waters lead to the Florida Bay and the Gulf of Mexico. It is a highly developed ecosystem with a large variety of plant and animal species and very complex biotic relationships, still not thoroughly understood.

A forest fire is destructive, but fire has always been a part of nature. Fires also clean and refertilize, and alter successional paths. (Courtesy of Forest Service, USDA.)

that used to grow there had been periodically consumed by fire. Fast-growing grasses quickly regained control, while forests were never able to survive even though the soil and weather conditions were favorable. These grasslands exhibited a **fire climax,** a condition in which the continuance of a given system is maintained by fire. During growth after a fire, the ecology of the area resembled that of an early successional stage, for the standing quantity of organic matter was low and production was faster than consumption. Eventually, animals and plants evolved which were able to accommodate to, or even depend on, the fires. Some grasses and trees developed seeds that sprouted only after being cracked open by fire. Those trees that did exist evolved thick, fire-resistant barks. Although fire returned fixed nitrogen to the atmosphere, this effect was counterbalanced because nitrogen-fixing legumes were common among the early successional plants.

It is interesting to consider what determines natural boundaries between climax systems. In many places, local weather patterns such as those observed around a large river or a mountain range result in rather rapidly changing climates with the resultant change in climax systems. In other places, soil structure or local factors such as human work will affect the change. Even under these conditions, one wouldn't expect an abrupt transition between ecosystems but rather some sort of border area that harbors representative species of each neighboring region. If you live near a place that is part farmland and part woods, take a walk in the country some day and direct your attention to the boundary between plowed fields and woodlots. In most places you will find a band, sometimes not more than 10 or 20 yards wide, in which birch, poplar, or aspen saplings mingle with various grasses and maybe a corn stalk or two. Flora from both the field and the woods will be present. Additionally, many types of bushes that aren't characteristic of either major ecosystem will be found. If you have the eye of a naturalist or of a hunter you will discover that more rabbits and upland game birds live here, where they can take advantage of the rich plant life, the good cover, and easy access to crops or woods. Such a border area is known as an **ecotone.** Ecotones generally support more life and numbers of species than either bordering climax area. An estuary is an example of an aquatic ecotone.

4.8 THE ROLE OF PEOPLE

People, too, occupy a position in the flow of energy through the biosphere and must necessarily interact with thousands of other species of plants and animals. Why, then, do we consider humans separately? The answer is that today people have unprecedented power at their disposal, and with it the ability to increase the productivity of or destroy the ecosystems of the earth. People have been so successful at reducing competition from other species that human population has risen precipitously in a time span that is insignificant on the evolutionary clock. Before civilization, species and ecosystems evolved together over long stretches of time, but human social and technological changes are orders of magnitude faster than evolution.

Nowhere is the rapid effect of human technological advancement more pronounced than on the North American continent. Three hundred years ago, this land mass housed a set of diverse, balanced ecosystems where many types of plants and animals, as well as people, lived in a manner which had changed little since the last ice age. Today, some of the native ecosystems such as the tall-grass prairie and the eastern hardwood forests are gone or altered almost beyond recognition. Those biomes that still exist, such as the northern forest or the Rocky Mountain tundras, are shrinking in the wake of progress, even as the original balances of species are being altered by predator control and hunting. In fact, every predator large enough to take a deer or a cow, with the exception of man himself, has been pushed into a few remote or bitterly cold areas. Even there the predators risk eradication. In place of the natural forests and ranges, millions of acres of intensely cultivated land produce large yields of foodstuffs to feed millions of non-farmers living in cities.

Human systems, like natural ones, are complex and often poorly understood. A wheatfield or a power plant may become unstable just as ecosystems naturally do. As we rely more and more on systems of our own design, we must ask some serious questions. Will our systems endure? Is the world that we are creating stable, or is it more vulnerable and more prone to catastrophic collapse than the natural world that is being displaced?

On the one hand, some argue that complex systems devised by humans over long periods of development operate quite reliably; transportation networks and electric power grids, for example, are both dependable. When breakdowns do occur, human intervention can improve the system still further. In modern jargon, such a remedy is called a **technological fix.** Others consider such positions to be not only unjustified but even arrogant — lacking in the humility with which we should confront problems of such great complexity. They point to the fact that human systems are replacing natural ones at an extremely rapid rate and that the environmental problems caused by these displacements are becoming increasingly complex and interconnected. If this trend continues, perhaps someday our technological fixes may not be adequate.

4.9 CASE HISTORY: *The Pine Beetle Epidemic on the Eastern Slope of the Colorado Rockies*

The pine beetle, *Dendroctonus ponderosae,* indigenous to western North America, is a natural parasite of the ponderosa pine. It attaches itself to a healthy tree and bores its way in toward the center, sucking out the tree's sap for nourishment. Certain species of fungi infect the beetle hole and further upset the tree's life support system. If enough beetles attack a given tree, the tree will die.

At the time this is being written (1976–1977), the ponderosa pine forests in several Rocky Mountain regions are being devastated by the pine beetle. Our case history will concentrate on the mountain ecosystems near Boulder, Colorado. The epidemic has been so severe that in some stands in this area nearly all of the mature trees have died. The pine beetle is not new to these hills, but rather has been a part of the ecosystem for many years. Why, then, did a population bloom recently occur?

The eastern slope of the Colorado Rockies has a relatively dry climate compared with other

Dry conditions on the eastern slope of the Colorado Rockies encourage growth of species of yucca.

A woman cutting dead timber from a stand of ponderosa pine killed by pine beetles.

forest systems. Only about 45 cm (18 in) of rain falls annually, and the dry air and hot sun of the region encourage rapid evaporation. Small cacti and yuccas grow on exposed hillsides, and the trees tend to thrive best in gullies or shaded areas, where water is more plentiful and the summer sun less intense. About 100 years ago, before gold and silver miners settled in large numbers in these regions, the mountainsides were only sparsely timbered. The Colorado gold rush brought an influx of settlers to the area. These miners cut so many trees to construct cabins and to timber the mines that the hillsides were virtually stripped and only the low-lying brush, grasses, and cacti remained. Within a few short years the forest-grassland system had been artificially converted to a grass-range system.

But we must remember that a great many pine cones must have lain on the ground. As the seeds sprouted, the young trees started to grow in a favorable environment. With the large trees gone there was little competition for space, sunlight, and water, so the saplings were healthy, and new forests began to grow quickly. Since several young trees require less space and water than is needed by one large tree, these

The native pine trees in many Rocky Mountain regions were cut by miners during the gold rush days to build log cabins.

Miners also cut trees to supply timbers for mine supports. (Courtesy of State Historical Society of Colorado.)

new forests grew to be much denser than the original system. By the 1960's, the young trees had grown to a mature size. The ecosystem appeared to be healthy. Yet, since the new forests were much denser than the original ones, competition for vital nutrients, especially water, was much more keen. The available water was now distributed among many more trees, so all the trees suffered from scarcity.

If a beetle attacks a vigorous, healthy, well-watered tree, the tree will protect itself by producing enough sap to force the beetle out of its burrow. The beetle, unable to penetrate the wood, will have to fly away and search for a more vulnerable tree. Thus the trees of a healthy forest can survive amidst a population of pine beetles. A weak tree, living with inadequate supplies of water, cannot produce excess quantities of sap, and is more vulnerable to infestation. In the dense weakened forests of the present day, there are large areas where there are no trees healthy enough to repel a beetle attack. Consequently the beetles have bored successfully into the wood of many trees, and the forests are dying.

What will happen in the future? Of course, we can't predict with certainty. In some regions all the trees may die, leaving a grassland similar to that left by the miners and loggers a hundred years ago. In that case, the cycle may repeat itself; a new dense forest may grow, only to be killed by beetles a century from now, leaving a barren grassland and establishing repetitive cycles of growth and destruction. In such a situation, true ecological balance will not be achieved. On the other hand, perhaps most, but not all, of the trees will die. The death of the weak individuals may reduce the competition for water sufficiently so that the remaining trees will become healthy enough to survive the onslaught of beetles. The beetle population would then diminish and a stable forest system would regenerate.

The final result of this outbreak is being affected by yet another poorly understood factor. Many of the people who live in the area infested by the beetles are actively cutting dead trees,

A healthy ponderosa pine.

thinning stands, spraying heavily infested areas, and in some instances planting species of trees that are less susceptible to the beetles. At the present time no one knows how effective these efforts really are, for the experiment is still in progress.

The infestation of beetles continues, and it may be many years, perhaps decades or centuries, before we will be able to determine whether the forests will regain their original balance.

TAKE-HOME EXPERIMENTS

1. **Population growth.**
   (a) Yeast
   Prepare a bread dough according to the following recipe: Heat $2\frac{1}{4}$ cups water to 85° F (29° C). Add:

   2 tablespoons sugar

1 tablespoon butter
$2\frac{1}{2}$ teaspoons salt
Stir in $\frac{1}{4}$ oz active dry yeast.
Sift together 3 cups of whole wheat flour and 3 cups of white flour. Add this mixture slowly to the liquid. Knead the dough. (See

Kitchen metrics (to the nearest ml or g)		
1 teaspoon	$= \frac{1}{3}$ tablespoon	$= 5$ ml
1 tablespoon	$= \frac{1}{16}$ cup	$= 15$ ml
1 cup	$= \frac{1}{4}$ qt	$= 237$ ml
1 fluid ounce	$= 30$ ml	
1 ounce	$= 28$ g	

any standard cookbook for directions on how to knead dough.)

Remove a small sample of the dough and set it in a measuring cup in a warm (85°F, 29°C) moist room. Record the volume of the dough at 10-minute intervals for 2 hours, then at ½ hour intervals for an additional 3 hours. Draw a graph of your results.

The dough expands because yeast organisms release carbon dioxide during their metabolism, and the gas is trapped in the dough. If the yeast population remained constant, the dough would rise at a constant rate. Did the dough in your experiment rise at a constant rate? If not, what was the shape of your graph? Interpret your results.

(b) Mold

Allow the bulk of the dough to rise, then bake it in a greased pan at 350°F (177°C) for approximately 45 minutes. You now have a warm loaf of bread with no preservatives added. Eat most of the bread, but save one slice. Place this slice in an open dish in a quiet place in the room. Within a few days, mold will start to grow on the bread. In this part of the experiment you will measure the growth of the mold as a function of time.

Take a piece of wax paper and draw a series of horizontal and vertical lines on it about a centimeter apart. You now have a grid with a series of squares that are each 1 cm on a side. When the mold first starts to form, place the grid over the bread and use it to estimate (a) the total surface area of the bread and (b) the surface area covered by mold. Remove the paper. Repeat this measurement once a day for 10 days. Draw a graph showing the growth of the mold population as a function of time. What type of function do you observe?

2. **Interspecies interactions.** In this experiment you will observe and record one or more interspecies interactions. The observations can be made in the field or in the laboratory; there is no limitation to the type of study that can be conducted. Two examples are given below, but you are encouraged to use your own imagination.

(a) Set a bird feeder in a convenient location and keep the feeder well stocked with bread and seeds. Observe the behavior of the birds. Do some individuals chase others away? Do some species of birds dominate the feeder? Describe your observations.

(b) Take a slow walk in the woods or in a park. Can you observe any instances where two plants appear to be competing for light, space, or water? Can you prove that competition is occurring, or would a further experiment be necessary? Can you observe the growth of any plant parasites? Are any insects present on the plants? Can you see the insects eating the plants; each other? Describe your observations.

PROBLEMS

1. **Definitions.** Define biotic potential; environmental resistance; carrying capacity.

2. **Sigmoid growth function.** Examine the sigmoid curve shown in Figure 4.1 and mark the points where the rate of increase is largest and smallest. What is the smallest rate of increase on the curve?

3. **Geometric growth rates.** Imagine that the population of a given species quadrupled (increased by multiples of four) every 10 years. If there were 10 individuals in 1950, how many would there be in 1960, 1970, 1980, 2000? Draw a graph showing the number of individuals as a function of time.

4. **Unstable growth function.** In some areas of the world, human populations increase rapidly during years of abundance, only to be faced with famine and starvation during years of drought. During these periods of famine, population growth slows down

somewhat, and occasionally the population decreases, but the long-range trend is toward increasing populations. Draw a hypothetical curve of human population size, illustrating these events. Show which points correspond to years of abundance and which to years of famine. How does your graph compare with the population graphs for yeasts, reindeer on St. Paul Island, and sheep in Tasmania?

5. **Carrying capacity.** The worldwide human population has been increasing continuously for the past few hundred years. Can this trend continue indefinitely? Are human populations subject to the constraints of a worldwide carrying capacity? Explain.

6. **Time lag.** In Chapter 3, Section 3.5, the effects of drought on a grass-grazer-tiger system in a hypothetical valley in Nepal were discussed. Draw a graph of the imaginary population levels as a function of time. Do the population maxima and minima of the organisms in the three trophic levels coincide? Discuss briefly.

7. **Lemming cycles.** (a) The numbers of many of the lemming predators fluctuate much less dramatically than do the numbers of the lemmings themselves. Can you think of a reason for this? (b) The snowy owl is a predator which specializes in catching arctic rodents. Snowy owl populations fluctuate dramatically with lemming populations. Explain.

8. **Law of Limiting Factors.** Animal migrations can be considered to be an adaptation in response to the Law of Limiting Factors. Explain.

9. **Law of Limiting Factors.** Do you think that a human being living in a cold climate could survive as well on a severely limited diet as a person living in a moderate climate could? Explain.

10. **Environmental resistance.** Select a non-domestic plant or animal with which you are familiar and list the primary components of the environmental resistance that affects it.

11. **Environmental resistance.** The physical components of the environmental resistance are not always completely independ-

ent of the biological ones. Discuss this statement and give examples to support it.

12. **Niche.** Define ecological niche. Are you living in your optimum niche? Defend your answer.

13. **Niche.** Decay organisms living in flowing streams often attach themselves to rocks. One species may predominate on the lee side and another on the current side. What factors might be involved in establishing such a relationship?

14. **Niche.** Can two individuals of the same species occupy exactly the same niche? Explain.

15. **Niche.** Discuss some differences between the home territory of the nomadic hunters (p. 80) and the niche concept of animals.

16. **Two-species interactions.** List the eight major types of two-species interactions. Include a brief explanation of each.

17. **Diversity.** Explain how niche competition promotes diversity.

18. **Competition.** Competition can be interspecific or intraspecific. Explain and give examples.

19. **Predation.** The ecologist Elton contends that predators live on capital, while parasites live on interest. Explain. Is this true from an individual or from a community viewpoint? Explain.

20. **Predation.** Would you expect a buffalo herd to be healthier after years of being hunted by men with bows and arrows or by men with guns? Explain.

21. **Predation.** After the removal of starfish from an intertidal ecosystem, the total number of species decreased. (See page 87.) Do you think that the total number of individuals also decreased? Defend your answer.

22. **Two-species interactions.** Refer to the eight types of two-species interactions and categorize each of the following examples: (a) mistletoe sucks the sap from the pine tree on which it grows; (b) a barnacle gains mobility by attaching itself to a whale, but does not consume any of the

whale's tissue; (c) a small bird eats the meat caught between a crocodile's teeth—the bird gets fed and the crocodile has free dental care; (d) a paramecium eats a bacterium; (e) an elephant steps on an ant; (f) two trees growing side by side reach out for light.

23. **Community interaction.** Do you think that it might be economically profitable to raise deer along with cattle in some areas of North America? Explain.

24. **Ecosystem stability.** Which do you expect to be better able to survive a drought: a cornfield or a natural prairie? Explain.

25. **Natural balance.** Discuss the statement, "Natural systems are perfect because they are always in harmonious balance."

26. **Ecosystems.** Would you say that the Earth includes many ecosystems that are relatively independent of each other, or that it contains only one ecosystem that occupies the entire biosphere, or that both statements are true? Present arguments in favor of your position.

27. **Climax systems.** It has been observed that large, complex plants and animals are more characteristic of climax situa-

tions than of early successional stages. (a) Name some organisms and their habitats that serve as examples of this observation. (b) To explain this observation, it has been suggested that species with long life cycles have evolved away from unstable environments. It has also been suggested that large, complex animal species cannot find proper sustenance from plants that grow in simple systems such as marshes. Argue for or against these hypotheses.

28. **Succession.** Define natural succession. What factors bring about changes in an ecosystem? What is the climax of an ecosystem? Cite three examples of a climax ecosystem.

29. **Succession.** Imagine that a new island just arose in the South Pacific. Trace the succession that would be expected to occur. Estimate the time span required for the climax to be reached. Compare the energy balance of the early ecosystems of the island to the energy balance of a marsh. Compare the types of species present.

30. **Imbalance.** What do you think would happen to the Everglades if a set of dikes were built to ensure constant water levels all year round?

BIBLIOGRAPHY

The three basic textbooks cited in the bibliography of Chapter 3 all contain material germane to this chapter as well. Several books dealing specifically with population biology are:

Arthur S. Boughey: *Ecology of Populations.* 2nd Ed. New York, Macmillan Co., 1973. 182 pp.
Arthur S. Boughey: *Contemporary Readings in Ecology.* Belmont, Calif., Dickenson Publishing Co., 1969. 390 pp.
Kenneth E. F. Watt: *Ecology and Resource Management.* New York, McGraw-Hill, 1968. 450 pp.

Edward O. Wilson and William H. Bossert: *A Primer of Population Biology.* Stanford, Conn., Sinauer Assoc., 1971. 192 pp.

Two comprehensive books dealing with the distribution of species on islands and continents are:
Robert H. MacArthur: *Geographic Ecology—Patterns in the Distribution of Species.* New York, Harper & Row, 1972. 269 pp.
Robert H. MacArthur and E. O. Wilson: *The Theory of Island Biogeography.* Princeton, N. J., Princeton University Press, 1967.

5

THE EXTINCTION
OF SPECIES

5.1 WHAT IS A SPECIES?

No two individual organisms are exactly alike; on the other hand, some groups of organisms are similar enough so that we recognize their common characteristics. Biologists who classify organisms into discrete groups recognize levels of similarity. Thus, they observe that any given plant is similar in many respects to all other plants and is basically different from animals, whereas all animals, in turn, have common characteristics. Subgroups within these two large groups also exist; for example, all animals with backbones form one such smaller unit. Within the larger subgroup of animals with backbones, all those that are fur-bearing and suckle their young form a distinct class called mammals. Of course there are many different orders of mammals, such as rodents, ungulates, and primates. As these groups become more numerous, the distinctions among them become finer and finer. A pertinent question is, When do we stop classifying? If we stopped too soon (for example, if we did not subclassify mammals), our distinctions would be too gross to be useful. On the other hand, if we stopped too late (for example, if we classified humans according to the curliness of their hair and the color of their eyes), our categories would be too numerous to be very useful.

Since individual animals or plants breed only with similar animals and plants, and not with dissimilar ones, it is convenient to classify organisms that breed together into discrete groups. A **species** is most satisfactorily defined as a group of animals or plants that exhibits reproductive isolation; that is, a group whose members form an interbreeding population but do *not* reproduce outside the group. The number of ways in which groups of organisms can isolate themselves reproductively from other groups is unexpectedly large. For example, potential mates may have different breeding seasons or spawning grounds and hence may never meet during the time that they are interested in mating. Characteristic behavior at the time of mating (for example, courtship and display behavior) may not be recognized. In some cases, mating may be physically impossible owing to incompatibility of reproductive organs. Mating may take place, but the egg may not be fertilized; or if it is fertilized, it may die before many cell divisions. An egg may even develop into an adult organism, but one with greatly reduced fertility, as is the case with the mule (which is the result of a cross between a horse and donkey). All these mechanisms operate in nature on various groups, singly or in combination, to ensure their identities as separate species.

The species is a fundamental unit of classification because parental characteristics are always passed on to members of succeeding generations. As members of these succeeding generations reproduce, the group characteristics become intermingled, but do not intermingle with characteristics outside the group.

Classification of Humans

Kingdom	Animalia
Phylum	Chordata
Subphylum	Vertebrata
Class	Mammalia
Order	Primates
Family	Hominidae
Genus	*Homo*
Species	*sapiens*

5.2 EVOLUTION OF SPECIES

Since the beginning of life on this planet, new species have continually been formed and existing species have continually been driven to extinction. The number of species on the Earth today represents the total number of new species that have evolved minus the total number of extinctions.

Individual plants and animals face many pressures, such as intraspecies competition, interspecies competition at a given trophic level, interspecies encounters in the form of predation or parasitism, and climatic variation. Now, if we have a varied population subject to certain environmental stresses, some of the individuals may be more successful at reproducing than other members of the species. Obviously, individuals who produce the most viable offspring will make a greater contribution to the next generation than those who produce fewer or weaker offspring. Those individuals who enjoy a reproductive advantage over others are said to have greater **fitness.** The process by which environmental stresses give certain members of a species a reproductive advantage over others was termed **natural selection** by Charles Darwin, who was chiefly responsible for suggesting the overwhelming importance of this mechanism in evolution.

Evolution of mammals. The original unspecialized ancestor is thought to have looked like the animal in the center—not too different from today's shrews. Each of the various descendants has become adapted to a specific habitat. (After C. A. Villee and V. G. Dethier: *Biological Principles and Processes*. Philadelphia, W. B. Saunders Co., 1971.)

What determines which individuals have greater fitness and hence which will be preserved? This question is a reworded form of the ecological question: what niche adaptations enhance the survival of a species? There is no single answer to this question, for there are many ways in which different species adapt. Some individuals, like antelopes, may survive predation because they can run fast, and others, like elephants, because they are strong enough to beat off their attackers. But strength and speed are only two of many types of attributes that are selected for. Recall the example of the two species of Scottish barnacles from Chapter 4. One species survived because it was better able to compete for the underwater niche; the other species gained fitness by its ability to grow in spite of periodic desiccation. The nature of this interspecies competition is quite complex and goes beyond conventional concepts of strength and speed. Ability to survive is also manifested in resistance to disease and parasites, which is a powerful selection pressure. Some species compensate for weaknesses by various forms of cooperation, such as the mutual relationship between algae and fungi that form lichens, or the herd and troop defenses of many animals.

Natural selection explains how a species evolves but does not explain how a new species is formed. As long as there is sexual contact among members of a species, new traits, if beneficial, are dispersed throughout the popula-

Figure 5.1 Theoretical stages in the evolution of new species from a single common ancestor. Prolonged geographic isolation leads to permanent reproductive isolation. Interbreeding cannot occur, even if the new species again overlap. (From M. E. Clark: *Contemporary Biology.* Philadelphia, W. B. Saunders Co., 1973, p. 570.)

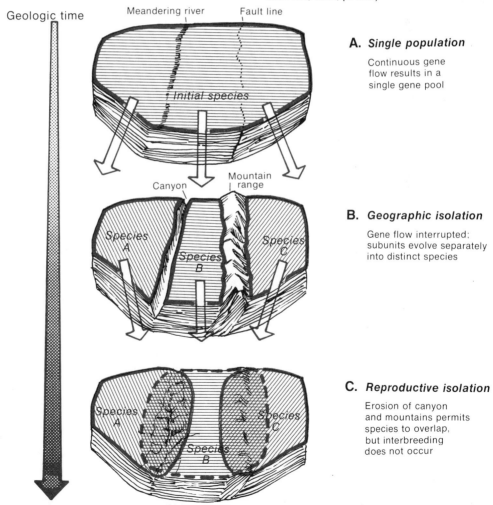

Geologic time

Meandering river   Fault line

Initial species

A. Single population

Continuous gene flow results in a single gene pool

Canyon   Mountain range

Species A   Species B   Species C

B. Geographic isolation

Gene flow interrupted; subunits evolve separately into distinct species

Species A   Species B   Species C

C. Reproductive isolation

Erosion of canyon and mountains permits species to overlap, but interbreeding does not occur

Figure 5.2 Different species of birds of paradise which evolved by isolation on New Guinea.

tion. For two species to diverge, however, populations must be isolated reproductively from each other for a long period of time, say 2000 to 100,000 generations. During the course of this isolation, changes occur in each population independently, and since the environmental pressures are, in general, different on opposite sides of some natural barrier, eventually two distinct species may arise (Figs. 5.1 and 5.2). While only a large barrier, such as an ocean, a jungle, or a high mountain range, can separate groups of birds, different species of rodents have evolved even on opposite sides of a large river.

5.3 THE EXTINCTION OF SPECIES

Extinction and evolution may sometimes proceed more or less in concert, so that the loss of one species is balanced by the introduction of another. For example, an early form of the horse, called *eohippus,* or dawn horse, stood 25 to 50 cm (10 to 20 in) high at the shoulder and lived in dense semitropical forests. As the climate and the vegetation of the world gradually changed, horses evolved along with the

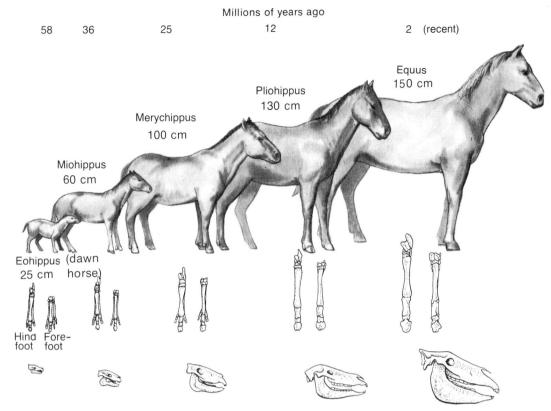

Millions of years ago

58   36   25   12   2 (recent)

Equus
150 cm

Pliohippus
130 cm

Merychippus
100 cm

Miohippus
60 cm

Eohippus (dawn
25 cm   horse)

Hind Fore-
foot   foot

Figure 5.3 Some stages in the evolution of the horse.

environment, and the dawn horse gradually developed as shown in Figure 5.3. The point is that the dawn horse never died out but rather changed in the course of millennia.

When we study the succession of ecosystems, however, we find that many extinctions occurred that cannot be classified as development. Throughout geological history, certain species have developed gradually, flourished, and then suddenly vanished, leaving behind them an empty niche. For example, approximately 500 million years ago, enormous numbers and varieties of primitive sponges populated the seas. Although for 30 or 40 million years these sponges dominated the seas from pole to pole, most species later disappeared, and different ecosystems developed. In more recent times, the rise and fall of the dinosaurs was certainly one of the most outstanding events in the history of the Earth. Small dinosaurs first evolved some 225 to 250 million years ago. For perhaps 25 to 50 millions years they slowly developed and grew in numbers and variety until they dominated the Earth. The dinosaurs reigned for 100 million years and then they all died, whereupon the established ecosystems and the regulatory mechanisms of the food web collapsed. Dinosaurs have been often belittled in popular literature as being stupid and clumsy. But in fact the great reptiles were outstanding evolutionary successes; they dominated the Earth longer than any other order has.

Figure 5.4 *Top*, Triceratops. *Bottom*, Stegosaurus.

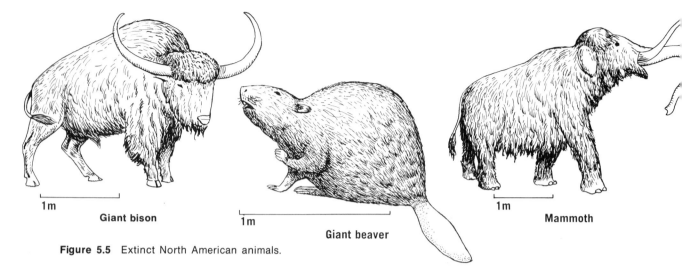

Figure 5.5 Extinct North American animals.

1 m
Giant bison

1 m
Giant beaver

1 m
Mammoth

"You're being recalled—He's going to try mammals."

(From *Saturday Review,* March 25, 1972. Copyright 1972 by Saturday Review Co. Used with permission.)

Figure 5.6 Hunting techniques of early humans. Artist's conception of how the Tule Springs site, Nevada, may have looked 12,000 to 10,000 years ago. (Courtesy of John Hackmaster.)

The extinction of the dinosaurs reminds us that seemingly stable biological systems can collapse.

Of course, the period following the extinction of dinosaurs was not devoid of life, and during this time, the previously slow evolution of mammals accelerated. Many species, including human beings, arose. The next significant wave of extinctions occurred about 10,000 years ago, at the end of the Pleistocene Age. Before this time the mammalian fauna of the world were significantly more varied than at present. Mammoths, mastodons, camels, horses, wild pigs, giant ground sloths, giant long-horned bison, woodland musk oxen, tapirs, bear-sized beavers, and many different kinds of now-extinct deer roamed the North American continent (Fig. 5.5). They were hunted in part by sabre-toothed tigers, giant jaguars, and dire wolves. All these animals have become extinct, for the most part without ecological replacement. In fact, estimates are that 95 percent of all large animal species in North America were lost, and mass extinctions occurred simultaneously in what are now South America, northern Asia, Australia, and Africa.

What caused these extinctions? There are many uncertainties. Perhaps climatic changes had an effect. We know that rapid alterations in weather patterns occurred at this time, but we also know that the same animal species that suffered extinction had already lived through several advances and retreats of the glaciers. Also, few small animals followed their larger cousins into oblivion, and it is hard to explain why climatic variation affected large animals differently from small ones. It seems certain that the hunting activities of primitive people played some role in the drama. We know, for example, that all the herbivores that are now extinct were pursued by early humans. However, no major extinctions preceded the development of advanced hunting techniques such as throwing spears, shooting arrows, and setting fires to drive animals over a cliff or into a trap (Fig. 5.6). If large herbivores were hunted to extinction, the specialized carnivores like the sabre-toothed tiger must have perished soon after.

For a period of five to six millennia, very few additional species became extinct, and then suddenly, in recent times, a new age of species destruction has begun. Not only have extinction rates been increasing rapidly in the last half century (Table 5.1) but also a great many more animal and plant species are seriously endangered today. Unless current trends reverse, extinction rates will continue to accelerate in the near future.

TABLE 5.1 Past and Present Extinction Rates

YEAR SPAN	AVERAGE NUMBER OF YEARS REQUIRED FOR EXTINCTION OF ONE MAMMALIAN SPECIES
1–1800	50
1801–1850	25
1851–1900	1.6
1901–1950	1.1

5.4 FACTORS LEADING TO THE DESTRUCTION OF SPECIES IN MODERN TIMES

We know that climate has been relatively constant in recent years; therefore it wasn't any change of weather that expelled grizzly bears from California and passenger pigeons from Wisconsin. We also know that devastating disease epidemics have not occurred recently in wild animal populations, and that vegetation hasn't changed. We know, in short, that people are the major agents in the extinction of species today. The following paragraphs discuss four major destructive mechanisms: (1) Destruction of habitats; (2) Introduction of foreign species; (3) Extermination of predators; and (4) Hunting by people for food or fashion.

DESTRUCTION OF HABITAT

The relationship between human beings and nature is no longer a simple predator-in-the-forest system. If we try to outline why a great many species of plants and animals have become endangered or extinct in recent years, we find that the ax, the chain saw, the tractor, and the bulldozer have in many cases been more detrimental than the hunter and the gun, for as people destroy natural systems, many organisms cannot survive. For example, the trailing arbutus and the ivory-billed woodpecker are two specialized species that have adapted to conditions in certain stable, climax forests of North America. As these forests have been cut and replaced with stands of second-growth timber, the ecological niche of the arbutus and the ivory-billed woodpecker has been gradually destroyed. Neither species has been able to adapt to new conditions, and both are facing extinction.

As cities, suburbs, farms, roads, and harbors are displacing native prairies, marshlands, forests, deserts, and waterways, an increasing variety of plants and animals cannot survive. In North America, thousands of species of prairie grasses and flowers are disappearing as the prairies

A

B

A, Trailing arbutus. (Photo by Alvin E. Steffan, National Audubon Society.) *B,* Giant white pine tree—a reminder of the primeval American forest. (From Rutherford Platt: *The Great American Forest.* © 1965. Published by Prentice-Hall, Inc., Englewood Cliffs, New Jersey.)

Ivory-billed woodpecker. (Photo by James T. Tanner, from National Audubon Society.)

Whooping crane. (Photo by Allan D. Cruickshank, from National Audubon Society.)

114

are destroyed; nor can the majestic whooping crane population survive when the marshes they need are being drained. In Asia, many species of plants and animals, including the Bengal tiger, are endangered as the jungles in which they range are being cut. The list of destruction of habitats could go on and on. Current reports estimate that over one hundred species of large animals face extinction, along with tens of thousands of species of spiders, mites, crustaceans, insects, and plants, and many of these are endangered, at least in part, by loss of habitat.

Habitat is not just a single specific place. It encompasses an entire ecosystem, with its complex food webs, sources of water, and energy. A habitat is the sum total of all the conditions necessary for life. For example, many animals such as the North American cougar and the grizzly bear normally hunt food throughout a large territory. If a new and impassable barrier such as a superhighway bisects that territory, the predators may not be able to survive even though the highway accounts for only a tiny fraction of the forest.

Migratory animals are particularly vulnerable to destruction of habitat. As mentioned in Chapter 3, many ocean fish spend most of their life in deep water, but reproduce in shallow, nutrient-rich coastal bays and estuaries. Although each fish may spend only a small fraction of its life near the coast, destruction of the coastal environment may endanger the survival of the species.

Sometimes a habitat may be destroyed without its physical appearance being significantly changed. For example, the California condor, the largest flying bird alive today, is a scavenger. Many problems beset the condor population. These birds breed slowly and the young are highly dependent on their parents for a long time. Condors are shot, trapped, and poisoned despite laws that are supposed to protect them. Their traditional range has been preempted by farming communities, and condors appear to avoid places where people congregate. But the condors might survive all these obstacles and yet succumb to the most threatening problem of all—the loss of their food supply. When an animal dies in a natural environment, its carcass remains above ground, where scavengers like the condor can feed. In a farming community, dead and diseased sheep and cattle are often removed and buried, depriving the condor of its food supply and further endangering the survival of the population.

Chemicals may also alter habitats. In particular, the proliferation of pesticides has been a significant factor leading to the decline of many species of animals, including the bald eagle, varieties of terns, and many species of hawks. Chapter 12 discusses the mechanisms by which pesticides aimed at insects are destructive to birds and other animals.

When a climax ecosystem is altered or destroyed, the number of individual organisms living within the system

California condor. (Photo by Carl Koford, from National Audubon Society.)

does not necessarily decrease. In fact, in many cases the number of individuals may actually *increase*. White-tailed deer and cottontail rabbits, for example, do not thrive well in established stands of old trees, for there are little brush and few low-lying trees to provide food and shelter. Therefore, when a climax forest is logged, the deer and rabbit population actually increases. But other species, such as the ivory-billed woodpecker and the trailing arbutus, cannot survive. If all the climax forests are destroyed, the total productivity of the planet may temporarily increase, but some species will become extinct, and the overall diversity of our world will be severely diminished.

INTRODUCTION OF FOREIGN SPECIES

In prehistoric times, migration of species from continent to continent or even from watershed to watershed often required thousands of years and in many instances did not occur at all. Consequently, evolution and speciation have occurred independently in various areas of the world, and each continent has given birth to its own native species. The results of this independent evolution are dramatically shown by the contrasts among herbivores in each of various sections of the globe: the bison in North America, the llama in South America, the yak in Himalayan

Important grazers that have evolved on different continents.

Caribou

Yak

Bison

Llama

Elephant

Kangaroo

116

regions, the elephant in Africa, and the kangaroo in Australia. Many plants also originated in specific areas: corn in North America, wheat in the Middle Eastern region, and rice in Eastern Asia.

During modern times, however, myriads of species have migrated with unprecedented ease as passengers on trucks, trains, automobiles, airplanes, boats, and in some cases, human bodies. Many migrating organisms have been unable to compete in foreign lands, and colonizations have been unsuccessful. By contrast, some organisms have succeeded so well in their new environments that they have disrupted the ecological balance and have endangered other species.

Some efficient invading predators and parasites have upset local ecosystems by decimating populations of prey. For example, the American chestnut tree used to forest much of the middle and southern Atlantic coastal areas of the United States, but now the chestnuts are seriously endangered because a parasitic tree fungus imported from China is lethal to the American trees (Fig. 5.7). Both American and Chinese trees had developed resistance to the parasites that grew in their own habitats but not to foreign parasites. As a result, all large American chestnut

Figure 5.7 Chestnut blight. (From Odum: *Fundamentals of Ecology,* 3rd ed. Philadelphia, W. B. Saunders Co., 1971.)

From
THE VILLAGE BLACKSMITH
by Henry Wadsworth Longfellow

Under a spreading chestnut tree
 The village smithy stands;
The smith, a mighty man is he,
 With large and sinewy hands;
And the muscles of his brawny arms
 Are strong as iron bands.

A healthy chestnut tree. (From Julia E. Rogers: *The Tree Book.* New York, Doubleday, Page & Co., 1905.)

trees on the East Coast died within 50 years of the arrival of the Chinese fungus. Perhaps some young trees will survive and provide a new breeding stock for the species, but the outcome is uncertain.

Successful ecological invasions also present economic problems. In many cases, invaders quickly become pests and are responsible for large financial losses. The sea lamprey, a predator-parasite that attacks various species of fish, migrated into the upper Great Lakes when a ship canal was built to bypass Niagara Falls. Shortly after gaining access to Lake Huron and Lake Michigan, lampreys virtually eliminated a lake trout industry of 8.5 million pounds (3.8 million kg) per year. Many such examples abound: the Japanese beetle, imported from the Orient, feeds on many crops, such as soybeans, clover, apples, and peaches; the American vine aphid, imported to France from the United States, was responsible for destroying three million acres of French vineyards. In fact, over half of the major insect pests in many areas are imports.

The invasions of foreign plants and animals are also responsible for more subtle perturbations. A few hundred years ago, individual continents possessed their own unique types of ecosystems. Thus, for example, temperate grasslands in America, Asia, and Australia all contain different groups of organisms. With the rapid immigration of some species and the destruction of others, the differences between ecosystems are becoming smaller, and the biological world is becoming simpler.

These incidents show that predators do have the potential to destroy a species of prey. The often realize this potential if they find themselves in a new and hospitable environment.

Figure 5.8 Young sea lampreys, *Petromyzon marinus*, attacking brook trout in an aquarium. (From Lennon, R. E.: "Feeding Mechanism of the Sea Lamprey and its Effect on Host Fishes." Fish Bulletin: U.S. Fish and Wildlife Service 56 #98: 247–1293, 1954.)

EXTERMINATION OF PREDATORS

Humans have always shared the planet with other predators, and during the millennia of coevolution, delicate relationships have developed. The most striking aspect of the interactions between people and their carnivorous competitors is that the large terrestrial predators of the earth habitually avoid attacking humans. This does not mean that people are never killed by other predators, but just that confrontation has been minimal. For example, the grizzly bear is one of the largest and strongest carnivorous animals alive today (Fig. 5.9). These giants weigh from 200 to 400 kg when mature, and measure up to 1⅓ meters high at the shoulder. They can run at about 40 to 50 kilometers per hour on flat ground and are strong enough to crush the skull of a large cow with one blow. Our fear of this formidable animal is reflected by its scientific name, *Ursus horribilis*. Yet the bear's record is far less fearsome than its name. One of the last significant ranges of the grizzly south of Canada and Alaska is in the Rocky Mountains of northern Montana in and near Glacier National Park. In the past 50 years, millions of hikers have toured the park and adjacent wilderness areas, thousands of unarmed people have sighted grizzlies, thousands of armed people have hunted and sometimes killed the bears, and on only a few occasions have the giant animals killed or maimed anyone.

The record of the relationship between humans and wolves is even more remarkable. European literature is filled with horror stories of wolves, and English-speaking children learn the lesson early when they listen to the tale of Little Red Riding Hood. Yet, there have been *no* documented cases of wolves attacking people in North America during the past 50 years.

The list could continue: American alligators do not attack unless cornered, lions generally avoid humans, and cheetahs have *never* been known to attack a person unless pursued. Other species, such as the Bengal tiger, the polar bear, and the Nile crocodile, attack humans more often, but even for these animals, people are only a rare prey. Yet, though they do not constitute any real threat to human life, predators have been hunted, sometimes vigorously, in the name of securing safety for humans (Fig. 5.10).

A second major source of conflict between predators and people is competition for food. Wild carnivores do kill game and domestic stock, and consequently hunters and farmers have often advocated the extermination of predators. But elimination of all predators does not mean more food for humanity. As an example, let us examine the relationships among ranchers, grass, sheep, and coyotes in the western United States and Canada. A coyote's diet includes many different items, among which are field mice, rabbits, moles, pack rats, sage hens, carrion of lambs that have died

Figure 5.9 Alaskan bear with salmon. (Photo by Leonard Lee Rue, National Audubon Society.)

Figure 5.10 Mountain lion cornered by bounty hunters in Colorado. (Photo by Carl Iwasaki, *Life* Magazine.)

of natural causes, and freshly killed lambs. If the coyotes were exterminated, the rodents would flourish, and since rodents eat grass, the quantity of forage available for the sheep would decrease. On the other hand, a spiraling coyote population would prey on range lambs and inflict financial losses on the ranchers. Thus, though coyote control is defensible, extermination is ecologically unsound. Even limited control programs have been questioned by some study groups, who contend that the cost of the anti-predator campaigns does not warrant the meager savings in livestock that are realized. Unfortunately, mass poisoning programs are being advocated by many ranchers who may not appreciate the true complexities of ecosystems.

The coyote is not now an endangered species, so why use this particular example here? Why not, instead, focus our attention on other animals that used to roam the rangelands of the West, such as the grizzly bear and the wolf, but that are now on the brink of extinction? The reason is that in the case of the coyote there is still time to initiate a balanced program of control and thus to avoid a crash extermination campaign that would later be followed by a crash conservation program to save this species.

HUMAN HUNTERS

After the wave of Pleistocene extinctions, primitive hunting societies seemed to have existed in ecological balance with their prey. Yet today, many systems are so seriously disturbed that once again some species are being eliminated by food gatherers.

The pigmy hippopotamus has coexisted with tribal hunting societies in central Africa for a long time, but as industrialization reduces the animal's range and brings hunger to many people in the area, the hippo is being hunted more systematically today, and extinction may result. This type of situation poses a serious dilemma. If an animal is hunted to extinction, the future food-producing capabilities of the area may be decreased, but a hunter with hungry children at home understandably hunts whatever prey is easiest to find.

Perhaps one of the saddest and least excusable types of environmental disruption caused by people in the world today is the wholesale killing of endangered species to satisfy the sport or fashion demands of a few individuals.

Conservationists estimate that between 500 and 700 wild giant sable antelopes live today in west central Africa (Fig. 5.11). These animals are endowed with a set of magnificent curved horns, sometimes five feet long, which may easily be the animal's downfall, for hunters come from many parts of the world to shoot a big bull and bring home a trophy. Additionally, many hunters and tourists who are unable or unwilling to shoot a giant sable purchase their trophy from native hunters and return home with an expensive wall decoration. Hope for the survival of this species is slim.

Figure 5.11 Giant sable antelope. (Courtesy of Richard D. Estes.)

The snow leopard is one of the few large predators living in the high country of the Himalayas. Leopard skins are considered to be elegant, and snow leopard skins are warm as well. As a result, there are fewer than 500 snow leopards left alive. Snow leopard coats can still be purchased, even though some major furriers no longer sell them.

These two examples were chosen out of several hundred cases where sport and fashion hunters are responsible for large-scale killing of animals and the destruction of natural systems. Other well-known examples include Nile crocodiles and American alligators for handbags and shoes, many jungle cats for furs, and several species of birds for their elaborate feathers. In recent years, national and international law enforcement organizations have banned the transportation and sale of the hides, feathers, and bone of many endangered species. These laws are a significant step forward, but they have not eliminated the slaughter, for smugglers have been able to maintain a steady trade in these items.

5.5 CURRENT TRENDS IN SPECIES EXTINCTION

There are three important categories of risk that endanger the survival of a species:

(a) Highly specialized or immobile animals and plants are particularly vulnerable to pressures from loss of habitat.

121

Bison, a large native grazer of the North American plains, were nearly driven to extinction in the early 1900's.

The trailing arbutus and ivory-billed woodpecker were already cited as examples of many plants and animals that are dying out while their more versatile relatives, capable of surviving either in early successional or in climax systems, can make successful adaptations.

(b) Among more mobile and adaptable creatures, large animals seem to be particularly at risk for several reasons. First, their habitat is most easily destroyed. Compare two major grazers of the Great Plains of North America—the bison and the field mouse. The bison population that now exists in Wyoming and Montana is not allowed to expand because the pastures and wheatfields of the West are not compatible with wild herds. Because it is easy to fence them out, the bison have few places left to roam free. In contrast, mice, which cannot be barred by conventional fences, still range throughout the plains.

Second, hunting pressures are generally more severe against large animals than small, for large game animals are more desirable for food and trophies. Also, relatively small grazers such as rodents have traditionally fled from their non-human predators, so they naturally flee from human predators, too. A mature moose or bison, on the other hand, is strong enough to stand ground against wolves or mountain lions. Consequently, these animals often try to defend themselves rather than flee. When people with modern firearms reached North America, they found that larger game animals were far easier prey than the smaller ones. Consequently, the strong have fallen in proportionally greater numbers than the quick.

Third, small animals such as rodents and insects usually have a higher reproductive rate than larger animals. For example, a female bank vole is independent of

Populations of field mice have not been threatened by the rise of civilization. (Photo by Eric Hosking, from National Audubon Society.)

her parents at $2^{1}/_{2}$ weeks, reaches sexual maturity in 4 to 5 weeks, and in one year can give birth to five litters of five infants each. Insects lay several thousand eggs a year. These species have evolved amidst powerful predators and appear to adapt well to the addition of one more. By contrast, the mighty creatures have traditionally bred more slowly and maintained population growth through maternal care or defense by the herd. These protective measures are not effective against the repeating rifle.

(c) Many *predator* species are seriously endangered for reasons that should be clear by now. Large predators have traditionally known almost no external enemies and are even more poorly adapted to predation than are large herbivores. In addition, their niches are at the top of the food chain, and they often need a large hunting range. Thus, as destruction of their habitat constricts their natural range so that the wild lands can no longer support them, the hunting animals often roam toward farmers' fields and prey on cattle. As a result, they are hunted extensively in extermination programs. Also, because environmental poisons tend to concentrate at the top of food chains, predators are often victims of pesticides. This phenomenon will be discussed in Chapter 12.

5.6 HOW SPECIES BECOME EXTINCT — THE CRITICAL LEVEL

The pressures discussed in the preceding section reduce the populations of a number of species, but rarely eliminate the last few surviving individuals. What factors lead to the final extinction of a species, or, in other words, how does an endangered species become an extinct species?

Let us consider the case of the passenger pigeon (Fig. 5.12). In the late 1880's, approximately two billion birds flew over the North American continent in flocks that blackened the skies. Commercial hunters, shooting indiscriminately into the flocks, killed millions for food and many more for fun, for the species was thought to be indestructible. As hunting pressures increased, the pigeon populations naturally suffered, and by the early 1900's market hunting was no longer profitable. Yet thousands of pigeons survived. Then suddenly they all vanished. Poof! Ducks, geese, doves, and swans had all been hunted and they survived in reduced numbers; why did the passenger pigeons succumb?

The naturalist Aldo Leopold could offer none of the conventional ecological arguments and said of the pigeon,

The pigeon was a biological storm. He was the lightning that played between two opposing potentials of intolerable intensity: the fat of the land and the oxygen of the air. Yearly the feathered tempest roared up, down, and across the continent,

Figure 5.12 The passenger pigeon—a lesson learned too late.

sucking up the laden fruits of forest and prairie, burning them in a traveling blast of life. Like any other chain reaction, the pigeon could survive no diminution of his own furious intensity. When pigeoners subtracted from his numbers, and the pioneers chopped gaps in the continuity of his fuel, his flame guttered out with hardly a sputter or even a wisp of smoke.*

In modern biological language, we say that the pigeon population was reduced below its **critical level** and thus could not survive even though many mating pairs remained alive. The concept of the critical level is a general one and applies to nuclear reactors as well as to pigeons. However, it is often difficult to determine accurately the critical level for a given system. To understand how critical levels operate and why they differ quantitatively from species to species, the following aspects must be considered:

(a) When animals face certain types of stress, many of them fail to reproduce normally. This phenomenon is fairly common in the animal kingdom. Female rabbits reabsorb unborn fetuses in the bloodstream during drought, and many animals such as the Javanese rhinoceros have never bred successfully in captivity. It even extends to people. For example, an abnormally high percentage of American males became impotent during the Depression of the 1930's. We just don't know enough about behavior to predict how a certain stress will affect fertility of a given animal, but we do know that in certain cases destruction of range and harassment by hunters alter the reproductive potential of the remaining stock. A sudden decrease in the reproductive rates in the face of mounting pressures would certainly lead to a rapid population decline.

(b) Often the pressures that act on one species of an ecosystem do not affect the entire ecosystem equally, and severe imbalance can result which might lead to extinction. The original North American population of a few billion passenger pigeons must have supported a large and varied population of predators. When commercial hunters slaughtered a significant number of pigeons, they did not shoot a proportional number of predators, and it is possible that the pigeon was exterminated because the ratio of predator to prey was so unfavorable.

(c) The population density may decline to a point where members of the opposite sex have a hard time finding each other.

(d) When the population of a species becomes low, that population is particularly subject to bad luck, and a few unfortunate events can lead to extinction. The situation is analogous to that of a gambler who is playing a game that is rigged so that the odds are in his favor. If he has a large cash reserve and can play for a long time, he will win, but if he starts off with a small stake, a few unlucky hands in the beginning can end the game. So it is with species. Any large population living in its own natural surroundings would be expected to survive. However, if only a small

*From Aldo Leopold: *A Sand County Almanac.* New York, Sierra Club/Ballantine Books, 1970.

breeding stock exists, chance occurrences might lead to extinction.

An unlucky example is the Steller's albatross. Despite decimation by hunters, a few viable flocks survived and bred. Then, in 1933 as one flock was nesting peacefully on an offshore island, a volcano erupted, killing most of the adults and all of the young. Again, in 1941, another volcano erupted and destroyed a second nesting population. The species survived for another 20 years, and reproductivity had just begun to accelerate when the last sizable flock was caught in a typhoon and destroyed. The future of the species is in question.

Sometimes, however, chance favors survival. The sea otter was extensively hunted for fur from about 1750 to 1900 (Fig. 5.13). By 1910 the otter was believed to be extinct. No one knows how many individuals actually survived the fur trade, or where they lived, but around 1930 this "extinct" animal reappeared and established itself in several locations. Obviously, a small breeding stock had survived and increased in spite of two decades of hard times.

Biological luck is more complicated than the presence or absence of a typhoon, volcano, landslide, avalanche, or other natural disaster. When breeding populations are very low, genetic misfortunes can be disastrous. For example, appearance of a rare deleterious mutation in one offspring of a large population will have no effect on the survival of the population. But if the breeding population is tiny, such a mutation may threaten continuation of the species simply because the tiny community can ill afford to lose even one potentially reproductive member.

(e) **Inbreeding** (mating between closely related individuals such as brother and sister or cousin and cousin) poses another danger to the existence of reduced population. Geneticists have shown that, in general, unions between brother and sister or cousin and cousin have a high probability of producing weakened offspring. In a small population such mating must necessarily occur quite frequently. This inbreeding increases the probability that a species will become genetically weak and perhaps extinct.

Figure 5.13 Sea otter surfacing with sea urchin (*top*) and abalone (*bottom*). (Courtesy of James A. Mattison, Jr., Salinas, California.)

5.7 THE NATURE OF THE LOSS

What will happen to the biosphere if these current trends continue? Ecosystems and food webs will become simplified, but the pattern of primary, secondary, and tertiary consumption will continue. If the deer becomes the largest grazer and the coyote becomes the largest non-human predator, grazing and predation will continue, so why worry about extinction?

There are several reasons to worry. In some ways the

Musk ox have traditionally defended themselves by banding together. This type of defense is effective against most natural predators but is ineffective against hunters with rifles. (Leonard Lee Rue, from National Audubon Society.)

Drosophila, magnification about 6×. (Courtesy of Dr. E. Novitski, Department of Biology, University of Oregon.)

Thomas Hunt Morgan (1866–1945) was an American geneticist whose experiments formed the basis for his theory that paired elements or "genes" within the chromosome are responsible for the inherited characteristics of the individual. He won the Nobel prize in medicine in 1933.

most compelling is the aesthetic and religious argument. Different individuals may express their feelings in different ways. To some, species and the wilderness should be preserved simply because they exist. Others might reformulate this in saying that humans have no right to exterminate what God has created. To still others, an unobtrusive passage through an untouched wilderness area is a source of enormous aesthetic gratification, as valid and moving an experience as a great play or string quartet. As many of the greatest and noblest creatures of the earth fall prey to the thoughtless acts of people, so too, the richness, variety, and fascination of life on this planet diminish with their passing.

A second reason is concerned with developments in medicine and the biological sciences, which have always been dependent on various plants and animals as experimental subjects. The range of species used in different experiments has traditionally been very wide. For example, when Thomas Hunt Morgan initiated his famous studies of genetics, he needed an animal that was easy to breed in large numbers and had few, large, and accessible chromosomes. He chose the fruit fly, *Drosophila,* not because the life style of these insects was of particular interest in itself, but because the animals were easy to study and the lessons learned from them could be generalized to studies of other creatures. Another serious scientific loss in this category is that of wild populations of plants and animals that have traditionally been used in breeding hybrid species for agriculture. For example, domestic corns are often susceptible to various diseases. Geneticists have periodically crossed high-yielding susceptible corns with hardy wild maize in an effort to develop high-yielding resistant corn. However, natural maize growing along fence lines and roadways is often considered to be a weed and has been combated across North America with herbicides. The loss of this species would be a severe blow to modern agriculture.

A third category of species destruction that may prove detrimental is occurring among the herbivores on the African savannahs. A simplified picture of the savannah ecosystem, with its migrating species, was discussed in Chapter 4, page 83. In many places natural grazers are being displaced to make way for cattle ranches. But cows are unable to utilize grass as efficiently as the heterogeneous African herds. Domestic stocks tend to overgraze certain species of plants selectively, thus upsetting the water table. Moreover, cattle are susceptible to many tropical diseases, especially sleeping sickness. The result is that an untouched savannah is capable of an annual production of 22 to 34 metric tons of meat per square kilometer in the form of wild animals, while the best pasture-cattle systems in Africa can yield only 7 metric tons of beef per square kilometer per year. Yet in the name of agricultural progress, many ungulates are being threatened with extinction, and other herd sizes are being substantially reduced.

The final reason, and perhaps the most compelling of all, concerns the richness and variety of life on the planet. The Earth is constantly changing. Climates change, environments change, products and foodstuffs change. Traditionally, species of plants and animals have always evolved along with their changing environment. If a mountain range uplifted, certain organisms evolved to live at higher altitude; if a species of small, slow-moving dawn horse developed into large, fast-moving horses, then quicker,

Impalas in Africa. (In T. A. Vaughan: *Mammalogy.* Philadelphia, W. B. Saunders Co., 1972, p. 249. Courtesy of W. Leslie Robinette.)

smarter, stronger predators evolved to pursue these horses. Our world is now developing and changing rapidly. Most scientists feel that if our systems are to remain stable, plants and animals must evolve to meet the change. But evolution is slow. If we allow species to become extinct now or in a few short years, the total global potential for adaptation and change will be greatly diminished, and the stability of future systems may be endangered.

5.8 CASE HISTORY: *The Blue Whale*

Antarctica is a cold, frozen continent, covered with ice and nearly devoid of vegetation. By contrast, the southern oceans that surround this continent are rich, fertile, and teeming with life. In fact, life is more abundant in the Antarctic Ocean than in many central tropical oceans, for deep-sea currents carrying valuable nutrients rise in the lower latitudes and fertilize the surface layers of ocean. The nutrient-rich euphotic zone supports large populations of phytoplankton, the primary producers of the ocean. In turn, there are a great many zooplankton that feed on the tiny plant species. The zooplankton population consists, in part, of various species of small, shrimplike crustacea, known collectively as **krill.** An individual krill organism is only about a centimeter or two long, but these animals congregate in schools that can be 10 meters thick and cover a surface area of several square kilometers. Schools of krill this size naturally represent a significant food supply; in fact the schools are large and plentiful enough to feed the largest animal that has ever lived on this planet—the blue whale. A blue whale is about 30 meters (100 ft) long and weighs 90,000 kg (100 tons). It has no true teeth, but rather a set of elastic, horny plates called **baleen** that forms a sieve-like region in the whale's mouth. A blue whale feeds by swimming through a school of krill with its mouth open, allowing the krill to pass through the baleen but excluding any larger objects.

Little is actually known about the life and habits of the blue whale. We know that they are mammals; they are born alive and breastfeed for some time after birth. Scientists believe that a young animal becomes sexually mature when it is about six years old and that a mature cow gives birth to a single infant every two or three years. Other species of whales that have been studied more intensively have proved to be highly intelligent.

Blue whales appear to be well adapted to life in their natural environment. Their bodies are streamlined and their skin, smooth and nearly hairless, offers little resistance to water flow. Body heat is conserved with the aid of a thick insulating layer of fat, called blubber. A mature blue whale has no significant non-human predators. Thus, the low reproductive rate of these animals has insured that they do not overpopulate their range.

The Antarctic is one of the most dangerous and tempestuous oceans in the world. Early whalers, therefore, did not venture south after these giants. But gradually, as whaling grounds in the more central latitudes became depleted, and as ships were built to be stronger and more seaworthy, people ventured south in search of the blue whale. At about the turn of the century,

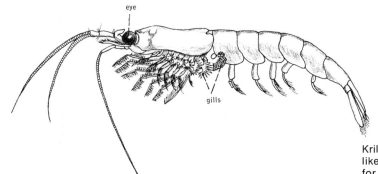

Krill (*Euphausia superba*) are small (5 cm) shrimplike animals that provide the major food source for most Antarctic whales.

Skeleton of an Atlantic right whale. Note the numerous sieve-like baleen in the whale's mouth. The blue whale is a plankton eater and also has baleen rather than teeth. (Courtesy of the American Museum of Natural History.)

it was estimated that the blue whale population numbered about 200,000 individuals. In 1920, over 29,000 whales were killed. The blubber was boiled down to oil to be used directly as a lighting fluid or as a raw material for the manufacture of cosmetics, medicinals, lubricants, shoe polish, paint, and other products. Some of the meat was eaten by humans, some thrown away, and in later years some was ground into food for dogs and domestic mink. Whaling was a dangerous but profitable business. The whales, having evolved in a predator-free environment, could not adapt to the new predatory pressure. The population started to decline precipitously. By the middle 1950's the species had been decimated. Marine biologists warned the whaling communities that not only would blue whales be threatened with extinction if current practices continued but also that the indiscriminate slaughter of whales was economically unsound. If whalers could practice reasonable conservation methods and allow a large breeding population to survive, they would be able to harvest several thousand individuals annually on a prolonged and sustained basis. On the other hand, if the wholesale slaughter were continued, and the breeding population were destroyed, then in a very short time there would be no more whales. Both whales and whalers would suffer. Unfortunately, the open ocean is

False killer whale leaping. Photograph taken in the Sea of Cortez, Mexico, by Jen and Des Bartlett.

subject to neither national nor international control. Each owner of a whaling vessel is able to establish his own guidelines, which are frequently based on immediate economic gain rather than on any long-range ecological or economic consideration. So the slaughter continues.

The International Whaling Committee (IWC) was established in 1946 to regulate whaling practices. The Committee ignored the recommendations of most knowledgeable marine biologists and set limits that greatly exceeded the sustained catch levels. The whalers ignored even these generous guidelines and killed more whales than the IWC recommended. Predictably, the whale population declined so that the total catch decreased steadily. In 1964–65, only 20 whales were harvested, and in 1974–75 the catch dropped still further. With the blue whale population nearly extinct, whalers turn to hunting other smaller species—fin, sei, and brydes whales.

At the present time, the blue whale population is estimated at a few thousand. Some experts feel that the critical level may have been reached already and that the blue whale, largest of all animals ever to live on this planet, is destined for extinction. At the same time, the total yield, whether we count that yield in number of animals killed, or in kilograms of food harvested, or in dollars, or in rubles, or in yen, has declined well below the theoretical productivity of the southern ocean. As a result, the world community has directly suffered from the slaughter of the whales. Yet whaling continues virtually uncontrolled even though it is now illegal to import products of the blue whale into many nations, such as the United States, Great Britain, and Canada. The problem is that a few nations, including Japan and Russia, continue to hunt Antarctic whales. If current practices continue, and species of whales become extinct, not only will we suffer a great aesthetic and scientific loss but also a hitherto productive region of the Earth's surface will produce less protein and fewer raw materials for the human population.

TAKE-HOME EXPERIMENTS

1. **Habitats.** Prepare a list of the various types of environments that exist within a 10 km radius of your home. Typical entries might be: residential; commercial; public park; woodlot; farm; lake; stream or river; undeveloped mountain. What types of wild plants and animals live in each of the sections? Discuss the effect of civilization on species distribution in your area.

2. **Foreign species.** Go to local farmers or nursery owners and ask them to list the major insect pests in your region. How many of these pests are native to your area? How many have been imported from foreign lands?

PROBLEMS

1. **Evolution.** Outline briefly the types of environmental pressures to which an animal must adapt. Using this outline as a guide, discuss Darwin's concept of "survival of the fittest."

2. **Speciation.** Many zoos are finding it advantageous to breed specimens in captivity. If this practice continues, do you feel that there is a possibility that separate "zoo species" will evolve?

3. **Speciation.** It is believed that new species of insects may evolve within a few hundred years, whereas many thousands of years are required for a new species of mammals to evolve. Explain.

4. **Pleistocene extinctions.** Discuss the role of people in the Pleistocene extinctions. Do you feel that people, alone, caused these extinctions? Defend your answer.

5. **Pleistocene extinctions.** If the sabre-toothed tiger could be reintroduced into North America today, do you think that it would survive? Defend your answer.

6. **Habitat destruction.** List the major factors which lead to habitat destruction in the modern world. What factors would you list as being most disruptive to wild animals? Explain.

7. **Habitat destruction.** Elk normally feed in the high mountains in summer and travel to lower valleys during the winter months. Imagine that a developer was planning to build a housing complex in a mountain valley in Montana. The developer claims that since only 10 percent of the elk's annual range is being preempted, the herd will not be seriously affected. Do you agree or disagree with this argument? Defend your position.

8. **Introduction of foreign species.** Australia has been separated from Asia, Europe, and Africa for millions of years, and consequently many species of plants and animals unique to this continent have evolved. When European settlers immigrated to Australia in the nineteenth century, they brought with them many new species of plants and animals, such as sheep, dogs, rabbits, and various cereal grains. In many cases these new organisms displaced native ones, and many native species have recently become extinct. Suggest some reasons why these substitutions have occurred.

9. **Environmental impact of major construction.** When ships pass from ocean to ocean through the Panama Canal they are raised through a series of locks, sail across a freshwater lake, and then are lowered through a second series of locks. Passage through the canal would be facilitated if a deep trench were dug to connect the Atlantic and Pacific oceans directly. Would such a canal affect the survival of aquatic species? Discuss.

10. **Predators and people.** Considering the grizzly bears' record, do you feel that it is safe to feed park bears despite regulations forbidding it?

11. **Critical level.** Define critical level. Give two examples.

12. **Critical level.** Discuss how the critical level might differ greatly from species to species. In your answer, briefly explain how each of the following factors could affect different species differently: (a) reproductive failure, (b) ecosystem imbalance, (c) ability of males to find females, (d) luck, and (e) inbreeding.

13. **Survival in the twentieth century.** Give some reasons why the bison herd in North America has been virtually eliminated while the white-tailed deer population has actually increased during the past century.

14. **Survival in the twentieth century.** Female polar bears give birth to one or two infants every three years. Discuss the survival potential of this low reproductive rate in past ages and at present.

15. **Survival in the twentieth century.** Consider the following fictitious species and comment on the survival potential of each in the twentieth century: (a) A mouse-sized rodent that gives birth to 40 young per year and cares for them well. This creature burrows deeply and eats the roots of mature hickory trees as its staple food. (b) An omnivore about the size of a pinhead. Females lay 100,000 eggs per year. This animal had evolved in a certain tropical area and can survive only in air temperatures ranging from 80° to 100° F (27° to 38° C). (c) A herbivore about twice the size of a cow adapted to northern temperate climates. This animal can either graze in open fields or browse in forests. It is a powerful jumper and can clear a 15-foot fence. Females give birth to twins every spring.

16. **Justification for the preservation of species.** Discuss the importance of wild grasses in the modern world.

17. **Extinction of species.** Referring to Table 5.1, calculate the average numbers of extinct species of mammals per year in each of the four time periods given.

BIBLIOGRAPHY

Four books which deal specifically with the destruction of animal species are listed below:

Roger A. Caras: *Last Chance on Earth.* New York, Schocken Books, 1972. 207 pp.

Kai Curry-Lindahl: *Let Them Live.* New York, William Morrow & Co., 1972. 394 pp.

H. R. H. Prince Philip, Duke of Edinburgh, and James Fisher: *Wildlife Crisis.* Chicago, Cowles Book Company, 1970. 256 pp.

Vinzenz Ziswiler: *Extinct and Vanishing Animals.* New York, Springer-Verlag, 1967.

Two valuable books which deal more specifically with ecosystem alterations are:

David W. Ehrenfeld: *Biological Conservation.* New York, Holt, Rinehart & Winston, 1970. 226 pp.

Charles Elton: *The Ecology of Invasions by Animals and Plants.* New York, John Wiley & Sons, 1958. 181 pp.

A detailed, fascinating, and advanced discussion of the Pleistocene extinctions is given in:

Paul S. Martin and H. E. Wright, Jr., eds.: *Pleistocene Extinctions.* New Haven, Yale University Press, 1967. 453 pp.

An interesting book that discusses the relationship between wild animals and humans is:

Roger A. Caras: *Dangerous to Man — The Definitive Story of Wildlife's Reputed Dangers.* New York, Holt, Rinehart and Winston, 1975. 422 pp.

Unit III

Human Population

6

THE GROWTH AND CONTROL OF HUMAN POPULATIONS

6.1 INTRODUCTION

Think of the place where you grew up. Picture in your mind how it looked when you were young and what it looks like now. If your childhood home was a suburb like Bethesda, Maryland, some of your favorite secret wooded haunts may now exist only in your memory. Today, these woodlands are the driveways of large apartment complexes or the sites of huge shopping centers. Destruction of scenic, quiet areas is one of the effects of rapid population growth. Perhaps you grew up in an old, inner city area. Many buildings in which your childhood friends once lived are now abandoned, boarded up, and the insides gutted. Urban decay is an unhappy effect of unplanned population decline. Perhaps you lived in a rural area, like Ridgeway, Colorado. The landscape is still as lovely as your fantasies of childhood recall. The same families who used to be your neighbors are still there; few newcomers have moved in; few have moved out. Life is still hard for your neighbors; they work long hours, and can afford few of the luxuries common to more affluent parts of the country. Economic stagnation often coincides with population constancy. These and other changes you can recall have occurred within two decades. Ask your parents to compare the childhood appearance of the place in which *they* grew up to its appearance now. Their memories will probably provide sharper contrasts than your own. If your parents are from New York City, they may remember farms in the Bronx. Californians will remember vast expanses of uninhabited land. Southerners will recall small cities and very rural areas. If you were to ask your grandparents to recall their childhood environments, they might tell you about wagon trains,

An abandoned building in a residental district of Manhattan.

135

Downtown in Ridgeway, Colorado, on a Thursday afternoon.

View of Harlem River High Bridge looking northwest toward Manhattan from the Bronx, 1861. (From the collection of the New York Historical Society.)

buffalo herds, and Indian children riding ponies across the prairies.

Dramatic changes in local populations have been the human condition since prehistoric times. When *Homo sapiens* first evolved, about 100,000 years ago, the total human population of the Earth must have fluctuated widely. In some years there were more deaths than births, causing human population to decrease temporarily. By the first century AD, however, world population had already established its present pattern of almost uninterrupted growth. Rates of growth were then very slow and extremely variable. The Black Plague of the fourteenth century is believed to have caused a temporary decrease in world population, while the European Renaissance marked the

beginning of a rapid rise in world population. At the time of the discovery of America, there were about one-quarter billion people alive on Earth. In 1650, about a century and a half later, world population had doubled to one-half billion. In another 300 years, world population multiplied fivefold to 2.5 billion persons. During the 1950's, the population increased almost another one-half billion. By 1975, world population was approximately 4 billion persons. In other words, the *increase* in world population from 1950 to 1975 was more than twice the *size* of world population in 1650. Or consider another comparison. Today there are more people in China than there were people on Earth in 1650. Or yet another: more than two thirds of all people born since 1500 are alive today.

Figure 6.1 presents a schematic graph of world population size since the emergence of *Homo sapiens*. The curve looks very smooth. Of course, population has not grown this evenly for 100,000 years. The smoothness of the curve reflects ignorance of the details of population change. Moreover, not all parts of the world have grown at the same rate. An extreme example is provided by Mexico. Before the Spanish landed, there were roughly 30 million people living in what is now Mexico. The Spanish soldiers brought European civilization, European religion, and European disease, especially smallpox. Within a century the population had plummeted to 1.6 million persons. Today, 60 million people live in Mexico.

Figure 6.1, then, does not reflect the experience of any particular region or nation, but provides a rough global picture. A glance at the curve shows that world population growth is becoming more and more rapid. Indeed, Figure 6.1 may well cause the reader to panic. If the population continues to grow ever more rapidly, or even if it continues to grow at its current rate, very soon there will be too many people for Earth to support. If poverty and starvation exist now, how can economic and agricultural development

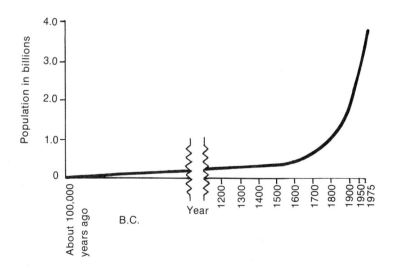

Figure 6–1 World population size from emergence of *Homo sapiens* to 1975.

be expected to keep pace with an exploding population? Destruction of land, depletion of natural resources, production of waste, and pollution of air and water can all be expected to increase with increased population. You have probably read dire predictions based on extrapolation of the growth curve of Figure 6.1. One projection which has gained a certain currency in both lay magazines and scientific journals says that if the present rate of world population increase were to continue, there would be one person for every square foot of the Earth's surface in less than 700 years. If we accept the premise, the conclusion is necessary. But we known that the conclusion must be false. It is impossible for people to stand elbow to elbow on this planet. One square foot of Earth's surface cannot feed, clothe, and shelter a person. Thus the premise must be untenable. In other words, human population cannot continue to grow at current rates indefinitely. Indeed, similar reasoning should convince you that *population size cannot grow forever,* even very slowly. It will be checked by such factors as limits of space and food, by explicit decisions of families and nations, by famine and diseases, and by complicated interrelated social forces.

Clearly, though, it is of great importance to know where the curve in Figure 6.1 is going. One reason such estimates are needed is to plan for the future. How much food must be produced during the next decades? How many schools should be built? Where should roads be placed? Parks? Power plants? Without some methods for projecting future population growth, planners would have insurmountable difficulties. Lest these questions appear to imply that population growth acts as an inexorable and independent force, remember that social and economic factors help to determine population size just as population size is one determinant of social and economic situations. For example, a well-educated and well-fed society typically tends to grow slowly, for people are both aware of population control and

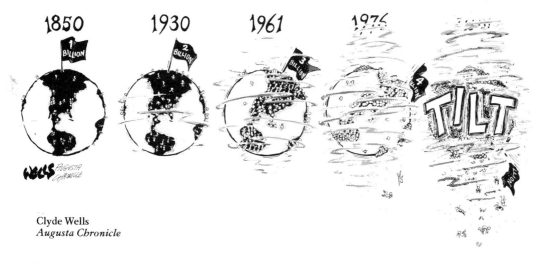

Clyde Wells
Augusta Chronicle

138

"Excuse me, sir I am prepared to make you a rather attractive offer for your square." (Drawing by Weber; © 1971 The New Yorker Magazine, Inc.)

motivated to practice it. Conversely, a society that grows slowly is more likely to be well fed. These are not firm rules. After World War II, for instance, American population growth was quite rapid.

A second reason for knowing how to predict population growth is of a political nature. A reasonably accurate estimate of population size at some future date is a numerical weapon with which to fight for the implementation of population control measures.

In order to evaluate for yourself the validity of statements, predictions, and proposals concerning population size, you must understand the mechanism of population growth and the terms used to express it.

6.2 EXTRAPOLATION OF POPULATION GROWTH CURVES

The most obvious method for predicting population growth is to construct a graph that plots population size against time and to guess how the curve will continue. Guessing points on a curve outside the range of observation is called **extrapolation.** Extrapolation is a subtle art. Look at Figure 6.1. If you didn't know the labels of the axes and were asked to continue the curve, what would you do? Such an exercise is frivolous. One person might think the curve will continue to go up indefinitely; another might try to draw a curve that peaks and then falls below zero. Still another might finish it by continuing up for a while and then leveling off. Someone else might think that the curve will become wiggly and erratic. If, however, you knew you were graphing population, your historical knowledge would allow you to exclude several kinds of curves. Infinite and negative populations would be impossible, a zero population highly improbable for many years, and wide fluctuations unlikely. Exclusions, however, wouldn't construct your curve. On the other hand, if you had a theory of population growth, you would have a basis for extrapolation.

A model for a mechanism of population growth was introduced in 1798 by the Reverend Thomas Robert Malthus, who claimed that generally populations grow much faster than food supplies. Therefore, he predicted, any uncontrolled population would eventually outstrip its food supply. Mathus stated:*

> By that law of our nature which makes food necessary to the life of man, the effects of these two unequal powers must be kept equal.
>
> This implies a strong and constantly operating check on population from the difficulty of subsistence. This difficulty must fall somewhere and must necessarily be severely felt by a large portion of mankind.
>
> ... The race of plants and the race of animals shrink under this great restrictive law. And the race of man cannot, by any efforts of reason, escape from it. Among plants and animals its effects are waste of seed, sickness, and premature death. Among mankind, misery and vice. The former, misery, is an absolutely necessary consequence of it. Vice is a highly probable consequence, and we therefore see it abundantly prevail, but it ought not, perhaps, to be called an absolutely necessary consequence. The ordeal of virtue is to resist all temptation to evil.

Nearly 200 years have passed since Malthus wrote his famous essay. World population has indeed risen rapidly. In many places, conditions have been getting harsher, and people are suffering on a world wide basis; however, Malthus's pessimistic predictions have so far proved inaccurate. Instead, the development of new agricultural techniques during the nineteenth and twentieth centuries has helped to improve economic conditions in much of the world.

No one can predict with certainty the shape of the curve of human population for the years to come. Perhaps humans will act with foresight and will gradually control population to coincide with the carrying capacity of the biosphere. However, the core of the Malthusian argument

Thomas Robert Malthus (1766–1834) was an English economist best known for his *Essay on Population,* published originally in 1798 and republished in 1803, considerably revised and enlarged. In his first edition, Malthus asserted that war, famine, pestilence, misery, and vice prevent population from increasing beyond the limits of subsistence. In the second and later editions of his work, he suggested that "moral restraint" (postponement of marriage and strict sexual continence) acts as a further check on population growth. Throughout his life, Malthus remained pessimistic in his assessment of the future progress of mankind.

*Thomas Robert Malthus: *An Essay on The Principle of Population.* Published originally in 1798. Republished in 1970 by Penguin Books Ltd., Harmondsworth, Middlesex, England, pp. 71–72.

140

cannot be ignored, for there are limits to the number of persons Earth can support, and unless growth is checked rationally, something akin to the Malthusian's "misery and vice" *will* afflict humankind.

6.3 AN INTRODUCTION TO DEMOGRAPHY

In order to investigate the population dynamics, or the **population ecology,** of a species of animal or plant, the ecologist studies the geography of the region of interest, the food supplies, climatic fluctuations, predation by other species, and competition between and within species. For human populations, however, social and economic forces are more important determinants of population growth. People have no important predators except, of course, other people; intraspecies competition in the form of exploitation, subjugation, and war is at least as old as recorded history. If an animal species grows too plentiful for its habitat, its members often starve; happier options are available for overpopulated human societies. The usual choices are emigration, or importation of food, or even the development of more effective modes of agriculture.

Recall from Chapter 4 that population curves for some species are **sigmoid,** or S-shaped, while the curves for other species may oscillate, or even decline to zero. If the human population curve eventually becomes sigmoid, examination of Figure 6.1 gives no clue as to when the top of the S will begin to form. A reliable method of predicting population is to look not at changes in total population size with time but at patterns of changes in **rates of growth.** For this purpose, the reader needs to understand how birth and death rates affect the size of populations and how current trends affect future patterns. Because of the uniqueness of the human pattern of growth, we shall consider it in demographic, rather than ecological, terms.

Demography is that branch of sociology or anthropology which deals with the statistical characteristics of human populations, with reference to total size, density, number of deaths, diseases, and migrations, and so forth. The demographer attempts to construct a numerical profile of the population viewed as groups of people, not as individuals. For this purpose, the demographer needs to know facts concerning the size and composition of populations, such as the number of females alive at a given time or the number of infants born in a given year.

The subject may sound terribly dry to those of you who are uncomfortable with numbers and computations, but demographic data often reflect in fascinating ways the history of the country studied, the trends in medical care, and the occurrence of social changes.

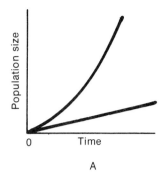

A

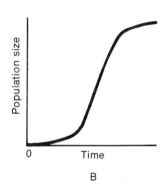

B

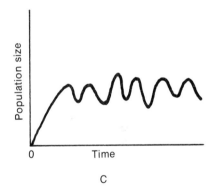

C

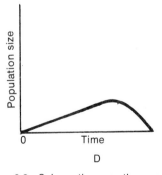

D

Figure 6.2 Schematic growth curves. *A,* Arithmetic (straight-line) and geometric (curved-line) patterns of growth. *B,* Sigmoid curve of growth. *C,* Oscillating population curve. *D,* Growth curve of a population that becomes extinct.

141

In addition to studying the composition of populations, the demographer is interested in how populations change in time. This is studied by counting the number of **vital events** — births, deaths, marriages, and migrations. If we knew the composition of a population at any given time and the number of vital events occurring between that time and another, we would know the composition of the population at the end of the period. For example, suppose that in 1924 the population of some village was 732. In the next two years, if there were 28 births and 15 deaths and if 4 people moved in and 1 moved out, there would be $732 + 28 - 15 + 4 - 1 = 748$ people at the end of 1926.

How does a population grow? Imagine you decided to deposit $100 in a bank that offered 3 percent interest per year. Suppose you started walking to the bank carrying three dollars in change, seven $1 bills, ten $5 bills, and four $10 bills, for a total of $100. If you deposited the entire $100, you would expect to have $103 at the end of the year. But on the way to the bank you bought an irresistible ice cream sundae for one dollar. Thus, you had only $99 to deposit when you arrived at the bank. No matter how you paid for your sundae, whether you used coins, or a dollar bill, or a bill of higher denomination and received change, your $99 would grow at a rate of 3 percent. In one year you would have $99 + $99(.03) = $101.97. This kind of growth, where the increase is proportional to the initial size, is called **geometric** or **exponential growth.**

How different a population is! Imagine a population of 100 people — 3 infants, 7 children, 50 adults under 65, and 40 people at least 65 years old. Suppose that during an entire year, no one moved in or out of the population, five of the women under 65 had babies, and two of the people over 65 died. These were the only vital events. At the end of the year, the population would be $100 + 5 - 3 = 103$, for an annual rate of growth of 3 percent.

Now suppose the population had contained only 99 people at the beginning of the year. What would the rate of increase have been? If the population grew in the same way that money in the bank grows, the rate of growth would be 3 percent no matter which person in the original population were no longer there. People, however, are not interchangeable like dollar bills. If the population had been missing an infant, there still would have been five births and two deaths. There would have been $99 + 5 - 2 = 102$ persons at the end of the year. The annual rate of growth would have been:

$$\frac{102 - 99}{99} \times 100\% = 3.03\%$$

On the other hand, if the population had been missing one of the women who had a child, only four births would have occurred, and the rate of growth would have been

142

$$\frac{101 - 99}{99} \times 100\% = 2.02\%$$

If one of the elderly persons who died had been missing from the population, the population would be $99 + 5 - 1 = 103$ persons at the end of a year, for an annual growth rate of

$$\frac{103 - 99}{99} \times 100\% = 4.04\%$$

This very simple example has pointed out some of the difficulties confronting the student of population size, but it also leads to three important insights that are necessary for an effective approach to the investigation of growth:

1. An overall "rate of growth" is really the difference between a rate of addition (by birth or immigration) and a rate of subtraction (by death or emigration). The rate of growth is positive only when there are more additions than subtractions.

2. The probability of dying or of giving birth within any given year varies with age and sex.

3. The age-sex composition, or **distribution,** of the population has a profound effect upon a country's birth rate, its death rate, and hence its growth rate.

6.4 MEASUREMENTS OF GROWTH

The simplest way to measure growth is by subtracting the lower value from the higher one. The following example does this for the populations of India and the United States during the same quarter-century, 1950 to 1975.

POPULATION (millions)		
	India	U.S.
1975	600	213
1950	360	152
Increase	240	61

These differences show that the United States, with many times the wealth of India and about three times the land area, had about one quarter the population increase.

Unfortunately, these population differences give little feeling about how fast a country is growing relative to other countries. More useful are rates of growth. For the data from India, the average annual rate of growth from 1950 through 1975 was approximately:

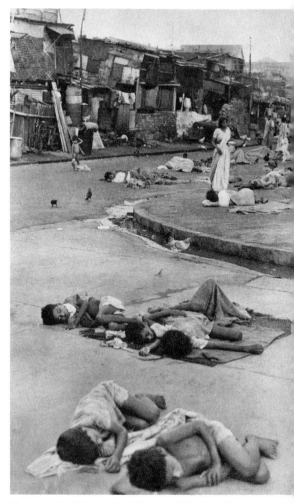

Thousands of homeless sleep on the streets of Bombay. (Courtesy of Sygma.)

$$\text{Average annual growth rate (India, 1950–1975)} = \frac{(600 \times 10^6 \text{ persons} - 360 \times 10^6 \text{ persons}) \times 100\%}{360 \times 10^6 \text{ persons} \qquad \times 25 \text{ years}} = 2.7\%/\text{year}.$$

For the United States in the same period the average growth rate was:

$$\text{Average annual growth rate (U.S., 1950–1975)} = \frac{(213 \times 10^6 \text{ persons} - 152 \times 10^6 \text{ persons}) \times 100\%}{152 \times 10^6 \text{ persons}} \times 25 \text{ years} = 1.6\%/\text{year}.$$

These rates are averages and do not imply, for example, that the population of India in fact grew by 2.7 percent each year. For if it had, the population in 1975 would have been $360(1.027)^{25}$ or 700 million persons.

A common popular measure of population growth is the **doubling time** (t_d) of the population. The doubling time is defined as the length of time a population would take to double if its annual growth rate (r) were to remain constant. (See Appendix C6 for calculation of doubling times.)

Two very basic measures of population growth are the **crude birth rate** and the **crude death rate**. Subtracting these two rates yields the **rate of natural increase**. For any geographical area or ethnic group being studied, these rates are:

(a) Crude birth rate in year X $= \dfrac{\text{Number of live children born in year X}}{\text{Midyear population in year X}} \times 1000$

(b) Crude death rate in year X $= \dfrac{\text{Number of deaths in year X}}{\text{Midyear population in year X}} \times 1000$

(c) Rate of natural increase in year X $=$ Crude birth rate $-$ Crude death rate

For long-term assessment of historical trends, these three crude rates are concise, useful, and graphic. Figure 6.3 and Table 6.1 show birth and death rates for the developed and the developing parts of the world. These growth rates provide interesting measures of long-term historical trends, but not much more. The growth rates of real populations depend very strongly on the ages and sexes of the people they contain. Therefore, beware of conclusions made

TABLE 6.1 Vital Rates* for the Developing and the Developed Nations 1900–1970

PERIOD	DEVELOPING NATIONS			DEVELOPED NATIONS		
	Birth Rate	Death Rate	Rate of Natural Increase	Birth Rate	Death Rate	Rate of Natural Increase
1900–1950	41	32	9	26	18	8
1950–1960	43	22	21	22	10	12
1960–1970	41	17	24	20	9	11

*Per 1000 population.

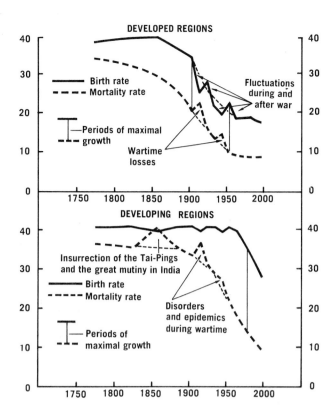

Figure 6.3 Estimated and predicted crude birth rates and crude death rates for the period 1750–2000 in the developed and developing regions of the world. The United Nations defines the developed regions as Europe, the United States, the Soviet Union, Japan, temperate South America, Australia, and New Zealand. The rest of the world is called developing. (Adapted from *The Demographic Situation of the World in 1970*, United Nations, 1971.)

on the basis of projecting doubling times. Remember that although the term "doubling time" sounds like a projection, the measure uses only the crudest available population data.

Patterns of migration do not greatly affect growth except when a large proportion of the population migrates, especially when the migrants are predominantly of one sex. For example, in the last century many Irish young men emigrated, leaving a surplus of women behind. Consequently there were many childless women. Conversely, at the beginning of this century, there were many more men than women in the United States. The differential can be attributed in large part to the fact that the heavy immigration of the period was male-dominated. In general, groups of migrants are usually quite different with respect to age, sex, and vital rates from the inhabitants of the country either from which they leave or to which they go. In particular, migrants are often healthy males of reproductive age.

The current age-sex distribution of an area should be used in predicting future population sizes. In the hypothetical age-sex distribution of Figure 6.4, each age group has the same number of males as females. In particular, there are 500 boys and 500 girls under 10, and 50 men and 50 women between 90 and 100 years of age. Furthermore, there are exactly 50 fewer men and 50 fewer women at each succeeding age decade. Thus, while there are 300 of

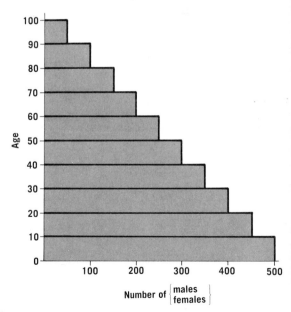

Figure 6.4 Age-sex distribution of an ideal population.

145

each sex between the ages of 40 and 50, there are only 250 of each sex between 50 and 60.

Do human age-sex distributions look like Figure 6.4? Not at all! Figure 6.4 would represent a population in which (a) boys and girls were born with equal frequency; (b) the same number of persons were born every year for over a century; (c) everyone died by the age of 100; and (d) any person, at birth, had an equal chance of dying throughout each year of his life span. However, in real human populations, about 106 boys are born for every 100 girls.* Nor is the probability of dying constant throughout one's life span. Instead, a relatively large proportion of people die when they are very young, comparatively few die between the ages of 10 and 50, and the proportion of people dying each year after 50 increases rapidly. In addition, there are marked sex differences in **mortality,** the number of deaths which occur in a given period. Women have a higher probability of surviving from one year to the next throughout the life span except during the childbearing years in areas without modern medical care.

Consider the effects of realistic patterns of vital events on a group of people born in the same five-year period. The greater survivorship of women over men means that even though more boys are born than girls, the ratio of women to men increases as the group grows older. By the time the group is elderly, there are considerably more women than men. Also, data are usually collected in such a way that we know only the total population of each sex over 70, 80, or 85. Therefore, the graphs can be only approximate for the very old age groups. Figure 6.5 presents an age-sex dis-

*This figure represents a worldwide average. There is considerable geographic variation.

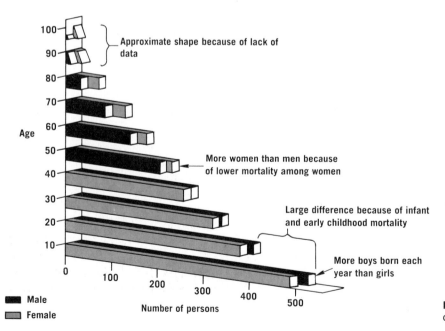

Figure 6.5 Typical age-sex distribution.

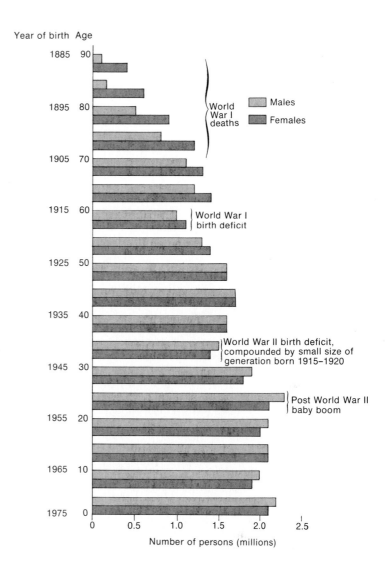

Year of birth Age

World War I deaths

Females

World War I birth deficit

World War II birth deficit, compounded by small size of generation born 1915–1920

Post World War II baby boom

0   0.5   1.0   1.5   2.0   2.5

Number of persons (millions)

Figure 6.6 Population of France in 1975.

tribution that more nearly reflects these demographic characteristics.

In addition to these reasonably predictable phenomena, many changes that are less predictable can occur. Population growth is affected by such events as war, famine, medical advances, and changes in social custom. For example, Figure 6.6 presents the age-sex distribution of France in 1975. This is a very bumpy curve. To interpret it we need knowledge from many fields. For example, the graph shows that in 1975 only one-third of the population of France between 80 and 84 were males. Why so few? Part of the explanation is that men die earlier than women, but the observed discrepancy is much larger than can be explained solely by natural differences in rates. History provides an answer. The people who are now 80 to 84 years old were born between 1890 and 1894. At the outbreak of World War I, in 1914, therefore, they were between 19 and 23 years of age. Many of these men fought and died in World War I. Therefore, much of the difference

147

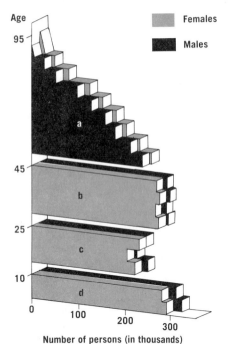

Age

Females
Males

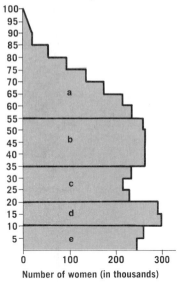

Number of persons (in thousands)

Figure 6.7 Age-sex distribution of Sweden in 1950.

Number of women (in thousands)

Figure 6.8 Female age distribution of Sweden in 1960.

in numbers of men and women in 1975 reflects the mortality of French men during World War I. The distribution curves can aid in predicting future population growth. Since many males born between 1890 and 1894 were away from home during some of their reproductive years, and since so many were killed, there should have been relatively few births about one generation later. Indeed, the generation born during World War I, especially in the years 1915 to 1920, is very small. Examination of Figure 6.6 shows a similar pattern for World War II.

To illustrate another example of predicting population growth from age distributions, examine the population distribution of Sweden in 1950 (Fig. 6.7). The total population in 1950 was nearly 7 million persons. First, note that the male and female age distributions are very similar. Therefore, only one sex is needed to study population growth. Note that there are relatively few people in the reproductive age group, 15 to 35 years. (Refer to groups b and c in Figure 6.7.) Therefore we would expect that 10 years later, there would be relatively few infants. The shape of the age-sex distribution curve in 1950 enables demographers to predict the age-sex distribution in 1960. In fact, the shape of the actual distribution curve in 1960, shown in Figure 6.8, is just about what one would expect. The male age distribution is nearly the same shape.

The shapes of many population curves tell much about the growth of a population. Rapidly growing populations have very large bases; populations that are remaining constant have relatively narrow bases, and declining populations have pinched bases (Fig. 6.9).

The change in population distribution over time has been likened to the digestion of a mouse by a snake. A snake swallows a mouse whole and thereby gets a big lump in its throat. The body of the mouse then moves slowly through the digestive system of the snake. The bump in the snake gets smaller and smaller as the mouse is slowly digested. The movement from head to tail can be considered analogous to the aging of a generation, and digestion to its gradual dying. The problem with guessing what a snake will look like tomorrow is that the observer doesn't know when, or what, the snake will eat next. Once a mouse

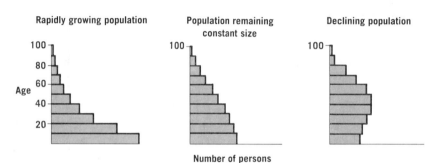

Rapidly growing population

Population remaining constant size

Declining population

Age

Number of persons

Figure 6.9 Schematic population age distributions.

148

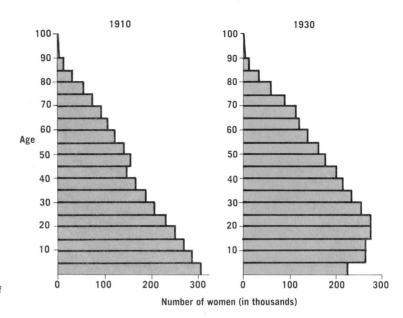

Figure 6.10 Female age distributions of Sweden in 1910 and 1930.

Number of women (in thousands)

is in the snake's body, it is easy to predict what will happen to the shape of the snake. So it is with population distributions. Once a generation has been born, demographers can predict quite accurately how that generation will change. However, accurate prediction of the size of the coming generation is extremely difficult. The size of the generation of childbearing age give clues, but not definitive information. For example, Figure 6.10 shows that the distribution for Sweden looked like a triangle in 1910. Not knowing anything about changes in vital rates, one would predict a triangle in 1930. However, in 1930 the base of the age distribution was pinched. Reexamination of Figure 6.7 shows that the pinched base persisted through the births of 1940. A demographer would have had to predict World War I as well as its profound economic and social effects on all of Europe to have projected accurately the population of Sweden from 1910 to 1930.

6.5 SUMMARY MEASURES OF MORTALITY AND NATALITY: EXPECTATION OF LIFE AND TOTAL FERTILITY RATE

To compare mortality and natality from country to country, measures are needed that are independent of population distribution. Much, but not all, information about the status of mortality in a given country during a given year can be summarized by a single measure—the **expectation of life at birth, e_0,** or the **life expectancy at birth.**

The expectation of life at birth measures the average number of years a baby born in a given year and a given country would live if the death rates prevailing at the time

149

of the baby's birth were to remain constant for a generation. For example, Table 6.2 shows that a hypothetical baby boy born in 1970 could be expected to live 59.4 years if conditions of mortality in Mexico remained constant for about 100 years. A real baby boy may not necessarily live exactly 59.4 years because, first, expectation of life reflects an average experience, and second, death rates in Mexico will not remain constant for a century.

A measure of the fertility in a country is the **total fertility rate (TFR),** that is, the total number of children a woman in a given country can be expected to bear during the course of her life if birth rates remain constant for at least one generation. There is considerable variability in TFR. In Mexico in 1965, for instance, women were producing babies so quickly that if those current rates were to

TABLE 6.2   Expectation of Life for Several Selected Countries*

COUNTRY	LATEST AVAILABLE YEAR	CRUDE DEATH RATE PER 1000	EXPECTATION OF LIFE	
			Male	Female
Denmark	1971–72	10.2	70.7	76.1
Iceland	1966–70	7.0	70.7	76.3
Netherlands	1973	8.0	70.8	76.8
Sweden	1973	10.6	72.1	77.7
Australia	1965–67	8.8	67.6	74.2
Bulgaria	1965–67	8.5	68.8	72.7
Canada	1970–72	7.4	69.3	76.4
Czechoslovakia	1970	11.0	66.2	72.9
England and Wales	1970–72	12.1	68.9	75.1
France	1972	10.4	68.6	76.4
Germany (East)	1969–70	13.3	68.9	74.2
Germany (West)	1970–72	11.7	67.4	73.8
Israel	1973	7.2	70.2	73.2
Japan	1972	6.6	70.5	75.9
Austria	1973	12.5	67.4	74.7
Hong Kong	1971	5.2	67.4	75.0
Poland	1970–72	8.2	66.8	73.8
United States	1972	9.1	67.4	75.1
Soviet Union	1967–68	7.6	65.0	74.0
Argentina	1965–70	9.4	64.1	70.2
Taiwan	1965	5.5	65.8	70.4
Yugoslavia	1971–72	8.5	65.6	70.4
Mexico	1970	8.9	59.4	63.4
Guatemala	1963–65	15	48.3	49.7
Pakistan	1962	20	53.7	48.8
Syria	1970	15	54.5	58.7
India	1951–60	20	41.9	40.6
Burundi	1965	25	35.0	38.5
Chad	1963–64	25	29.0	35.0
Nigeria	1965–66	25	37.2	36.7
Togo	1961	29	31.6	38.5

*Data from U.N. Demographic Yearbooks (1970 and 1974).

continue, the average woman would give birth to 6.8 babies. By contrast, in Hungary, the TFR was only 1.8. In the developed countries, a TFR of roughly 2.1 is the **replacement level,** that is, the value of the TFR which corresponds to a population exactly replacing itself.

6.6 ACCURACY OF THE DATA

The process of collecting data for the purpose of constructing age-sex distributions of a country is quite complicated. People must be identified, and ages must be known. Many opportunities for error arise. The reported information may be incorrect, and the procedures used for tabulating the data may introduce serious inaccuracies. Accurate data-gathering techniques are expensive and require considerable expertise.

Even when data appear reliable, the cautious demographer must question their validity. Figure 6.11 depicts the age distribution by five-year age groups for Mexico in 1960. The data were collected in the decennial (tenth annual) **census,** or count, of the population. Each person was asked to report his age. The graph appears odd, especially between the ages of 30 and 60. There are more 36- to 40-year-olds than 31- to 35-year-olds; more 46- to 50-year-olds

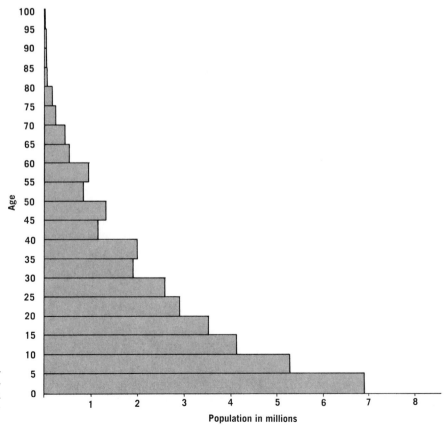

Figure 6.11 Population distribution of Mexico by age categories in the 1960 census. (Data from *U.N. Demographic Yearbook,* 1962.)

151

than 41- to 45-year-olds; more 56- to 60-year-olds than 51- to 55-year-olds. The dips in population cannot be attributed to successive birth deficits because they are too close together. Replotting the data by single year of age (Fig. 6.12) yields shocking results. Accordingly to the 1960 census of Mexico, people were much more likely to report an age divisible by 10 than to report the next year. If the reported data are true, there were three times as many 30-year-olds as 31-year-olds; five and a half times as many 40-year-olds as 41-year-olds; six times as many 50-year-olds as 51-year-olds; 10 times as many 60-, 70-, 80-, and 90-year-olds as 61-, 71-, 81-, and 91-year-olds, respectively. What demographic process could explain the overabundance of people whose age ended in five, and the lack of people whose age ended in seven? The answer, of course, is that in Mexico people tend to report their ages in some round form, for instance, to the nearest five, or to some even number close to the true age. Such marked **digit preference** is not uncommon in developing countries. In such cases, use of data as reported leads to unreliable age-specific measures.

Other sources of error abound. For instance, the definitions of live birth, immigrants, and emigrants differ from country to country. Another problem arises because conflicting political claims over territory cause some people to be counted twice—once as subjects of one country, once as subjects of another. Other difficulties result from the habits of the group being counted. Nomadic bands, for example, are notoriously hard to count.

Even in developed countries, demographic data have errors. Mothers often forget to report to the census taker the presence in the family of a newborn, especially if the baby is sleeping peacefully in another room at the time the interviewer arrives. In the United States, there is considerable inaccuracy in reporting age at about 65, because many people have been dishonest in reporting their age to the Social Security Administration. Some people are loath to tell the census taker their true age for fear of seeming "too old." Another serious problem in obtaining accurate age-sex distributions for the United States arises from welfare laws, especially Aid to Dependent Children (ADC). ADC is available only to families without fathers. Often the father, or a male, is present in the family, but his existence has been denied to the Welfare Department. When the census taker arrives, the father's existence is again denied, for

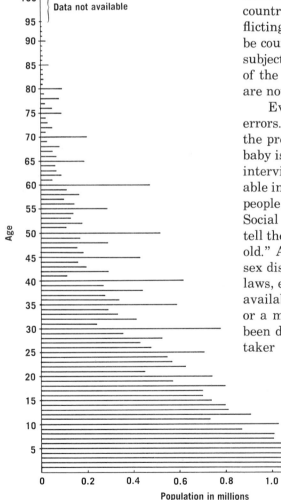

Figure 6.12 Population distribution of Mexico by single year of age in the 1960 census. (Data from *U.N. Demographic Yearbook*, 1962.)

fear that the Census Bureau will report his presence to the Welfare Department. (The fear is needless.)

All the foregoing is to warn against hasty uncritical acceptance of a demographic "fact." It is not a plea to ignore data, for even if data are imprecise or sketchy, general trends in birth or death rates will almost always appear. In using demographic data, always consider what sources of error might be present, and that such error might change the interpretation of the subject under investigation.

6.7 THE DEMOGRAPHIC TRANSITION

Population grows when the birth rate plus the in-migration rate exceeds the sum of the death rate plus the out-migration rate. For thousands of years, population grew slowly. Suddenly, in the fifteenth century, population began to rise rapidly. What happened? Have family sizes grown larger and larger during the past 500 years? The answer is no; indeed, the average number of births per family has become smaller. The change has occurred because people live longer, on the average, than they used to.

When nutrition is poor, water unclean, and infectious disease rampant, relatively few people live to adulthood. Many children are born, but many die; perhaps 50 percent of all live-born infants do not reach their fifth birthday. As modern principles of medical care have developed or are introduced into a society, death rates start to fall. In the developed countries, the death rates have been steadily dropping for over a century. Changes in sanitation practices, introduction of pasteurization, immunization against disease, and social welfare programs to alleviate abject poverty and malnutrition have slowly been introduced and have contributed to a gradually declining death rate. As the death rate has decreased, population growth has increased. Note in Figure 6.3 the large difference between birth and death rates for the developed countries in 1900.

In the developing areas of the world, modern medicine arrived suddenly in the twentieth century. Armed with a century of knowledge and experience, medical personnel from the developed countries introduced and helped to implement profound changes quickly. Death rates plummeted in the twentieth century, leading to very rapid population growth.

Why has this been so? An improvement in medical practice may lead to a sudden drop in death rates; changes in birth rates reflect changing human patterns of behavior. In the developed nations, medical technology has been advancing steadily for many years. During this time, people have gradually understood that it is possible to raise a family without having many babies. Consequently, the birth rate has been declining gradually for almost a century. People in the developing nations have not been exposed to such gradual change. Modern medical technology was introduced suddenly, and people's patterns of behavior have not had time to adjust, so birth rates have not dropped so quickly.

Demographers summarize the changing profile of a country in the following way. Medically primitive societies are characterized by both high birth and high death rates; since the difference is small, there is little or no growth. When modern medicine is introduced, mortality among children drops. Birth rates, however, remain relatively constant. The combined effect of these two trends—falling death rates and constant birth rates—ushers in a period of very rapid population growth, known as the demographic transition. Finally, birth rates drop, so that developed countries have low growth because birth and death rates are both low. Table 6.3 lists some countries at various demographic stages.

Recent research has indicated that this picture of the demographic transition may be oversimplified. In several European countries, for instance, birth and death rates declined simultaneously and at nearly the same rates in the nineteenth century. In France, examination of records of births and deaths indicates that in many parishes the decline in birth rate actually *preceded* the decline in death rates. Further study of transition is an area for research in demography.

TABLE 6.3 Some Typical Vital Rates Before, During, and After Demographic Transition (latest available data)*

		CRUDE BIRTH RATE/1000	CRUDE DEATH RATE/1000
Before Transition	*Very high birth and death rates*		
	Afghanistan (1965–70)	50.5	26.5
	Angola (1965–70)	50.1	30.2
Transition Period	*High birth and death rates*		
	Bolivia (1965–70)	44.0	19.1
	Indonesia (1965–70)	48.3	19.4
	High birth rates, moderate death rates		
	India (1965–70)	42.8	16.7
	Morocco (1965–70)	49.5	16.5
	Moderate birth rate, low death rate		
	Argentina (1970)	22.9	9.4
	Spain (1974)	19.3	8.4
	United States (1974)	15.0	9.1
After Transition	*Low birth rate, low death rate*		
	Switzerland (1974)	12.9	8.5
	Netherlands (1974)	13.8	8.0

*Source: U.N. Demographic Year Book (1974).

6.8 LAWS AND PUBLIC POLICY CONCERNING POPULATION GROWTH

In the past, laws concerning population have been in favor of growth. The Declaration of Independence of the United States accused King George III of England as follows:

He has endeavored to prevent the population of these States: for that purpose obstructing the Laws for Naturalization of Foreigners; refusing to pass others to encourage their migrations hither, and raising the conditions of new Appropriations of Lands.

Even the ancient codes of Hammurabi of Babylonia and Emperor Augustus of Rome contained provisions encouraging births. Only since World War II has the desirability of population growth been seriously questioned by many governments.

Today, public policy concerning population growth depends in large part upon the demographic stage of a country. Countries before transition (that is, those with both high birth and death rates), are justifiably concerned with the difficulties of life and strive to improve medical conditions. This improvement leads, of course, to transition and rapid population growth. Many nations undergoing transition strive to reduce birth rates. Most remedies are noncoercive. Governments issue statements of public policy, reinforce them with propaganda, and offer information on methods of family planning. Some governments even provide supplies for contraception.

In general, improvements in economic conditions tend to be associated with declining birth rates. The process is circular. As economic conditions improve, educational opportunities expand, and more and more families rise in social status. A particularly important social pattern is an economic structure that encourages women to work. These conditions all lead to smaller family sizes. Conversely, a small family size, with fewer mouths to feed and backs to clothe, helps to keep the modern family economically sound. Thus, a small family size leads to improved economic conditions, both in the individual family and in the country as a whole.

In recent years, several nations have found that these noncoercive methods do not lead to a rapid enough decline in birth rate. As a result, stronger measures have been adopted, such as economic penalties and even frank legislation against large families. In Singapore, for example, the state provides paid maternity leave for only two births. Also, only the first four children in a family have access to choice primary education. Such discrimination against latter-born children is an effective economic disincentive. Families are loath to have a child who will not be educated, and so, since many families are unable to pay for education, family size has decreased rapidly. In India, sterilization is now compulsory for all fathers of two or more

A jungle clearance in Brazil. (Photo by Leo deWys, In

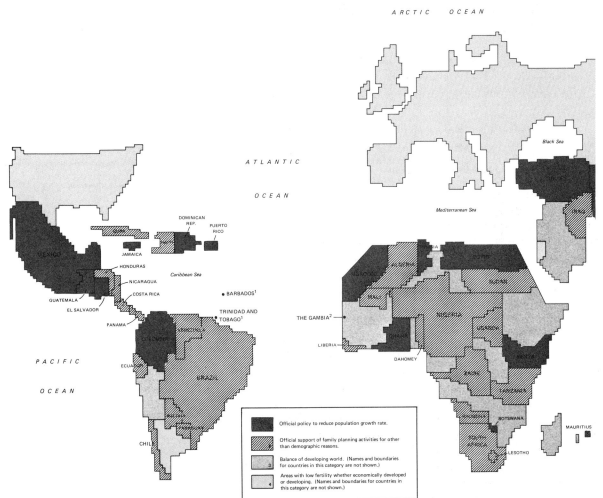

Figure 6.13 Government position on population growth and family planning among developing countries. Areas of designated countries and regions are proportional to their respective annual number of births; locations are in general accordance with geographical relations.

Legend (in map):
- Official policy to reduce population growth rate.
- Official support of family planning activities for other than demographic reasons.
- Balance of developing world. (Names and boundaries for countries in this category are not shown.)
- Areas with low fertility whether economically developed or developing. (Names and boundaries for countries in this category are not shown.)

children. If this program is effective, the birth rate in India will decline dramatically. These admittedly effective methods are very harsh; compulsory sterilization infringes upon individual liberty, and legislation directed against large families harms innocent children, who are made to suffer for "the sins of their fathers."

Some nations in transition are still encouraging births. The prime example is Brazil, a nation of vast area, much of which is sparsely populated, and abundant natural resources.

After transition, population growth is more or less stable. Most governments of post-transition countries are rather neutral toward population growth. Contraceptive methods are widely known and easily available, even in countries where religion frowns upon their use. Some of these nations actively encourage births in order to stimulate economic growth. Rumania, for instance, limits access to contraceptive devices in order to increase population. The French government, also, is encouraging births.

Figure 6.13 is a map of the world showing the relationship between the annual number of births and official policy toward family planning.

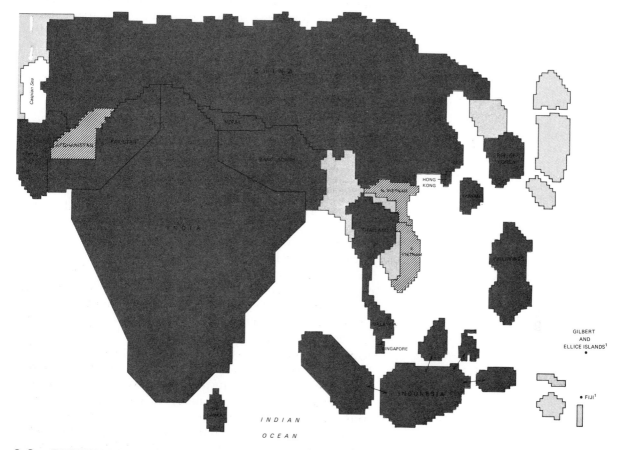

6.9 TECHNIQUES OF POPULATION CONTROL

Social trends leading to decreased fertility, such as the increasing role of the woman outside the home, will be discussed in Chapter 7. Such trends, of course, do not in themselves prevent births but lead to the desire to control family size.

Three important approaches to prevention of birth are available—**contraception, sterilization,** and **induced abortion.** Sterilization procedures render a person incapable of fathering or conceiving a child. Examples of contraceptives (devices or techniques which prevent conception) are the condom, the intrauterine device (IUD), contraceptive pills, the diaphragm, and spermicidal agents. Induced abortion is the artificial termination of pregnancy. Sterilization and induced abortion are, by their very nature, effective methods of population control. The effectiveness of various contraceptive methods is a function of both the theoretical efficacy of the method and the motivation of those who practice it. For example, the rapid decline in birth rate in nineteenth-century Europe has been credited in large part to the widespread use of *coitus interruptus,* intercourse with withdrawal before ejaculation. The historical success of this error-prone method has been attributed to the strong motivation of its users. By contrast, vaginal foam should theoretically be a highly effective method of birth control, yet in studies of contraceptives, many babies are born to women using foam. A likely explanation for its relative ineffectiveness is that highly motivated women chose other methods of conception control, and therefore, the less highly motivated women relying on foam use it improperly or often neglect to use it at all.

In addition to modern methods of contraception, many traditional practices tend to limit births. Perhaps the most important is breast-feeding, for fertility declines during lactation.

Most countries have laws concerning the teaching, advertising, dissemination, and use of various birth-control methods. The laws are often contradictory and widely disobeyed. In many countries which forbid birth-control methods, the well-to-do members of the society are able to circumvent the law by leaving their country or state for an abortion or sterilization, by purchasing smuggled contraceptives, or by going to a private physician who prescribes some contraceptive method as "therapy" or as a "disease preventive." The poor in any country have fewer opportunities to violate population-control laws. Until the recent Supreme Court decision allowing abortions throughout the United States, an unmarried pregnant American woman of moderate means could have gone to New York to have a safe, legal abortion, while many of her poor counterparts who wanted an abortion were forced to choose an illegal abortionist, where the operating conditions were often quite unsanitary. Out of every 100,000 American women who are pregnant and do not have an abortion, about 20 die of direct complications of pregnancy. Of every 100,000 who have a legal abortion in a hospital, only 3 die. By contrast, there are 100 deaths for every 100,000 women who undergo an illegal abortion; moreover, many of the survivors are rendered infertile.

In some countries, such as Colombia and the Republic of Ireland, abortion is illegal under all circumstances. In others, abortion is legal only to save the life of the mother (e.g., Guatemala and Indonesia) or to preserve her health (e.g., Australia and West Germany). Some countries permit abortion for eugenic reasons; that is, to prevent the birth of an infant with a severe birth defect (e.g., Czechoslovakia and Norway). Some allow abortion in cases of rape or incest (e.g., Argentina and Poland). A few countries, for example Denmark, the United States, and the U.S.S.R., permit abortion at the request of the mother. Abortion is prohibited in many eastern European nations except under special circumstances. In Bulgaria, for instance, where the state is trying to foster an increased rate of population growth, women with at least two living children and women over 40 with one living child are routinely allowed abortions; other women may be denied their requests.

Laws concerning contraceptives are varied as well. They are forbidden in Ireland and Spain but are required to be stocked by pharmacies in Sweden, and by pharmacies, department stores, and rural cooperatives in the People's Republic of China. Until July, 1970, when the law was declared unconstitutional by the U.S. Court of Appeals of the First Circuit, Massachusetts forbade the sale of contraceptives and the dissemination of information about birth control to unmarried women. In Poland, information about contraception is mandatory before an abortion; in Denmark it is required after childbirth and abortions. In some North African countries, the manufacture and importation of contraceptive devices are illegal.

Sterilization laws are perhaps the most ambiguous. In the United States, most states permit compulsory sterilization to be performed on mentally infirm patients maintained at state institutions. Only five states specifically allow voluntary sterilization for therapeutic or socioeconomic reasons. In the many states without any law concerning sterilization, the resulting legal ambiguity causes many doctors to be reluctant to perform sterilizations.

In 1968, the United Nations Conference on Human Rights unanimously proclaimed that family planning is a basic human right and declared that governments should abolish laws conflicting with the implementation of such rights and adopt new laws to further such rights. The legalization of contraception, abortion, and sterilization throughout the world would help reduce population growth and curtail maternal deaths from unsanitary, illegal abortions. It should be noted, however, that this proclamation of the United Nations is a recommendation which carries no legal weight anywhere in the world.

Among people of the developed countries, the choice of various methods of birth control is largely individual. Several methods may be tried and the most personally satisfactory chosen. In developing countries, the problem is quite different. Usually, a team of health workers makes a decision about the type of birth-control techniques to introduce. The wrong choice can lead to a rejection of family planning. Family planners in developing countries have learned that introduction of a highly reliable method of birth control to a relatively small group of couples is often more effective in reducing births than a moderately effective method introduced to many. Methods such as diaphragms, foams, and other techniques that require careful use are often not effective in the developing countries.

The acceptability of contraceptives varies by culture. The highly successful birth-control program in Puerto Rico is due to the widespread acceptability of sterilization there. The program in Taiwan has used the IUD, whereas Japan and eastern Europe depend on abortion for limiting population size. However, no family planning program can be successful unless the families involved are motivated to practice birth control.

Taiwan (formerly called Formosa) is an island about 150 km off the southeast coast of mainland China. Its area, 36,000 km², is about the size of Delaware and Maryland combined. Nearly half of the island is mountainous and sparsely inhabited. The remainder is densely populated. The total population is slightly over 15,000,000 persons, or three times the combined populations of Delaware and Maryland.

The aboriginal inhabitants of Taiwan are probably of Malaysian origin. The population was very low until the Dutch colonized the island in 1624 and encouraged migration of Chinese from the mainland to increase production of sugar and rice. After the expulsion of the Dutch by the Chinese in 1662, migration from the mainland of China continued, so that by the 1660's, Taiwan had an estimated 50,000 Chinese inhabitants. The number increased to about 120,000 by 1690 and reached 2 million by 1810.

In 1895, the island was ceded to Japan. The population rose rapidly during the next half-century, reaching the 3 million mark around 1905, and exceeding 6.5 million by 1943. When Japan surrendered in 1945 at the end of World War II, Taiwan was restored to the Republic of China, and about one-half million Japanese left the island.

After World War II the Nationalist and the Communist Chinese struggled for the control of the mainland of China. When the Communists prevailed, the Nationalists established a government on Taiwan. During the period from 1947 through 1964, more than one million people migrated from the mainland to Taiwan. At the same time, the death rate on Taiwan declined sharply. By 1964, the population had exceeded 12 million; by 1975, the population had reached 15 million. Thus, the population of Taiwan increased fivefold in the 70 years from 1905 to 1975! (See Figure 6.14.)

How has this extreme population pressure been felt? Taiwan has not experienced a food shortage; in fact, it is still a net exporter of food. The nation has maintained one of the highest standards of living in Asia; real per capita income has been steadily increasing. In fact, the large supply of cheap labor has provided Taiwan the manpower for rapid industrialization. The people are among the healthiest in Asia; the infant mortality rate is low, the life

Terraced farms in the mountains of Taiwan. (Photo courtesy of the Chinese Information Service.)

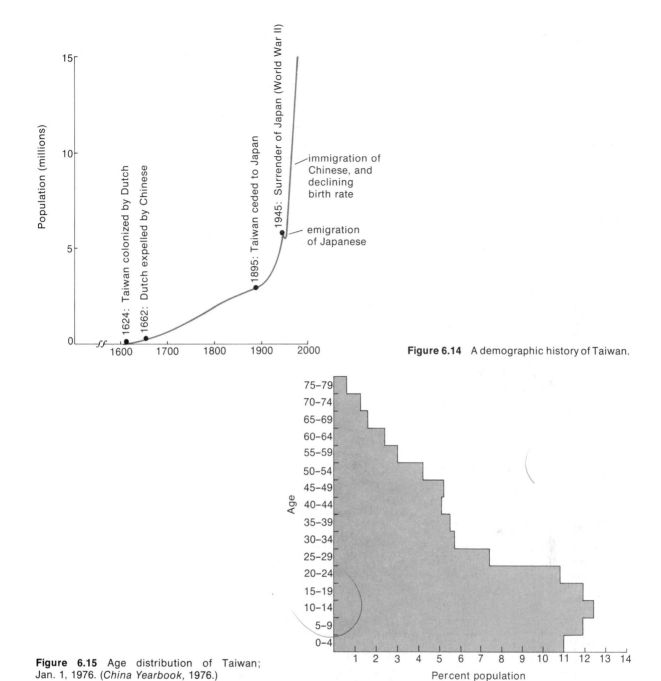

Figure 6.14 A demographic history of Taiwan.

Figure 6.15 Age distribution of Taiwan; Jan. 1, 1976. (*China Yearbook*, 1976.)

expectancy high, and the level of nutrition more than adequate. Literacy rates are among the highest in the developing world. Clearly, the effects of rapid population growth in Taiwan have not been disastrous.

However, rapid growth has led to an age-sex distribution (Figure 6.15) with many children. In 1968, 42 percent of the population was under 15 years old. In the developed countries, the comparable proportion was 22 to 31 percent. Such a large proportion of children in a

country that values education so highly has made primary education extremely costly to the government. Moreover, population growth has been outstripping the rate of increase of physicians, dentists, nurses, and health technicians.

Other social costs of rapid population growth have been apparent. Farms have become fragmented. Cities have grown noisy and polluted. Housing is inadequate, and streets are congested with traffic.

These changes have led to a conviction both

in the government and among the general public that population size should be limited. The Taiwanese are eager to provide education for their children, and this eagerness has been credited with discouraging large family size which may cause a financial impediment to further education.

The religion of the island is a mixture of Buddhism, Taoism, and Confucianism. Confucianist teaching values large families, specifically urging families to have at least one son. However, none of the religions has specific doctrines against family planning.

Taiwan, then, affords an example of a country ripe for an active and successful family planning program. The population growth has been very fast, so that the problem of population is readily apparent. The people are literate, healthy, and well fed. As mentioned in Section 6.8, all these factors are important for the success of family planning. In spite of this, the initial introduction of family planning met with resistance. During the period from 1920 to 1950, the government's official position was to encourage growth. Population during these years was checked by three natural regulators—disease, famine, and war. Suddenly,

around 1950, the situation changed dramatically. Political stability, improvements in agriculture, and modernization of medicine and public health measures led to a jump in the rate of natural increase from 20 per 1000 in 1947 to 38 per 1000 in 1951. In 1950, responding to this sudden spurt in growth, the Sino-American Joint Commission on Rural Reconstruction issued a pamphlet on the rhythm method of birth control. Official reaction was unfavorable. The leaders of government, fresh from several years of bitter war with the Communists, viewed the pamphlet as a Communist plot to weaken the military by limiting the number of future soldiers.

In less than a decade, however, the governor of Taiwan became convinced that a reduction in population growth rate would not reduce the size of the army for at least 20 years and would be a stimulus to the economy. A national program for teaching women about contraception was begun.

In 1962, the Population Council began a study to determine an effective approach to family planning in Taiwan. The study concluded that, in Taiwan, the best method would be to make visits to individual homes to teach

High-rise apartments were introduced to Taipei in the 1960's. Today there are hundreds of apartment buildings in the city. (Photo courtesy of the Chinese Information Service.)

A district health station in Taipei with a poster urging family planning. (Photo by Daniel Turk.)

This balloon, flying over Taipei, says "Childbirth needs planning for a happy family!" (Photo by Daniel Turk.)

about contraception. The contraceptive of choice would be the Lippes loop, an IUD. The government, in 1968, officially inaugurated a program for population control. The government's intention was to reduce the rate of natural increase from 3 percent to 2 percent within a decade. A year later, therapeutic abortion and sterilization were legalized for some conditions. The program has enlisted the help of many government agencies and private organizations. Field workers and private doctors assist under government contracts. Family planning is promoted by the mass media, including radio and television spots, slide shows in movies theaters, posters in buses, newspaper releases, even advertisements on matchboxes. Field workers visit new mothers in their homes to offer free family planning assistance.

The program has been remarkably effective. In 1967, about 38 percent of wives in Taiwan knew nothing about the loop, and 53 percent knew nothing about the contraceptive pill. Two years later, roughly one third of all wives had tried the loop. By 1972, the rate of natural increase had fallen to 1.9 percent. Forty-five percent of wives ages 20 to 44 were using family planning services (Fig. 6.16). The government hopes to reduce the rate still further by enrolling more participants.

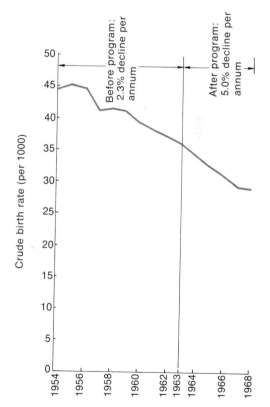

Figure 6.16 Crude birth rate in Taiwan before and after establishment of family-planning program in 1963. (Adopted from Country Profiles: Taiwan, February 1970.)

PROBLEMS

1. **Population ecology.** Discuss some important factors relating to the population growth of species other than man. Contrast these factors with those important to man's population growth.

2. **Vocabulary.** Define the following terms: rate of growth; rate of change.

3. **Exponential growth.** Explain why world population size cannot grow forever.

4. **Population prediction.** Discuss some uses of predictions of population size.

5. **Growth curves.** Outline Malthus' theory of population growth.

6. **Growth curves.** What is a sigmoid curve? Discuss its relevance as a model for human population growth.

7. **Vocabulatory.** Define demography; vital event; vital rate.

8. **Migration.** Immigration and emigration data are defined in terms of the stated intention of the migrant at the time he or she applies for the proper visa. Do you think such data are reliable measures of long-term migration? Define migration in terms that would be more demographically useful. Why is your definition not employed by data-collection agencies?

9. **Family size.** What would happen to human population size if no family had more than two children? If the average family size were two children?

10. **Accuracy of data.** In Table 16 of the United Nations Demographic Yearbook (1970), the infant mortality rate in Angola for 1968 is cited as 15.9. What are the units of infant mortality rates? A footnote states, "Rates computed on number of baptisms recorded in Roman Catholic Church records." Comment on the accuracy of the quoted rate.

11. **Accuracy of data.** The U.N. cites that the crude birth rate in the Netherlands Antilles in 1968 was 23.0 and that the infant mortality rate was 20.2. These data ex-

clude most live-born infants dying before registration of birth. How do these definitions affect international comparisons?

12. **Population distribution.** Explain why some declining animal populations have age distributions with large bases. (Consider populations where there is heavy predation of adults.)

13. **Population distribution.** How are age-sex distribution curves useful in predicting future growth? What are some of their limitations?

14. **Population distribution.** How would you expect each of the following to affect population growth? Consider which age groups are most likely to be affected by each event and the interrelationships between the event and the population change: (a) famine; (b) war; (c) lowering of marital age; (d) development of an effective method of birth control; (e) outbreak of a cholera epidemic; (f) severe and chronic air pollution; (g) lowering of infant mortality; (h) institution of a social security system; (i) economic depression; (j) economic boom; (k) institution of child labor laws; (l) expansion of employment opportunities for women.

15. **TFR.** Explain why the replacement value of the TFR must be greater than 2.0.

16. **Birth control.** The following measures have been proposed to measure contraceptive effectiveness; (a) number of failures (pregnancies) to 1000 women on contraceptives; (b) number of failures per year to 1000 women on contraceptives; (c) birth rate to population on specific contraceptives; (d) completed family size to women on specific contraceptives. Discuss their relative merits.

17. **Birth control.** What factors are most important in introducing birth-control methods to a developing nation?

18. **Birth control.** Suggest reasons why a highly reliable method of birth control introduced to a relatively small group of couples may be more effective in reducing

births than a moderately effective method introduced to many.

19. **Social policies.** In the United States, families are allowed an income-tax deduction for each child. In most developed countries, each family is allotted an annual grant for each child. How do you think these tax laws affect family size?

20. **Social policy.** If you had the responsibility of discouraging population growth, would you consider curtailing income tax deductions if a family has more than four children? curtailing health benefits? Whom would such policies harm?

21. **Population control.** Joan P. Mencher, in an article entitled, "Socioeconomic constraints to development: The case of south India" (Transactions of the New York Academy of Sciences, 1973, pp. 155–167), points out that poor people are often treated shabbily in medical clinics. In south India, she says, "To have a baby does not require contact with hospital people, but to *avoid* having a baby requires contact with maternity assistants, doctors, etc., all of whom tend to treat the poor and low-caste people as 'animals.'" How do you think such treatment affects population control among the poor of south India?

BIBLIOGRAPHY

This chapter has introduced demographic techniques for analyzing population growth. There are several valuable texts available for those interested in further study of demography.
Two excellent introductory texts are:
Peter R. Cox: *Demography.* 4th Ed., Cambridge, England, Cambridge University Press, 1970. 469 pp.
Donald J. Bogue: *Principles of Demography.* New York, John Wiley & Sons, 1969. 899 pp.

More mathematical introductions are:
Mortimer Spiegelman: *Introduction to Demography.* Rev. Ed. Cambridge, Mass., Harvard University Press, 1968. 514 pp. (Spiegelman includes an extremely large bibliography covering a wide range of topics related to population size, control, measurement, and so forth.)
Nathan Keyfitz: *Introduction to the Mathematics of Population.* Reading, Mass., Addison-Wesley Publishing Co., 1968. 450 pp. (This highly technical and mathematical text is especially careful in its presentation of interrelationships among various measures of population composition and vital rates.)

A very complete source is the two-volume work:
Henry S. Shryock and Jacob S. Siegel: *The Methods and Materials of Demography.* Washington, D.C., Bureau of the Census, 1973. 888 pp.

The most useful and complete source of world population data is the United Nations Demographic Yearbook, *published annually since 1948. For many nations and areas of the world the* Yearbook *includes the most recent available information on population sizes, vital rates, and many more specialized demographic statistics.*

For up-to-date trends in population studies, refer to Studies in Family Planning, *a monthly bulletin published by* The Population Council, *New York City.*

A recent book discussing the background and the causes of overpopulation is: Thomas McKeown: *The Modern Rise of Population.* New York, Academic Press, 1976. 168 pp.

7

THE SOCIOLOGY OF HUMAN POPULATIONS

7.1 SCOPE OF THE PROBLEM

Suppose you desired to study the population dynamics of the North American beaver. Armed with a description of its ecological niche—the beaver lives along streams and lakes in wooded areas of Canada and northern United States; it eats the barks and twigs of aspen and poplar trees; it builds dams, thus permanently altering its environment; and it constructs dens of sticks and grasses where it lives with its family—you would know where to look for beavers, and you could devise a methodology for studying them. By contrast, imagine that you are assigned to study the population dynamics of a much more versatile mammal, whose habitat extends from the icy Arctic through the temperate Northern Hemisphere to the steamy rain forests of South America and Africa to the temperate Southern Hemisphere. The species, though nonmarsupial, has outposts in Australia as well. Small bands live in such hostile terrain as the Sahara and the Andes Mountains. Some spend a large part of their adult life on the surface of the sea. Most members of the species are omnivores whose diet varies depending on where they live. Some are strictly herbivorous even when animal flesh is readily available. One of the most remarkable features of the species is its wide variety of dens. A few compete with other large mammals for natural caves; some carry their dens on their backs; many build dwelling places of sticks and grasses; some use ice or stones; many fashion special building materials for their homes. Some individuals or families live in near-isolation; others live in special dens physically separate from each other, but close enough for regular social contact. Still others live in elaborate structures with many inhabitants, not necessarily of the same family. Although generally members of the species cooperate with each other, occasionally members of one troupe will viciously at-

C

D

tack another troupe. Moreover, the litter size of this species varies by habitat. Finally, the species is well known for its propensity to alter its environment radically.

Studying such a species would be terribly difficult. The only effective mechanism would be to separate the problem into subproblems. Rather than investigating this species as a whole, the ecologist must investigate the species habitat by habitat.

The species described is, of course, the human being. Studying effects of population growth, or change, on people as a whole is fruitless and meaningless. The problem must be more sharply delineated, not only because of the wide variety of human habitat, food, and dwelling, but also because of the strong social and political forces that operate in human population dynamics.

This chapter focuses attention on different spheres of human experience. It regards population change as it affects the family, the community, the larger political unit, and finally, the Earth as a whole. This view starts from the

People live in many harsh environments. Some examples are: (A) the Sahara desert (courtesy of Sygma); (B) the Arctic (courtesy of the American Museum of Natural History); (C) the woods of Colorado; (D) by the edge of the water. (Photo courtesy of the Hong Kong Tourist Association.)

THE MILWAUKEE JOURNAL
Dist. Field Newspaper Syndicate, 1976

premise that an individual birth or death is felt most keenly by the small social unit most directly involved, that is, the family. The community is next most strongly affected. Addition, or deletion, of a member affects the availability of food, space, schooling, and so forth. The effect of the birth or death of an individual is attenuated as the social unit gets larger. When one studies a nation, the birth or death of an individual is nearly irrelevant (except when the individual is one who wields power). Yet, the occurrence of many births, or of many deaths, is felt by even large political units. Population dynamics, then, operates on many levels simultaneously.

7.2 POPULATION CHANGE AT THE FAMILY LEVEL

Many forces—biological, social, economic, and political—operate to influence family size. Among hunter-gatherers, for example, very young children must be car-

Children in agricultural and industrial societies work to help support the family. *A*, Mexico; *B*, Southeast Asia. (Photos courtesy Ken Heyman.)

ried when the band moves from one camp site to another. The ideal interval between births is therefore long enough so that a mother has to carry no more than one child. Hunter-gatherer societies typically have relatively small family sizes; the total population of a such a society is usually well below the biological carrying capacity of the area it occupies. Anthropologists believe that hunter-gatherers have practiced fertility control by infanticide, abortion, and primitive forms of contraception.

In agricultural societies, families can have several very young children without great inconvenience. The babies can stay near the mother or be cared for by older children. Moreover, when children grow older they can assist the family on the farm. Even quite young children can help—they can shoo birds away from newly sown hay or weed the crops. Slightly older children can care for the animals. Family size in such agricultural societies tends to be large.

Even larger families are found among peasants in feudal societies. Peasants must not only produce enough food for themselves, but also their agricultural output must support a large ruling class. Children in these societies are true economic assets, for they produce more food than they eat. Similar conditions prevail in colonized societies.

Among some groups, religious beliefs encourage large families. Traditional Hindus, for instance, believe that only a son can perform certain religious rites after the death of a father. The father is doomed to a particular kind of hell if these rites are not performed. In order to be assured that he will be outlived by at least one son, a Hindu man tends to marry early and to have a large family.

In modern societies, having a child becomes more and more a conscious parental choice. In fact, paralleling society's demographic transition is a dramatic microdemographic transition within the family. Consider the situation in England:

> The typical working class mother of the 1890's, married in her teens or early twenties and experiencing ten pregnancies, spent about fifteen years in a state of pregnancy and in nursing a child for the first years of its life. She was tied, for this period of time, to the wheel of childbearing. Today, for the typical mother, the time so spent would be about four years.*

This revolutionary reduction occurred in only two generations. How did these remarkable changes take place, and what factors lead to similar changes in other countries?

One of the first steps in modernization is marked improvement in public health and in medical care. These measures lead to healthier parents and healthier babies. The healthier a woman is, the more likely she is to be biologically capable of conceiving a child and of bearing a normal infant. The healthier an infant is, the more likely it is to reach childhood and adulthood. Perhaps the single most profound demographic effect of modern medicine has been

The cathedral at Rouen. Begun about 1200; completed about 1500. Feudal European agricultural society produced so much food that it could support a large class of artisans who built magnificent structures. (Photo courtesy Morton Shor.)

---

*R. M. Titmuss: *Essays on the Welfare State.* London: Unwin University Press, 1966, p. 91.

170

its influence on infant mortality. In medically primitive societies, nearly half of the babies born do not survive until the age of five; in the developed world, well over 90 percent of infants reach the age of fifteen. Thus, improvements in health and medicine lead at first to an increase in individual family size.

For countries now undergoing demographic transition, infant mortality is rapidly dropping. Families that are producing many babies will find that most will now survive to adulthood. The consequences within the family will be multiple. Maternal and child health, nutrition, education, and even the intelligence of the child may all be adversely affected by large family sizes.

Demographic theorists contend that when parents realize that babies are likely to survive, couples have fewer children. Until that realization, family size rises. The European experience lends support for this theory. Thus, in the nineteenth century, when England was undergoing its demographic transition, average family size increased temporarily. The explanation given is that until families realized that their babies would survive, they continued to reproduce at a high rate. Once the effects of the fall in infant mortality became manifest, fewer babies were produced, and family size returned to its former level.

Another important change that modernization brings to a society is an increase in the educational level of the population. One of the first things that people learn is better health practices. Specifically, they learn to sterilize water, and they become eager to bring their children for immunization and for medical care. The result is that more children live, and family size increases.

Education eventually also leads to exposure to books and to other cultures. Such new awareness often breaks down traditional beliefs and practices. For some societies, the change leads to increased fertility; for most, exposure to other practices leads to decreased fertility.

Education also gives people knowledge of fertility control, thus contributing further to the reduction of family size.

The most long-lasting effect of education appears to be a shift in taste. A small family is considered optimal among most educated peoples. Educated families have high expectations for each child. They demand high quality education and child care and better food. All this is costly. The change in taste, coupled with knowledge of contraception, leads to declining family size.

Urbanization is central to modernization. Families moving to a city find that food is much more costly than in rural areas. Urban mothers often must work outside the home to help support the family, and child care is expensive. Infant mortality is generally higher in urban than in rural areas. Housing is more cramped and much more expensive. All these factors encourage small family sizes. Moreover,

Children in a modern society.

knowledge and techniques of fertility control are more readily available than in rural places. Finally, cultural tastes in urban centers often favor small families.

With modern development come new goods, increased per capita income and, eventually, leisure time. Tastes become more modern. Things — refrigerators, cars, radios, televisions, air conditioners — that were luxuries in one generation become necessities in the next. All these factors are economic disincentives to a large family size.

Political forces also influence family size. Most countries have been **pro-natalist,** that is, favoring births. If all the neighbors of a couple have many children, and if the government urges large families, that couple is likely to produce more babies than one living in a country where most families are small and where the government discourages births.

The detailed demographic history of each country is unique, so that not all the forces described above are relevant to the families of all societies. But a generalization can be made: when children are economic or social assets, a family is influenced to have more babies. When children are costly, either in terms of economic expense or social disapproval, family size tends to be small. Public policies that reward families with many children encourage births;

policies penalizing children discourage births. Refer to Chapter 6 for a discussion of political and legal approaches to population control.

7.3 LAND USE, URBANIZATION AND RURALIZATION

Most of the Earth is sparsely populated. Few people live in high mountains, in tropical rain forests, in deserts or steppes, in tundra or taiga. Antarctica, Greenland, Canada, the U.S.S.R., Australia, and most of the countries of Africa and South America have fewer than 15 persons per square kilometer. (By contrast, Western Europe is much more densely populated, with over 150 persons per square kilometer.) When one reads that the Earth is over-populated, the problem referred to is not one of lack of space but rather of the limited availability of resources, the inefficient utilization of land, and the psychological sensation of being "crowded." Subsequent chapters deal with resources, energy, and food and with pollution caused by human activity. This section discusses the settlement patterns of people, with particular reference to modern trends.

What is striking about settlement patterns is their variability. People live in large cities, in remote rural areas, in small villages, and in towns. A good climate, adequate rainfall, and favorable agricultural conditions are usually preconditions for dense populations. Sometimes, however, an otherwise inhospitable area may attract settlers by its richness of resources—Alaska for its minerals and oil, the Congo for its ivory. Figure 7.1 is a schematic picture of various types of human settlement.

In the twentieth century, there has been a worldwide trend towards **urbanization,** the process characterized by the movement of people from rural to urban settlements,

Urban village in Nigeria. (Photo courtesy Ken Heyman.)

Coastal fishing village, La Conner, Washington.

173

I. Dispersed agricultural settlement (e.g., midwestern United States and Canada)

II. Agricultural villages (e.g., European farming villages)

III. Coastal settlements (e.g., fishing villages)

IV. Settlements along a strip (e.g., along banks of a river, tracks of a railroad, or by a highway)

V. Centralized urban places (e.g., Mexico City)

VI. Suburbanized urban places (e.g., Boston)

VII. Urban clusters (e.g., megalopolis of the northeast corridor of the United States: Boston-NYC-Philadelphia-Wilmington-Baltimore and Washington, D.C.)

Figure 7.1 Schematic diagram of patterns of human settlement.

WHAT IS AN URBAN PLACE?

No one quarrels with classifying New York City as "urban" nor with labeling Quintana Roo in Mexico as "rural." However, when one tries to classify places of intermediate size as urban or rural, problems of definition arise.

The Statistical Office of the United Nations has recognized five major methods of classifying places along the urban-rural axis:

(a) A place may be called urban if its government says it is! This is admittedly a circular definition ("A city is a city"). It is based on administrative, historical, and political considerations rather than on statistical criteria. Jamaica, for example, simply defines Kingston and its metropolitan area as urban. All other parts of the island are classified as rural.

(b) A place may be called urban if it contains a certain minimum number of people. This minimum varies from country to country and shifts in time. In particular, highly industrialized nations tend to have a higher minimum size than predominantly agricultural countries. For example, Ghana defines a place as urban if it has more than 5000 inhabitants, while Japanese places need 30,000 inhabitants to qualify as urban. This definition begs the question of how large an area is to contain the population.

(c) A place may be called urban if it possesses some particular form of local government. Thus, in El Salvador, administrative centers of departments, districts, and municipios are called urban.

(d) A place may be called urban if it possesses one or more of the above characteristics and it has certain specific urban characteristics, for example, established street patterns, public services, and/or contiguously aligned buildings. Pakistan, for example, classifies as urban "all municipalities, civil lines, cantonments not included in municipal limits, any other continuous collection of houses inhabited by not fewer than 5000 persons and having urban characteristics, and also a few areas having urban characteristics but fewer than 5000 population."[*]

(e) A place may be called urban if a specified proportion of the economically active population is engaged in nonagricultural occupations. This definitional type is usually used in connection with one of the previous kinds of classifications. Israel, for example, defines as urban all settlements of more than 2000 inhabitants except those where at least two thirds of the heads of households participating in the labor force earn their living from agriculture.

The United Nations, recognizing the wide variety of societal types in the world, recommends that each nation's statistical bureau define "urban" in the manner most useful to that country's particular purpose. The UN, then, does not promulgate any uniform system of definition. Thus, it is extremely difficult to compare degree of urbanization from country to country. When one refers to the process of urbanization, that process reflects the increase in urban populations in the sense defined by the individual countries.

---

[*]Henry S. Shryock and Jacob S. Siegel: *The Methods and Materials of Demography.* U.S. Department of Commerce, Bureau of the Census, 1973, p. 155.

from small towns to large cities, and from large cities to their suburbs. This migration is caused by, and in turn leads to, profound societal changes. Rural areas are depopulated, while urban areas—towns, cities, and urban zones—become more densely populated. Figure 7.2 diagrams several types of urban places.

The first cities arose along the Tigris and Euphrates rivers roughly 6000 years ago. Since then, great cities have grown and fallen in many areas of the world. Most people, however, have traditionally lived in rural areas. In Europe, historically the most urbanized continent, only 1.6 percent of the population lived in cities of over 100,000 in 1600; the figure rose to only 1.9 percent by 1700 and 2.2 percent by 1800. In fact, before 1800, no country was predominantly urban. Between the time of the fall of the Roman Empire and the beginning of the nineteenth century, no European city had one million inhabitants. Thus, on the eve of the Industrial Revolution, Europe was essentially an agrarian continent. Outside of Europe, settlement patterns were even more rural.*

In the hundred years from 1800 to 1900, cities grew

---

*An exception may have been East Asia. Archeological evidence suggests that the first city to exceed one million persons may have been the capital of the Khmer Republic (Cambodia). Tokyo and Shanghai may have had populations of over one million by 1800. Moreover, by 1800, many parts of East Asia boasted very large central cities surrounded by highly productive agricultural areas.

I. City   II. Urban region with city at center

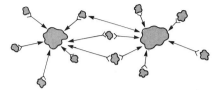

III. Urban region with two cities at center

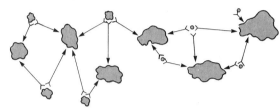

IV. Urban megalopolis

Figure 7.2 Schematic diagram of types of urban places. Direction of arrows shows major commuting directions. Tail represents place of residence; head, place of work.

A view of Ankhor Wat, ancient capital of the Khmer Republic. (Photo courtesy Picon A.P.)

175

rapidly. In 1900, at least twelve cities had populations over one million – London, Paris, Berlin, Vienna, Moscow, St. Petersburg (now Leningrad), New York, Chicago, Philadelphia, Tokyo, Shanghai, and Calcutta. By 1960, more than 30 percent of the inhabitants of the world lived in urban areas, and the proportion has steadily grown since then. Worldwide, about 140 cities had populations over 1,000,000. By the year 2000, over 50 percent of the world's population will probably live in urban places, and there will probably be more than 250 cities of over one million inhabitants.

If population in the twentieth century has grown prodigiously, urban growth, particularly in the developing countries, has exploded. World population tripled in the period from 1800 to 1960; in the same time, the population living in localities of 20,000 people or more increased more than fortyfold!

It is important to realize that the extraordinary growth of cities during the twentieth century has been primarily due to migration from rural areas to urban places and not to high birth rates in urban areas. The rapid influx of migrants leads to serious and in many cases unsolved problems. Municipalities are often unable to provide clean water, adequate housing, transportation, education, and other services to newcomers.

Two types of cities should be distinguished. In the industrialized countries, people flock to the cities for available jobs. As economic standards rise, people demand more space for housing, and many eventually move from the central city to less densely populated suburban areas. This outflow creates a demand for improved transportation facilities – specifically, new roads. The typical city in the developed world, then, has a densely populated inner core with businesses and housing, both luxury and slum. Surrounding the central city is a rapidly expanding peripheral suburban region, growing both in population and in area. Roads and other transportation systems needed to maintain the periphery require a large amount of space; road-building encourages new suburban growth, which in turn creates a demand for more roads. The geographical range of the city expands. The combined effect of these patterns can lead to **urban sprawl.**

As an example of the changing patterns of land use in this century, consider the United States. In 1920, 4 million hectares of land had been urbanized. Highways, railroad beds, and airports used 8 million hectares. By 1969, 14 million hectares had been urbanized, and 10 million were used for transportation systems. Most of this increased area had been farmland in 1920.

In the developing world, cities also expand. Indeed, urbanization is proceeding at a faster rate in the developing world than it is in the industrialized nations. The developing world, however, is still much less urbanized than

Dense dwelling in urban Brazil. (Photo courtesy Ken Heyman.)

Suburban region in Boulder, Colorado.

Europe, North America, Australia, New Zealand, and East Asia. The land area of a city in the developing world grows slowly in relation to the increase in population. The reasons are several. First, jobs are often not available, so that people do not earn enough money to pay for adequate housing. Second, systems of transportation are not developed, so that the inhabitants must live close to the center of the city. As a result, these cities become overurbanized; that is, there are too many people for the economic base. As a result, the city suffers from poor sanitation and increasing social problems. The fringes of cities in the developing countries are often characterized by dismal living conditions.

Even though the living conditions of many twentieth century cities are deplorable, people migrate there voluntarily. The reason must be that conditions in rural areas are even worse. Migration to the city has been called a push-pull phenomenon—people are pushed out of rural areas by hopeless poverty and pulled by the lure of the city. India is characterized by what is termed a push-back phenomenon: conditions in the city are even worse than in many rural areas, so that many migrants return to their homes.

In the USSR, Europe, and North America, rural areas have actually *declined* in population since 1950 and are expected to decline still further.* In the developing regions,

*The fact that rural areas decline in population does not necessarily mean that areas become depopulated; it may mean that a small town previously characterized as "rural" grows large enough to be called "urban."

rural areas are expected to continue to grow, but not so fast as urban areas.

Until the twentieth century, the size of a city was limited by modes of transportation. A worker had to live close to his place of work. Ancient Rome, for example, was a city of slightly over a million people. Geographically tight-knit, it had apartment houses of six to eight stories and narrow streets, and people usually worked close to their homes. Because of geographical and technological limitations, it could not grow much larger.

Today, modern methods of transportation permit people to travel long distances to get to work and to places of recreation. Elevators permit the city to grow tall; people can live and work in high, multi-storied buildings. As a result, modern cities have grown haphazardly. If an omnipotent genie could be assigned to rearrange the people, houses, schools, hospitals, and power plants of Los Angeles without reducing the number of buildings nor increasing the area of the city, he could place people closer to their places of work and could provide larger expanses of open parkland closer to where most people live. The result would be a reduction in air pollution, because cars would be used for smaller trips. Less energy would be used; more leisure

The ruins of Pompeii, an ancient Roman city. (Photo courtesy of the Italian Government Travel Office.)

time would be available. All in all, the city would be a much more pleasant place in which to live.

No one, however, has the power to level Los Angeles or any other city and start anew. What, then, can be done to ease the problems caused by urbanization? Solutions are frustratingly difficult, almost impossible. The approaches have been different in the developed and the developing world. In the former, many older cities are undergoing a process of "urban development." Old sectors are being renovated, or destroyed and replaced. The new areas, at best, incorporate sound principles of urban design. At worst, they force the former inhabitants into ever more crowded slums and substitute sterile new buildings for vibrant but poor old neighborhoods.

Another approach is the so-called planned or new town. A town is designed and built, and then residents move in. Such construction requires huge capital outlay and often needs tax advantages and other governmental support. Well-planned new towns are often quite lovely and designed for convenient living.

Few developing nations have the capital to execute planned urban development. Masses flock to the cities because of dismal rural poverty. The urbanization process is much too rapid to permit orderly development. It would be cheaper to provide economic assistance to rural areas. Even this approach, however, is too expensive for most Third World nations.

Various settlement patterns prevail in rural areas. In the United States and Canada, single families live isolated from others. The single greatest problem with this pattern lies in the difficulty of receiving emergency medical care. More typical rural patterns are small clusters of houses surrounded by farms. This pattern is found in Europe,

An aerial view of Arad, a new town of the Negev desert. (Photo courtesy of the Consulate General of Israel in New York.)

A planned community of luxury row houses in the United States.

Urban growth in Brazil. In the developing world, cities often grow in a haphazard fashion. (Photo courtesy Ken Heyman.)

Asia, Africa, and South America, and was typical of early settlements in eastern North America. Often, a larger town serves as a nucleus for several small farming villages.

The trend towards urbanization has serious consequences for rural areas. Often, the migrants to the city are young people in their economically active years. Some-

A farmer's house in the United States. A single family lives isolated from others. (Photo courtesy Ken Heyman.)

times, so few people of working age remain behind that there is insufficient labor to tend the fields. Some agriculturalists in the developed countries fear that even if public policies designed to encourage migration back to rural areas could be implemented, the new migrants would lack agricultural expertise.

7.4 CONSEQUENCES OF POPULATION DENSITY

As the population density in a given area increases, each person's share of the available supplies of land, water, fuels, wood, metals, and other resources must necessarily decrease. In the past, people in many parts of the world have raised their standard of living despite a rising population by using the available resources with increased efficiency. However, there are eventual limits to population growth. Since we cannot predict the future advances of technology accurately, we cannot predict the maximum possible human population. The total world supplies of food, energy, and resources are discussed in Chapter 8, 9, 10, and 11.

Before humans are eliminated by death through starvation and thirst, it is certain that the quality of life on Earth will change. Many forests and wild places will disappear and be replaced by cities and indoor environments. Some individuals will welcome the change, some will be adversely affected; but what will happen to society as a whole as populations continue to increase?

Violence, disunity, political upheavals, and personal unhappiness are attributed by some to population density. In a series of experiments with strains of rodents, John Calhoun* studied the effects of extreme crowding. He constructed cells supplied with enough food and water for many more rodents than the space would normally hold. A few animals were placed in each cell and allowed to breed.

*See John B. Calhoun in *Scientific American*, **206**:139, February, 1962, and "Environment and Society in Transition," *Annals of the New York Academy of Sciences*, June, 1971.

A barrel organ in Amsterdam. Note the racing pigeon loft on the top of the center house. Amsterdam is one of the most densely populated cities in Europe.

The population and the density grew quickly, and the animals began to act bizarrely. The females lost their ability to build proper nests or to care for their infants. Some of the males became sexually aggressive; most retreated from communication with others. In short, the normal processes of socialization were destroyed. Other literature presents evidence of decreased fertility and strange behavior in many animal species under conditions of overcrowding. However, we do not know in what way conclusions from animal experiments apply to human populations.

Available data do not support the hypothesis that high population density of a nation is necessarily associated with social upheavals, violence, or poverty. Research into the nature of the relationship between population density and social problems has yielded a morass of conflicting conclusions. No consistent patterns emerge when national population densities are defined as population size divided by total national area. Such densities are grossly misleading, for they do not measure population densities in populated areas. For example, the Netherlands, one of the most densely populated areas of the world, had 340 persons per square kilometer in 1976. By contrast, the population density in India was only 180/km² and in Algeria, only 7/km.² Since, however, nearly all of the Netherlands is inhabitable, while much of India is jungle and most of Algeria is desert, the fact that the density of the Netherlands is so high and its society so stable does not by itself disprove a hypothesis that high population densities are socially detrimental. These examples are not isolated instances, for the 62 percent of the Earth's surface that is semi-arid, taiga, tropical jungle, arctic, tundra, or desert holds only 1 percent of its population.

The relation between population density on arable

TABLE 7.1 Density of Population in 1960 and 1975 and 1960–1975 Gain in Population Density in Major Areas of the World*

REGION	INHABITANTS/km² 1960	1975	GAIN IN DENSITY 1960–1975
World total	22.1	29.4	7.3
More developed regions	16.0	18.6	2.6
Less developed regions	27.0	39.5	12.5
Europe	86.1	96.0	9.9
USSR	9.6	11.4	1.8
North America	9.2	11.0	1.8
Oceania	1.9	2.5	0.6
South Asia	54.9	80.4	25.5
East Asia	67.0	85.5	18.5
Africa	9.0	13.2	4.2
Latin America	10.5	15.8	5.3

*Source: *UN Demographic Yearbooks.*

A view of Hong Kong, the most densely populated city in the world. (Photo courtesy Ken Heyman.)

land and social problems is also confusing. Japan, which supports 1700 persons per square kilometer of arable land, is an example of a very densely populated country that maintains a prosperous and relatively crime-free society.

Some psychologists and sociologists claim that density by itself is not relevant to the feeling of being crowded. A more important factor, they say, is the amount of space available in the individual's dwelling unit. Even here the evidence is ambiguous. Hong Kong has probably the highest residential density ever known in the world. Nearly 40 percent of residents of Hong Kong share their dwelling unit with nonrelatives; almost 30 percent sleep three or more to a bed, and 13 percent sleep four or more to a bed. Most of the population lives in a dwelling unit of a single room; most dwelling units are homes to more than nine persons; most dwelling units are homes to two or more unrelated families.* Even under these conditions of extreme crowding, there is little or no evidence of antisocial behavior attributable to the crowding itself.

The degree to which density is an important factor in breeding antisocial behavior is still unknown. One difficulty in studying the effects of population density on humans is that spatial requirements may in part be culturally determined. Much sociological research is needed to gain an understanding of the factors causing members of different societies to feel crowded.

*R. E. Mitchell: Some social implications of high density housing. *American Sociological Review*, **36**:18–29, 1971.

7.5 EFFECTS OF POPULATION CHANGE IN THE DEVELOPING COUNTRIES

Rapid population growth, especially when it leads to increasing urbanization and rising unemployment, may be detrimental to the economic and social aspirations of a developing country. A developing country is often defined, arbitrarily, as one in which the average per capita annual income is under $600. Birth rates in such countries are generally well above 30 per thousand and often more than 40 per thousand. If such a country is growing rapidly, much of the economic effort that could have been expended in development to improve the per capita income is instead necessarily diverted to efforts for increasing agricultural production. Many schools are needed, because a very large proportion of the population is under age 15. But the development of better schools, as well as improved health care, greater industrialization, more modern housing, and pollution control can only be undertaken after an initial effort is made to assure minimal standards of nutrition. Although an adequate diet is often not attained, rapid population growth precludes raising the per capita caloric or protein content of the diet.

Problems caused by rapid urbanization are especially acute in the developing world. In the West, urbanization has often been associated with economic, industrial, artistic, and intellectual productivity. Migration in the West from rural areas to the cities has, in the past, been accompanied by fundamental changes in styles of life. Cities have been wealthier and more modern than the rural areas and have been characterized by a considerable occupational differentiation. By contrast, the patterns of urbanization in many developing countries are not associated with productivity. The modern Indian, African, or South American city is often culturally rural. Migrants to the city may not be the upwardly mobile but rather the farmer whose crops have failed. Cities in the developing nations are often poor and agricultural; the literacy rate may be very low; housing is often shoddy; food is scarce. Dakarta, India, for example, is a coastal city that has five times more people now than it did before World War II. For food, it depends upon imported grain. Since India has not dispensed food efficiently in the past, the hungry flock to the seacoast cities to be near sources of food. To transform these poor cities to modern ones would require assured supplies of food, the introduction of profitable industry, and population control.

The economic success of some developing nations, even at great odds, is truly remarkable. An example of such success is Ghana. If there were no external investment, and if fertility were constant, the per capita income would probably drop 8 percent between 1960 and 1985. If, however, the fertility were to decline only 1 percent annually, per capita

income would rise 9 percent; a 2 percent decline in fertility should lead to a 24 percent rise in income. These figures mean that the per capita growth in productivity is very high. In most developing countries, the rapid rise in productivity rates has not been sufficient to prevent per capita incomes from falling farther and farther below the incomes in the developed nations.

The elderly are often not economically independent. The sign this Mexican beggar wears says, "I am 80 years old. Therefore, no one will give me work. To live, I beg your charity. Thank you." (Photo by Anton Turok.)

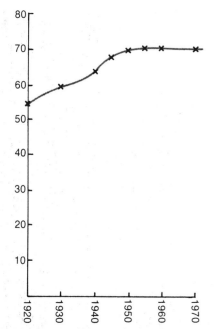

Figure 7.3 Expectation of life at birth, 1920–1970, United States (males and females combined). (From U. S. Public Health Service, *Vital Statistics of the United States.*)

Economic and social pressures resulting from rapidly increasing population growth lead to shortages in food, energy, and resources, as well as to rapid environmental deterioration. The following chapters discuss many environmental problems. The student should realize that all these problems are exacerbated by population growth: the more people, the greater the need for materials, and the greater the amount of pollution produced.

7.6 DEMOGRAPHIC TRENDS IN THE UNITED STATES

The demographic history of the United States is largely the history of rapidly changing patterns of fertility; by contrast, mortality has remained relatively constant. Fertility reflects social and economic events: In the 1920's, a period of economic boom and the Great Gatsby, fertility was quite high. The total fertility rate (TFR) was then about 3.2. The TFR fell sharply throughout the years of the Great Depression. It reached a low of about 2.2 in 1936. During World War II, in spite of the fact that many men were overseas, the fertility rate increased steadily. In 1946, the year after the war ended, the TFR was about 2.5 At that time, demographers and social scientists believed that the trend was due to the return of war veterans. They predicted that fertility would soon decrease. Instead, it rose

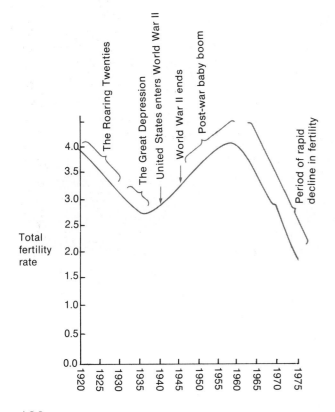

Figure 7.4 Total fertility rate for the United States, 1925–1975. (Source: Bureau of the Census.)

rapidly to a peak of 3.7 in the late 1950's. Analysts then attributed the rise to economic and social well-being. Businesses that specialized in baby products flourished; the middle class began in earnest its flight from the cities to provide their children with "fresh air" and places to play. Schools became overcrowded; communities couldn't build schools fast enough to keep pace with the growing numbers of children. Experts predicted a continued pattern of high births. Then, suddenly and inexplicably, the TFR began to drop quite rapidly. In 1971, it fell below the replacement level of 2.1 and continued to fall. By 1975, it had fallen to 1.8. Population experts were so surprised by the continued decline in TFR that in 1975 the Census Bureau drastically revised its assumptions concerning U.S. population growth. In the 1970's communities were forced to close many of the new schools they had built less than a decade before, because the number of school-age children had dropped so precipitously. Businesses that had specialized in baby goods were forced to diversify. Older teachers were fired and many young teachers were unable to find jobs.

When the TFR is below replacement level, the population may continue to grow if there is a disproportionate number of young persons. (Refer back to Section 6.3.) In the United States in 1977, 37 percent of the female population was at the childbearing age (18–44), and another 26 percent was under 15 years old. If the TFR remains at its 1975 level, the population of the United States will continue to increase for about 70 years before it stops growing. At that time, the population will be approximately 265 million. Figure 7.5 shows the age-sex distribution of the United States for 1975 and the projected distribution in the year 2000 if the fertility patterns of the mid-1970's prevail.

The projected age-sex distribution for the year 2000 is markedly different from conditions in 1975. First, the projected population is older. The percent of the population over 65 will increase from 10.0 in 1975 to 11.5 in 2000, a

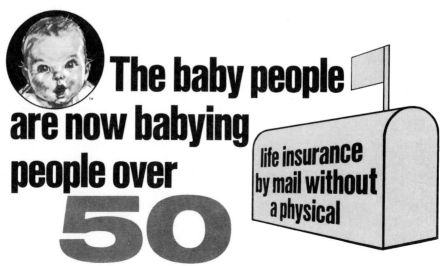

Gerber, long associated with baby products, also sells services to older people.

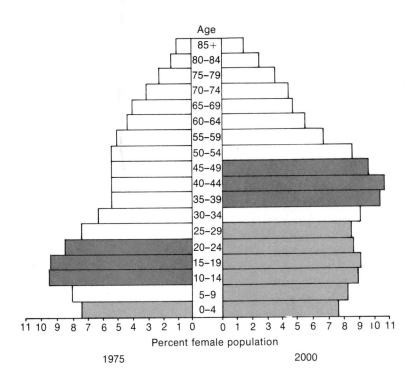

Age

Percent female population

1975   2000

Figure 7.5 U. S. female age distributions, 1975 and 2000. Note that in 2000, the postwar baby boom children will be of working age. (Data for 1975 from U. S. Census Bureau. Data for 2000 from the Fifth Annual Report of the Council on Environmental Quality.)

Post World War II baby boom

TFR below replacement level

negligible amount. However, the absolute increase will be considerable. There were 22 million people in the United States over age 65 in 1975; there will be a projected 29 million elderly in 2000. More housing and facilities for health care for the elderly will be necessary. The ratio of retired persons to people of working age will increase slightly. (If 1975 conditions prevail even longer, the ratio will increase dramatically by the middle of the twenty-

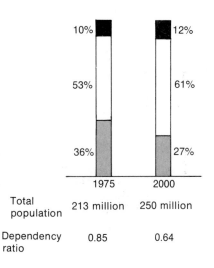

	1975	2000
Total population	213 million	250 million
Dependency ratio	0.85	0.64

Under 20

20–64

65 and over

Figure 7.6 U. S. age distribution, 1975 and 2000. (Data for 1975 from U. S. Census Bureau. Data for 2000 from the Fifth Annual Report of the Council on Environmental Quality.)

TABLE 7.2 Population Change for the 11 Cities in the United States with Populations Greater than 750,000 in 1950

CITY	POPULATION (000's) 1950	POPULATION (000's) 1970	PERCENT CHANGE
New York City	7892	7895	+0.0
Chicago	3621	3367	−7.0
Philadelphia	2071	1949	−5.9
Los Angeles	1970	2816	+42.9
Detroit	1850	1511	−18.3
Baltimore	950	906	−4.6
Cleveland	915	751	−17.9
St. Louis	857	622	−27.4
Washington, D.C.	802	757	−5.6
Boston	801	641	−20.0
San Francisco	775	716	−7.6

first century.) One effect will be on the social security system. The system finances benefits from current income. Therefore, a higher ratio of retired people to working people will increase the amount of money that workers will be required to contribute. This effect, however, will not be noticed for about 20 years and may be temporarily offset by an increase in the employment of women.

The percentage of the population under 20 years of age will fall from 35 to 26 percent from 1975 to 2000. Municipal expenditures for schools and teachers will be substantially reduced.

Young and old people depend upon the working age population for support. The **total dependency ratio**—the ratio of the elderly plus the young to the total number of working age people—is projected to fall rapidly from 0.85 in 1975 to 0.64 in 2000. Thus, there will be a larger percentage of working people with relatively fewer people to support. The economy will have to expand considerably to provide enough jobs for all who want them.

Not only is the age-sex distribution changing but also the relative size of the urban and rural population is undergoing modifications. In spite of an overall trend toward urbanization, there has been considerable movement away from the large cities into suburban areas. (See Table 7.2.)

7.7 CURRENT WORLDWIDE POPULATION TRENDS

Predictions of the future are hazardous. Dramatic upsets—major wars, cataclysmic geological changes, major climatic shifts—are, by their very nature, unforeseeable. Prediction, then, is based on the assumption that at least some social order and control will be maintained in the world, that people will not be totally frustrated in their attempt to maintain an acceptable quality of life, and perhaps even improve it, and that the geological and meteor-

189

ological conditions of the world in the next century will be nearly the same as they are now. If all these hold true, then population growth will necessarily be contained in the future, and this containment will be the result of decreasing fertility, rather than increasing mortality. Two questions arise: When will population stop growing, and how large will the population of the world become?

Questions are easier to ask than to answer. Perhaps population will grow quickly in some areas of the Earth and decline in others; perhaps growth will be slow for many years; perhaps growth will be rapid only for a century; perhaps growth will stop, then start again later. The prediction in this section is based on projections made by the United Nations in 1974. The first assumption made by the UN is that, Ponce de Leon notwithstanding, people will not achieve immortality; there is no Fountain of Youth. The UN assumes that the expectation of life will rise to 74.8 years. This ultimate life expectancy will be reached by various countries at various times in the future.

If expectation of life is 74.8 years, mortality in childhood and adolescence will be so low that 2.08 children per woman would ensure long-run population replacement or zero population growth. The date when fertility will fall to 2.08 will vary by region. Even if the TFR falls to 2.08, population will continue to grow until age distribution is stabilized.

Figure 7.7 diagrams the changes in demographic structure for eight major regions of the world as estimated by the United Nations. The current world population is about 4.5 billion. The United Nations projects that world popula-

Figure 7.7 The changing demographic patterns of eight major geographic regions. Data and projections, 1925–2100.

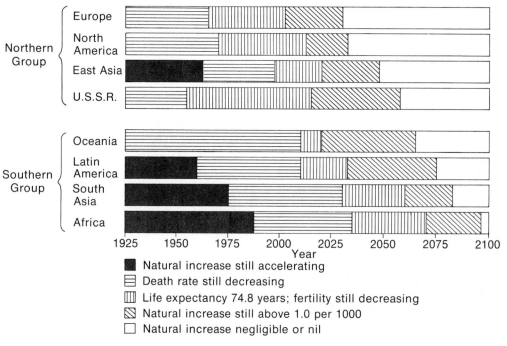

190

tion will reach 12 billion by 2075. If fertility is higher than projected, the total population may be nearly 16 billion, while if the fertility declines very rapidly, the total population may be slightly less than 10 billion (Fig. 7.8).

Two striking projections emerge: first, the population of the world is likely to triple in the next century, and second, much of the growth is likely to occur in the countries of the Southern Hemisphere—Africa, Latin America, and South Asia. In 1975, for instance, about half the world's population lived in the south; in a century, roughly three fourths will live there. The difference will be due to a very rapid rate of natural increase. As the needs of the southern portions of the globe become more pressing, economic development and efficient use of resources of those regions will become imperative if humans are to live in comfort and dignity.

Along with rapid population growth will come even more urbanization, posing technological, administrative, economic, and social problems as well as problems of environmental protection.

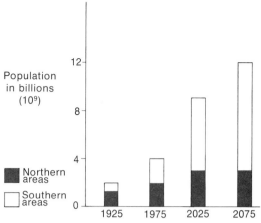

Figure 7.8 Estimated and Projected World Population 1925–2075.

7.8 Case History: *Pakistan*

It is simple arithmetic, simple economics, to realize that something has to be done to curtail the growth of populations. . . . We realize the urgency.
Prime Minister Bhutto of Pakistan, as quoted by Milton Viorst, *Science,* **191**:52, 1976.

The Pakistani government has been experimenting with programs for population control for the past quarter of a century, for Pakistan's population growth has been prodigious. Its population of 20 million in 1911 grew to 40 million by 1952 and to over 70 million by 1976. The growth has been due to a rapid decline in mortality. In spite of the decline, one Pakistani child in four dies before its fifth birthday. Life expectancy is only about 50 years. More than one half the population is under 15 years old, so that even if the average Pakistani family dropped to two children, Pakistan's population would be nearly 150 million people by 2025.

In the 1960's, Ayub Khan, then president of Pakistan, included a family planning program into the government's five-year plan. The intrauterine device was the method most highly recommended. The government claimed considerable success. When Ayub was overthrown, villagers attacked and burned several family planning centers. It is not known whether the hostility was directed against family planning

or against Ayub. Later, analysts established that the claimed success was based on doubtful, perhaps falsified, statistics.

Ayub's successor, Yahya Khan, believed that the population control programs were a political liability and discarded them. Some bureau-

Even densely populated countries have vast areas of uninhabited land. The Khunjerab Pass in Pakistan. (Photo courtesy George B. Schaller.)

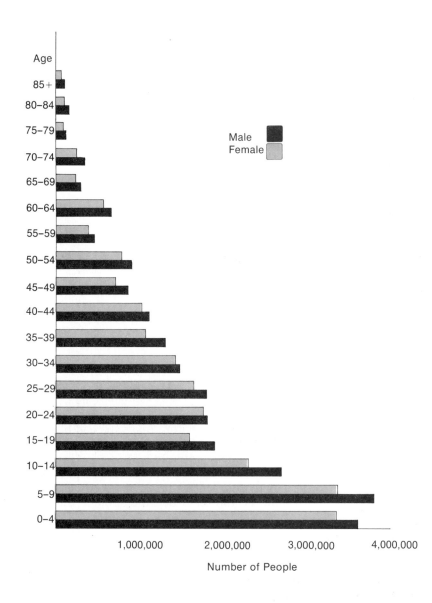

Figure 7.9 Pakistan, age-sex distribution, 1968.

crats, however, continued to work on developing programs for population control.

Yahya was succeeded by Bhutto, who, with his wife, has initiated programs for population control. In 1975, a new experiment was introduced in Pakistan. The program has two parts, a "continuous motivation system" developed by the Pakistani government and a "contraceptive inundation program" supported almost totally by the U.S. Agency for International Development (AID).

In the continuous motivation system, trained teams of health workers, both male and female, are assigned to groups of 10,000 to 15,000 people. In each such group, demographers calculate there are about 1500 couples of childbearing age. The team's responsibility is to visit all couples four times a year to urge them to limit their families and to provide them with information about birth control.

The "contraceptive inundation" phase is just what it sounds like — the countryside is flooded with condoms and birth control pills at a price (a few cents a month) that even the poorest peasant can afford. This program has reached nearly all of Pakistan's 43,000 villages. Pakistani men have shown little resistance to condoms; they are now the preferred method of contraception. Use of the pill has increased dramatically since the start of the program. The Pakistani government permits sale of the pill without medical or paramedical approval. Such an approach would be unthinkable in the United States, where the populace is more and

Balcony-lined houses in a main street of Peshawar, Pakistan. (Photo courtesy Helmut Schadt.)

more aware of the possible deleterious side effects of the pill. In Pakistan and in other developing nations, the cost of *not* providing the pill is too great. Health services in Pakistan are in such short supply that they must be used for more pressing needs than examinations of the possible side effects of contraception. Moreover, the mortality rate in pregnancy and childbirth is five or six times higher than in the United States, so that the risk attendant on the use of the pill is relatively unimportant. Women in rural areas of Pakistan seem to be much more willing to use the pill than the IUD.

The theory behind the experiment was as follows: No one understands why couples choose to have, or not to have, a child. No matter what the social or economic state of a family, some would like to limit the number of children and should not be deprived by lack of knowledge of or material. This approach flies in the face of conventional thought, which states that parents have many babies because they want and need them to work in the fields, and expect them to be providers of social status, of old-age security, and even of a measure of immortality.

Thus far, the program appears to be successful—villagers are responding positively to the program, team morale is high, and the central government is more and more supportive.

Does the apparent Pakistani success refute conventional thought? Probably not, for Pakistan has been undergoing sociological changes. With the Green Revolution (see Chapter 12) came new methods of agriculture. Old methods of cultivation were changed; fewer hands were needed to produce the same amount of grain. Farmers who used to be considered traditionalists now seek new technology, for they are beginning to recognize that they have control over their destiny. Some resist subdividing their farms among their sons, because they see that larger farms are more amenable to modern agricultural practices. Many peasants are now reluctant to have their daughters marry at the earliest possible age. Perhaps most important, the peasantry has the transistor radio, which provides a glimpse of the outside world and the promises of modernization. The radio also presents birth control propaganda. Thus, in spite of some doubts about the birth control program, early successes have been realized. In Pakistan, as in other regions of the world, economic and social improvements may be developing hand in hand with social change.

193

A Pakistani school (Photo courtesy Leo de Wys, Inc.)

Loading a boat in Pakistan. (Photo courtesy Günter R. Reitz.)

PROBLEMS

1. **Family size.** Explain why agricultural societies tend to have larger families on the average than hunter-gatherer societies.

2. **Infant mortality.** Explain why reduction in infant mortality leads to marked changes in population distributions. Comment on the assertion that modern medicine, by prolonging the lives of elderly persons, has contributed greatly to the population explosion.

3. **Modernization.** Outline several changes that modernization brings to a society and discuss their relevance to population growth or decline. Specifically, describe how education tends to lead first to an increase and later to a decrease in family size.

4. **Urbanization.** New York, Chicago, and London have been called cities ecologically dependent upon the railway and the steamship; Los Angeles and the Boston-New York-Washington megalopolis are said to be ecological consequences of the automobile. Discuss.

5. **Urbanization.** Briefly contrast the process of urbanization in the developed and the developing worlds.

6. **Population density.** Describe some unpleasant effects of high population densities.

7. **Population density.** Describe some unpleasant effects of very low population densities.

8. **Population density.** Population density is usually defined in terms of people per unit of land area. Density is often cited as an index of poverty. How useful is such a definition in a city with many luxury high-rise buildings and large areas of decaying tenements? What alternative definitions would you suggest?

9. **Settlement patterns.** Describe the settlement pattern of the town or city in which you live. Does the pattern in your locale resemble any of the patterns of Figure 7.1 or 7.2? Discuss.

10. **Urban places.** Explain why the problem of defining "urban" renders international comparisons of urbanization difficult. Do you think an international definition of "urban place" would be useful? Defend your answer.

11. **Rapid population growth.** In times of rapid population growth from excesses of births over deaths, what types of professions and services must become increasingly available very rapidly?

12. **Trends in the United States.** What factors will tend to raise American birth rates? lower them?

13. **Total dependency ratio.** Define the total dependency ratio. Discuss its relevance in assessing the demographic and economic health of a nation.

14. **Population projection.** The United Nations bases its population projections on the prediction that expectation of life will not exceed 74.8 years. What would happen to human population size if expectation of life rose to 75 years? 80 years? 200 years? Do you think governments should continue to spend money on research into aging, or should the fear of finding a secret to longevity serve as a damper to such research?

15. **The developing world.** Why do you suppose that many Third World politicians suspect the call for population control is a devious ploy by the developed nations to maintain the current economic imbalance among the nations? Do you believe their distrust is well founded? Why or why not?

16. **Technology.** Discuss the use and importance of the radio as an agent of population limitation.

17. **Social policy.** What arguments and programs might allay the fears that population limitation programs are designed to preserve economic inequalities?

BIBLIOGRAPHY

An introduction to population distribution and settlement patterns is found in:
Emrys Jones: *Human Geography.* New York, Frederick A. Praeger, 1966. 240 pp.

A concise summary of the effects of changes in population size and distribution is presented in:
Lennart Levi and Lars Andersson: *Psychosocial Stress: Population, Environment, and Quality of Life.* New York, Spectrum Publications, Inc., 1975, 142 pp.

Several books sound a tocsin for our crowded planet. One of the most popular of these is:
Paul R. Ehrlich: *The Population Bomb.* New York, Ballantine Books, 1968. 201 pp. (Ehrlich includes a bibliography of similar discussions.)

On the other hand, there is an important argument for encouraging moderate population growth expressed in a very provocative work:
Alfred Sauvy: *General Theory of Population.* (Translated by Christophe Compos.) New York, Basic Books, 1969. 550 pp.

A very interesting group of papers on population growth is presented in the two-volume work:
National Academy of Sciences: *Rapid Population Growth.* Baltimore, Johns Hopkins Press, 1971. (Vol. 1, 105 pp; vol. 2, 690 pp.)

The United Nations publishes several volumes on population. Among the most useful are:
The Population Debate: Dimensions and Perspectives. New York, The United Nations, 1975. (Vol. 1, 676 pp.; vol. 2, 726 pp.)
The Determinants and Consequences of Population Trends. New York, The United Nations, 1973. 661 pp.

For a discussion of population policies, see:
World Bank Staff Report: *Population Policies and Economic Development.* Baltimore, The Johns Hopkins University Press, 1974. 214 pp.

Unit IV

Resources and Energy

OUR GEOLOGICAL ENVIRONMENT

8.1 THE FORMATION OF MINERAL AND FUEL DEPOSITS

STRUCTURE OF THE EARTH

Ancient philosophers theorized that the Earth was hollow, rather like a tennis ball, with a thin outer shell and a void in the center. This belief persisted in some circles even up to modern times. In the 1960's the Congress of the United States was considering legislation to finance the drilling of a test well under the ocean in an effort to penetrate the Earth's crust and sample the rock of the mantle. Several concerned people wrote letters telling their senators and representatives that if such a hole were bored it would unplug the stop, so to speak, and all the oceans' water would drain away into the middle of the Earth. No scientists believe this theory anymore. We now know that the Earth is composed of several distinct layers of different kinds of solid or liquid matter.

Scientists believe that billions of years ago, natural decay of radioactive elements within the Earth released enough heat to melt vast volumes of primordial rock. Most of the dense elements such as iron and nickel that were found in this mass of liquid rock gravitated toward the center. These were then surrounded by a **mantle** of lighter rock and a surface **crust** that is generally of comparatively low density. The present structure of the Earth is shown in Figure 8.1.

Although we picture distinct layers of rock inside the Earth, the boundaries between them are not rigid and impenetrable. Rather, rock flows upward and downward in slow but continuous and dynamic exchange. Thus the Earth is believed to have formed a hard solid crust about 3.5 to 4 billion years ago, but virtually none of that original crust remains. Almost all of the original rock has been

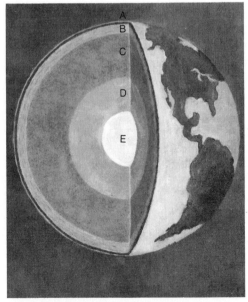

A   Crust
B   Mantle, upper layer   C Mantle, lower layer
D   Outer core, probably liquid iron or iron and nickel
E   Inner core, may be solid iron and nickel

Figure 8.1 The structure of the Earth.

pushed downward into the mantle, reabsorbed, and replaced by material coming from deep within the globe.

Careful studies have shown that the mantle itself is separated into several layers, each with its own unique physical characteristics. Some mantle rock is brittle. Other regions of the mantle are hot and semifluid and behave somewhat like heavy, viscous putty. Various forces within the mantle create large stresses, pushing the rock in one direction or another. When the brittle layers are stressed

Figure 8.2 Continental drift.

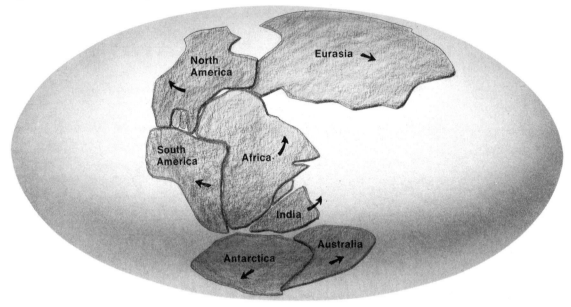

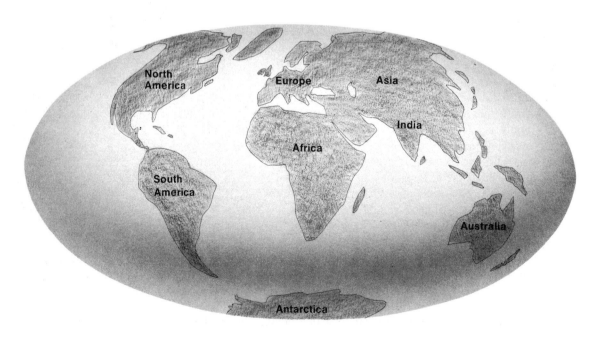

severely they may crack sharply, causing earthquakes. On the other hand, the puttylike "plastic" layers tend to flow gradually when stressed and do not break sharply. There are certain regions within the plastic mantle layer that contain a much more fluid material. This fluid, called **magma,** consists of melted rock mixed with various gases such as steam. Magma flowing quickly to the surface of the Earth produces a volcano; the outpouring magma is called **lava.** If magma protrudes into the crust slowly through a crack or fissure in the rock but does not travel all the way to the surface, it will cool gradually deep within the crust.

For most practical purposes it may safely be assumed that the Earth's crust is a rigid mass of rock lying on the surface of our planet. Thus we expect that the distance between any two cities in the world will remain constant from year to year, and that the continents will always lie in the same relationship to each other as they now do. However, according to modern geological theory, the continents are not immobile and rigidly fixed in position. Geologists now believe that the Earth's crust is composed of several large continent-sized pieces of solid rock and that each piece floats about on the semifluid plastic mantle. The continents and ocean basins float on the denser mantle fluid, much as tightly packed icebergs might float on water.

This concept of "floating" continents explains a great many observations. Look at a map of the world as shown in Figure 8.2A. If you were to cut out the continents and try to piece them together as part of a jigsaw puzzle, you find that they fit together amazingly well, as shown in Figure 8.2B. From this evidence alone early scientists deduced that perhaps there once existed one or perhaps two large supercontinents. The supercontinent(s) then broke apart, and the pieces slowly drifted away from each other to their present position. This idea is called the theory of **continental drift.** Of course, as continents move, they carry their mineral deposits with them.

FORMATION OF MINERAL DEPOSITS

As we go from place to place on the Earth's surface, not only does the topography of the land change, but also the chemical composition of the rock and soil varies. These changes often occur abruptly, especially in mountainous regions. For example, outside of Boulder, Colorado, at the foothills of the Rocky Mountains, a series of uplifted sedimentary rock marks the landscape (Fig. 8.3). Just a few miles away, however, many of the imposing cliffs of Boulder Canyon are of granite. Rock chemistry may change even more abruptly, and two or more distinct types of minerals may appear in a single formation. Figure 8.4 shows a small segment of one type of rock, called a **vein,**

Figure 8.3 The third flatiron, a hard sandstone formation near Boulder, Colorado.

201

This rock outcrop, located only a few miles from the rock shown in Figure 8.3, is composed primarily of granite.

embedded in a dominant rock formation. Geologists try to understand the processes that cause local concentrations of one mineral or another. Aside from the theoretical interest, this problem has considerable practical importance, for people are always searching for concentrated deposits of metals, fuels, and fertilizers.

The magma lying in the Earth's mantle varies from place to place. The chemical composition of the lava coming from Mauna Loa in Hawaii is likely to be different from that coming from Vesuvius in the Mediterranean. But only a small fraction of the Earth's crust is composed of lava that shot rapidly out of a volcanic opening. Great quantities of rock have oozed up slowly through cracks in the crust, cooling gradually during its travel. If a layer of older rock is cracked apart and invaded by upflowing magma, the final formation will contain a vein of foreign rock, as shown in Figure 8.4. A vein may be a few centimeters or several kilometers wide.

Erosion also plays an important role in developing heterogeneity. If several different types of rock lie in the same region, the softer rock will erode away more quickly, leaving exposed layers of hard rock rising over a valley or plain.

Many other chemical and physical processes are responsible for separation of rock into distinct formations. In reviewing some of these, we will emphasize the development of economically significant ore, fuel, and fertilizer deposits.

Suppose that two minerals were mixed together within a molten lava. If the magma were agitated and pushed upward to the surface by a volcanic eruption and then frozen quickly, the minerals would be more or less evenly dispersed in the newly formed rock. But suppose that the magma had started to move up slowly through a fissure in the crust and cooled gradually while still kilometers under the surface. Say for example that one of the minerals solidifed when cooled to 1500° C while the other solidified at 1200° C. As the total mixture cooled, the mineral with the 1500° C melting point would start to solidify while the rest of the magma remained liquid. Remember that the cooling process occurs slowly; sometimes it requires many thousands of years. During this time, all the tiny pieces of solid minerals would settle downward through the lighter liquid until there existed a concentration of one mineral at the bottom of the uplifting magma. This deposit may lie deep in the crust, unreachable by modern mining techniques, or alternatively it may be uplifted by geological processes and exposed by erosion.

Minerals may also become concentrated inside the Earth by differential solubility. To understand how this process works, take a little bit of salt and mix it up with a lot of sawdust. It would be physically difficult to pick out the salt grains from the sawdust. But suppose you put the entire mixture in a glass of water and stirred it up. The

Figure 8.4 Vein structure in rock. (Photo by Ken Brewer.)

salt would dissolve into the water, and the sawdust would float to the top. You could then skim off the sawdust, evaporate the water, and collect the salt.

Similar processes can occur at the Earth's surface. If a mineral is either more or less soluble than the adjacent rock, moving water may separate the two.

Once a mineral deposit is formed, it may be displaced by geological processes. Many mineral deposits on the Earth are concentrated in geographical strips, or belts, on the various continents. Such beltlike formations of tin deposits, for example, are shown in Figure 8.5. As seen on the map, many of these belts appear to end abruptly at the ocean. But if we fit the continents together, many tin belts line up. Belts of coal seams, salt deposits, and gypsum concentrations also seem to disappear into the ocean but align with each other when continents are pieced together. It is believed that these belts were formed many millions of years ago on a primordial supercontinent by individual processes such as sedimentation in a large river valley, or by the movement of large masses of ore-bearing magma through massive fissures in the Earth's crust. When the supercontinent broke up, the belts separated into the patterns we now observe. Thus, the theory of continental drift has proved to be a useful tool in helping geologists find various mineral deposits.

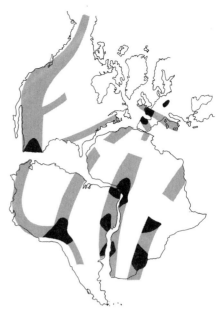

Figure 8.5 Tin belts of the world.

FOSSIL FUEL DEPOSITS

Coal is a valuable geological deposit that was formed from organic debris such as decayed animal and vegetable matter (fossils). As each layer of this ancient organic debris started to decompose, it was gradually covered by successive layers, and so the decomposition was interrupted before it was complete. Thus some of each layer was preserved under new layers of sediment and debris. The combined heat and pressure from accumulating layers initiated a series of transformations that changed the underground plant tissue into **coal.**

Go to a swamp or marsh and dig up a shovelful of the muck on the bottom. If you examine it closely, you will find that there may be very little dirt in your sample; it is mostly decayed plant matter. If such a swamp bottom is covered with inorganic sediment and compressed for hundreds of thousands of years, a small coal deposit will develop. However, most modern swamps are poor coal producers. It is estimated that a layer of compressed organic debris 12 meters (39 ft) thick is required to produce a 1-m layer of coal. For a layer this large to accumulate, conditions must be favorable and stable in a region for a great many years. Coal deposits probably are being formed today in many areas, notably in the Ganges River delta in India, but the process is extremely slow—much, much slower than the exploitation of existing reserves. Therefore, since we cannot expect formation of new deposits to keep pace with use, we have all the more reason to conserve our present reserves.

Oil and gas are also organic deposits, but these fuels were probably formed from tiny marine microorganisms rather than from the debris of large plants. As microscopic sea creatures settled to the bottom of the ocean and were later covered with mineral sediment, tiny droplets of bodily oils were squeezed out of each organism. If the rock formation was favorable, this oil was trapped into large deposits and altered chemically by heat and pressure until petroleum was formed.

8.2   MATERIALS AND ENERGY

A natural ecosystem is stable when the various processes that take place in it are in balance. Birth and death, growth and decay, the absorption of nutrients and the elimination of wastes—all of these processes involve the cycling of materials and the flow of energy. When any one material becomes depleted, as for example when the water supplies are exhausted by pine trees on the eastern slope of the Colorado Rockies (see Section 4.9), the entire system is upset. Similarly, if the flow of energy is disrupted, as when phytoplankton in a body of water are destroyed by an oil spill, the system is again thrown out of balance. Human systems are also delicately dependent on the maintenance of an orderly supply of materials and energy. To survive in a developed society, we need food, fuel, clothing, shelter, and networks of transportation. Each of these needs can be supplied only if *both* energy and materials are widely available. Most of the food consumed by people in industrial nations is grown on commercial farms. The farmers use tractors made of steel, a structural material, and powered by gasoline, a fuel. Agricultural fertilizers are usually materials that are mined or chemically manufactured, transported, and spread on the fields with machines powered by fossil fuels.

Materials and energy are intimately interrelated in our modern world. Think of an oil well. Drilling for oil requires specialized materials. For example, steel is used for the drilling rig, diamond or carborundum chips for the augers, and copper or aluminum wire for the electrical system. Products made from oil help mine the metals needed to build the equipment to drill for oil. The drilling rig itself needs fuel (which comes from oil) to drive the drill. If any material or fuel is unavailable and no substitutes are found, the entire system will fail. Therefore both the developed and the developing world must ask, "How stable is

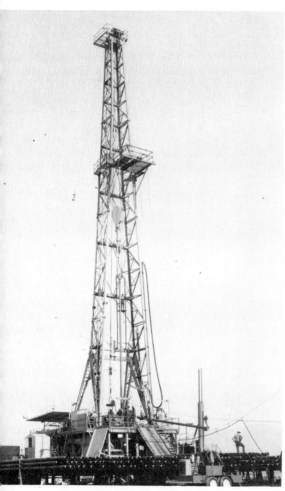

Oil is needed to power this drilling rig to search for oil.

our cycle of materials, and how reliable is our energy flow, and how long can they be expected to last?"

First, we must understand how energy differs from matter, or material substances. Wood, sand, and iron are materials, and they consist of atoms that retain their identity despite physical or chemical transformations. Thus, one can take a bar of iron, beat it, roll it into a sheet, shred it into pieces, melt it, and allow it to solidify again, and one would still have iron. If the bar of iron is set outside in damp air, it will react chemically with oxygen to produce iron oxide, commonly called rust, but even this chemical conversion does not destroy iron atoms. The iron can be easily recovered from its oxide and reconverted to the pure metal. Moreover, chemical and physical operations can be repeated again and again; the iron atoms never get "tired," or destroyed. At the end of any conceivable set of chemical processes, the original supply of iron atoms remains.

Energy, on the other hand, is not a material. Energy is defined as *the capacity to perform work or transfer heat.* Where does energy go after it has been used? The carbon atoms in a lump of coal unite with oxygen as the coal burns, and its inorganic content turns to ash, but what becomes of its energy? This nonsubstance, called "the capacity to do work," is elusive and hard to keep track of. If we can find used iron and reuse it, why can't we find used energy and reuse it? Or better yet, if energy isn't really matter, could we find some for nothing? These two questions plagued scientists for a long time and the search for answers led to the study of heat-motion, or **thermodynamics.**

A lump of coal burns and turns to ash, but what becomes of its energy?

8.3 THE FIRST LAW OF THERMODYNAMICS (OR "YOU CAN'T WIN")

To understand the first two laws of thermodynamics and the answers to the hopeful questions posed at the end of the previous section, let us search for solutions to the following imaginary work problem:

A landslide causes a large boulder to roll downhill; it comes to rest in a position where it blocks the entrance to a cave. Work is required to remove the boulder in order to gain access to the cave. How can this work be done? A primitive caveman might have tried to push it, roll it, or drag it. If these were unsuccessful, he might have cooperated with a group of his friends to remove the boulder.

After people had learned to domesticate large animals and to use them as beasts of burden, animals might have been enlisted to help move the rock. Although cooperation with other people or with animals represents a great advance over solitary action, it is obviously a method of *shar-*

Caveman moving boulder.

ing the work, not reducing it. A much more far-reaching development for doing work was the invention of machines. The lever, the wheel, the roller, the screw, the inclined plane, and later, the gear and the block and tackle were all devices that could increase the *force* that a person or an animal could exert. Scientists now understand that such devices do not actually reduce the amount of work required, but rather spread it out over a longer time and thus smooth out the effort. However, this difference can easily be overlooked, and a device such as a lever can therefore be mistakenly thought of as a work-saver. As machines became more and more effective in extending the force of living muscle, such erroneous impressions were reinforced, and inventors saw no reason to doubt that they could provide machines which could produce work indefinitely just because their mechanisms were clever. Such hope was a powerful stimulus to invention because it promised that continued improvement in design would eventually free people and their animals entirely from their labor. If our caveman had had such a device (which is called a perpetual motion machine of the first kind), he could have moved all the boulders he wished at his leisure. It may be difficult for the modern reader to appreciate the fact that the search for such a perpetual motion machine seemed entirely reasonable, apparently requiring only continued progress along the lines that had already been so successful. However, all attempts failed. The failures have been so consistent that we are now convinced that the effort is hopeless. This conviction has been stated as a law of nature and is called the **First Law of Thermodynamics.**

This law can also be expressed in terms of conservation of energy by the statement "energy cannot be created or destroyed."

Don't ask for a proof of the First Law. There is none. The First Law is simply a concise statement about human experience with energy. If it is impossible to create energy, then it is hopeless to try to invent a perpetual motion machine, and we may as well turn to some other method of doing work.

8.4 HEAT ENGINES AND THE SECOND LAW OF THERMODYNAMICS (OR "YOU CAN'T BREAK EVEN")

Modern technology demands far more energy than can be supplied by men and beasts. Instead, people use heat engines, which consume fuel to produce heat, and convert the heat into work.* The idea that heat could be converted into work was far from obvious. In fact, heat engines have been used successfully only during the past 200 years or so. (James Watt developed a steam engine in 1769.) The first

*Of course, a person (or an ox) is also a heat engine. People consume food, which is their fuel, and convert its energy into the work of muscle contraction. Mechanical heat engines, however, can be much larger than animals and can consume fuels such as gas, oil, and coal, which animals do not eat. The result is that the total amount of work done and heat produced is increased to an extent that has new effects on the environment.

experimental proof that energy can be converted, without gain or loss, from one form to another was supplied by James Joule in 1849. Since heat and work are two forms of energy, it is possible to convert heat into work or work into heat. Thus one can create heat by rubbing two sticks together, and the heat produced is exactly equivalent in energy to the work required to rub the sticks.

Fuels "contain" internal energy. Thus, a kilogram of coal can release heat when it burns in air. Energy is also stored in ordinary substances that are not fuels, such as water. If you heat water, it becomes hot; therefore, hot water must have energy. But even cool water loses energy when it turns to ice; therefore, cool water must also have energy. Why not, then, use this energy to drive a machine to do work? Such a machine, although seemingly not quite so miraculous as the perpetual motion machine of the first kind, would extract energy from its surroundings (for example from the air or from the ocean) and convert it into useful work. The air or water would then be cooled by the extraction of energy from it, and could be returned to the environment. Automobiles could then run on air, and the exhaust would be cooler air. A power plant located on a river would cool the river while it lighted the city. Such a machine would *not* violate the First Law, because energy would be conserved. The work would come from the energy extracted from the air or water, not from an impossibly profitable creation of energy. Such a device is called a perpetual motion machine of the second kind; alas, it too has never been made and never will be.

Let us return now from the impossible engine to a heat engine that uses fuel. Even here the situation continues to be discouraging. Experiments have shown that the potential energy inherent in a fuel is never completely converted into work; some is always lost to the surroundings. This energy is "lost" only in the sense that it is no longer available to do work; instead the energy warms the environment. Ingenious inventors did try to design heat engines that would convert *all* the energy of a fuel into work, but they always failed. They found, instead, that a heat engine could be made to work *only* by the following two sets of processes: (a) Heat must be absorbed by the working parts from some hot source. The hot source is generally provided when some substance such as water or air (called the "working substance") is heated by the energy obtained from a fuel, such as wood, coal, oil, or uranium. (b) Waste heat must be rejected to an external reservoir at a lower temperature.

A heat engine cannot work any other way. The original form of this negative statement, as made by Lord Kelvin (1824–1907) is, "It is impossible by any means ... to derive mechanical effect from any portion of matter by cooling it below the temperature of the coldest of the surrounding objects." This is an expression of the **Second Law of Thermodynamics.**

James Watt (1736–1819) was a Scottish inventor, and James Joule (1818–1889) was an English physicist. Standard units of energy and power have been named after these two men. A joule is a unit of energy. A watt is a unit of power and is defined as the amount of work done at a rate of one joule per second.

William Thomson (later Lord Kelvin) was a British physicist who proposed the absolute scale of temperature (the "Kelvin" scale) in 1848. His major contributions to science were in the field of thermodynamics.

Steam locomotive.

To gain further insight into this very fundamental concept, imagine a few simple experiments with gases. It is known that gases push against the walls of their containers when they are heated, and that this property can be used to convert heat to work. Consider a cylinder full of gas with a freely sliding piston in the center (Fig. 8.6A). Suppose that heat is transferred to the system until the gas on the left side of the piston reaches 300°C, while the gas on the right stays at 0°C. The heated gas on the left would expand and force the piston to the right, as shown in Figure 8.6B. This, then, is a simple heat engine. Heat is converted to motion of the piston. Now suppose that you started over again with the piston in the center and heated both sides of the cylinder equally (Fig. 8.6C). The gas in the left side would warm and exert a pressure against the piston, but this effect would be equally balanced by the pressure of the hot gas in the right-hand chamber. Since the two pressures would be equal, the piston would not move and the heat would not be converted to work. The piston moves only if the temperature on one side of the cylinder is greater than the temperature on the other. The amount of work performed depends on the *difference* in temperature between the two sides. What does this have to do with heat engines and energy? If one side of the apparatus is heated to 300°C while the other side is maintained at 0°C, the piston will move a large distance, whereas if we heat one side to 300°C and the other to 250°C (Fig. 8.6C), much more heat energy is used but less motion is realized. Since the work performed by the piston is proportional to the distance that it moves, little work is realized in this

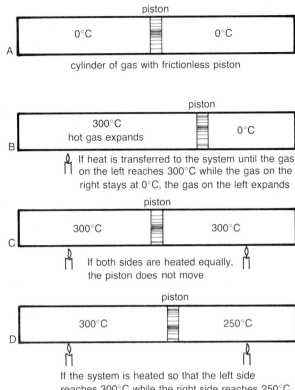

Figure 8.6 A piston moves in response to the temperature difference between the two sides.

latter example. We immediately see that the *work performed by a heat engine depends on the temperature difference between the hot reservoir and the cold reservoir.*

Heat energy is useful only if large differences in temperature can be realized. Suppose a lump of coal is burned in a steam locomotive. The heat is converted to useful work and the engine travels up a hill. Now imagine that the fire burns out and the engine rolls back down the hill. There is friction in the wheel bearings, between the wheels and the track, and between the locomotive and the air. As a result, a large mass of air and metal is heated slightly. When the locomotive has come to rest, the energy of the original lump has been conserved, but because no large temperature differences have been created, not much useful work can be extracted from this energy. Thus, the energy from the coal cannot be recycled. This observation is a general one, and explains why energy, once used, cannot be reused to perform work efficiently. In brief, materials can be recycled but energy cannot.

8.5 OUR FOSSIL FUEL SUPPLY

In 1976, 93 percent of the total energy used in the United States was derived from fossil fuels—coal, oil, and

209

natural gas. Three percent was from nuclear fission. Only four percent was from a renewable supply—mostly power from falling water, plus some solar energy and energy from wood and geothermal sources. The Second Law assures us that someday our reserves will be depleted. How much time do we have? For a realistic estimate of the number of years before people have used all the Earth's energy reserves, we must forecast human population growth (see Chapter 6), estimate the quantity of the remaining reserves, and predict accurately our future rates of consumption. All such forecasts are subject to large errors.

Geologists have used several methods to estimate the remaining reserves of fossil fuels. Fuels that have been positively located and identified are called *proved* reserves. Moreover, from seismic recordings and other data, the location and size of *probable* sources have been recorded. Educated hunches and preliminary data give us a third category, *possible* deposits. By adding to the proved sources a reasonable fraction of the probable ones, and assuming that new, as yet unthought-of sources will balance disappointments among the possible deposits, we can get a rough guess at the world's supply of fuel.

To estimate the energy requirements from the 1970's into the early twenty-first century, we start by plotting the past energy consumption, and then try to guess how the curve will continue in the future. Figure 8.7 shows the production of energy in the United States from 1850 to the present, with predictions to the year 2000. From 1850 to 1915, the average growth in energy production was 6.9 percent per year, compared to a population growth of only 4 percent. During this period, then, the quantity of energy used doubled every 10 years. Between 1915 and 1955 the consumption rate increased only 1.85 percent per year, corresponding to a doubling in use roughly every 40 years. In the period of time between 1961 and 1975, consumption has been increasing about 4.5 percent per year, corresponding to a 15- to 16-year doubling period. During the same period the population was rising at a rate of 2 percent. Total world use of energy is increasing at a rate of roughly 7 percent per year. If that rate remains constant, world energy requirements will double every 10 years.

It is obvious that the world cannot double its consumption of a depletable resource every 10 years for very long, for each doubling corresponds to an increasingly large growth. Think of it this way: If energy consumption is doubling every 10 years, then in the past 10 years we have used as much energy as in our whole previous history. Or another way: If energy consumption is doubling every 10 years, then by the time we have used up half our total reserves, we will have only enough left for another 10 years. Even if there were a very large reserve of fuel, the environmental side effects of energy production would eventually limit expansion. In that case, a reasonable

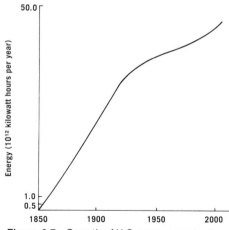

Figure 8.7 Growth of U.S. energy production. (Adapted from M. King Hubbert: The energy resources of the earth. In: *Energy and Power, A Scientific American Book.* San Francisco, W. H. Freeman & Co., 1971.)

210

prediction for growth with time is depicted in Figure 8.8. This resulting curve, called a **sigmoid function,*** shows an initial growth rate, followed by a rapid rise, and finally a leveling off. This level portion would correspond to a condition in which the environmental cost of increased use would be higher than the expected gain.

In the real world it is reasonable to suspect that the combination of environmental side effects and depletion of resources will lead to the leveling off and eventual decline of consumption of a depletable fuel. We can expect that consumption of coal, gas, and oil will increase rapidly for a while as technology improves the methods of exploration, mining, and transmission. As poorer reserves must be tapped and pollution control becomes more expensive, the cost of power will rise and people will consume less fossil fuel. Eventually, when the reserves are depleted—that is, when more energy is needed to mine and refine a fuel than is retrieved when the fuel is burned—consumption will drop to zero. A generalized fuel consumption curve is shown in Figure 8.9.

Although scientists can draw the general shape of the curve showing the consumption of the fuel, they cannot accurately predict how long a given reserve will last. Sometimes predictions by various experts differ widely. Moreover, a prediction offered one year may be retracted the next. For example, in the early twentieth century most exploration for oil was performed on land. Gradually, geologists began to search the continental shelves under the oceans for petroleum, and as a result, wells were dug along many coastal regions. The search for oil under the sea has recently accelerated so that during the decade between 1965 and 1975 many new deposits have been found in such places as the North Sea in Europe, off the East Coast of North America, and in Southeast Asia. No one really knows how many new deposits will be found or how large some of the recently discovered ones are. This uncertainty makes it hard to predict future patterns of consumption. Of course, our fossil fuels are finite and therefore cannot supply our energy needs indefinitely.

Most authorities believe that natural gas is our least abundant fossil fuel. The peak consumption rate will probably be reached about 1985, or before most of the readers of this book have reached middle age. Already, contractors in many areas are unable to supply gas lines for heating new houses because the industry claims it will not have enough fuel for new customers. The scarcity of natural gas is environmentally unfortunate, since it is less polluting than any other widely used fuel.

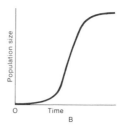

Figure 8.8 Sigmoid growth curve.

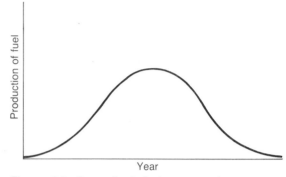

Figure 8.9 Generalized fuel consumption curve.

The search for oil under the sea.

---

*See Chapter 6 for a discussion of the relationship between growth rates and doubling times and for a further discussion of curves and projection.

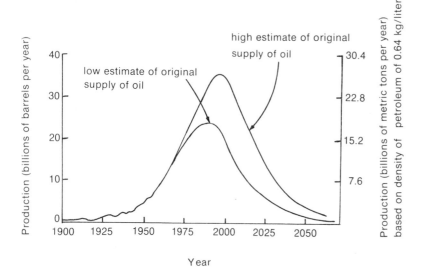

Figure 8.10 Past and predicted world oil production based on two different estimates of initial supply. (Adapted from Hubbert, ibid.)

Figures 8.10 and 8.11 combine estimates of world supply with predicted rates of use to display past and predicted oil and coal production.

Petroleum, the most widely used fuel today, can be easily converted to a variety of free-flowing liquid forms, such as gasoline, kerosene, and fuel oil. Liquid fuels are particularly well suited for use in both small internal combustion engines and large industrial boilers. Unfortunately, plentiful supplies of petroleum cannot be expected to last for more than a century or so. Even if enough new reserves were discovered to double our present supply, they would probably be depleted within 100 to 150 years.

As Figure 8.11 demonstrates, we have barely begun to utilize our coal reserves. As gasoline and oil become more scarce, consumption of coal will rise precipitously for a while. Coal alone, however, will not be able to satisfy all of our energy needs. By the year 2300, for example, the consumption rate of two to twenty metric tons per year predicted by Figure 8.11 may not be enough to satisfy our energy needs. Therefore, it is assumed that other sources will be exploited.

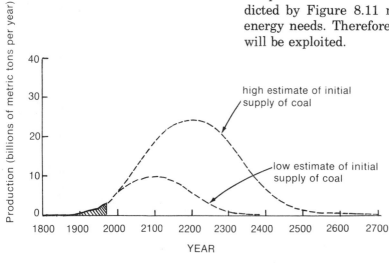

Figure 8.11 Past and predicted world coal production based on two different estimates of initial supply. (Adapted from Hubbert, ibid.)

Will people be driving horses when the oil runs out?

What will happen when the fossil fuels are depleted? Will people all learn to drive teams of horses to town and rediscover the use of wood as a fuel? The answer is almost certainly no, for several stopgap measures are available, and a few more long-range solutions will be found.

Until now, most of our petroleum has been extracted from underground wells, or "lakes," made up of a viscous but pumpable liquid. Underground deposits of shale in Colorado, Wyoming, and Utah have been found to contain large quantities of petroleum impregnated in the pores of the rock, and sand fields, laden with oil, have been discovered in Africa, the United States, and Canada. The quantity of oil that can be obtained from oil shales and tar sands is unknown. The richest deposits, containing about 400 liters of oil per metric ton of ore, are believed to add another 25 percent to the world's usable supply of liquid petroleum. Today, importation of oil is usually more economical than mining the shales, although a Canadian firm is currently extracting oil from tar sands, and several pilot plants for handling oil shale are in operation in western Colorado. As imported oil becomes more costly and as the technology needed for extraction becomes more highly developed, shale deposits may be used more. Many shales have relatively small quantities of fuel per ton of rock. If all the

(From: The Energy Crisis. Science & Public Affairs Book Bulletin of the Atomic Scientists, p. 63.)

"Be practical—people always need coal."

Oil shale country in western Colorado. (Courtesy of Atlantic Richfield Co.)

shales containing from 20 to 400 liters per metric ton are to be recoverable for fuel, the total reserves are equal to 100 *times* the estimated supply of oil. We do not know how much of this can in fact be used. Present technology would require more than 20 liters of fuel to mine, transport, extract, and refine a metric ton of shale. Thus it is now clearly uneconomical to mine some shales. Furthermore, experts believe that many low grade shales will never be economical to mine. But what about the shale with 200 liters of oil per metric ton? One hundred? As the cost of fuel from other sources increases and as the technology of retrieving oil from shale advances, exploitation of oil shale will probably increase, for lower-grade ores will become relatively more economical. Another problem arises because large quantities of water are needed to extract oil from shale, and in the semiarid regions where the shale is found, the water is needed to irrigate farmlands. Thus it is impossible to predict how much shale will actually be mined.

Another method of increasing the supply of oil and gas is to convert coal to liquid fuel. The theory of such conversions is well understood, and the practice has been worked out both in the laboratory and in the industrial plant. In fact, gasification of coal was commonly practiced in the 1920's, and the conversion of coal to gasoline was carried

A pilot plant that produces synthetic gasoline from coal. (Courtesy of U. S. Energy Research and Development Administration.)

out by the Germans during World War II. By 1975, there were over 50 plants in different parts of the world that produced gaseous or liquid fuels from coal. Most of these were either pilot or small industrial operations, and large-scale conversion of coal has not yet proved feasible. The major problem is one of economics. In the United States the most expensive imported gas cost $3.00 per billion joules* in 1976, while gas from coal was estimated to cost $3.80 per billion joules. Undoubtedly, gas from coal will become relatively cheaper than other gas in future years. In fact, conversion of solid to liquid and gaseous fuel will probably be common before the end of the century.

8.6 NUCLEAR FUELS

Oil shales, tar sands, and chemical conversions will undoubtedly extend the life of the Fossil Fuel Age, but such stopgaps cannot be the basis for a long-range continuation of our technological existence. Instead, we must turn either to renewable resources or to nuclear energy.

In 1975, only three percent of the energy used in the United States was provided by nuclear fuels, but this fraction is expected to increase rapidly in the near future. By the year 2000, 25 percent of the total energy of this country may be supplied by nuclear plants, and the percentage may continue to rise. We must therefore consider the availability of uranium, from which the fissionable uranium-235

*As given in Appendix A, 4.184 joules = 1 calorie. Therefore one billion joules = 2.4×10^8 calories or 240,000 kilocalories. This is equivalent to about 280 kilowatt hours.

Mining uranium ore undergound near Grants, New Mexico. (Courtesy of Ranchers Exploration and Development Corporation.)

isotope, now used as a fuel in nuclear reactors, is extracted (see Chapter 10). Once again, we are concerned with both abundance and concentration in natural ores, since both factors determine cost. At present, uranium costs about $18.00 per kilogram, but the higher grade ores which make a low price possible are scarce. Table 8.1 shows the amounts of ore available at various prices.

Uranium ores costing less than $65/kg will probably be exhausted during this century. Then the price of uranium will rise to $110/kg or more (over six times the present price). At this time our uranium supply will not be exhausted, but the price of electricity generated in this manner will be so high that we cannot expect conventional nuclear fission to be a source of inexpensive energy. In recent years breeder reactors, which use plentiful, nonfissionable uranium-238 or thorium-232 to synthesize fissionable materials, have been developed (see Chapter 10). Breeders can produce many times as much energy per kilogram of uranium ore as a conventional fission reactor. The proponents of nuclear energy predict that the anticipated environmental and security problems can be solved and that breeders can therefore extend the life of our fissionable fuel supply for many thousands of years. However, not everyone agrees; in fact, the question is a matter of considerable controversy.

Another possibility can be considered. Scientists may someday harness the hydrogen fusion reaction. The technological problems yet to be solved are so difficult that some authorities expect fusion never to be functional, while at the other extreme some optimists predict the first demonstration of feasibility within 10 years. Most members of the scientific community feel that development of such a power plant is highly unlikely before the year 2000. If fusion* is found to be practical, the energy available to us will be enormous. Probably the first process to be developed will require lithium for the synthesis of tritium, the raw

---

*The development of fission and fusion reactors will be discussed further in Chapter 10.

TABLE 8.1 Estimated Uranium Reserves (As Uranium Oxide, U_3O_8)

COST OF URANIUM OXIDE, $/kg*	THOUSANDS OF METRIC TONS AVAILABLE	YEAR OF EXHAUSTION
18	600	1986
22	700	1988
33	900	1990
65	1000	1992
110	4000	?
220	7000	?
450–1100	more than 5,000,000	?

*Dollar value in 1974.

material needed for fusion. We have only enough lithium to provide energy equal to our original supply of fossil fuel, but if scientists can control fusion reactions at higher temperatures, fuel can be obtained from ocean waters, and energy will be available for the long-term future.

8.7 RENEWABLE ENERGY SOURCES

In the early nineteenth century, wood was the world's most widely used source of energy. If a forest is managed properly and loggers cut trees only as fast as they regrow, fuel can be harvested indefinitely. Our present rate of energy consumption is so great that the world's forests cannot meet the demand. However, several other renewable sources are available. Energy can be obtained from the Sun, the power of falling water (hydropower), the winds, and the tides. Large quantities of heat are also available from the internal layers of the Earth (although strictly speaking this energy supply is not renewable). If all the renewable sources of energy were exploited efficiently, they could provide enough power for our civilization and could replace fossil and fission fuels. Many of the technological problems needed to harness these sources have already been solved. However, the development of renewable energy sources has been quite slow, and their full potential is still very far from being realized. Some of the social, economic, and technical problems involved in harnessing renewable energy sources will be discussed in the next chapter.

Wood is still used as a heat source in many rural regions.

8.8 ORES AND FERTILIZERS

The Second Law of Thermodynamics assures us that since energy cannot be recycled, our fossil fuel deposits will someday be depleted. There are no such restrictions on metals or ores. As mentioned earlier, these materials are never used up; they can be used over and over again indefinitely.

What happens, then to "used" iron in our modern world? Some is recycled and recast, while the rest is thrown into garbage dumps and dispersed through land and water. This far-flung material is theoretically recoverable, but the actual cost of searching it all out, separating it from other garbage, and gathering it all together would be prohibitive.

Actually, most minerals are widely distributed in the soil and water all about us. Does this mean that you could open an iron mine by digging a hole in your back yard? Probably, but the processing costs would be far greater than the value of ore collected. Much iron is dissolved in the oceans as well, but it is so diluted that one could extract only about 11 g of metal from one million liters of liquid. It certainly is possible to build a factory that would extract iron and other metals from sea water. The major problem would be that a tremendous quantity of energy would be needed. Considering the cost and scarcity of fuel today, it is simply impractical to collect any material that is so widely dispersed. Thus, when scientists speak of depletion of a mineral deposit such as iron, they do not mean that all the iron in the Earth will disappear, but rather that it will be spread about so diffusely that it would be too expensive and use too much energy to collect. The Earth's iron reserves will be considered to be depleted when the concentrated deposits have been mined, processed, and spread about.

To date, people have not exhausted the supply of any essential mineral, and most forecasters predict adequate reserves until at least the end of this century. Since forecasting more than 25 years into the future is extremely difficult, disagreement among experts is so great that estimates of long-term availability of metals often differ by a

Iron can exist in many forms.

factor of ten. Several factors are relevant to understanding of the problems inherent in predicting a time scale for mineral depletion.

ORE DEPOSITS

Prospectors or geologists seldom find pockets or veins of pure metal; rather they find valuable minerals mixed with other types of rock. An **ore** is considered to be a rock mixture that contains enough valuable minerals to be mined profitably with currently available technology. A rare and valuable mineral such as uranium may be mined commercially if it exists in concentrations as low as 0.1 percent by weight in the rock, whereas for an iron deposit to be considered an ore the rock must contain at least 30 percent metal. The **mineral reserves** of a region are defined as the estimated supply of ore in the ground. Reserves are depleted when they are dug up, but our reserve supply may be augmented by either of two processes. First, new deposits may be discovered. In addition, there are many known deposits that are now uneconomical to mine. For example, a deposit containing 20 percent iron is not considered an ore, because it is so expensive to extract the metal from the rock that it would not be profitable under current market conditions. If technology improves so that the materials can be refined cheaply, or if the market price of iron increases, the deposit will suddenly become an ore reserve.

Many of the very high grade, concentrated, and easily accessible ores, such as the 50 percent iron deposit of the Mesabi Range in the north central United States, are being used up rapidly and either have been depleted or will be

A small-time mining operation.

A large open pit mine. (Courtesy of Bethlehem Steel.)

219

depleted in the near future. These mines are essentially nonrenewable, and once they are gone, our civilization will have suffered an irreplaceable loss. But our technological life will not end with the exhaustion of these rich reserves, for less concentrated deposits are still available in great abundance. For example, there has been recent concern that phosphorus reserves needed for the manufacture of fertilizers will be depleted soon. At the present time it is economical to exploit a mine containing about 32 percent or more of phosphates. These reserves are being exploited heavily and will be depleted in the near future. There are, however, many deposits containing 29 percent phosphate, perhaps 100 times as much as the original supply of 32 percent ore. Much of this 29 percent phosphate lies under the sea and therefore would be expensive to extract, but it is available.

The situation is similar for a great many other minerals as well. For example, there are greater volumes of deposits containing 0.5 percent copper than 0.6 percent copper, and more ores containing 35 percent iron than 40 percent iron. Some of these lower grade deposits are easy to reach; others exist in harsh environments such as under the sea or in the Arctic. The question arises: Will the technology be developed to extract the lower grade ores at reasonable prices? Naturally, there are differences of opinion in this matter. Optimistic geologists say yes, the technology and machinery for refining ores are rapidly becoming more efficient, and progress will continue in the future. These people point to several phenomenal success stories. Shortly after World War II it became obvious that high-grade iron ore deposits in the United States were being depleted rapidly, and consequently there was a great deal of alarm about possible iron shortages. Fortunately metallurgical engineers discovered a new refining process to extract iron from low-grade ores at competitive prices. Thus the impending crisis was averted, and the price of iron and steel did not rise precipitously. As technology improves, many well-known mineral deposits become ores. In addition, new deposits continue to be discovered, with the result that the reserves of many important minerals have recently *increased* in some parts of the world.

Other geologists contend that such trends are not likely to continue for long. They point out that a finite supply of concentrated mineral deposits cannot last forever. Moreover, as increasingly lower grade ores are sought, the technological problems inherent in all aspects of mining and refining rise sharply. Dependence on technology to solve all problems may lead to grave disappointment. These scientists emphasize that the future availability of metal depends on many factors besides the quantity of ore in the ground and the state of refining technology. Some of these are discussed in the following paragraphs.

Availability of Energy

To extract metal from ore, the dirt and rock must be dug up and crushed, the ore itself must be separated and chemically reduced to the metal, and the metal must finally be refined to purify it. Each step, especially the chemical reduction, requires energy. Moreover, low-grade ores require much more energy to process than do high-grade ores. For example, if you are mining ore containing 29 percent phosphorus, more material must be dug, transported, and crushed to obtain a given yield of product than if you are mining ore containing 32 percent phosphorus. Some low-grade ores differ from high-grade ores not only in concentration but in chemical composition as well. Some chemicals are easier to purify than others. For example, it is energetically advantageous to extract lead from its sulfide ore (PbS), according to the generalized reaction:

$$PbS + O_2 \longrightarrow Pb + SO_2$$
$$\text{lead sulfide} + \text{oxygen} \longrightarrow \text{lead} + \text{sulfur dioxide}$$

Steel is manufactured from iron ore in large open hearth furnaces. (Courtesy of Bethlehem Steel.)

Lead also occurs in ores of other chemical compositions, but the overall energy requirements for processing them are much greater.

Pollution and Land Use

Most mining processes cause significant pollution of land, water, and air. For example, sulfur is found in large quantities in many ore deposits. This sulfur, bound in forms such as copper sulfide (CuS and Cu_2S), reacts with water in the presence of air to produce sulfuric acid (H_2SO_4), which runs off into the streams below the mine. This pollution, known as **acid mine drainage,** kills fish and disrupts normal aquatic life cycles. When sulfur accompanies other chemicals through refining processes, it is often converted to gaseous air pollutants such as hydrogen sulfide and sulfur dioxide. Sulfur, of course, is not the only polluting chemical from mining operations. Many other mine pollutants cause serious air and water pollution.

Just as more energy is required to handle low-grade ores than high-grade ones, more pollution generally results from processing these impure materials. The pollution can be controlled with highly specialized pollution-abatement equipment, but such measures are expensive and add to the total cost of refining ore.

The world is running short of food, energy, and recreational areas as well as of high-grade mineral deposits. What should our policy be if a valuable ore or fuel lies under fertile farmland or a beautiful mountain? Which resource takes precedence? At present this question is being raised principally with respect to exploitation of fuel reserves, for vast coal seams lie under the fertile wheat fields of Montana and the Dakotas in the United States, and southern Saskatchewan in Canada. If large areas of low-grade ore must be exploited, the problem will extend to metal reserves as well. These difficult questions will be discussed in more detail in the next chapter with reference to fossil fuel supplies.

The Future Demands For Metals

In order to determine how long current mineral resources will last, estimates of how much will be used in the future must be available. Unfortunately, it is extremely hard to predict future levels of demand. No one knows how much industrialization will occur in developing nations. Will most of the inhabitants of India own automobiles and televisions someday? If so, it will not be easy to satisfy future demands for minerals.

Engineers speculate about the future role of substitutes for metals. Reinforced concrete, fibrous glass, and plastics are now used in many applications that once called

222

Concrete bridge (above); steel bridge (below).

for steel or other metals, and the use of concrete and synthetics is rising rapidly. If metals can be replaced by nonmetals for many uses, current reserves will last longer. Concrete is an especially desirable substitute because it is relatively cheap and comparatively little energy is used to mine, process, and mix the ingredients. Many other substitutes for metal require much energy. The reason is twofold. First, the actual consumption of energy in an industrial process may be high, as in the manufacture of glass. Second, the raw material may itself be a potential fuel whose energy content is not utilized. For example, synthetic resins and plastics are made from coal and oil, and their manufacture therefore depletes our fossil fuel supply just as if these raw materials were burned in a furnace. Because of the difficulty in predicting future trends in metallurgy, in plastics technology, and in the cost of energy, it is difficult to assess the extent to which glass and plastics will replace metals in the future.

Rates of Recycling

Valuable metals and fertilizers are often dispersed and discarded rather than recycled. More efficient recycling

TABLE 8.2 Uses of Copper in Some Common Alloys

ALLOY	COMPOSITION
Bronze	90 % copper, 10 % tin
Red brass	90 % copper, 10 % zinc
Yellow brass	67 % copper, 33 % zinc
Coinage gold	90 % gold, 10 % copper
Coinage silver	90 % silver, 10 % copper
Silver solder	63 % silver, 30 % copper, 7 % zinc

procedures would certainly extend the life of our concentrated mineral reserves. Chapter 15 discusses the complex scientific, economic, social, and political aspects of recycling.

What will happen when a metal such as copper becomes prohibitively expensive, an eventuality expected sometime in the twenty-first century? Copper is now widely used. It is an excellent conductor of electricity and is therefore fabricated into wire and electrical components. Copper is a major component of bronze and brass as well as a minor component of many other alloys. When the price of copper rises precipitously, substitutes for this metal will be needed. Aluminum is also a good conductor of electricity, though not quite so effective as copper. In the future, aluminum wire is likely to replace copper. In fact, even today aluminum is commonly used for household wiring. Bronze and brass will become obsolete, but other alloys will be substituted. In short, economic and technological accommodations will produce some hardships, but the civilized age will not die. This guardedly optimistic outlook should not be interpreted to mean that it is somehow acceptable to use copper wastefully today, because more energy is needed, for example, to replace aluminum with copper. Remember, too, that energy supplies are limited.

8.9 CASE HISTORY: *The Alaska Pipeline*

Every stage in the production and utilization of a fuel or metal is a potential source of pollution. What are the environmental effects of exploiting all these resources? The answer to this question requires not only an estimation of the total quantity of resources available, but also an assessment of the pollution problems inherent in their recovery and use. The Alaska pipeline serves as an illustration of these complexities.

Most of the petroleum that has been drilled in the United States has been found in stable, temperate ecosystems, where the oil is relatively easy to extract, transport, and refine without disrupting large regions of the land surface. As the more accessible sources become exhausted, however, fuels are increasingly sought in more inhospitable places, where all phases of extraction and transportation will cause more severe environmental problems.

The general public has been made aware of these difficulties as a result of the discovery of large quantities of oil and gas near Prudhoe Bay on the north slope of Alaska. Once this oil

Oil rigs in Prudhoe Bay.

of heated oil across Alaska. The northern sections of Alaska through which the pipeline runs is a tundra ecosystem. The land of the tundra consists of a thin layer of topsoil over a layer of permanent frozen subsoil, or permafrost. Plants grow quite slowly in these regions. The soil, which rests on ice, is easily disrupted and slow to be rebuilt by natural processes. The primary large herbivore of the tundra is the caribou, an animal that travels in large herds. The caribou winters in the southern forest and migrates across the tundra to feed on moss and lichens in the spring and summer. The southern section of the pipeline travels through the Alaskan forests.

A careful study of the potential environmental impact of any technical enterprise, such as a dam, a nuclear generating station, or the Alaska pipeline, should consider (a) the adverse effects that occur even if all operations proceed according to plan and (b) the possibility that more severe environmental disturbances may occur as a result of human carelessness or mechanical failures.

The pipeline construction companies have made considerable effort toward preserving the Alaskan wilderness, but the unavoidable bustle and activity associated with construction have already been disruptive. The presence of thousands of workers in an area that was previously almost uninhabited has upset the behavior of wildlife. In addition to unavoidable disturbances, construction workers have shot caribou indiscriminately or run them to death with airplanes or snowmobiles, thereby intensifying the problem.

A new road has been built along the route of the pipeline to facilitate its construction. This first road to bisect Alaska has the potential to provide easy access for hunters and tourists to move across vast wilderness regions previously inaccessible to cars and trucks. Present plans require that when construction of the pipeline is completed, pipeline maintenance personnel shall be transported primarily by helicopter and the road shall be given to the state. The state government now plans to close the road to all vehicular traffic. But laws change, politicians are voted in and out of office, and a road once built is easily reopened.

When construction is completed and the workers return to their homes, activity along the pipeline route will decrease dramatically. Severe problems will remain, however. The pipe itself may pose a barrier to continued caribou migration. The hot oil flowing through the pipe may melt regions of permafrost and cause

is brought to the surface, it must be transported more than 1300 kilometers over land and nearly 3200 kilometers by sea, across some of the largest land and aquatic wildernesses of the world, before reaching the urban centers where it is needed. Will the nature of these wildernesses be disrupted? If so, how severe will that disruption be?

The Alaska pipeline is one of the most ambitious single construction projects in history. Every day it will carry about 200 million liters

225

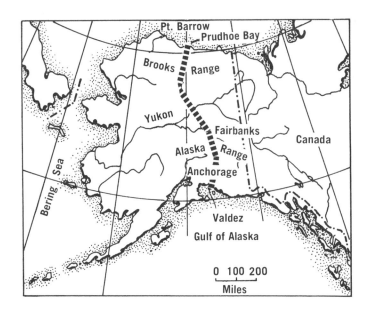

Pipeline route.

the soil to shift and settle. Large sections of pipe have been elevated on trestles to minimize these possibilities. While such precautions will reduce environmental disruptions, they will not eliminate them.

Even worse, the pipe itself may leak or break open. Engineers and contractors have tried to build a system that will not rupture. Sections of pipe have been connected together in a series of S turns (page 227) so the pipe will be able to expand and contract in response to temperature changes or soil movement. The pipe is welded together, and construction plans have directed that all welds are to be subject to

Elevated sections of the Alaskan Pipeline. (Wide World Photos.)

226

a detailed x-ray analysis. Furthermore, a series of pumping stations and shutoff valves have been installed along the route. If the pipe does break, operators should be able to detect the loss of pressure and close the pipeline. Then only the oil in the pipe between the point of rupture and the valve immediately upstream will flow out across the land. But this quantity of oil could be considerable. At the very least, hundreds of hectares of delicate tundra might be destroyed, and rehabilitation would be quite slow; still, the economic or environmental perturbations would be localized rather than widespread. At worst, however, a break in the pipeline could spill millions of liters of oil into the Gulkana River, a commercially valuable salmon spawning ground. The Gulkana flows into the Copper River, which now provides a runway for salmon and supports large summer populations of wild ducks and geese.

Despite a great number of precautions, a variety of human and natural factors combine to make a rupture of the pipeline a real possibility. Most important of these is the fact that the route traverses an active earthquake region. About 30 major earthquakes have struck within the area of the pipeline during the past 75 years, and the probability is high that a large quake will occur sometime during its operating lifetime. If a significant movement of earth occurred during such a natural

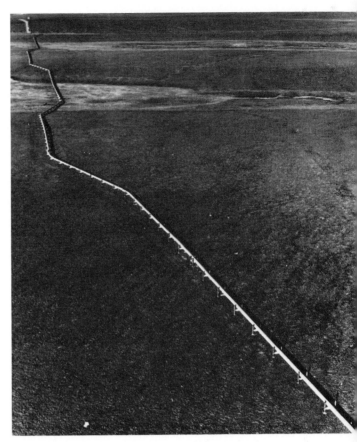

Pipeline zigzags over the tundra so that the pipe will resist fracture if the earth moves. (Wide World Photos.)

The 2300-foot long Yukon River Bridge was opened to truck traffic for the trans-Alaska pipeline project. (Alyeska Pipeline Service Company.) (World Wide Photos.)

227

Icebergs present a hazard to shipping off the coast of Alaska. (Photo by Ken Brewer.)

catastrophe, the pipeline could easily be fractured.

Many of the planned inspection systems required by law have been ignored or the records falsified. People hired to search for flaws in welds have, in some instances, neglected their duties and approved sections of pipe without examining them. Government auditors have uncovered nearly 4000 cases of fraudulent or questionable x-ray examinations, and nearly one quarter of them were located in critical locations such as underneath rivers or flood plains. In one case, pipe buried under a river broke apart and floated to the surface even before it was subject to the stress of flowing oil.

The environmental threat does not end at the pipeline itself. The oil is piped to the port of Valdez (see map on page 226), from which it is to be shipped south in supertankers. The entrance to Valdez harbor is narrow, treacherous, and subject to rapid tidal currents that make navigation difficult. In addition, the Gulf of Alaska just south of the port is a foggy, tempestuous stretch of ocean. Supertankers are cum-

The harbor at Valdez. The first oil arrived at the terminal at 11:02 PM, July 28, 1977. (Wide World Photos.)

bersome, clumsy ships with poor steering and turning abilities. They have been involved in many accidents in the past (see Chapter 14), and most probably more accidents will occur in the future. If a ship is grounded or broken apart it may release large quantities of oil into a beautiful, productive, and delicate ecosystem.

The pipeline may be only the beginning of the development of the Alaska oilfields. Already exploratory crews are finding oil and gas in many other delicate Alaskan ecosystems—mountain ranges, tundra, and ocean bottoms. Even now, new and expanded pipeline routes are being discussed.

The destruction of an ecosystem can be likened to the extinction of a species; in both cases something stable and beautiful, a product of eons of evolution, is lost.

Very few people have traveled in the northern wilderness and paddled by canoe along the Kobuk, Gulkana, or Copper rivers and seen moose in the northern forests, millions of geese nesting in the muskeg country, bald eagles diving for salmon, or caribou darkening the tundra. Yet, with the encroachment of human activity, with the construction of projects like the Alaska pipeline, these areas are being altered and are suffering as a result of the search for increasingly large quantities of fuel. Most of the large wilderness areas in this world have been destroyed. In North America, for example, the herds of bison in the Great Plains have been all but wiped out, the eastern forests have been largely cut, and many species of plants and animals have become extinct. Will the Alaska wilderness meet a similar fate?

TAKE-HOME EXPERIMENTS

1. **Flow.** There are several readily available materials that appear to be solid but tend to flow slowly as a liquid. In this respect their behavior is analogous to that of the material in the plastic layer of the Earth's mantle. Two such solids are road tar and Silly Putty, a commercial product available in many toy stores. Take some tar or Silly Putty and place it in a small pan so that you have a layer about 5 cm thick. Now find a piece of wood and a piece of iron that each weigh about 150 g. Place each weight on your fluid and observe what happens after five minutes, five days, two weeks. Explain.

2. **Work and heat.** Fill a small plastic bucket half full of water. Collect a pile of small pebbles and let the pebbles and the water sit in the same room for several hours so that all the materials are at the same temperature. Record the temperature of the water carefully. Standing as high above the bucket as you can, drop the stones into the water. Record the temperature of the water after about a kilogram of stones have been dropped into it. Can you detect any change of temperature? Explain.

3. **Work and heat.** Take a bowl of cold water and record the temperature. Beat the water vigorously with a fork, or if available, an electric egg beater. Record the

temperature again. What do you observe? Explain.

4. **Soluble minerals in soil.** Mix 25 to 30 g (about an ounce) of soil with enough warm water so that you can pour the mixture into a filter. A coffee filter will do if laboratory filter paper is not available. The filtered liquid should be clear; if it is not, simply pour it back through the same filter and let it run through again. Now evaporate all the water by warming it on the *lowest* setting of a hot plate, or by leaving it to stand overnight on a warm

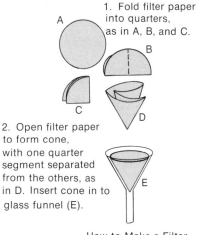

1. Fold filter paper into quarters, as in A, B, and C.

2. Open filter paper to form cone, with one quarter segment separated from the others, as in D. Insert cone in to glass funnel (E).

How to Make a Filter

object such as a radiator. The residue is the soluble mineral content of your soil sample. Note the color. A yellow-brown tint may be due to iron.

To make a further test, dissolve some washing soda (not bicarbonate of soda or baking soda; it must be washing soda, which is a form of sodium carbonate) in a little water. Dissolve your soil minerals in 2 or 3 drops of water and add a few drops of the carbonate solution. The mixture should become cloudy. This material is the insoluble carbonates of the metallic elements in the soil.

PROBLEMS

1. **Structure of the Earth.** Briefly outline the interior structure of the Earth. What regions are solid and brittle? plastic? molten?

2. **Continental drift.** What is continental drift? What evidence supports the theory of continental drift? Discuss.

3. **Continental drift.** Defend or criticize the following statement: Since coal deposits were formed from vegetation of tropical swamp systems, it would be useless to search for coal in polar regions.

4. **Formation of mineral deposits.** It is common for a single mine to contain fairly high concentrations of two or more minerals. Discuss how geological processes might favor the deposition of two similar minerals in a single location.

5. **Formation of mineral deposits.** If one compound is to be separated from a complex mixture, it must somehow be transported away from the rest of the material. How are ores moved out of a mixture in each of the following processes: (a) separation by gravity, (b) separation by differential solubility?

6. **Fossil fuels.** Discuss briefly the formation of both coal and petroleum.

7. **Materials and energy.** Explain why we need petroleum to produce steel, and steel to produce petroleum.

8. **Energy.** What is energy? Which of the following processes involve a transfer of energy? (a) A ball rolls down a hill. (b) A spring is held in a compressed condition for a year. (c) A mouse eats seeds and metabolizes them to keep its body warm. (d) A frictionless wheel rotates around its axle 1000 times. (e) A discarded sandwich rots in the garbage can.

9. **First Law.** Write three statements of the First Law of Thermodynamics.

10. **First Law.** The accompanying sketch shows a "perpetual motion" machine based on osmosis. The two identical membranes are permeable to water but not to sugar. Water from the reservoir passes up through membrane 2 to a height that is determined by the osmotic pressure of the solution. Water also permeates membrane 1, falling back to the reservoir and doing work on the way down. Do you think this machine will work? If not, why not?

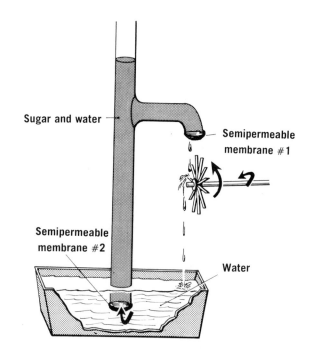

11. **Second Law.** Write four statements of the Second Law of Thermodynamics.

12. **Second Law.** Problem 12 in Chapter 3 reads, "We speak about nutrient cycles and energy flow. Explain why the words cycle and flow are used in their present context." Armed with the message of the Second Law, can you better explain the meaning of energy flow?

13. **Second Law.** Explain why materials can be recycled but energy cannot.

14. **Second Law.** Suppose you had two power plants, one that used $500°\,C$ steam as its working substance and the other that used $150°\,C$ steam. In all other respects both were identical. Each was cooled to $30°\,C$, and both produced the same amount of electricity. Which power plant is more efficient? Which would add more waste heat to the environment? Explain.

15. **Resources.** The curves showing the past and predicted production levels for oil, coal, and gas all rise slowly at first, then rapidly, and then decrease to zero. Explain why all three curves behave in this manner. Also discuss the differences among the three graphs. For example, explain why coal production will continue long after natural gas resources have been depleted.

16. **Oil shales.** Predict some environmental side effects of mining oil shales.

17. **Liquid fuels from coal.** Recent reports indicate that many carcinogenic (cancer-producing) compounds are produced when liquid fuels are synthesized from coal. In addition, over one third of the fuel content of the coal is consumed during the conversion. Using this information to supplement the material in the text, list the categories of direct costs and external environmental costs required to produce liquid fuel from coal. Make a similar list for the production of liquid fuel from petroleum. Which costs do you think will undergo the greatest changes within the next 25 years?

18. **Uranium reserves.** Two geologists, estimating the availability of uranium for fission reactors, arrive at different answers. Explain how such discrepancies can occur.

19. **Uranium mining.** Predict some environmental side effects of mining Sierra Nevada granites for their uranium and thorium content. Compare your answer with the answer to Problem 16.

20. **Energy supply.** Predict some problems of energy production, supply, and use in the year 2200.

21. **Ores.** What is an ore? Explain how an ore deposit can be depleted. How can it be augmented?

22. **Ore deposits.** Briefly discuss five factors that govern the future supplies of metals.

23. **Recycling.** It is often uneconomical to recycle certain metals or fertilizers. Do you feel that economic priorities should be reordered to encourage recycling? If so, can you suggest methods for changing these priorities?

24. **Mineral reserves.** Petroleum is generally burned as a fuel, but in many applications the chemical compounds in the oil are used for the manufacture of plastics and other materials. Explain why petroleum cannot be economically recycled after it is burned, but if it is used for the synthesis of plastics it can be recycled many times.

25. **Depletion of mineral resources.** A noted environmental scientist reported that the world tin reserves may be depleted in 1990. In making this prediction, he assumed that (a) mining technology and world economic activity will remain constant, (b) consumption levels and population will remain constant, and (c) no new deposits will be discovered. From this information alone, do you feel that you can agree with his conclusion? Defend your answer.

26. **Recycling.** Does recycling of automobiles really conserve iron, or does it only conserve energy? Defend your answer.

27. **Recycling.** Sand and bauxite, which are the raw materials for glass and aluminum, respectively, are plentiful in the Earth's crust. If we are in no danger of depleting these resources in the near future, why should we concern ourselves with recycling glass bottles and aluminum cans?

28. **Alaska pipeline.** Analyze critically the statements below.

"[The Department of] Interior estimates that if performance of the oil tankers on the Valdez run were no better than the worldwide average, we could anticipate spills averaging 384 barrels a day." (From "The Living Wilderness," Summer 1970, page 8.)

"The Copper River Basin has the highest density [of waterfowl]; however these high densities occur a mile or more from the route." (From "The Environment," published by the Alyeska Pipeline Service Co.)

29. **Alaska pipeline.** Explain why a pipeline across the tundra is more difficult to construct and more disruptive to the local ecosystem than a pipeline across Texas.

30. **Alaska pipeline.** Do you feel that it would be possible to exploit the Prudhoe Bay oil fields without disrupting the Alaskan wilderness ecosystems? Defend your answer.

BIBLIOGRAPHY

Several books that discuss energy resources include:

Philip H. Abelson (ed.): *Energy: Use, Conservation and Supply.* Washington, D.C., American Association for the Advancement of Science, 1974. 154 pp.

Ford Foundation Report: *A Time to Choose, America's Energy Future.* Cambridge, Mass., Ballinger Publishing Co., 1974. 511 pp.

S. S. Penner and L. Icerman: *Energy—Demands, Resources, Impact, Technology, and Policy.* Reading, Mass., Addison-Wesley, 1974. 373 pp.

Marion Shepard, Jack Chaddock, Franklin H. Cocks, and Charles M. Harmon: *Introduction to Energy Technology.* Ann Arbor, Mich., Ann Arbor Science Publishers, 1976. 300 pp.

A few books dealing with the mineral resources of the world include:

David N. Cargo and Bob F. Mallory: *Man and His Geologic Environment.* Reading, Mass., Addison-Wesley, 1974. 548 pp.

Eugene N. Cameron (ed.): *The Mineral Position of the United States, 1975–2000.* Madison, The University of Wisconsin Press, 1973. 159 pp.

Gary D. McKenzie and Russell O. Utgart (eds.): *Man and His Physical Environment.* Minneapolis, Minn., Burgess Publishing Company, 1975. 387 pp.

Ronald Tank (ed.): *Focus on Environmental Geology.* New York, Oxford University Press, 1973. 474 pp.

National Academy of Sciences Report: *Mineral Resources and the Environment.* Washington, D.C., National Academy of Sciences, 1975. 348 pp.

9

ENERGY:
CONSUMPTION AND
POLLUTION

9.1 ENERGY

Whenever an object is moved, heated, cooled, or chemically altered, energy exchanges are involved. A growing plant, a howling gale, a diesel locomotive laboring uphill or a skier sliding down, a nuclear power plant generating electricity, and a caterpillar crawling along a limb all utilize and release energy in some form. Energy transformations are studied in all disciplines. The environmental scientist is concerned with the economic, social, and ecological consequences of such transformations. As an example, think about a bulldozer consuming gasoline as it pushes aside a pile of dirt to clear an area for the foundation of a house. The prospective occupants of the new

Even a simple act such as digging a foundation for a house raises important environmental questions.

233

house will probably feel that the work is beneficial and productive, for bulldozing is the easiest and cheapest way to push dirt. A neighbor may disagree and see only that the machine is cutting an ugly slash across a previously flowering hillside and squandering precious fuel to do it. The neighbor might argue that the bulldozer pollutes the air, creates too much noise, and exposes the hillside to erosion. Furthermore, the argument might continue, an architect could design a beautiful house that did not require such defacement. Both the newcomer and the neighbor would be able to supply strong arguments to support their positions.

Even in this simple case, it is difficult to judge the relative merits of the two viewpoints. In more complex situations, such as the construction of a nuclear power plant or a new highway system, positions become even more polarized, and the "answers" become even more uncertain. On the one hand, our entire developed society depends on the continued supply of large quantities of energy, while on the other hand we face the depletion of resources, pollution of the environment, and the loss of some of the amenities of life. How do we see our way through the maze? How do we judge, when all factors are considered, whether a given change is valuable or detrimental to society? Of course, no textbook can tell you how to establish values for your own life, but you can begin to learn how to marshal the facts on which to base your decisions.

9.2 ENVIRONMENTAL PROBLEMS CAUSED BY MINING AND DRILLING FOR FUEL

Extracting fossil fuels and transporting them to urban centers always cause some pollution. Coal can be mined either in traditional underground tunnels or in open pits. **Tunnel mines** lie unseen below the surface of the ground, but they are not innocuous. Perhaps most significantly, tunnel mining is dangerous and unhealthy for the miners. Fires, explosions, and cave-ins claim many lives. In addition, fine coal dust suspended in the air enters the miner's lungs, disrupting normal respiration and giving rise to a series of debilitating and often fatal illnesses known collectively as **black lung disease.** Tunnel mines disrupt the flow of ground water and pollute underground streams. They are also often more expensive and less efficient than open pit mines, for coal seams lying between tunnels are often left untouched. However, the surface contours of the land are left undisturbed, and the topsoil is not removed, although occasionally buildings situated above old coal mines have collapsed as the land above the tunnels settles.

To improve the efficiency of mining, coal companies are switching an increasingly large portion of their operations to open pits, otherwise called **strip mines.** In the operation

1. Man without fire
(2000 kcal/day)

2. Primitive agriculture
(12,000 kcal/day)

3. ca. 1860
(70,000 kcal/day)

4. ca. 1970
(230,000 kcal/day)

Energy consumption by people.

of a strip mine, the surface layers of topsoil, subsoil, and rock are first stripped off by huge power shovels or draglines to expose the underlying coal seam. The soil is piled alongside the cut or into an adjacent one that has already been thoroughly mined. The coal is then removed, and a new cut is started. By 1975, 800,000 hectares (2 million acres) of land in the United States had been mined in this unsightly way.

Even though strip mines are generally cheaper for the mining company to operate than tunnel mines, the external environmental costs often run quite high. Unreclaimed strip mines are a multifaceted environmental insult. Neither facts nor figures are needed to convince a discerning person that open, rootless, dirt piles are uglier, less useful, and more liable to water erosion than a natural forest or prairie The uglification is an aesthetic loss to all of us. Erosion silts streams and reservoirs, and kills fish. Fur-

Reclaimed farmland in previously stripmined coal area of central Illinois. The former agricultural productivity of such farms is not always completely recovered.

thermore, sulfur deposits are often associated with coal seams. This sulfur, generally present as iron sulfide, FeS_2, reacts with water in the presence of air to produce sulfuric acid (a highly corrosive and poisonous substance), which runs off, together with the silt, into the stream below.

Perhaps even more crucial is the fact that millions of acres of coal lie under farmland in many parts of the world, especially the fertile wheat fields of central regions of North America. Is the land more valuable as a wheat field that has the potential to produce food indefinitely or as a coal mine? Our present mode of calculating costs favors the extraction of coal over the growing of wheat, for a mining company can buy a farm that lies over a coal seam, mine the coal, destroy the land, and still realize a large profit. Many people feel that this monetary structure, by ignoring economic externalities, does not recognize long-term human needs. A coal company can produce coal at a lower price if a depleted mine is simply abandoned when operations are complete. But the land can be reclaimed at a cost. If the pit is refilled with subsoil, regraded, and topped with fertile soil, argicultural productivity can be regained in as few as five years. Complete soil fertility is not usually achieved, however. When the topsoil is dug up and exposed to air, the organic matter may oxidize. Thus, land that was rich before exposure becomes less fertile Despite these problems, crops can be grown on reclaimed strip mines. The price of coal in 1977 was about $20 to $25 per metric ton. If strip mined lands were restored adequately, the price of coal would rise by about $1 to $1.50 per metric ton.

Strip mining operation. (Photograph by Arthur Sirdofsky.)

Strip mined land.

Are users of coal willing to pay this price? If it is not paid directly, the cost is paid ultimately in the form of rising agricultural prices.

The United States has roughly 16 million hectares (40 million acres) of coal deposits that can be strip mined in the near future. Such vast areas magnify the problems outlined above. A much greater effort at land reclamation should be made. If the true environmental cost of ruined streams and land is levied against the price of coal, an incentive would be provided to rework and replant the land.

In contrast to the disruption caused by coal mining, oil drilling in an area of flat land, such as the plains of Texas, is relatively innocuous. However, as our easily accessible oil fields are being pumped to depletion, the oil industry has begun to invade the delicate ecosystems of the

Mining coal in an underground mine. (Courtesy of Exxon.)

Oil pumper near Bakersfield, California.

Air pollution from burning of fossil fuels.

Arctic and the continental shelves for more petroleum. (See the Case History in Chapter 8.)

Despite careful precautions, accidents seem to occur periodically in all industrial operations, and drilling for oil is no exception. Broken drill pipes, excess pressure, or difficulties in capping a new well have repeatedly led to blowouts, spills, and oil fires. When these have occurred on land, the problems have been locally contained, and the environmental disruptions have been minimized. However, when they occur on offshore rigs such as those currently in operation in the coastal waters of Louisiana and southern California, the result is disastrous. Refer to Chapter 14 for a more detailed discussion of the ecological consequences of oceanic oil spills.

9.3 ENERGY CONSUMPTION AND POLLUTION—AN OVERVIEW

Every barrel of oil, ton of coal, and kilogram of uranium consumed represents multiple sources of environmental disruption. Land is preempted and defaced in mining and in the establishment of transportation routes. The refining of petroleum and the consumption of fuel cause serious air pollution, and the waste heat produced has the potential to disrupt ecosystems or alter world climate. Is it necessary to pollute the environment so severely? Would it be possible to preserve the amenities of civilization and simultaneously improve the quality and cleanliness of the planet? The answer is definitely yes, for there are a great many ways to reduce pollution without undue sacrifice of personal conveniences or comforts.

One approach to reducing pollution is to reduce the source—that is, to reduce energy consumption. We could accomplish this goal by a concerted, widely practiced conservation program. If a large proportion of the population lowered their thermostats to 20°C (68°F) in winter and raised their air conditioner settings to 25°C (77°F) in summer, chose to live in smaller houses that were easier to heat, drove fewer miles ("Is this trip necessary?" or "Could you ride a bicycle instead?"), turned off unused lights, and recycled valuable products, significant quantities of fuel could be saved. Unfortunately, these altruistic solutions have not been particularly effective in recent years. After the fuel shortage of 1973 was alleviated, people in the United States started driving more than ever before, demands for home heating fuel increased, the number of air conditioners and the total power consumed by the average air conditioner increased, and industrial use of energy climbed.

In the developing nations, the rate of increase of energy use has risen even faster than it has in the United

States. Such trends are alarming. However, it is unrealistic to exhort the affluent to do with less, and it is unfair to ask the poor to accept their lot. Therefore, planners must search for other types of answers to the problems of energy and pollution.

A second approach, which would allow us to maintain or even increase our lavish dependency on machines, would be to improve and extend technological methods of reducing energy consumption and controlling pollution. Is it reasonable to expect success from such efforts? How difficult and controversial this question is! Some people feel that technological solutions, alone, can solve environmental problems. They argue, for example, that scientists and engineers can build large, comfortable, pollution-free automobiles that use energy efficiently. Such solutions, they say, simply require time.

These technical arguments may be valid, but most scientists do not believe that we have the time to wait for their realization. Automotive smog is a serious health problem right now, and priceless energy supplies are dwindling at an alarming rate. Perhaps the most promising solution to energy conservation and reduction of pollution involves some compromise — a union of technological, social, and economic changes. If some significant technical improvements can be realized, and if people will accept some changes in their patterns of living, only a small fraction of the energy used today will be needed and only a small fraction of the pollution will be produced, with *very little loss of comfort or convenience.* For example, small cars can operate with one-third the energy consumption of large ones. If people used carpools with three passengers to a car, a small private automobile would become as efficient as a bus. Or consider fuel consumption for home heating. A well-built house with solar collectors installed to carry some of the heating load can easily be built to consume only one quarter of the fuel used in the average home built today. No one would even have to turn the thermostat down. Moreover, the savings in fuel bills would easily pay for the additional construction cost within a short period of time. But few solar heated houses are being built, partly because people wish to avoid high initial costs, and partly because social attitudes are not responsive to these changes.

Throughout the rest of this chapter, we will examine present patterns of energy consumption and study some technologies that can be used to conserve valuable resources while simultaneously reducing pollution.

The bicycle is an energy-efficient means of transportation. (From D. Plowder: *Farewell to Steam.* Brattleboro, Vermont, The Stephen Greene Press, 1966.)

9.4 USE OF SOLAR ENERGY

Every 17 minutes the solar energy incident on our planet is equivalent to human energy needs for a year at the 1975 consumption level. In the *least* sunny portions of the United States (excluding Alaska), an area of only 80 square meters (a square approximately 9 m, or 29 ft, on a side) receives enough sunlight to supply the total energy demands of the average American family. In addition, the technology is available *today* to trap and utilize much of this energy to produce usable power without creating a pollution problem. Yet solar energy is not in common use.

PRODUCTION OF HOT WATER—SOLAR COLLECTORS

A few years ago, one of the authors (Jon) lived in the mountains in a cabin without running water. I set a black plastic pipe in a spring on the hillside above the cabin, and extended the pipe along the sunny surface of the hillside toward the cabin. At the bottom of the pipe I installed a simple faucet. The water which filled the pipe sat in the sun all day and became warm, so that by evening there was always a plentiful supply of hot water for bathing and washing. You can see, therefore, that a solar hot water heater is not necessarily a complex device. Even a bucket of water sitting in the sun all day will become warm. A **solar collector** improves the efficiency of the design. One type of collector consists of a coil of copper pipe welded to a blackened metal base. The whole assembly is covered by a transparent layer of glass or plastic. The operating principle is uncomplicated. Sunlight travels through the glass and is absorbed by the blackened surface. Metal conducts heat readily, so the water in the pipe gets hot. The glass serves as a barrier so that heat trapped within the collector will not readily escape back into the atmosphere (Fig. 9.1).

A solar collector of this type can be easily connected to a home heating system. The hot water produced in the collector is stored in an insulated tank and then piped through radiators to heat the house. Of course, sunshine is not available at night or on cloudy days. Hot water can conveniently be stored to heat a house overnight, but it is expensive to build a system large enough to heat and store enough water to supply heat during several days of cloudy weather. Therefore, most solar systems are installed in conjunction with a conventional furnace that supplies heat during prolonged overcast periods. Naturally, such a dual system is initially more expensive than a simple furnace, but if one calculates the total cost of a solar collector, a furnace, and fuel over a 15-year period, considering all operating costs and including the interest on initial investment, one finds that in nearly all cities of the United States solar heating is economically competitive with other forms of heat. Since the price of fuel is almost certain to rise in the future, the economic advantage of solar heating will increase still further. In addition, solar heat is clean, nonpolluting, and completely renewable. Why, then, aren't solar heaters widely used? Perhaps many people simply are

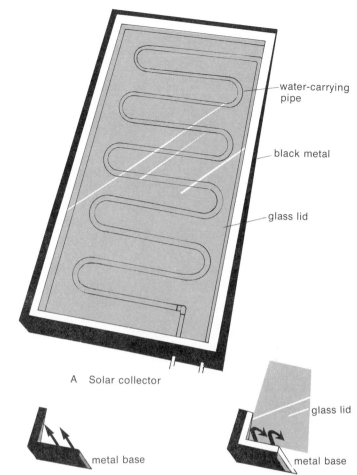

Figure 9.1 Solar collector.

water-carrying pipe

black metal

glass lid

A   Solar collector

metal base

glass lid

metal base

B Without the glass lid, heat escapes rapidly into space.
With the glass lid, much of the heat is retained.

Solar collectors on a residential home. (Courtesy of Energy Systems Division of the Grumman Corporation, manufacturers of Sunstream Solar Collectors.)

not aware that the technology is well established and that units are commercially available. Perhaps the high capital investment outweighs the prospects of low fuel bills. Perhaps people just don't think about it. But a trend does appear to be starting. Several companies are marketing various types of solar units, and solar water heating is finding widespread popularity in Japan, Australia, and Israel.

PRODUCTION OF HIGH-TEMPERATURE STEAM

If you take an ordinary magnifying glass, 10 to 15 cm in diameter, and focus sunlight onto a piece of paper, you can easily burn a hole in the paper. The lens concentrates the solar energy from a large area to a small one so that high temperatures can be realized. Sunlight can also be concentrated through the use of special curved mirrors. A large solar furnace of this type, completed in France in 1970, reaches temperatures up to 3500° C, hot enough to melt any metal. Such a device can be used to make hot steam to drive a turbine and produce electricity. At the present time, it is cheaper to generate electricity in a fossil fuel plant, but again, as the price of fuels rises, the relative costs will certainly change.

SOLAR GENERATION OF ELECTRICITY

In the late 1800's scientists discovered that if light of the proper frequency were shined on certain types of metals, electrons could be knocked off the surface of the metal and ejected into space. Thus, light energy could be converted into electron motion, or electrical energy. Early photoelectric devices were extremely inefficient and were not considered for the production of electric power. However, with the development of transistor technology in the 1950's and 1960's, practical, efficient, **solar cells** were built. These devices are commonly used today to convert sunlight to electricity in spacecraft. They are quiet and troublefree, they do not emit pollution, and they appear to have long life expectancy. Today, it would be entirely possible to build an electric generating station on Earth using solar cells. A 1000-megawatt plant (equivalent to a large fuel-burning facility) would occupy only 10 square kilometers if it were built in the southwestern American desert. The major problem at the present time is that the capital costs are so high that a solar plant is uneconomical even though the fuel is free. But the situation is changing. New processes are being developed to manufacture solar cells more cheaply, and many experts believe that economical units could be in production by the mid-1980's.

Solar furnace, Centre National de la Recherche Scientifique Solar Energy Laboratory, Font-Romen Odeillo.

Of course, solar generating stations alone could never entirely replace fossil or nuclear plants, for the Sun does not shine on cloudy days or during the night. But they could be incorporated into a power system and reduce our dependence on other energy sources.

If techniques are developed to store energy for times when the Sun is not shining, it is conceivable that solar energy will completely replace fossil fuels in the future. Many proposals have been suggested. Perhaps the most practical idea is to use electricity from a solar generating station to dissociate water into hydrogen and oxygen according to the following equation:

$$2H_2O \xrightarrow[\text{energy}]{\text{electrical}} 2H_2 + O_2$$
$$\text{water} \qquad\qquad \text{hydrogen} \quad \text{oxygen}$$

Hydrogen is a versatile and useful fuel that can burn in air and could be used as a replacement for gasoline and other fluid fuels. The only product of combustion is water.

Unfortunately, rapid development of solar cell technology is being hampered by a shortage of money allocated for research. In the United States, in 1975, the solar energy research budget amounted to $50 million, which is only about six percent of the money spent on research for nuclear fission.

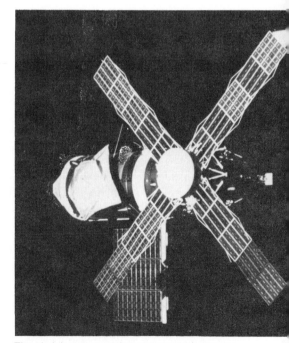

The skylab space station, powered by a group of photoelectric cells mounted on "windmill" arms. (Courtesy of NASA.)

9.5 OTHER RENEWABLE ENERGY SOURCES

HYDROELECTRIC ENERGY

Today, only four percent of the power used in the United States is derived from hydroelectric sources. The Sun evaporates water at sea level, and the winds bring it to the mountains, where it collects into streams with potential to fall and release energy.

Thus, hydroelectric energy is an indirect form of solar energy. As more sites are exploited, more power will be tapped by harnessing falling water. Even if the overall consumption of energy continues to rise, however, the relative importance of hydroelectric energy will probably not increase much. There simply are not enough good locations for building large dams, and those that are available are often in recreational areas. Furthermore, damming the Grand Canyon or the Snake River Gorge would be a great aesthetic loss.

Energy in falling water.

GEOTHERMAL ENERGY

Energy derived from the heat of the Earth's crust, called **geothermal energy,** has gained attention in the past few years. In various places on the globe, such as in the hot springs and geysers of Yellowstone National Park in Wyoming, hot water is produced near the surface. Although no one suggests harnessing Old Faithful for generating electricity, there are several places where hot (250° C, 482° F) underground steam is available for the price of a deep well. The Pacific Gas and Electric Company is producing some electricity from a generator connected to wells in central California. This project is new in the United States, but an Italian facility at Larderello has been in operation since 1913 and works at a capacity of 400,000 kw. Even optimistic supporters, however, do not expect a large portion of our power to be supplied in this manner, for there are not enough hot springs at or near the surface of the earth. Additionally, continuous exploitation for more than a century or two is expected to exhaust the water or heat content of these wet wells. Furthermore, geothermal energy is not always free from pollution. Underground steam or hot water is often contaminated with sulfur compounds, which must be removed before they are discharged to the air or to a lake or river.

A more ambitious proposal is to drill 1.5 to 6 kilometers through the Earth's crust into its molten core. New types of drills have been devised that melt their way through bedrock. If the core could be reached, water could be poured down the hole and the resulting steam withdrawn. If successful, geothermal tappings of this type could satisfy

Section of the geothermal steam field where Pacific Gas and Electric Company generates 396,000 kilowatts of electricity from underground steam.

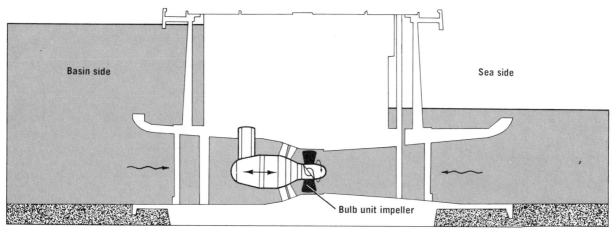

Tidal dam and turbine.

requirements for energy far into the future. Great care must be exercised in such drillings, for the deep wells and large quantities of water flowing through the holes might cause slippage of rock layers, and hence earthquakes.

ENERGY FROM THE TIDES

The force of the tides has long been an intriguing source of power. Imagine a turbine which is immersed in the sea near shore. Electricity could be generated as the tides turn the turbine blades, but the output from a single generator would be small. Moreover, lining the coast with millions of small generators would be uneconomical.

An alternative practical solution is to build a tidal dam across a bay or estuary where there is a large rise and fall of the tides. By damming and funneling the water, significant concentrations and economical power output can be realized. A tidal generating plant in France on the coast of Brittany generates 240,000 kw, about the output of a small fossil-fueled plant. If we should exploit this resource further, however, we shall have to reckon with the repercussions on aquatic life that would inevitably result from the construction and maintenance of dams across the estuaries. Also, there are not enough properly oriented bays and estuaries to provide more than a small percentage of our energy requirements.

ENERGY FROM THE WIND

The power of wind has been used since antiquity to pump water and grind grain. Construction of large-scale electric generating stations of this type would pose serious engineering problems because of their huge size, but small windmills as an auxiliary source of home electricity would be economical and would conserve other sources of power.

"I mean, like, don't you think it kind of *weird* to get electricity from our windmill-powered generator and then use it to watch reruns of 'I Love Lucy'?"

245

Wind generators.

Of course, the wind does not always blow, and therefore some form of energy storage would be needed to provide power during calm periods.

ENERGY FROM THE BURNING OF GARBAGE

Another practical proposal is to burn garbage as a renewable energy source. We shall see in Chapter 15 that half of the solid waste of the United States is paper. Burning it would both provide energy and significantly reduce the problem of solid waste disposal. In fact, a Parisian company has been burning solid wastes to generate electricity for over 50 years. At the present level of operation, a total of 1.5 million metric tons of garbage is consumed per year as an auxiliary fuel source. In the United States in 1975, there was an estimated 900 million metric tons of dry solid waste, of which only 140 million metric tons was actually recoverable for use as a fuel, the remainder being difficult and uneconomical to collect and transport. Even this 140 million metric tons could provide energy equivalent to 28 billion liters of oil (15 percent of the 1975 consumption in the United States) or 38 billion cubic meters of natural gas (38 percent of the total). Saving 28 billion liters of oil in a year while simultaneously reducing solid waste pollution is a healthy approach to environmental problems. Once again, our failure to exploit a valuable resource is a measure of societal inertia. Individual and industrial patterns are slow to change, even when such change is beneficial to all.

Clearly, the burning of garbage is not a panacea. But if garbage, solar, geothermal, tidal, and wind energies were all exploited, along with the implementation of various social remedies outlined later in this book, the total contribution could be very substantial. In fact, these energy sources, if exploited extensively and used conservatively could altogether eliminate the need for fossil fuels.

9.6 ENERGY CONSUMPTION AND CONSERVATION FOR TRANSPORTATION

The two *least* efficient modes of transportation, the automobile and the airplane, are the two fastest-growing industries in the transportation field. Consider seven means of moving people: bicycles, walking, buses, trains, automobiles, "jumbo" jet passenger planes, and supersonic transport (SST) planes. Table 9.1A shows the energy required to drive each of these; the units are expressed in passenger kilometers per liter (pkm/l) of petroleum fuel.

Consider also various means of moving freight. These are expressed in relative units that are dimensionally similar to those cited above, namely the number of kilometers that a given amount of freight can be transported by a liter of fuel. The data are shown in Table 9.1B. The implications of these data are clear: Reliance on inefficient methods of transportation is contributing to the rapidly accelerating depletion of our fossil fuels.

It is relatively easy to outline a series of proposals that would reverse this dependency on inefficient modes of transportation. Millions of barrels of fuel would be con-

TABLE 9.1 Energy Requirements for Transporting People and Freight

A. PEOPLE*

Bicycle	425 pkm/l
Walking	280 pkm/l
Intercity bus	35 pkm/l
Train	20 to 85 pkm/l, ranging from poorly patronized intercity routes to commuter trains
Automobile	4n to 16n pkm/l, depending on model and condition of car, where n = number of passengers
Jumbo jet	8 to 10 pkm/l, for estimated average passenger loads
SST	5.7 pkm/l, for estimated passenger loads

B. FREIGHT**

Pipeline	93 km/l
Railroad	63 km/l
Waterway	62 km/l
Truck	11 km/l
Airplane	1 km/l

*In passenger-kilometers per liter of fuel (pkm/l).
**The value of 1 km/l for air freight is used as a standard for reference.

Noonday traffic in Boulder, Colorado.

served *daily* if people used **mass transit** more frequently, drove smaller, more efficient cars, used carpools, changed their living habits so that homes were located near places of work, built and used safe, attractive bicycle paths, and placed increased reliance on railroads for freight. Outlining such a plan is easy; implementation is difficult.

In many places, especially in the United States, the automobile has modified our styles of living. Not only do personal habits embrace automobile use but also the very layout of most suburbs, towns, and cities encourages use of the private car. For example, consider two localities in Colorado: Boulder, an expanding city, and Telluride, a

There is only one paved street in Telluride, Colorado, and even here traffic is sparse.

small mining and ski town, which has seen few major structural changes in the last quarter century. Telluride is centralized. The grocery store, clothing store, hardware store, post office, bank, and in fact the entire town is concentrated within an area of a few square blocks. Thus, it is easy to do all one's shopping on foot. In Boulder, on the other hand, there are no food stores in the central part of the city; a person shopping in the downtown mall area must travel nearly 1.5 kilometers to buy groceries. Bus service, though available, is slow. Since most people do not care to wait long or to walk far burdened with packages, they prefer to drive automobiles. Moreover, any change of the social order would involve a major, very expensive reorganization of the city.

Many people live in suburbs and commute to large cities to work. Billions of dollars have been spent on superhighway systems to link the two areas. We cannot simply move the factories or housing developments, nor can we change highways to railroad tracks without a tremendous outlay of capital, so once again, one of the least efficient modes of transportation—the automobile—carries the largest portion of the traffic.

The situation is so well established that significant improvements are extremely slow. Approximately 97 percent of urban traffic is now carried by the automobile, and only 3 percent is carried by public transportation systems—buses, subways, and trolleys. Thus, even a significant increase in mass transit has little effect on the total fuel consumption in a city. For example, in Atlanta (Georgia) city officials lowered bus fares and increased the number of routes served by the municipal bus lines. As a result, ridership increased by 28 percent, a significant amount; but that 28 percent increase was based on only 3 percent of the total traffic. Meanwhile, over 96 percent of the traffic was still carried by automobiles, and the total fuel consumption for transportation was reduced by only

Not only are present transportation systems well entrenched in people's habits, but they are set in concrete as well.

BART, a modern mass transit system in San Francisco.

one half of one percent. These data are not meant to discourage the construction of mass transit systems but rather serve to emphasize the difficulty of changing well-established patterns. Significant effort must be made *now* toward building mass transit systems so that in 25 years, when fuel shortages become even more acute, people's habits and the structures of cities will be geared toward less waste.

Not only is the automobile inefficient in fuel consumption but also it uses space inefficiently. A two-track local subway uses a roadbed 11 meters wide and can carry 80,000 passengers per hour On the other hand, an eight-lane superhighway is 38 meters wide and carries only 20,000 people per hour under normal traffic conditions. A superhighway capable of carrying 80,000 people per hour

Cable car in San Francisco.

Rush hour traffic in Denver, Colorado.

would have to be 152 meters wide (approximately 1½ times as wide as the *length* of a football field).

Many modern cities are becoming so congested that the automobile is no longer an efficient means of transportation. An automobile is designed to provide direct, quick, comfortable, and private transportation. There is no need to walk to a bus stop, wait for the bus, and then mix with a heterogeneous group of commuters and shoppers. However, in recent years, expressways have become so crowded that commuting by automobile has lost much of its advantage. During rush hours, highways are choked with automotive smog, traffic is often stalled for long periods of time, and accidents are common. Moreover, parking spaces are often hard to find in urban centers so people often cannot drive directly to their destination anyway. If quiet, efficient, comfortable mass transit systems were built in most cities, and if roadbuilding projects were curtailed, mass transit might become more popular.

9.7 ENERGY CONSUMPTION AND CONSERVATION FOR INDUSTRY

Industrial power consumption involves a wide variety of products and processes. In Table 9.2 industrial consumers are listed in decreasing order of their energy demands. Manufacturing processes are using increasingly

**TABLE 9.2 Industrial Energy Consumers
in the United States**

TYPE OF INDUSTRY	TOTAL ENERGY PER YEAR ($\times 10^{18}$ joules)
Primary metals	5.6
Chemicals and chemical products	5.2
Petroleum refining	2.9
Food and kindred products	1.4
Paper and allied products	1.4
Stone, clay, glass, and concrete	1.3
All other industries	8.5

large amounts of energy because both the demand for goods and the energy needed per item are increasing. One reason that the energy needed per item is so high today is the trend toward increased automation to offset labor costs. Another factor, however, is more foreboding. As a natural resource like iron ore becomes depleted. lower-grade ores become commercially more attractive. The mining, purification, and metallurgy of low-grade ores requires more energy than the same operations performed on high-quality deposits.

One way to reduce industrial energy consumption is to increase recycling of material. In general, production of manufactured goods from recycled wastes consumes less energy than the production from raw materials. However, total manufacturing costs reflect the price of labor, transportation, and capital investment as well as fuel bills, and unfortunately, when economic externalities are neglected, recycling operations are often more expensive than primary production. Unless industries are offered economic incentives or threatened with penalties, we can expect little immediate increase in industrial recycling.

Many aspects of the problem of industrial energy consumption are closely linked with the problems of solid waste disposal, for most manufactured goods are ultimately discarded. Thus, reevaluation of planned obsolescence, the reduction of the amount of paper used in packaging, and generally reduced consumption would greatly alleviate environmental stresses caused by manufacturing and waste disposal.

9.8 ENERGY CONSUMPTION AND CONSERVATION FOR HOUSEHOLD AND COMMERCIAL USE

Most of the residential and commercial uses of power involve the burning of fossil fuels to heat air and water, and the consumption of electricity for other purposes. Of course, solar energy could be easily used for many of these applications.

A modern petroleum refinery.

Fire for heating is the most ancient use of fuel. Without space heating, people could not live in some areas they now inhabit. Today, space heating, which accounts for only 12 percent of our total energy consumption (compared to 16 percent for the automobile, for example) remains one of the most efficient uses of fuel. Home furnaces can be built to operate at 65 to 85 percent efficiency, a figure well above that for electric generating stations or automobiles (Fig. 9.2). However, these efficiencies are seldom realized in the home because most furnaces are poorly adjusted. Moreover, most homes are poorly insulated. Building in most areas in the United States is regulated by strict legal codes. Yet the insulation requirements of the Federal Housing Authority are below the recommendations of the insulation manufacturers. Compliance with the manufacturers' suggestions would reduce fuel consumption, and hence fuel bills, by about 33 percent. The increased investment in insulation would be recovered in the first or second year.

Despite the high efficiency of home furnaces, gas and oil units do pollute the air, and the control of pollution on a house-by-house basis would be very costly. Therefore, there is an increasing trend toward the use of "clean, efficient, electric heat," as expressed by electric utility advertising slogans. If "electric heat" is taken to mean heat generated by electrical resistance, as in a giant electric oven, then the process in the home is, in fact, clean and 100 percent efficient. Of course, the pollution is simply released at the site of the electric generating station instead of in the home. Section 9.10 shows that commercial electric generating stations are only 41 per cent efficient at best. Including losses in transmission, we cannot expect more than about 33 percent efficiency as delivered to the home.

The technology is available to make improvements toward the realization of clean, efficient electric heat, if we are willing to pay the price. Consider the household refrigerator, which we normally think of as being only a cooling machine but which actually is a device that uses fuel to pump heat from the cold interior to the warmer exterior. Therefore, the refrigerator is just as much a warming machine as a cooling machine. Fuel is needed because, according to the Second Law of Thermodynamics, heat would normally flow the other way, from the hot outside room into the cold refrigerator, as in fact it does when the electricity fails. If one stands behind a refrigerator while it is operating, one can feel the warm air being pumped out. Heat pumps that can heat a room by cooling the outside are, in fact, commercially available, and can be designed to achieve efficiencies three times as great as electric resistance heaters. Thus, the overall efficiency of an electric heat pump, including calculation of the inefficiencies of the power plant, is competitive with home furnaces. However, high installation costs have discouraged the wide application of heat pumps.

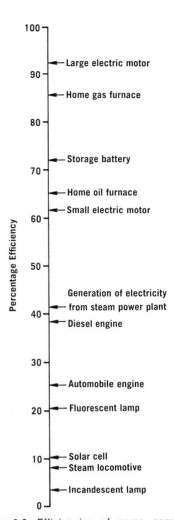

Figure 9.2 Efficiencies of some common machines, devices, and processes. The values for electrical devices do not include the energy losses at the generating station.

Personal comfort depends not only on temperature but also on the humidity and air flow in a room. Additionally, certain forms of radiation can produce comfort even if the ambient temperature is low. Thus, if you sit in bright sunlight on a winter day you can feel warm even at temperatures at which the moisture from your breath condenses. Likewise, a person may feel as comfortable at 20° C in front of a cast-iron stove as at 25° C in a room warmed by hot air. If the best available engineering were used in home heating, fuel bills could be lowered.

The relationship between temperature and comfort is also dependent on habits and customs. According to research carried out by The American Society of Heating and Ventilating Engineers in 1932, the preferred room temperature during the winter for a majority of subjects was 19° C. Similar research at later dates showed that the comfort range had risen to 19.5° C in 1941 and to 20° C by 1945. In 1975 most homes, schools, and offices were heated to 21 or 22° C. Part of the change is due to the fact that 35 years ago people wore sweaters and long underwear indoors. If one lives through several winters at 19° C, then 21° C seems uncomfortably warm. In many areas of the world, fashions are responsible for increased energy consumption. For example, men feel constrained to wear jackets, ties, and long pants in business offices during the summer and are then forced to turn up the air conditioners, and similarly women wear dresses in winter, and turn up the heat to warm their legs!

Such needless consumption is encouraged by advertising campaigns and even by the price structure for electric power. According to current pricing practice, electricity becomes cheaper per kilowatt hour as demand is increased. In a typical example, the first 40 kwh cost ten cents per kilowatt hour, and any consumption above 900 kwh costs only two cents per kilowatt hour. This rate struc-

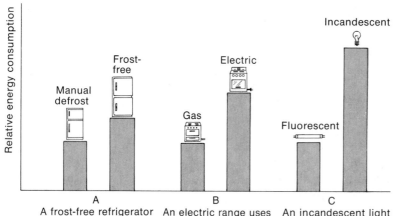

A
A frost-free refrigerator uses 1½ times as much energy as a convention-al one

B
An electric range uses twice as much energy as a gas range

C
An incandescent light uses three times as much energy as a fluorescent light

Figure 9.3 Fuel consumption in the home could be reduced without any loss of comfort if more efficient appliances were used.

ture naturally encourages additional use and must be re-evaluated. If the system were reversed, and prices per kilo-watt hour rose with increased demand, people would be more reluctant to use electricity to perform tasks that were done manually 10 years ago. As a result, such a system would undoubtedly slow the rapid rise of per capita consumption. Of course, different rate structures might be necessary for industrial consumers, or many manufactured goods might become too expensive to produce. On the other hand, one could argue that such goods *should* be more expensive; otherwise, nonbuyers are subsidizing them.

From fuel mining to fuel transportation to fuel consumption to energy transmission, the environment always loses when energy is consumed. The problems are magnified by population density and increased demand. An electrical generating facility in Great Falls, Montana, in 1920 would have produced all the pollutants inherent in a large modern facility, but space, air, and clean water were so plentiful, and the population was so sparse, that it would have been difficult to detect any environmental effects. No one would have believed that a serious problem could exist. A cowboy standing on a butte two days out of town could scan a circle a hundred miles in diameter and see almost no sign of people. This concept of inexhaustibility gave rise to what economist Kenneth Boulding calls "cowboy economics." Early Americans thought that since you can neither damage nor deplete the environment, you may as well take the most convenient route to your goal. Today we

TABLE 9.3   Homes with Selected Electrical Appliances: 1952 to 1971*

ITEM	1952 NUM-BER	1952 PER-CENT	1960 NUM-BER	1960 PER-CENT	1965 NUM-BER	1965 PER-CENT	1970 NUM-BER	1970 PER-CENT	1971 NUM-BER	1971 PER-CENT
Total number of wired homes	42.3	100.0	51.7	100.0	57.6	100.0	64.0	100.0	65.6	100.0
Air-conditioners, room	0.6	1.3	7.8	15.1	13.9	24.2	26.0	40.6	29.2	44.5
Bed coverings	3.6	8.6	12.2	23.6	20.0	34.7	31.7	49.5	33.5	51.1
Can openers	(NA)**	(NA)	2.5	4.8	14.2	24.7	29.1	45.5	31.5	48.1
Dishwashers	1.3	3.0	3.7	7.1	7.8	13.5	17.0	26.5	19.4	29.6
Dryers, clothes (includes gas)	1.5	3.6	10.1	19.6	15.2	26.4	28.6	44.6	31.2	47.6
Freezers, home	4.9	11.5	12.1	23.4	15.7	27.2	20.0	31.2	21.4	32.7
Mixers	12.6	29.7	29.0	56.0	41.9	72.8	52.8	82.4	55.3	84.4
Refrigerators	37.8	89.2	50.8	98.2	57.3	99.5	63.9	99.8	65.4	99.8
Television: Black and White	19.8	46.7	46.2	89.4	55.9	97.1	63.2	98.7	65.4	99.8
Color	(x)†	(x)	(NA)	(NA)	5.5	9.5	27.2	42.5	33.5	51.1
Washers, clothes	32.2	76.2	44.1	85.4	50.3	87.4	59.0	92.1	61.8	94.3
Water heaters	5.8	13.8	9.8	18.9	13.5	23.4	20.2	31.6	22.2	33.8

*Wired homes in millions, as of December 31, 1971. Percentages based on total number of homes wired for electricity. (From *Merchandising Week,* annual statistical issues. New York, Billboard Publications, Inc., 1972.)

**Not available.

†Not applicable.

'Congressman, you forgot your energy policy!'

know that we can damage and deplete our planet. Old attitudes must therefore be re-evaluated.

9.9 ELECTRICITY

Few issues in the field of environmental science have received so much attention as the problem of generating electrical power. The production of electricity leads to water pollution, air pollution, solid wastes, radioactive wastes, land uglification, and thermal pollution. Yet our present society could not operate without electric power, because lighting, entertainment, home conveniences, many industrial processes, and thousands of cheap, relatively silent and trouble-free motors are dependent on it.

Electrical consumption has risen faster than any other form of energy consumption in the past few years and will probably continue to accelerate rapidly. One cannot reap the benefits of electricity without some environmental side effects, but the use of innovative technology can certainly reduce the environmental insult.

Most of our electricity is now produced in steam turbines. The operating principle here is uncomplicated. Some power source, such as coal, gas, oil, or nuclear fuel, heats water in a boiler to produce hot, high-pressure steam. The steam expands against the blades of a turbine, and the spinning turbine operates a generator which produces electricity. The steam, its useful energy now spent, flows into a condenser where it is cooled and liquefied, and the water

"The egg timer is pinging. The toaster is popping. The coffeepot is perking. Is this it, Alice? Is this the great American dream?" (Drawing by H. Martin; © 1973 The New Yorker Magazine, Inc.)

is returned to the boiler to be reused. The cooling action of the condenser is essential to the whole generating process.

Recall from Chapter 8 that whenever heat is converted to work, the amount of work produced always depends on the temperature difference between the hot and cold parts of the engine. Maximum efficiency is therefore realized when a very hot working substance can discharge its waste heat to a very cold sink. Therefore, some provision *must* be made to maintain a cool environment at one end of the engine. These relationships are shown schematically in Figure 9.5. In practice, cooling is accomplished by circulating water around the condenser. Maximum efficiency is obviously reached with very hot steam and a very cool con-

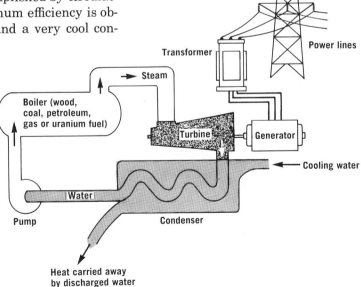

Figure 9.4 Power generator (schematic), showing discharge of waste heat.

denser. In practice, the nature of the metals in the turbine limits the temperature to a maximum of about 540° C (1000° F). The low temperature is limited by the cheapest coolant, generally river, lake, or ocean water. Within the constraints of these two limits, an efficiency of 60 percent is theoretically possible, but uncontrollable variations in steam temperature, and miscellaneous heat losses, reduce the efficiency to about 40 percent, even in the best installations. Thus, for every 100 units of potential energy in the form of fuel, 40 units of electrical energy are available as useful work and 60 units of energy are dissipated to the surroundings as heat. Moreover, because refrigeration of the condenser is too expensive, further increases in efficiency depend on the use of hotter steam. Since the maximum upper temperature is limited by the ability of metals to withstand the heat stress, efficiency cannot be expected to increase appreciably unless some new breakthrough in metallurgy takes place.

An electric generating system that converts the internal energy of a fuel into work is, in effect, a heat engine. Although a modern generator is only 40 percent efficient, it is more efficient than other common heat engines, such as a large diesel (38 percent), an automobile (25 percent), or a steam locomotive (9 percent). If electricity is used to perform mechanical work, say to drive an electric motor,

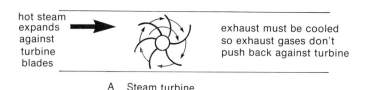

A   Steam turbine

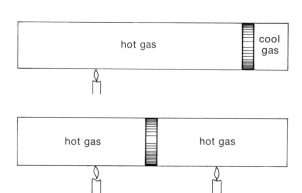

B   Piston analogy. Recall from Figure 8.6 that the piston moves in response to a *difference* in temperature between the two sides

Figure 9.5 The exhaust gases in a turbine must be cooled!

the overall efficiency of the total system is higher than the efficiency of a small internal combustion engine. Thus an electric chainsaw is more efficient than a gasoline-powered one. But if electricity is used for heat, as in an electric stove or space heater, the overall efficiency is lower than if a fuel is burned directly. A propane stove is twice as efficient as an electric stove because the propane does not convert heat to work. (Recall that the efficiency of all heat-to-work conversions is constrained by the Second Law of Thermodynamics.) Yet many home electrical appliances are heating devices. Considerable quantities of energy could be conserved if electricity were used only in situations where it is the most efficient source of energy, and if fuels were used directly when heat is needed.

9.10 THERMAL POLLUTION

The amount of heat that must be removed from an electrical generating facility is quite large. A one-million-kilowatt plant running at 40 percent efficiency would heat 10 million liters of water by 35° C (63° F) every *hour*. It is not surprising that such large quantities of heat, added to aquatic systems, cause ecological disruptions. The term **thermal pollution** has been used to describe these heat effects.

The processes of life involve chemical reactions, and the rates of chemical reactions are very sensitive to changes in temperature. As a rough approximation, the rate of a chemical reaction doubles for every rise in temperature of 10°C (18°F).

We know that if our own body temperature rises by as much as 5°C (or 9°F), which would make a body temperature of 42°C (107.6°F), the fever may be fatal. What then, happens to our system when the outside air temperature rises or falls by about 10°C? We adjust by internal regulatory mechanisms that maintain a constant body temperature. This ability is characteristic of warm-blooded animals, such as mammals and birds. Thus, the body temperature of a child or a canary in a room cannot be determined by reading the wall thermometer. In contrast, non-mammalian aquatic organisms such as fish are cold-blooded: that is, they are unable to regulate their body temperatures as efficiently as warm-blooded animals.

How, then, does a fish respond to increases in temperature? All its body processes (its metabolism) speed up, and its need for oxygen and its rate of respiration therefore rise. The increased need for oxygen is especially serious, since hot water has a smaller capacity for holding dissolved oxygen than cold water. Above some maximum tolerable temperature, death occurs from failure of the nervous sys-

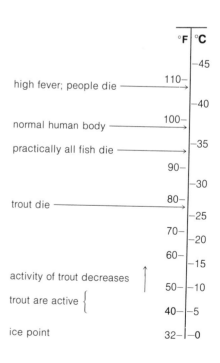

For many years, the word pollute has meant "to impair the purity of," either morally* or physically.** The terms air pollution and water pollution refer to the impairment of the normal compositions of air and water by the addition of foreign matter, such as sulfuric acid. Within the past few years two new expressions, thermal pollution and noise pollution, have become common. Neither of these refers to the impairment of purity by the addition of foreign matter. Thermal pollution is the impairment of the quality of environmental air or water by raising its temperature. The relative intensity of thermal pollution cannot be assessed with a thermometer, because what is pleasantly warm water for a man can be death to a trout. Thermal pollution must therefore be appraised by observing the effect on an ecosystem of a rise in temperature. Similarly, noise pollution has nothing to do with purity: foul air can be quiet, and pure air can be noisy. Noise pollution (to be discussed in Chapter 16) is the impairment of the environmental quality of air by noise.

---

*(1857.) Buckle, *Civilization*. I., viii, p. 526: "The clergy . . . urging him to exterminate the heretics, whose presence they thought polluted France."
**(1585.) T. Washington. trans. *Nicholay's Voyage*, IV, ii; p. 115: "No drop of the bloud should fall into the water, least the same should thereby be polluted."

tem, the respiratory system, or essential cell processes. According to the Federal Water Pollution Control Administration, almost no species of fish common to the United States can survive in waters warmer than 34° C. The brook trout, for example, swims more rapidly and becomes generally more active as the temperature rises from 4 to 9° C. In the range from 9 to 15.5° C, however, activity and swimming speed decrease, with a consequent decline in the trout's ability to catch the minnows on which it feeds. This inactivity is more critical because the trout *needs* more food to maintain its higher metabolic rate in the warmer water. Outright death occurs at about 25° C. In addition, spawning and other reproductive mechanisms of fish are triggered by such temperature changes as the warming of waters in the spring. Abnormal changes, to which the fish is not adapted, can upset the reproductive cycle.

In general, not only the fish but also entire aquatic ecosystems are rather sensitively affected by temperature changes. Any disruption of the food chain, for example, may upset the entire system. If a change of temperature shifts the seasonal variations in the types and abundances of lower organisms, the fish may lack the right food at the right time. For example, immature fish (the "fry" stage) can eat only small organisms, such as immature copepods. If the development of these organisms has been advanced or retarded by a temperature change, they may be absent just at the time that the fry are totally dependent on them.

Higher temperatures often prove to be more hospitable for pathogenic organisms, and thermal pollution may therefore convert a low incidence of fish disease to a massive fish kill as the pathogens become more virulent and the fish less resistant. Such situations have long been known in the confined environments of farm and hatchery ponds, which can warm up easily because the total amount of water involved is small. As thermal pollution in larger bodies of water increases, so will the potential for increased loss of fish by disease.

Aquatic ecosystems near power facilities are subject not only to the effects of an elevated average temperature but also to the thermal shocks of unnaturally rapid temperature changes. Power generation and heat discharge vary considerably from a peak in the afternoon to a low in the hours between midnight and daybreak. Additionally, a complete shutdown of a day or longer occurs occasionally. Thus, the development of cold-water species is hindered by hot water, and the development of hot-water species is upset by the unpredictable flow of heat. In addition, both types of organisms are adversely affected by rapid temperature changes.

Additional disruptions can occur because hot water has a reduced oxygen content. Power plants are usually located near population centers, and many cities dump sewage into rivers. Since sewage decomposition is dependent on oxygen, hot rivers are less able to cleanse themselves than cold ones. The combination of thermal pollution with increased nutrients from undecomposed sewage can lead to rapid and excessive algal growth and eutrophication (see Chapter 14). Therefore, thermal pollution imposes the unhappy choice of dirtier rivers or more expensive sewage treatment plants.

Synthetic poisons, too, become more dangerous to fish as the water temperature rises. First of all, toxic effects are accelerated at higher temperatures. Second, as just mentioned, warm water favors increased growth of plant varieties such as algae. The algae tend to collect in the power-plant condensers and reduce water flow efficiency The electric company responds by periodically introducing chemical poisons into the cooling system to clean the pipes. These poisons are then mixed with the downstream effluent. Additionally, domestic and industrial water consumers are more apt to discharge treatment chemicals (such as copper sulfate) into water with high algae concentrations than into clean water. Thus, in warmer water, not only are fish less likely to resist poisons but they are also likely to be exposed to them more.

The Second Law of Thermodynamics assures us that we cannot invent a process to avoid the production of this excess heat. We can, however, reduce the amount of heat wasted or we can put it to good use. Some suggestions are described in the following paragraphs.

"It's not the humidity—it's the thermal pollution." (*American Scientist*, Sept.-Oct., 1971.)

HOW TO DECREASE THE THERMAL LOAD

Since solar energy imposes no additional heat on the environment, increased reliance on sunlight, especially by the construction of solar generating stations, would decrease thermal pollution (see Section 9.5).

As mentioned previously, mechanical losses account for a 15 to 20 percent loss of efficiency in conventional power plants. Thus, thermal pollution could be reduced if power plants operated at higher efficiency. A promising new design that increases efficiency is based on the principle of **magnetohydrodynamics** (MHD). In this system, air is heated directly and is seeded with metals such as potassium or sodium,

$$K + heat \longrightarrow K^+ + e^-$$

| potassium atom | potassium ion | negative electron |

which lose electrons at high temperatures. This hot, electrified air stream is allowed to travel through a large pipe that is ringed with magnets. The movement of these

An experimental MHD generator. (Courtesy of General Electric.)

charged particles through a magnetic field generates electricity. Furthermore, the exhausted hot air can operate a conventional turbine, thus producing additional electricity. The advantages of this system are twofold. First of all, the overall efficiency of the system is expected to reach 60 percent, and secondly, most of the waste heat is dissipated directly into the air rather than into aquatic ecosystems. The feasibility of the technique has been shown by the Russians, who now operate a 250,000-watt MHD generator near Moscow.

HOW TO PUT WASTE HEAT TO GOOD USE

Large quantities of hot water can be used for diverse purposes. Hot water can improve growth conditions in large greenhouses. The costs of the greenhouse and water handling systems are high, and the economic feasibility might ultimately depend on the accommodation that can be made between the farmer and the power companies. Unfortunately, owners of greenhouses don't need hot water in the summer, a time when power demands for air conditioning are high and the deleterious effects of thermal pollution on aquatic life are apt to be most severe.

Success in utilizing hot water for irrigating open field crops has been reported in the state of Washington. Alternatively, hot water circulating in closed pipes can be used to heat the soil without irrigating it; the resulting benefit is a higher agricultural yield, as shown by the data in Table 9.4.

The practice of adding chemical poisons to kill algae in the condensers is incompatible with the use of that water for irrigation, although not incompatible with closed-system soil heating. If a power plant were to shut down during a cold snap, the flow of hot water would suddenly halt and a whole crop could be lost.

Another possible use of excess hot water is in raising fish. "Agriculture" in the water, or **aquaculture,** is the science and practice of raising fish in artificially controlled ponds or pools. The growth rate of certain fish (not all) can be enhanced markedly by raising them under carefully controlled temperatures. Proponents of aquaculture feel that high yields of food can be realized if cheap hot water is available. (Japanese aquaculturists support up to three million kilograms of carp per hectare per year!) Thus, the heated discharge that can be detrimental to a whole ecosystem can be advantageous to a single species. However, fish farmers have the same sorts of problems as greenhouse operators. In addition, they must somehow dispose of considerable quantities of warm water heavily polluted by fish wastes.

Pumping hot water into home radiators is an attractive idea, but although it is done in some cities, it is economically impractical in others. First, many residents prefer not to live close to power plants, and the costs and heat losses involved in piping hot water any significant distance are prohibitive. Second, the installation of an underground steam system would be an extremely complex task in a large established city. Such a proposal would be practical only for newly built areas. Third, the maximum power demands do not coincide with maximum heating demands. Peak electric utilization is in the afternoon, while late-night use is low. Some storage facility or auxiliary steam generator would be needed to supply heat in the evening and at night. Finally, home heating would not remove waste heat during warm seasons or in warm climates. Despite these difficulties, the prospect of "free" heat is appealing, and the proposals are being studied in many cities.

Other proposals are to use the heat to speed the decomposition of sewage (without contact between the hot water and the sludge) or to desalinate seawater. Both these proposals require careful economic analysis.

An interesting example of the use of waste heat is provided by the relationship between the Bayway, New

TABLE 9.4 Effects of Soil Heating Without Irrigation on Vegetable Production in Muscle Shoals, Alabama

CROP	YIELD IN TONS PER ACRE	
	Heat	No Heat
String beans	6.9	2.7
Sweet corn	6.2	3.2
Summer squash	20.6	17.6

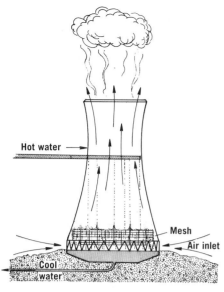

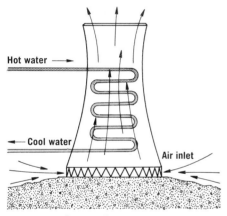

Wet cooling tower.

Dry cooling tower.

Jersey, refinery of the Humble Oil and Refining Company and the Linden, New Jersey, generating station. The Linden power plant is capable of producing electricity at 39 percent efficiency. For the past 15 years, this efficiency has been lowered by a less than optimum cooling of the condenser, and some of the waste heat has been sold as steam to Humble. If we consider the two-plant operation as a single energy unit, the overall efficiency of power production has been raised to a level of 54 percent. The process is beneficial to many: the companies save money, fuel reserves are conserved, and thermal pollution of waterways is reduced 15 percent. Similar operations are being planned or are in effect in other countries.

Though the technical problems of utilizing waste heat are not theoretically insurmountable, the solutions are often economically discouraging. Typically, the waste water is hot enough to damage an aquatic ecosystem but not hot enough to be attractive for commercial use. Perhaps if the total environmental cost of thermal pollution were considered, waste-heat utilization would become more feasible.

HOW TO USE THE ATMOSPHERE AS A SINK

Another approach to the problem of thermal insult to our waterways is to dispose of the heat into the air. Air has much less capacity per unit volume for absorbing heat than water does, so the direct action of air as the cooling medium in the condenser is not economically feasible. For this reason power plants must still be located near a source of water, the only other available coolant. However, the water can be made to lose some of its heat to the atmosphere and

This large cooling tower dwarfs the nuclear reactor at a reactor site in southern Washington.

then can be recycled into the condenser. There are various devices available that can effect such a transfer.

The two cheapest techniques are based on the fact that evaporation of water is a cooling process. Many power plants simply maintain their own shallow lakes, called **cooling ponds.** Hot water is pumped into the pond, where evaporation as well as direct contact with the air cools it, and the cool water is drawn into the condenser from some point distant from the discharge pipe. Water from outside sources must be added periodically to replenish evaporative losses. Cooling ponds are practical where land is cheap, but a one-million-kilowatt plant needs 400 to 800 hectares (1000 to 2000 acres) of surface, and the land costs can be prohibitive. A cooling tower, which can serve as a substitute for a cooling pond, is a large structure about 180 meters in diameter at the base and 150 meters high. Hot water is pumped into the tower near the top and sprayed onto a wooden mesh. Air is pulled into the tower either by large fans or convection currents, and flows through the water mist. Evaporative cooling occurs and the cool water is collected at the bottom. No hot water is introduced into aquatic ecosystems, but a large cooling tower loses over 3.8 million liters of water per day to evaporation. Thus, fogs and mists are common in the vicinity of these units, reducing the sunshine in nearby areas. Reaction of the water vapor with sulfur dioxide emissions from coal-fired power plants can cause the resultant air to carry sulfuric acid aerosols, as described in Chapter 13.

Environmental problems can be lessened if dry cooling towers are used instead of evaporative wet ones. A dry tower is nothing more than a huge version of an automobile radiator installed into a tower to promote a speedy flow of air past the cooling pipes. Dry towers are uneconomical because of the cost of the prodigious amount of piping required. The relative costs of the three cooling facilities are shown in Table 9.5.

An old Victorian house in Telluride, Colorado. High ceilings, poor insulation, and large single pane windows combine to cause houses of this type to be hard to heat

TABLE 9.5 Cost of Thermal Pollution Control for a Fossil-Fueled Plant

TYPE OF CONTROL	ESTIMATED AVERAGE COST mils per kwh*
Cooling pond	0.08
Wet tower, mechanical draft	0.10
natural draft	0.18
Dry tower, mechanical draft	0.81
natural draft	0.99

*The cost of electricity is about 2 cents per kwh for consumers who use 1000 kwh per month, assuming that non-recycled water is used for cooling. The figures in this column therefore represent the *added* cost for control of thermal pollution. One mil = one-tenth of a cent.

There are a great many different ways to conserve energy used for heating. The most obvious is for people to turn their thermostats down in the winter and up in the summer. However, this is a social rather than an engineering solution. An engineering answer is to design a structure that is inherently efficient so that fuel can be conserved without asking people to change their habits. The general strategy for conserving energy in buildings involves the following engineering features.

1. Minimize convection exchange with the outside.

   Any crack or hole in a house allows significant entry of cold air in the winter or hot air in the summer. While few houses have cracks in the walls, most have small openings around doors and windows. If every home owner sealed all leaky windows, millions of barrels of oil and thousands of metric tons of coal could be saved every day across the globe.

A large volume of outside air enters a house every time someone opens the door. Although there is no way to stop this exchange of air entirely, we can control its volume. Suppose that a small tightly sealed but unheated room is attached to the outside of the front door. A person entering the house opens the door to this anteroom and then (the architect hopes) shuts it. Now, when the person opens the door to the heated portion of the house, only a limited volume of cold air can enter.

2. Control interior convection currents.

   Since hot air rises, houses are warmer at ceiling level than at floor level. For example, the temperature of a house on a typical winter day is 16°C (61°F) at floor

---

If we heat one end of a metal bar with a candle, the other end will soon get hot, for heat travels directly through the metal. This type of heat transfer is called **conduction.** While all materials conduct heat, metals are many orders of magnitude more efficient than common insulators such as glass, brick, and wood.

Gases and liquids conduct heat, but they also permit heat transfer in other ways as well. Most materials expand when heated. A heater sitting on the floor in the corner of a room warms the air around itself. This warm air, being lighter than the air in the rest of the room, rises, displacing colder air downward. The cold air then comes in contact with the heater, becomes hot, and rises in turn. The continuous movement of hot air upward and cold air downward initiates a circulation in the room called a **convection current.**

Heat can also travel directly through a vacuum or through gases in the form of **radiation.** A pot-bellied stove and the Sun both radiate heat.

A Conduction

B Convection Currents in Room Heated by Potbellied Stove

C Radiation

level, 20°C (68°F) at shoulder level, and 22°C (72°F) near the ceiling. In a room with a high ceiling, a great deal of heat lies above the human occupants and is therefore wasted. Thus the airy feeling produced by high ceilings is paid for by excess heating costs.

3. Minimize conduction of heat through wall and roof surfaces.

Even if a house is tight and leakproof, conduction through wall and roof surfaces allows significant heat exchange. Thus a tight, well-sealed copper house would be expensive to heat because copper would conduct the heat out rapidly. Among common building materials, wood is the best insulator, concrete and brick are poorer, and all metals conduct heat very rapidly. Although wood is one of the best insulators among the collection of *structural* materials, many substances conduct heat much more slowly. Luckily for us, one fine insulator is absolutely free—air. Air that is not moving is 52 times more effective than soft wood and 338 times more effective than brick in providing insulation. Therefore a very good and cheap way to insulate a house is to trap a blanket of air around the outside. Thus, the walls of most buildings are essentially hollow, being built of inner and outer surfaces separated by the framework (Fig. 9.6). However, the enclosed space of a hollow wall is not so effective an insulator as we might think. Convection currents establish themselves inside the wall and carry heat from one surface to the other. To increase the insulating value of a trapped air space, we can add material such as glass fiber. The major function of fibrous glass insulation is to interrupt the flow of air in the wall. Therefore the insulation is sold as a collection of loosely spun fibers and not as a tightly compressed sheet. The fibers hold the air relatively motionless, and it is mainly the quiet air that insulates.

As discussed in Section 9.9, considerable quantities of fuel could be saved if homes were better insulated than many are at present.

4. Use radiation from the Sun.

Section 9.5 discussed systems for heating houses with solar energy. This section will consider the engineering aspects of more modest solar devices. A glass window, for example, is a device to utilize the lighting value of the Sun. Without windows people would consume large amounts of electricity for lighting while they lost a great aesthetic pleasure of being able to look outside. Most arrangements of windows are designed solely as sources of light, without consideration for heat transfer. Hence many windows tend to make a room cooler in winter and hotter in summer. They transmit radiant energy from the Sun on sunny days. Even on a cold winter day the sunlight streaming through a window can bring considerable warmth into a room. At the same time that glass admits radiant energy from the Sun, heat escapes back out through the glass by conduction. Not only is glass a poorer insulator than wood and other building materials, but also windows are generally quite thin, and heat escapes faster through thin barriers than through thick ones. Heat loss through windows can be reduced considerably by using two layers of glass separated by an air space. One way to accomplish this is to install storm windows. Alternatively, windows made of two panes of glass sealed around a trapped air space are commercially available. Of course, it is the air space more than the extra glass that provides the additional insulation.

Since energy can be transmitted through glass in either direction, the design and location of windows should favor heat gains during the winter months. A tightly fitted piece of doubled glass that faces the Sun on the south of the house allows more warmth to enter than to escape during the day. At night, however, since no light or

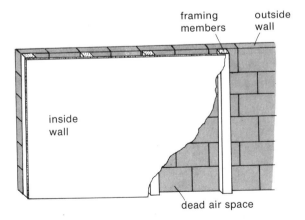

Figure 9.6 Standard wall construction.

A modern house with large glass surfaces. This glass wall faces north and is responsible for major heat losses.

radiant energy enters, the losses are greater than the gains. Over the course of an entire winter, even a good window allows more heat to escape than to enter and is therefore responsible for higher fuel bills. Heavy curtains drawn over the windows at night reduce heat losses significantly, but even greater conservation can be realized by building insulated panels, or shutters, that close against the windows at night. If a house is built with many south-facing windows and equipped with

panels for nighttime insulation, a 10 to 25 percent reduction in heat bills can be realized.

A building incorporating all the design features discussed in this section can be heated and cooled with considerably less than 50 percent of the fuel consumption of a poorly planned structure. A well-designed and tightly built house is more expensive, but the additional cost is quickly returned in fuel savings.

TAKE-HOME EXPERIMENTS

1. **Solar collectors.** Remove the inner dividers from each of four identical ice cube trays and fill each tray with the *same* amount of water, until it is about 2/3 full. (Between 1½ and 2 cups will do for the average-size tray. Use a measuring cup.) Set all four trays in the freezer compartment of a refrigerator until the water is frozen. Cover two of the trays with clear plastic wrap, sealed around the edges with string or a large rubber band. Replace the trays in the freezer and let them stay overnight. Next day, about midmorning, prop the four trays on a small box on a waterproof surface outdoors so that two of them (a covered one and an open one) slope toward the north and the other two slope south. Let them sit in the sun and note the rate at which they melt. Which design was the most efficient collector of solar energy? Why? Plan and carry out an experiment to test the effect of insulating the sides and bottom of your tray. (Use plastic foam or even dry crumpled newspaper as insulation.) Describe your results.

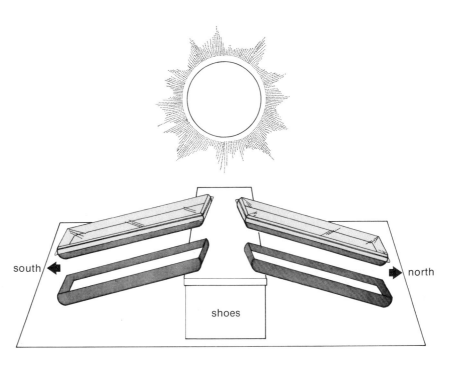

south   north

shoes

2. **Efficiency of Transportation.** Obtain a map of your local community. Mark the locations of your residence, your school, the grocery store, post office, bank, and the five other stores that you visit most frequently. Measure the distance from your home to each of these locations. What type of transportation do you usually use to travel from one to another? Discuss the efficiency in terms of fuel consumption, time, and cost of the transportation system you use. Would it be difficult to improve the fuel efficiency of this system? Discuss.

3. **Electric meter.** Find the electric meter in your house or apartment. If you live in a private house, the meter is probably located on the outside wall, whereas if you live in an apartment, it is probably in the basement. As shown in the photograph, a meter has several small dials and a horizontal disc. The disc spins when electric energy is being used, and the rate of spinning is proportional to the amount of electricity consumed. Thus, if many appliances are plugged in, the disc will spin rapidly, and if little electricity is being used in the house, the disc will barely move. Turn off all the electrical appliances in your house, and unplug the refrigerator and freezer. The disc should now be stationary. Plug in one 100-watt light bulb. How many seconds are required for the

disc to make one complete revolution? Plug in a 50-watt light bulb and measure the time required for one revolution. Now plug in an appliance whose power requirement you do not know. Record the time required for one revolution of the disc. How much power is consumed by the unknown appliance?

Household electric meter.

PROBLEMS

1. **Coal mining.** Discuss the relative advantages and disadvantages of strip and tunnel mining.

2. **Coal mining.** Explain how unrestrained strip mining of coal could cause the price of meat and bread to rise. Use this example to discuss the nature of economic externalities, introduced in Chapter 1.

3. **Energy and pollution.** In today's society, many businessmen wear jackets and ties indoors during the summer and then use air conditioners to keep cool. Discuss various social and economic solutions to the problem of excess fuel consumption for air conditioning. Which solution or combination of solutions would be most effective? Which would be easiest to implement? Defend your answer.

4. **Energy and pollution.** According to the present pricing structure in the United States, electricity becomes cheaper per kilowatt hour (kwh) as demand is increased. In a typical example, the first 40 kwh costs 10 cents per kwh, and any consumption above 900 kwh costs about 2 cents per kwh. Can you think of reasons why this rate structure was established? Discuss how this rate structure encourages needless energy consumption. Examine some social, political, and economic solutions to the problem.

5. **Solar energy.** Describe the construction and function of a flat plate collector.

6. **Solar heating.** Discuss the advantages and disadvantages of using solar energy for home heating. Explain what factors you would evaluate in deciding whether or not to install solar heat if you were building a new home.

7. **Solar generation of electricity.** Discuss the potential for using solar energy to generate electricity. Explain why this potential is not now being exploited.

8. **Public policy on solar energy.** A city-dwelling critic of governmental support of research on solar energy complained that (a) it would benefit only those who live in the suburbs or the country, and (b) if it is such a good idea it should attract private investment rather than use up tax dollars which come out of his pocket. How would you counter his argument?

9. **Hydrogen economy.** Explain how an economic system based on hydrogen fuel would function differently from a system based on petroleum.

10. **Alternative energy sources.** Discuss briefly the prospects for solar, geothermal, tidal, wind, and garbage energies. What problems do these methods entail?

11. **Tidal energy.** There are several sites in the Pacific Northwest that are excellent candidates for tidal generating stations. However, the tidal bays and inlets are vital to salmon migration. Discuss some of the environmental effects of placing tidal generating facilities in this region.

12. **Transportation.** Give some reasons why it is often difficult to alter patterns of surface transportation.

13. **Transportation.** Gasoline taxes have traditionally been used for the construction of new roads. This practice has been considered fair because the roads are paid for by those who use them most. Increasingly, economists and social philosophers feel that many of our traditional concepts of fairness must be reevaluated in the light of environmental problems. Do you feel that there should be a re-evaluation of road tax use? If so, how would you allocate funds? If not, explain.

14. **Transportation.** It takes a few hours to fly across the United States; a comparable journey by train takes several days. Yet the cost to the passenger is about the same. Discuss the impact of this economic structure on world fuel reserves.

15. **Transportation.** In 1975, 97 percent of urban traffic was carried by the automo-

bile, and only 3 percent was carried by mass transit systems. Thus, even if mass transit systems were to double the number of passengers they carry, there would be relatively little impact on overall fuel consumption rates. Should these figures be used as an argument against building new mass transit systems? Defend your answer.

16. **Energy conservation for industry.** Discuss some factors that might make recycling economically attractive.

17. **Space heating.** Why can a furnace be built to be more efficient than a steam engine?

18. **Heat pumps.** It has been suggested that small air conditioners be placed in the center of a room rather than in a window for more efficient central cooling. Do you think this is a good idea?

19. **Heat pumps.** What is the common name for a heat pump that heats the outdoors at the expense of the indoors?

20. **Electric power generation.** Explain the function of a condenser in a steam turbine. Is a condenser needed at a hydroelectric facility?

21. **Efficiency of energy conversion.** Discuss the efficiency of each of the following methods of warming the air in a room: (a) Oil is burned in a home furnace. (b) Coal is burned in a generating plant to make electricity, which is transmitted to the house to operate an electric heater. (c) Electricity from the same source is used to operate a heat pump that transfers energy from outdoors into the house. (d) Hydrogen is burned inside the room.

22. **Thermal pollution.** Define thermal pollution. How does it differ in principle from air or water pollution?

23. **Thermal pollution.** Dogs live on all parts of the Earth from the tropics to the Arctic. Trout, on the other hand, are confined to waters no warmer than about $15°$ C. Why are the permissible temperatures for trout so much more limited?

24. **Thermal pollution.** Since marine life is abundant in warm tropical waters, why should the warming of waters in temperate zones pose any threat to the environment?

25. **Thermal pollution.** Warm water carries less oxygen than cold water. This fact is responsible for a series of disturbances harmful to aquatic organisms. Explain.

26. **Thermal pollution.** On page 260 of this book it is stated that the heat discharged into aquatic systems is likely to produce serious ecological damage. Yet the discussion on page 263 describes uses of hot water for increasing fish yields in aquaculture. Is this a contradiction? Explain.

27. **Hot water.** Discuss some difficulties with use of hot water for agriculture; for aquaculture. Discuss the potential benefits.

28. **Solar energy.** Explain how a solar-operated steam generator can thermally pollute a river without having an effect on the total heat flux of the biosphere.

29. **Waste heat and climate.** What types of energy sources do not add any additional heat to the environment?

30. **MHD.** Explain the operation of an MHD generator. Why is it a desirable form of electrical generation?

31. **Cost of thermal pollution control.** A typical home consumes 1200 kwh per month. Use Table 9.5 to determine the monthly cost to a family of (a) a cooling pond, (b) a mechanical wet tower, and (c) a mechanical dry tower.

32. **Home heating.** List five ways to improve the heating efficiency in your home or apartment. If you live in a private home, estimate the cost of implementing these changes. If you live in an apartment house, do you think that such changes would be a worthwhile investment for the owner?

BIBLIOGRAPHY

A three-volume work that reviews many aspects of the relationship between energy use and human welfare is:

Barry Commoner, Howard Boksenbaum, and Michael Corr (eds.): *Energy and Human Welfare.* New York, Macmillan, 1975. Vol. 1, 217 pp.; vol. 2, 213 pp.; vol. 3, 185 pp.

Several other recent and comprehensive books on energy and the environment are:

Philip H. Abelson (ed.): *Energy: Use, Conservation, and Supply.* Washington, D.C., American Association for the Advancement of Science, 1974. 154 pp.

Barry Commoner: *The Poverty of Power.* New York, Alfred A. Knopf, 1976. 314 pp.

Ford Foundation Report: *A Time to Choose—America's Energy Future.* Cambridge, Mass., Ballinger, 1974. 511 pp.

Allen L. Hammond, William D. Metz, and Thomas H. Maugh, II: *Energy and the Future.* Washington, D.C., American Association for the Advancement of Science, 1973. 184 pp.

S. S. Penner and L. Icerman: *Energy.* Reading, Mass., Addison-Wesley, 1974. 373 pp.

Marion L. Shephard, Jack B. Chaddock, Franklin H. Cocks, and Charles M. Harmon: *Introduction to Energy Technology.* Ann Arbor, Mich., Ann Arbor Science Publishers Inc., 1976. 300 pp.

H. Stephen Stoker, Spencer L. Seager, and Robert L. Capener: *Energy from Source to Use.* Glenview, Ill., Scott, Foresman and Company, 1975. 337 pp.

Two books on special topics are:

J. Richard Williams: *Solar Energy Technology and Applications.* Ann Arbor, Mich., Ann Arbor Science Publishers, 1974. 120 pp.

Marvin M. Yarosh (ed.): *Waste Heat Utilization.* Springfield, Va., National Technical Information Service, 1971. 348 pp.

An excellent book on mass transit is:

Tabor R. Stone: *Beyond the Automobile.* Englewood Cliffs, N. J., Prentice-Hall, 1971. 148 pp.

NUCLEAR ENERGY AND THE ENVIRONMENT

10.1 THE NUCLEAR CONTROVERSY

Early in 1976, two articles about nuclear power appeared at the same time in popular journals, one by Barry Commoner in *The New Yorker,* the other by Hans Bethe in *Scientific American.* Both authors are well known: Commoner is a biologist and environmentalist; Bethe is a physicist who received the Nobel Prize in 1967 for his discovery of the nuclear reactions from which stars derive their energy. They could hardly have disagreed with each other more. Commoner concluded that "the entire nuclear program is headed for extinction. It will leave us with a monument, which people will need to care for with vigilance, if not affection, for thousands of years—huge stores of radioactive wastes, and the powerless, radioactive hulks of the reactors that produced them." Bethe, on the other hand, stated that if the United States must have sources of energy other than from fossil fuel, the only important source between now and the end of the century is nuclear fission. Bethe also minimized the various objections that have been raised against nuclear power. He pointed out that the risks involved (other than from nuclear weapons) "are statistically small compared with other risks that our society accepts."

The opposing sides in the controversy are by no means limited to two representatives. For example, under the heading "No Alternative to Nuclear Power," 32 scientists (one third of whom are Nobel laureates), headed by Bethe, fully supported his position.[*] One the other hand, another group of scientists, also including what seems to be a required component of Nobelists, submitted on August 6, 1975 (the thirtieth anniversary of the bombing of Hiro-

[*]*Bulletin of the Atomic Scientists,* March, 1975.

shima) a declaration against nuclear power,* which stated, in part, that the United States "must recognize that it now appears imprudent to move forward with a rapidly expanding nuclear power plant construction program. The risks of doing so are altogether too great. We, therefore, urge a drastic reduction in new nuclear power plant construction starts before major progress is achieved in the required research and in resolving present controversies about safety, waste disposal, and plutonium safeguards."

How can knowledgeable people differ so sharply? Part of the answer, at least, is apparent. They differ in the facts they believe to be true and select as a basis for their arguments; they differ in the assumptions they make about the probability of future events such as accidents; and finally, they differ in the judgments they make about the value of the expected benefits and the human cost of the expected damages. But perhaps the most significant gap that separates the two sides lies in their views of the human prospect. The proponents of a nuclear future assume that (a) we must have energy sources which can match what is now available and also provide for continued growth, (b) alternative sources such as solar energy are interesting but are not likely to contribute significantly within this century, (c) coal, which is available now, is environmentally a more damaging choice than nuclear power, and (d) conservation has its price. As Bethe's group expresses it, "One man's conservation may be another man's loss of job."

The opponents of nuclear power necessarily reject the central assumption that we must utilize energy at an ever-increasing rate. Other societies use less energy and provide good conditions of life. Even today's highly developed nations used much less energy only a few decades ago and were then hardly "primitive" by present standards. They argue further that alternatives such as solar energy could be used now with minimal social disruption. Of course, some adjustments would require societal changes, but so do our patterns of ever-increasing dependence on energy. Between the two sets of alternatives, the anti-nuclear argument continues, the more conservative one will be the better choice in the long run for preserving and improving the human condition.

10.2 ATOMS

During the nineteenth century, scientists reaffirmed the ancient Greek speculation that ordinary matter is a collection of fundamental units, called **atoms.** An atom consists of a small, dense, positively charged center called a **nucleus** surrounded by a diffuse cloud of negatively charged **electrons.** The atomic nucleus is composed of two

*Prepared under the auspices of the Union of Concerned Scientists, Cambridge, Massachusetts.

types of particles, positively charged **protons** and electrically neutral **neutrons.** Practically all of the mass of the atom is contained in the nucleus.

Chemical elements are considered to be the stuff of which all other substances are composed. About 104 elements are known; most are found in nature, but some have been created by transformation of other elements. Some common ones are hydrogen, carbon, nitrogen, iron, and uranium. The unique characteristic of a chemical element is that all its atoms have the same number of protons in the nucleus. Thus, all hydrogen atoms contain 1 proton, all carbon atoms have 6, nitrogen 7, iron 26, and uranium 92. The neutrons add mass to the nucleus but have relatively minor effects on the chemical behavior of the element. For example, some carbon atoms have 6 neutrons, while others contain 8. The sum of the number of protons plus neutrons is the **mass number.** A carbon atom that has 6 neutrons therefore has a mass number of 6 protons + 6 neutrons = 12, and is designated carbon-12. A carbon atom with 8 neutrons has a mass number of 6 protons + 8 neutrons = 14, and is designated carbon-14. Such elements, whose atoms have the same number of protons but different numbers of neutrons, are called **isotopes.**

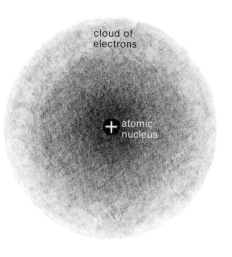

10.3 RADIOACTIVITY

It was just before the turn of the century, in 1896, that the French physicist Henri Becquerel accidentally discovered that uranium minerals spontaneously emit energy in the form of radiation. This fact is easy to accept now, but at that time the phenomenon seemed eerie; it even implied a threat to the law of conservation of energy. Several significant facts soon emerged from further studies of what came to be called **radioactivity:** The pure uranium compounds that were extracted from the mineral were *less* radioactive than the crude mineral itself. This difference implied that there were other more highly radioactive substances mixed with the uranium. A series of careful, tedious separations carried out by Marie and Pierre Curie (wife and husband) resulted in the discovery of new radioactive elements, the most important of which was **radium.**

The Curies also learned that the radioactivity of substances is associated with their elements, not with compounds. Thus, a gram of radium has the same radioactivity in the form of the pure metal as in the form of any of its compounds. Since chemical bonding involves the electrons of the atom but does not affect radioactivity, we are led to the conclusion that *radioactivity is associated with atomic nuclei.*

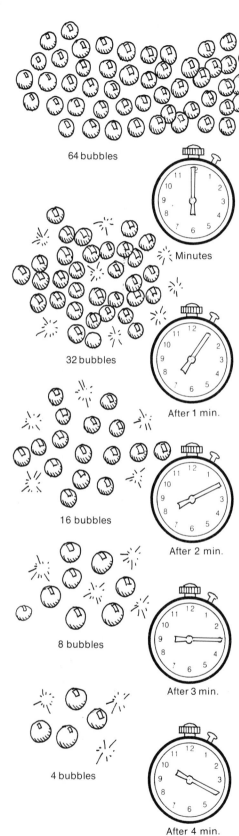

64 bubbles

Start,
64 bubbles

Minutes

32 bubbles

After 1 min.

16 bubbles

After 2 min.

8 bubbles

After 3 min.

4 bubbles

After 4 min.

Soap bubbles with one-minute half-life.

There are two aspects of radioactivity that must be emphasized:

(a) When an atom of a radioactive element emits energy, it also decomposes, thereby producing a new atom. The new element, called a **daughter** of the one that produced it, may also be radioactive, and produce another daughter, and so on, until a stable (nonradioactive) element terminates the sequence. For example, the radioactive disintegration series that includes radium progresses in nine additional steps to end with a stable form of lead. Although isotopes of the same element are chemically very similar, they may have very different nuclear properties; in fact, one form may be radioactive and another not. Radioactive forms are called **radioisotopes** These, too, are identified by their mass numbers, written, for example, as uranium-235, or radium-226.

(b) Consider, now, the *rate* of radioactive disintegration. If you observed just one atom of, say, radium-226, containing one nucleus, when would it decompose? This question cannot be answered. Think of the radium nucleus as an energetic bundle of electrically charged matter; it may or may not break apart at any time. But we do know what the *chances* are that the radium nucleus will decompose in any given minute, or day, or year, or century. To understand this idea better, leave the radium nucleus for a moment and consider a more familiar example. Imagine that you are blowing soap bubbles and watching them burst in air. How long will the first bubble last? You cannot predict its lifetime with certainty, but let us say that you know that, if you waited a minute, the bubble would have a 50–50 chance of surviving. Now imagine that you blew 64 such bubbles. After one minute, you would expect to find 32 bubbles intact; the other 32 would have burst. After another minute, the surviving 32 would be halved again, and only 16 would be expected to remain. After the third minute you would expect 8 to remain, and so on. Because half of the bubbles are expected to burst in any given minute, this time interval is called the **half-life.**

The half-life concept applies to radioisotopes. One nucleus of a radium-226 atom has a 50–50 chance of surviving in any given interval of 1600 years. This means that the half-life of radium-226 is 1600 years. Therefore, if one gram of radium-226 were placed in a container in 1980, there would be only one-half gram left after 1600 years (in the year 3580) and only one-quarter gram after another 1600 years (in the year 5180), and so on. This process is called **radioactive decay.**

The concept of half-life does not imply that after 1600 quiet years half of the radium will suddenly decompose. Think of the soap bubbles—they would not pop off in unison at one-minute intervals; rather, some popping would appear to be occurring more or less regularly. The half-life is an averaged value for all the bubbles, as it is for the radium nuclei. This means that there is a chance for some decompositions to occur *in any interval of time.* Since there are so many atoms in a sample of radium (about 2.65×10^{21} per gram), many will be decomposing every second, and any nearby Geiger counter will respond by clicking all the time.

The rate at which the radiation is emitted from a sample of radium-226 depends on its quantity. Since this quantity is constantly decreasing, the rate of disintegration is also decreasing. Remember, however, that radium produces other radioisotopes when it decomposes. Any sample that has been decomposing for some finite time will contain some of the original radium-226 plus some of each of its radioactive "daughters," as well as some of the stable final product, lead-206. These radioisotopes have different half-lives, ranging from fractions of a second to about 20 years. Therefore, the total radioactivity produced by a sample of radium together with its radioactive waste products is more than that produced by the radium alone.

A final question is: If nuclei of radioactive elements are unstable, why are there any left on Earth? The only conceivable answer is that these survivors are all daughters of radioisotopes with very long half-lives. The half-life of natural uranium-238, for example, is 4,500,000,000 years. The radiations from such materials plus the effect of radiation that comes to the Earth from outer space is called the **background radiation.**

In recent years humans have vastly increased the quantity of radioactive materials in various parts of the Earth. We cannot invent anything to stop this radioactivity. It slows down by radioactive decay at a rate determined by the half-lives of the radioisotopes involved.

10.4 BIOLOGICAL EFFECTS OF RADIATION

Energy is emitted by radioactive elements in three different forms; (1) highly penetrating radiation, similar to x-rays, (2) fast-moving electrons, and (3) small portions of the nucleus itself. All of these forms are capable of harming living organisms.

HOW RADIATION IS DELIVERED: EXTERNAL AND INTERNAL IRRADIATION

There are two ways in which an animal can be irradiated. The first is from an external source of radiation, for example an x-ray tube. The second is via the ingestion or inhalation of radioactive materials; this is called internal irradiation, since the sources of radiation are inside the body. It should be clear that this internal type poses the more important problem in connection with the health hazards of radioactive waste products. Whether or not a radioactive material poses a significant biological hazard depends on two quite independent properties, discussed below.

The Chemical Nature of the Substance

The chemical rather than the radioactive nature of a substance determines whether or not it can enter the food chain and be taken up by animals, plants, and ultimately people. For example, strontium-90, a common radioactive by-product of atmospheric nuclear weapons testing, is chemically similar to calcium. It is thus taken up by plants, where it is ingested by herbivores such as cows. Like calcium, it is then concentrated in the milk of the animals, and from there it becomes a part of the human diet. In people, it becomes a structural part of bone along with ingested calcium. Therefore, as radioactive decay continues, the cells in the bone and the bone marrow become the prime targets of the radiation from this source.

Another significant fallout isotope is cesium-137 (half-life = 30 years), which is chemically similar to potassium. Cesium-137 accumulates in muscle tissue, while strontium-90 accumulates in bone and milk. Cesium-137 is therefore the more significant source of radioactivity in the diet of Eskimos, who eat caribou muscle.

Remember also that the food web has the ability not only to transport but also to concentrate various materials. Concentration of radioisotopes in the food chain in Arctic regions is particularly efficient. The effect of this concentration was noted when it was found that Eskimos absorbed more radioactive fallout than did people who lived in other zones of the Earth, where more fallout actually occurred. The first step in this highly efficient concentration is provided by the Arctic lichen, a plant that gets its mineral nourishment directly from dust particles which settle on it. For this reason, the lichen collects fallout dust particularly efficiently. In summer, caribou migrate north to the tundra, where they wander over large areas in search of lichen. Lichen becomes an important part of their diet. The effect is as if someone sent out the caribou to collect and bring back the cesium-137, and they accomplish this task very well. Then, of course, the Eskimos eat the caribou, sometimes as their only food, and so, at the top of the food chain, they get the most concentrated radiation.

The Half-Life of the Radionuclide

It should be clear that even if the substance does get into the food chain in the manner described above, if the half-life is very short (for example, seconds or minutes), no hazard exists, since most of the material will have decayed before entry into the food chain. Similarly, if the half-life is measured in millions of years, no significant amount of radiation will be generated during the lifetime of the animal. In our previous example, the half-life of strontium-90 is about 28 years. Thus, this isotope has the ability to be incorporated into living tissue and, provided enough is absorbed, may generate significant radiation.

EFFECTS OF RADIATION ON LIVING CELLS

X-rays (and gamma rays, which are an even more energetic form of radiation) can knock electrons out of atoms that they strike. This damage may affect certain molecules that are necessary to the cell. One such molecule is deoxyribonucleic acid (DNA), which contains all the genetic information that is required for the growth and maintenance of the cell. DNA is a sensitive target for radiation; when a cell is irradiated, the DNA strands tend to break into fragments. If the rate of delivery of the radiation is low, the cell's repair mechanisms can seal the breaks in the strands, but above a certain dose rate, the repair process cannot keep up, and the DNA fragmentation becomes irreversible.

Cell types differ greatly in their sensitivity to radiation. In general, those cell types that divide most rapidly are the ones most easily killed by radiation. These include cells of the bone marrow, which make the red cells, white cells, and platelets of the blood; those that line the gastrointestinal tract and the hair follicles; and the sperm-producing cells of the testes. Conversely, muscle and nerve cells, which do not divide in the adult, are quite resistant even to large doses of radiation. The rule, however, is not invariable. Neither lymphocytes (the cells that control the immune response) nor the egg cells of the ovary divide under ordinary circumstances but both are exquisitely sensitive to killing by radiation.

What about nonlethal effects? You might expect, quite correctly, that anything as potent as radiation would exert some effect on cells, even if delivered in sublethal doses. It has been known for many years that radiation is a powerful inducer of mutations. Mutations are produced when the DNA is altered in some way. The alterations may sometimes be so gross that the chromosomes look abnormal when viewed under the microscope. Other changes may be far more subtle and occur over only a minute length of the DNA (point mutations). The importance of mutations, whichever type occurs, is that (1) they may cause changes in the function of the genes in which they occur; and (2) they are passed on to all daughter cells. A simple example of a mutation might be the loss of the cell's ability to transport a sugar such as glucose from the outside of the cell to the inside. A more complex example (but perhaps only because we do not yet understand the mechanism) is the change by which a normal cell becomes a cancer cell.

EFFECTS ON THE WHOLE ANIMAL

We now turn our attention to the effects of radiation not on single cells but on whole animals. It is convenient to divide these into **somatic effects,** or those which are confined to the population exposed to the radiation, and **genetic effects,** or those which are inherited by subsequent generations.

Early Somatic Effects: Radiation Sickness

On several occasions during the past 75 years, groups of people have been exposed to large doses of ionizing radiation over periods of time ranging from a few seconds to a few minutes. The holocausts at Hiroshima and Nagasaki, together with accidents at civilian nuclear installations, have provided much information about what radiation can do when a lot of it is administered to the entire body over a short period of time. Let us consider first the simplest and most drastic measure of radiation effect—death. Figure 10.1 shows the relation between the dose administered to a population of animals and the percentage of the population remaining alive three weeks or more after the exposure. Up to a dose of about 250 rads (see Table 10.1 for definition

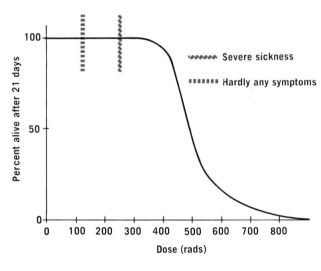

Figure 10.1 Curve showing the approximate relationship between the dose of radiation administered to a whole animal (such as a mouse or a man) and the percent of the treated population which survives three weeks afterward. Mice have, of course, been intensively studied in the laboratory; accidents in industry and the nuclear explosions in Japan have provided the approximate data for humans. (From *American Scientist* 57: 206, 1969. Copyright 1969 by Sigma Xi National Science Honorary.)

TABLE 10.1 Units Related to Radioactivity

UNIT	ABBREVIATION	DEFINITION AND APPLICATION
Disintegration per second	dps	A rate of radioactivity in which one nucleus disintegrates every second. The natural background radiation for a human body is about 2 to 3 dps. This does not include "fallout" from man-made sources such as atomic bombs.
Curie	Ci	Another measure of radioactivity. One Ci = 37 billion dps.
Microcurie	μCi	A millionth of a curie, or 37,000 dps.
Roentgen	R	A measure of the intensity of x-rays or gamma-rays, in terms of the energy of such radiation absorbed by a body. (One R delivers 84 ergs* of energy to 1 gram of air.) The roentgen may be considered a measure of the radioactive dose received by a body. The dose from natural radioactivity for a human being is 5 R during the first 30 years of life. A single dental x-ray gives about 1 R, a full mouth x-ray series, about 15 R.
Rad		Another measure of radiation dosage, equivalent to the absorption of 100 ergs per gram of biological tissue.
Rem		A measure of the effect on man of exposure to radiation. It takes into account both the radiation dosage and the potential for biological damage of the radiation. The damage potential is based on the following scale of factors: x-rays, gamma rays, electrons　　:　1 neutrons, protons, alpha particles: 10 high-speed heavy nuclei　　　　　: 20 The rem is then defined by the relationship: Rems = Rads × Biological damage factor. Therefore, 100 ergs per gram (x-rays) = 1 rad × 1 = 1 rem, but 100 ergs per gram (neutrons) = 1 rad × 10 = 10 rems.

*See Appendix A for a discussion of units of energy.

of a rad), virtually everyone survives. When the dose is increased above this point, survival begins to drop sharply, and above a dose of 700 rads, everyone dies.

Does this mean that below doses of 250 rads there is no observable effect? Not at all. For even if the exposed individuals do not die, they may become quite ill. At doses between 100 and 250 rads, most people develop fatigue, nausea, vomiting, diarrhea, and some loss of hair within a few days of the exposure; the vast majority, however, recover completely from the acute illness. For doses around 400 to 500 rads, however, the outlook is not so rosy. During the first few days, the illness is similar to that of the previous group. The symptoms may then go away almost completely for a time, but beginning about three weeks after the exposure, they will return again. In addition, because the radiation has impaired bone marrow function, the number of white cells and platelets in the blood will fall. This is of great significance, since without white cells the body cannot fight infection, and without platelets the blood will not clot. Looking at Figure 10.1 again, you can see that about 50 percent of those exposed in this dose range will die, and of these, most will die of either infection or bleeding.

If, instead, the dose administered is about 2000 rads, the first weeks of the illness will again be the same as in the previous groups, but rather than waiting three weeks for symptoms to return, these people become very ill in the second week, with severe diarrhea, dehydration, and infection leading to death. At these dose levels the cells of the gastrointestinal tract are affected before the bone marrow toxicity has a chance to become severe, and these patients may die even before their blood counts have dropped to life-threatening levels.

At doses above 10,000 rads, animal experiments have shown that death, which may occur within hours of administration of the dose, is due to injury of the brain and heart.

Delayed Somatic Effects

Of the late somatic effects of radiation (that is, those occurring months or years following the exposure), none is better studied or of more concern than the increased incidence of cancer in those with a history of exposure to radiation. Although the molecular mechanisms at work here are still largely obscure, the evidence that radiation does increase the incidence of cancer in exposed populations is overwhelming. Before the dangers of radiation were appreciated, early workers were careless in their handling of radioactive materials and suffered a greatly increased incidence of skin cancers. The famous case of the radium dial workers in the 1920's deserves mention. These women were responsible for painting the dials of watches with the phosphorescent radium paint in use at that time and routinely tipped the end of the brush in their mouths before applying the paint to the dial face. In later years this group experienced a very high incidence of bone tumors. (Ingested radium, like strontium, is preferentially incorporated into bone.) The survivors of the atomic attacks at Hiroshima and Nagasaki, within a decade of the attacks, had far more leukemia than one would have expected from a group of this size, and subsequently the incidence of a variety of malignancies other than leukemia seems to be rising as well.

Medical therapeutics itself also provides lessons. For example, evidence is accumulating that the children born to women who have had x-rays of their pelvis during pregnancy have a higher risk of developing leukemia than those whose mothers have not had such exposures.

Induction of malignancies is not the only late somatic effect. Irradiated animals also have a propensity for cataract formation in the lens of the eye. Also, for some reason, all irradiated species studied show a shortening of the lifespan. We shall discuss the implications of all these findings in a later paragraph.

Genetic Effects

We now consider those effects of radiation which do not manifest themselves in the irradiated individual himself but result in mutations in the genetic material of the **germ cells** (the sperm cells of the testis and the egg cells of the

"I hope you weren't planning on a big family, Miss Whipple."

From *Industrial Research,* September 1973.

282

ovary) that are passed on to succeeding generations. In every experimental system studied in the laboratory, in organisms as diverse as viruses, bacteria, fruit flies, and mice, radiation has been shown to be a potent inducer of mutations. Though both ethical and practical considerations militate against genetic experiments with humans, scientists have attempted to find evidence of an increased mutation rate in irradiated populations such as the atomic bomb survivors. These investigations have yielded only equivocal results. Nonetheless, there is no reason to believe that people should behave differently in this respect from every other well-studied species and, thus, scientists and policy-makers must assume that radiation is also mutagenic in humans.

SOME SOLUTIONS

If, then, one wishes to minimize the somatic and genetic effects of radiation, the task is clear—one must minimize unnecessary radiation exposure. To do this intelligently, one must first have some idea of the contribution of the various sources of radiation to people. By far the greatest source is natural background radiation; this amounts to about 0.125 rads to the gonads per person per year, and comes from sources in outer space, the Earth's crust, and building materials. Now that atmospheric testing of nuclear weapons has been largely curtailed, radioactive fallout accounts for very little increase over the background. In the western world, by far the largest addition to background comes from the diagnostic uses of medical x-rays; the best estimates are that, on the average, diagnostic x-ray studies increase the genetically significant radiation load to the population by about 50 percent of background. ("Genetically significant radiation" is that which reaches the gonads of people who are still in the reproductive age group.)

Since 1928 the International Commission on Radiological Protection,* a group of scientists from many countries, has been establishing radiation standards, usually in the form of maximum permissible doses for total body irradiation of members of a population. The general way in which this is done is that the results of animal experiments with high-dose radiation are extrapolated back to low doses to get an estimate of the likely effect of low-dose irradiation on whole animals or people. In doing so, the assumption is always made that there is no safe threshold; that is, there is no low-dose level of radiation below which radiation is completely harmless. Most, but not all, scientists agree that this "no threshold" assumption is valid; in any case, it is the safest assumption in the present state of our ignorance. Currently, for members of the general population (that is, people who do not work with radiation on a daily

*The ICRP has no legal status, but its recommendations are generally adopted by countries within which radiation is used.

basis), the ICRP recommends a maximum dose of five rads in a lifetime, which amounts to about 170 millirads a year. This is to include all radiation except natural background and medical sources. The ICRP has estimated that if the entire population of the United States were to be exposed to this maximum level, there would be about 2500 additional cases of cancer each year. Linus Pauling has calculated that the number is closer to 96,000 new cases. Drs. Arthur Tamplin and John Gofman, two radiation scientists who have been particularly vocal critics of the 170 millirad/year guideline, put the number at 30,000. The United States Atomic Energy Commission has been quick to point out that there is no obvious way that all the people in the country could be exposed to this level; rather, only those who live or work on the borders of a nuclear reactor site might approach this limit, and if they did, those further removed from the site would receive much less radiation.

It must be emphasized that all these calculations are based on assumptions which seem reasonable but are unproved, and they all involve large approximations in arriving at the final estimates. But the lesson is clear—we will not have nuclear reactors without some price, whether the number of additional cancer cases each year in the United States is 100 or 100,000.

We have seen that radiation poses many hazards to human health. Some are easily assayed (for example, acute radiation sickness) and some only with the greatest difficulty (genetic effects in future generations). The responsibility on those who use radiation is thus enormous, for the implications of what we do now extend far into the future to those yet unborn. If radiation were *only* dangerous, of course, there would be little difficulty in making policy decisions, but things are not so simple. The benefits and potential benefits of radiation are very great. Modern medical diagnosis is unthinkable without x-rays. Radiation can increase the incidence of cancer in exposed populations, but in many patients who already have cancer, therapy with radiation can sometimes cure the disease and can often provide substantial relief from suffering.

There is no way to balance known benefits in the present with unknowable dangers in the future. Nonetheless, there are reasonable steps to take. An x-ray examination should be undertaken only when truly necessary, particularly for children and for adults of child-bearing age. Whenever possible, gonadal shields should be provided to the patient undergoing the examination. Except in emergencies, women of child-bearing age should have x-ray examinations only during the first two weeks of their menstrual cycle, since pregnancy is most unlikely during this interval.

Of course, nuclear chain reactions provide the largest

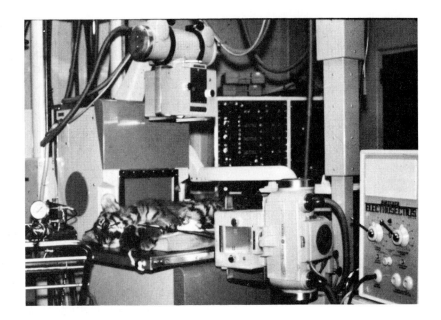

Diagnostic x-ray equipment.
(Photo by Patricia Weiner.)

potential source of radiation to which humanity could be exposed. The sections to follow will explore this threat in some detail.

10.5 NUCLEAR CHAIN REACTIONS

The first artificial nuclear reaction was carried out by Ernest Rutherford and his co-workers in 1919, when they bombarded ordinary nitrogen with energetic particles to produce oxygen-17, which is nonradioactive.

Fifteen years later, in 1934, Irène and Frédéric Joliot-Curie, Mme. Curie's daughter and son-in-law, converted boron to nitrogen-13, which is radioactive. This was the first artificially produced radioisotope, and its creation brought about the first artificial increase of radioactivity on Earth. However, the quantity of radioactivity produced by such a process has an inconsequential effect on the Earth, because only very small quantities of radioactive matter are involved.

The discovery of the **nuclear chain reaction,** which occurs in nuclear fission, led to the production of vast quantities of radioactive matter. It is important to understand what a chain reaction is so as to appreciate why it makes the production of radioactive wastes a real problem for life on Earth.

A chain is a series of links. Think of the process of making a chain; it involves the successive addition of links. The process of adding links to a chain is called **chain lengthening.** If the end of a chain links onto the beginning, it forms a cycle, and the chain ends. This is one form of **chain termination.** If more than one link is added to a

chain

terminated
chain

branched
chain

given link, various arms of the chain develop; this is called **chain branching.**

A series of steps in a process that occur one after the other, in sequence, each step being added to the preceding step like the links in a chain, is called a chain process or **chain reaction** (Fig. 10.2). Chemical chain reactions can also undergo branching. An example of a branching chemical chain reaction is a forest fire. The heat from one tree may initiate the reaction (burning) of two or three trees, each of which, in turn, may ignite several others. If the lengthening of one chain proceeds at a given rate, the production of 10 branches means that 10 reactions are going on at the same time, so that the rate has increased tenfold. A chemical chain reaction that continues to branch can produce an explosion. The condition under which a chain reaction just continues at a steady rate, neither accelerating nor slowing down, is called the **critical condition.**

The production of the atomic (fission) bomb and of nuclear reactors depends on branching nuclear chain reactions. The process is initiated when a neutron strikes a uranium-235 nucleus and can proceed in any of several different ways. Two examples are shown here:

$$\text{uranium-235} + 1 \text{ neutron} \longrightarrow \text{barium-142} + \text{krypton-91} + 3 \text{ neutrons} \quad (1)$$

$$\text{uranium-235} + 1 \text{ neutron} \longrightarrow \text{iodine-137} + \text{yttrium-97} + 2 \text{ neutrons} \quad (2)$$

Note the following important points about these equations:

(a) The reaction is started by one neutron but produces two or three neutrons. These neutrons can initiate two or three new reactions, which in turn produce more neutrons, and so forth. This is, therefore, a branching chain reaction.

(b) The uranium-235 nucleus is split in half (roughly) by these reactions. This is called **atomic** or **nuclear fission.** Fission releases energy because the uranium nuclei have more energy than their breakdown products. The amounts of energy involved are very large compared with those of chemical reactions. If the branching chain reaction continues very rapidly, it produces an atomic explosion. If the chain branching is carefully controlled, energy can be released slowly to form a nuclear reactor, which can be used for power production.

(c) Fission reactions produce radioactive wastes. Barium-142, krypton-91, iodine-137, and yttrium-97, the products shown in the preceding equations, are all radioactive. Furthermore, the reactions represented by these equations are only two out of many that occur in atomic fission. Many different radioisotopes are produced by atomic fission. Also, the fission products, as a group, are much more radioactive than the uranium from which they are produced. Recall that the half-life of the naturally abundant uranium istope* is 4½ billion years. But the half-lives of

*Uranium-238 is not itself a fissionable isotope but is the raw material for the production of plutonium-239, which is fissionable. This conversion is explained in Section 10.6.

the fission products are much shorter; some are measured in centuries, some in years, days, minutes, seconds, or fractions of a second. A compound with a short half-life decomposes quickly, but in its early stages it emits radiation at dangerously high levels. As an analogy, think of two logs, one rotting slowly and the other burning rapidly. The rotting log decomposes over a long period of time but always remains cool, whereas the burning one is hot during its short lifetime. Similarly, a radioactive element with a long half-life, such as uranium-238 (4½ billion years), emits low levels of radiation. But the fission products that have short half-lives, such as barium-142 (half-life, 11 minutes), are analogous to the burning log, for they emit dangerously high levels of radiation. These materials, produced in variety and abundance by nuclear chain reactions, are the atomic wastes that concern us.

10.6 NUCLEAR FISSION REACTORS

This section will describe the fundamental features of nuclear reactors. Its purpose is to provide a conceptual framework that will make it easier for you to confront the environmental problems associated with nuclear energy and to make evaluations concerning the social and economic issues that arise.

Nuclear fission reactors require fuel, and the fuel must be a substance whose nuclei can undergo fission. There are two significant nuclear fission fuels, uranium-235 and plutonium-239. These are not the only known fissionable isotopes, but they are the ones on which currently operating power plants and most designs for future plants are based.

Uranium-235 occurs in nature; it constitutes 0.7 percent of natural uranium. The remaining 99.3 percent is the heavier isotope, uranium-238, only very little of which undergoes fission in the reactor. The predominant fission reaction may be written as follows:

uranium-235 + 1 neutron $\longrightarrow$ fission products + 2 to 3 neutrons + energy   (3)

The second fuel, plutonium-239, does not occur in nature; it is produced by the reaction of uranium-238 with neutrons. The reactions may be written as follows:

uranium-238 + 1 neutron $\longrightarrow$ plutonium-239   (4)

plutonium-239 + 1 neutron $\longrightarrow$ fission products + 2 to 3 neutrons + energy   (5)

Thus, the two important naturally occurring sources of fission energy are uranium-235 (fissionable but not abundant) and uranium-238 (abundant but not fissionable until it is converted to plutonium-239).

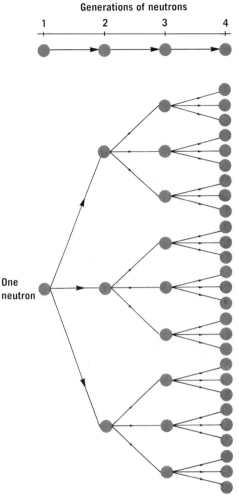

Generations of neutrons

One neutron

Figure 10.2 Neutron chain reactions. *Top:* unbranched chain. *Bottom:* three-for-one branching.

The Yankee nuclear power station at Rowe, Massachusetts. (From D. R. Inglis: *Nuclear Energy: Its Physics and Its Social Challenge.* Reading, Mass., Addison-Wesley, 1973.)

To *dampen* means to inhibit by the application of moisture. The usage is appropriate because a spreading fire is a branching chain reaction, and water slows down a fire by terminating many of the reaction branches.

It is apparent from the fission equations that nuclear reactors require another essential ingredient besides fuel, namely, neutrons. In fact, it is the behavior of neutrons, more than any other single factor, that determines the design of a nuclear power plant. Let us therefore consider what possible events neutrons can undergo, and how these events must be controlled to make the reactor work and to keep it safe. Let us assume that we start with natural uranium (99.3 percent uranium-238, 0.7 percent uranium-235) and make modifications as they are needed.

First, a neutron could undergo *fission capture* by uranium-235 (Equation 3). This would be fine; it provides more neutrons to branch the chain, and the reaction yields energy. But there are some problems. The uranium-235 nuclei capture only *slow* neutrons to undergo fission, but release *fast* neutrons. Some medium is therefore needed to slow down the emitted neutrons to make them more amenable to capture. Such a medium is called a neutron **moderator.** The first material used was graphite; water is also suitable. Another problem arises from the meager abundance of uranium-235 in natural uranium. The rate of fissions, and hence the production of energy, can be speeded up by increasing the proportion of uranium-235 in the fuel. Such enrichment must be limited, however, because the chain reaction must not be allowed to branch out of control.

Second, a neutron can undergo *non-fission capture by uranium-238,* thereby producing plutonium-239 (Equation 4), which can in turn undergo fission (Equation 5). This capture occurs only with fast neutrons, not slow ones, and will therefore be impeded by moderators used to slow down neutrons for uranium-235 fission capture. On the other hand, the production of plutonium-239 is, in effect, a "breeding" of new fuel and, therefore, is very attractive as a source of energy for human needs. These circumstances imply that the design of the reactor will depend to a great extent on the process we wish to favor.

Third, a neutron can undergo *non-fission capture by impurities.* This causes loss of neutrons and dampening of the chain, but it also offers a convenient means of controlling the reactions. Fission products accumulate as impurities (Equations 3 and 5), and because they soak up neutrons, the fuel elements that they contaminate must eventually be removed and purified. But it is necessary to have a controlled means of absorbing neutrons to regulate the reaction. The most direct method is to insert a stick of neutron-absorbing impurity. The more impurity that is pushed in, the more the reaction is slowed down. Devices that do this are called **control rods;** they usually contain cobalt or boron and other metals, and they can be inserted into or withdrawn from the reactor core so as to regulate the neutron flux with great precision.

Finally, a neutron, traveling as it does in a straight line, may simply miss the other nuclei in the reactor and *escape.* Obviously, the larger the reactor, the more atoms will stand in the way, and the greater will be the chance of neutron capture and the less the chance of escape. This circumstance imposes lower limits on reactor size—we shall never have pocket-sized fission generators, nor even fission engines for motorcycles. This escaping tendency also demands adequate shielding to prevent neutron leakage.

How, then, can a reactor be constructed so as to satisfy all these requirements? The first decision to be made is whether the reactor is to be designed only to generate energy from uranium-235 or whether it is also to "breed" plutonium-239. Let us start with the non-breeder.

NON-BREEDERS

Refer to Figure 10.3 as you read the following description. The reacting core contains fuel, moderator, and control rods. The fuel is typically a ceramic form of uranium dioxide. This compound is much better able than the pure metal to retain most fission products, even when overheated. The uranium fuel itself is prepared by conventional (non-nuclear) chemical treatment and consists of the natural nonfissionable isotope, uranium-238, enriched with the fissionable uranium-235 by a factor only three or four times above its naturally occurring level. This low level of enrichment provides an automatic protection against an increased fission rate if the temperature should rise accidentally, such as by failure of the coolant system. The fuel is inserted into the reactor core in the form of long, thin cartridges (Fig. 10.4). These fuel cartridges are covered (or "clad," as they say in the nuclear power plants) with stainless steel or other alloys.

The fuel elements are surrounded by the moderator,

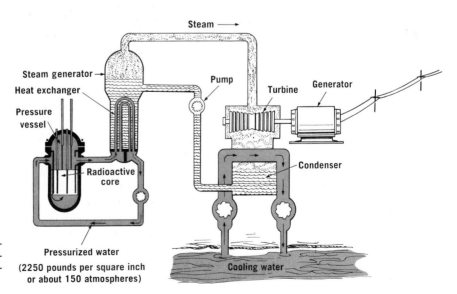

Figure 10.3 Schematic illustration of a nuclear plant powered by a pressurized water reactor.

Pressurized water
(2250 pounds per square inch or about 150 atmospheres)

Figure 10.4 A typical fuel assembly, consisting of many fuel rods, being lowered into place in a reactor. (From Inglis, *ibid.*)

whose function, remember, is to slow down the neutrons so that they will undergo fission-capture. Water is advantageous because it is both a moderator and a coolant. In the boiling water reactor, the fission heat converts the water to steam. In the pressurized-water reactor, shown in Figure 10.3, the water is maintained in the liquid state under high pressure. It then releases its energy to another body of water in a heat exchanger. The mode of generation of electricity is the same as in fossil fuel plants. The steam drives a turbine that powers an electric generator; the waste steam is then cooled and returned to the heat exchanger.

Interspersed into the matrix of fuel cartridges, moderator, and coolant are the control rods. They serve not only to

Figure 10.5 Control-rod driving mechanism of a pressurized water reactor (PWR), partially disassembled. (From Inglis, *ibid.*)

regulate neutron flow but also as an emergency shut-off system (Fig. 10.5). To speed up the chain reaction, the rods are partially withdrawn; to slow it down, they are inserted more deeply. In the event of malfunction, the rods are pushed rapidly all the way into the core to capture as many neutrons as possible and quench the chain reaction.

BREEDERS

Return now to page 287 for another look at Equations 4 and 5. This process produces fissionable plutonium-239 from abundant uranium-238, and, therefore, makes it possible for one reactor to provide fuel for another. (Hence the reactors "breed.") The uranium-238 is called a "fertile" material, to maintain the biological metaphor. The time required for a breeder to double its quantity of fissionable material depends on its design and operation; the values obtained by theoretical calculation range from 6 to 20 years. A doubling time about midway between these extremes is considered by many to be a reasonable expectation.

The breeding reactions require fast neutrons. This means that a moderator, which slows down neutrons, must be excluded. The reactor core, then, consists of uranium-238 that is highly enriched with uranium-235 or plutonium-239 (the sources of the fast neutrons), and no moderator. The space vacated by the moderator can accommodate additional enriched fuel; thus, the overall concentrations of *both* the fertile and fissionable materials in the breeder core are much greater than they are in a non-breeder. This situation is inherently more dangerous because energy is released in a more concentrated form. Thus, in case of a malfunction, there is greater danger of overheating and melting of the core, which would concentrate the fuel still further and release radioactive products still more rapidly.

The compactness of the breeder core demands a very rapid removal of heat. Water is disadvantageous because it is a neutron moderator, which must be avoided. Furthermore, water boils at relatively low temperatures even under high pressures, and steam is a poor heat conductor. The coolant of choice is liquid sodium. Sodium is a silvery, soft, chemically active metal. It reacts with water to produce hydrogen gas; if air is present, the heat of the reaction can spark the explosion of the hydrogen. The sodium becomes highly radioactive when exposed to the reactor core. But its saving virtue is its ability to carry heat away from the reactor core rapidly, since it is an excellent heat conductor and it remains in the liquid state over a very

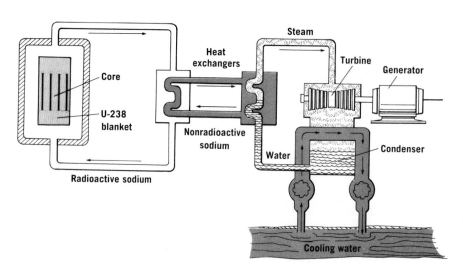

Figure 10.6 Schematic diagram of a fast breeder reactor.

292

wide temperature range, from 98° C to 890° C at normal atmospheric pressure.

The heat exchanger in which steam is produced to drive the turbine must be shielded from the radioactive sodium. This is accomplished by an intermediate loop of nonradioactive sodium. The entire arrangement is shown schematically in Figure 10.6, and details of the reactor building are pictured in Color Plate 7.

10.7 THE NUCLEAR FUEL CYCLES

The availability of nuclear fuels was discussed in Chapter 8. From mine to factory to disposal to ultimate death by radioactive decay, these materials will become more and more intimately involved with human activities if the nuclear industry continues to grow. It is therefore important to understand the steps involved in the processing, utilization, and disposal of nuclear fuels.

There are two different sequences to consider. The first applies to non-breeder reactors, and is really better characterized as a "once-through" process rather than a cycle (except that the wastes *do* eventually return to earth). The second applies to breeder reactors, which may represent the nuclear economy of the future, sometimes called the "plutonium economy." Both cycles are described below.

THE "ONCE-THROUGH" URANIUM CYCLE

Uranium ore is mined in various areas of the Earth as a black deposit containing perhaps 0.3 percent uranium. It is concentrated by a series of physical and chemical processes to a yellow mud that is about 80 percent uranium oxide, U_3O_8. After further processing, the uranium is obtained as a brilliant orange oxide,* UO_3.

Recall that only about 0.7 percent of natural uranium is the fissionable uranium-235 isotopes. The next step is therefore **enrichment,** which is the most difficult and costly portion of the fuel cycle. Since the isotopes are for all practical purposes chemically identical, the separation must be carried out by physical processes. The method, which is called **gaseous diffusion,** makes use of the fact that molecules of lighter gases move faster through porous barriers than do molecules of heavier gases. To take advantage of this phenomenon the uranium is chemically converted to a gaseous product (uranium hexafluoride, UF_6), and the molecules with the lighter uranium-235 pass through the barriers at a slightly higher rate than those with the heavier uranium-238. The enriched material is reconverted to an oxide form and fabricated into the fuel pellets used by the power plant.

*Many American homes contain old (pre-World War II) orange-colored kitchen pottery prepared from this uranium oxide pigment. The radiation level from such materials is low. However, it would be well to get them out of your kitchen. Donate them to the nearest university.

After a year or more, when an appreciable portion of the uranium-235 has been consumed and fission products have accumulated, the fuel assemblies are removed from the reactor. At this time the waste products are at their most intensely radioactive state, and they are too dangerous to ship. They are therefore stored underwater at a site on the plant premises for a few months to allow the most highly radioactive components to decay. The partially decayed fuel is then shipped to a fuel processing plant. Here the pellets are cut up, dissolved, and chemically processed to recover uranium and plutonium (which was produced during the time the fuel was in the reactor).

The uranium can be reconverted to UF_6 and recycled for enrichment, but this is not advantageous as long as rich uranium ores can be mined. Meanwhile, most of it is stored. Plutonium is also stored for possible future use in reactors or as an explosive. Some of the radioisotopes that have special applications in science, medicine, or industry are also separated and set aside for such uses. The remainder is a solution of **radioactive waste,** which must be boiled down to reduce its volume, and then stored somewhere for a length of time which, compared to the span of human lives or even political systems, seems like forever.

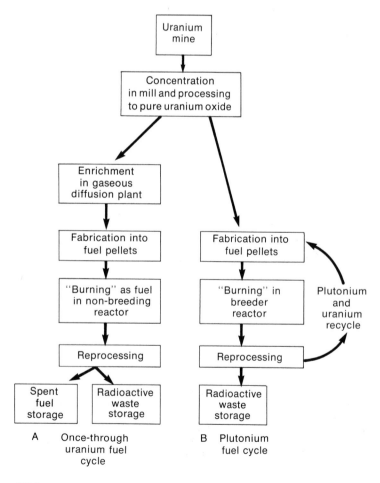

Figure 10.7 Nuclear fuel cycles.

294

The question of how and where to keep this material is the subject of Section 10.9.

THE PLUTONIUM BREEDER FUEL CYCLE

Review the earlier description of breeder reactors and recall that the breeder *produces* plutonium from uranium-238. This circumstance affects the fuel cycle in two important ways. First, the uranium enrichment step can be skipped because the reactor breeds its own fuel. Second, the fuel it breeds, which is plutonium, must be recovered and incorporated again into pellets. Since these operations do not all take place at the same location, the breeder fuel cycle has the effect of introducing plutonium in a highly enriched form into channels of commerce and transportation, with possible consequences that will be discussed later in this chapter. Figure 10.7 is a schematic diagram of the two cycles.

10.8 THE PRACTICE OF SAFETY IN NUCLEAR PLANTS

Questions about safety and hazards are at the heart of the nuclear controversy. It will be helpful, in approaching this very complex issue, to consider first some of the basic principles of industrial safety, and then to describe how they are applied in nuclear plants.

To become familiar with the general concepts of safe design, let us return to our familiar example, the automobile. You must first recognize that the purpose a car serves—moving people from place to place—is inherently dangerous, because other objects may get in its way and injure the passengers. Therefore, the first requirement is that *safe operation must be part of the original design.* For example, brakes should operate smoothly and reliably; the driver should be able to see the road clearly on all sides, even at night or during rain or snow; and the automobile body must be able to absorb shocks from irregularities of the road surface. But we all know that mechanical devices are far from perfect. Things can go wrong, and, given enough time, we may be confident that they will. There are two possible responses to such dangers. One of them is to *provide "back-up" or duplicating systems that will take over in cases of failure.* This approach is sometimes called **redundancy.** An example in the automobile is an independent braking system that will operate if the primary one fails. The other response is to *provide a warning,* so that the operator can act to avoid an impending accident. Thus a light or buzzer can indicate overheating or loss of oil. Since materials and components wear out, we must also *provide a schedule of inspection and maintenance.* Finally, if all these systems fail and an accident does occur, the design should *provide features that prevent or minimize injury to people.* Such features are safety belts, air bags, helmets, asbestos suits, self-regenerative bumpers, and roll bars.

Nuclear plants, like automobiles, serve an inherently dangerous function — they process materials that are exceedingly harmful to living organisms. Therefore the five principles of safe practice described above must be followed. Their applications in nuclear plants may be outlined as follows:

Safe Design. Recall that in a non-breeder reactor the uranium-238 is only modestly enriched with the fissionable uranium-235, so that the fuel is nothing like an atomic bomb. The control rods are inserted by pushing them down into the core, so that if power fails, they could simply fall. Ordinary water is both a coolant and a moderator. If excess heat should boil the water out, the loss of moderator would stop the chain reaction. Design specifications require that the materials of construction be of the highest engineering quality and be fully tested before use. As pointed out on page 292, the breeder reactor is inherently more dangerous, but this means only that safe design is, if anything, even more critical.

Redundancy. The system for which a back-up is most important is the one which cools the reactor core. If that should fail, there are generally at least *two* other independent cooling systems. If the power system on which the emergency measures depend should fail, an off-site source of power can be used. If *that* fails, on-site diesel generators or gas turbines can take over. Secondary systems of this type are quite complex and are interrelated in such a way that their responses are specifically appropriate to the nature of the emergency. Furthermore, these responses are fully automatic; they do not have to be initiated by a human operator.

Warning. The control room of a nuclear plant displays a panorama of gauges, dials, lights, buzzers, and bells. Individual workers are supplied with badges that are sensitive to radiation and that will monitor the degree to which the wearer has been exposed. Detection devices are distributed throughout the plant and are also set outdoors at various distances from the plant.

Inspection and Maintenance. Reactor operators must go through strict licensing procedures, with periodic renewal. The plants themselves are inspected several times each year, penalties are applied to violators of regulations, and listings are kept of any defects or failures.

Protection in the Event of Accident. The reactor vessel, made of thick steel, is itself surrounded by an anti-radiation shielding several feet thick (Fig. 10.8). As a final bar-

Control room of a nuclear plant.

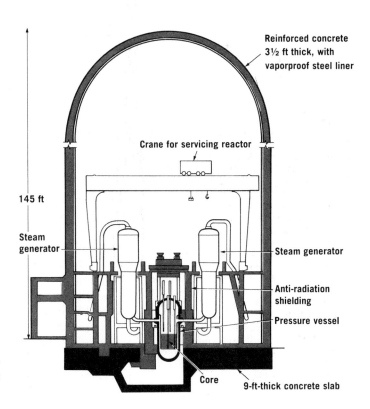

Reinforced concrete
3½ ft thick, with
vaporproof steel liner

Crane for servicing reactor

145 ft

Steam
generator

Steam generator

Anti-radiation
shielding

Pressure vessel

Core

9-ft-thick concrete slab

Figure 10.8 Containment structure for a nuclear power plant. (© 1971 by The New York Times Company. Reprinted by permission.)

rier, the entire system is surrounded by a vapor proof, steel-lined, reinforced concrete **containment structure.** This barrier is designed to withstand earthquakes and hurricanes and to contain all matter that might be released inside, even if the biggest primary piping system in the reactor were to shatter instantaneously. The Soviet Union and some other countries do not require this last barrier, a fact which nuclear proponents have cited to emphasize the high priority given to safety in the United States.

If all of the above sounds very comforting, remember that there are automobile accidents. Likewise, things go wrong in nuclear plants. Questions concerning such hazards are taken up in the next section.

10.9 ENVIRONMENTAL HAZARDS FROM FISSION REACTORS

Five sets of questions must be discussed in assessing the environmental hazards associated with nuclear fission plants: (a) What are the chances of a serious accident? (b) What are the extent and the environmental effect of the routine emissions of radioactive materials under normal conditions of operations? (c) What are the problems associated with the disposal of radioactive wastes? (d) What are the dangers of terrorism and sabotage? (e) What are the environmental effects of the waste heat released from the nuclear plants? The last of these effects, which is thermal pollution, was dealt with in Chapter 9. Here we consider the first four.

THE CHANCES OF A SERIOUS ACCIDENT

The nightmare of the reactor safety engineer is the possible overheating of the radioactive core, the melting of the reactor vessel, and the spread and possible escape of the molten mixture of fuel and fission products. How is it possible to judge the chances of such an accident and its possible consequences? A seemingly reasonable approach would be first to outline the steps that could lead to such a calamity, then to judge the probability of each event. Let us illustrate such a sequence, and then discuss how the probabilities can be assessed.

Imagine, first, that a pipe carrying cooling water to the reactor bursts. Such an occurrence is called a **loss of coolant accident (LOCA).** Since the water is also the neutron moderator, the chain reaction in the core would be terminated. However, there is another source of heat in the reactor that

A nuclear reactor is nothing like an atomic bomb. Opponents of nuclear energy have complained that no one ever made such an accusation, and therefore the statement diverts attention from more credible hazards. Nonetheless, the reader should recognize that a bomb contains highly concentrated fissionable materials, which leaves only two significant fates for neutrons: fission capture or escape. The factor that determines which of these two fates will predominate is size, or mass; the minimum mass required to support a self-sustaining chain reaction is called the **critical mass.** To set off an atomic bomb, therefore, subcritical masses of uranium-235 or plutonium-239 are slammed together by chemical high explosives to make a supercritical mass. The chain reaction instantly branches, and the mass explodes. The fuel in a nuclear reactor contains no such concentrations.

cannot be turned off—the energy release by the radioactivity of the accumulated fission products. In about 45 seconds, parts of the pressurized water reactor would heat up to 1480°C, at which temperature water would react with the fuel cladding to release hydrogen gas, which can explode in air.

To prevent such an occurrence, an **emergency core cooling system (ECCS)** automatically introduces another stream of water. Even if this works as expected, large quantities of water and steam carrying radioactivity will still spray out of the reactor at high pressure. The containment structure is designed to prevent any of this material from escaping to the outside. However, if the ECCS system should fail to operate—or, even if it does, if it should fail to hold the reactor below the 1480° C danger point—the fuel rods may buckle or even break and thus block the passage of water completely. In such an event, the temperature would continue to rise until, in about half an hour, the fuel would melt. Within a few hours the reactor vessel itself would be breached, and tons of white-hot radioactive material would melt its way into the ground. (This scenario is sometimes jokingly called the China Syndrome, meaning that the molten mass is moving toward the other side of the globe.) Large quantities of radioactive matter could be released from the ground to the air, to subsurface water or, if the containment structure itself failed, directly from the reactor core to the atmosphere.

Let us return now to the first step, the bursting of a water pipe. This is no ordinary household pipe that readily springs a leak during a cold snap when its water freezes because the furnace happens to go out and the occupants happen to be on vacation. Nonetheless, the pipe was manufactured by people in some other factory, and things could have gone wrong *there,* so its integrity is not absolute. It would do little good to stare at a pipe, scratch your head, and try to guess when it will burst. A much more reasonable approach would be to go where industrial pipes have been in service for many years and find out how many *have* burst. There are quite a few pipes in places like petroleum refineries, chemical manufacturing plants, and water distribution systems. With some effort, it is possible to survey the history of pipe ruptures in all such applications. This information may then be used as a basis for calculating the chance of a LOCA in a nuclear plant. Remember, however, that a LOCA triggers the ECCS, and that, too, would have to fail before a catastrophe could occur. Again, one can obtain a history of failures of pipes, valves, motors, and monitoring instruments (they are all susceptible) and, in like manner, apply the results to the nuclear reactor. Finally, if a radioactive cloud *does* escape, where will it go?

The numbers game now involves the chances that the wind will blow the cloud toward populated areas, or that it will be washed down by rain, and so on and on. Taken together, such procedures require considerable time and effort (not to mention guesswork). It is important to recognize, however, that the United States government estimates of nuclear safety are based on just such an approach. In fact, the estimates were prepared under the direction of Professor Norman C. Rasmussen of the Massachusetts Institute of Technology. The study, which required two years and cost three million dollars, was released by the Nuclear Regulatory Commission in 1975.

The key conclusions were as follows:

Event	Chance of Occurrence
Complete fuel meltdown of any one reactor in any one year.	1 in 17,000
An accident killing as many as 1000 people by acute radiation sickness in any one year.	1 in a million
The "worst" accident occurring to any one reactor in any one year, which would cause 3300 deaths, 45,000 "early illnesses," over 1500 latent fatal cancers, and an approximately equal number of genetic defects.	1 in a billion

"Of course it's perfectly safe. Any accident would be in complete violation of the guidelines established by the Nuclear Regulatory Commission."
(Sidney Harris, *American Scientist*, Sept.–Oct., 1976.)

It is very easy to show that, in Bethe's words (page 273), these risks "are statistically small compared with other risks that our society accepts." However, opponents of nuclear power reject the Rasmussen approach and its predictions on many levels. These objections may be categorized as follows:

- Various tests of small-scale ECCS operations (about 1/60 the size of full-sized units) have all resulted in failures. These findings are therefore very different from the Rasmussen predictions.
- The Rasmussen study assesses the risks to the people *within 25 miles* of nuclear reactors. It therefore neglects the possible effects of an accident on the entire United States population, let alone on the population of the world.
- Private insurance companies have been willing to insure nuclear power plants only for limited liabilities. Protection beyond that is available only from the government under the **Price-Anderson Act.** This circumstance has been cited to argue that insurance companies do not believe the Rasmussen predictions.
- The entire analysis is based on an unprovable series of assumptions (or guesses) about the chances of various malfunctions. Many scientists disagree with them. Some engineers have resigned from their jobs in nuclear plants in protest against what they assert are unacceptably great risks of which the public is not aware. Perhaps the most devastating criticism of all is the fact that the methodology of the Rasmussen report, when applied retroactively, failed to predict accidents *which had already occurred when the report was issued.* Many malfunctions (but no catastrophic accidents, up to the time of this writing) have plagued nuclear plants throughout the lifetime of the program. Rather than citing all these data, one incident will be described in some detail in the Case History at the end of this chapter.

> The limit of liability of owners of nuclear plants in the event of a nuclear accident, as set by the Price-Anderson Act, has been $560 million. This provision was declared unconstitutional in a decision handed down on March 31, 1977, by James B. McMillan, U.S. District Judge for the Western District of North Carolina. The occasion for the suit was a proposal by the Duke Power Company to build nuclear plants in North and South Carolina. The case will be appealed.

"ROUTINE" RADIOACTIVE EMISSIONS

Even with the best design and with accident-free operation, some radioactivity is routinely released to the air and water outside the plant. In the boiling-water reactor, for example, the water passes directly through the reactor core and thus circulates around the fuel elements. Some of the fuel claddings, which are very thin (about 0.05 cm) inevitably develop small leaks, which permit direct transfer of radioactive fission products to the water. This situation has been somewhat improved by substitution of zirconium-alloy cladding for the stainless steel previously used. Even in the absence of leaks, however, some neutrons do get through to the water and make some of its impurities radioactive, and this effect, too, is a source of "routine" emissions to the watercourses that serve as the ultimate coolants for the power plant.

Radioactive material can also be gaseous. Krypton-85, for example, is a radioactive fission product (half-life, nine years) that is insoluble in water and escapes to the atmosphere through a tall stack. These emissions, taken together, are so small that their effect may be compared with the background radiation from cosmic rays and naturally radioactive materials in the Earth's crust to which all life is subject. This fact is often cited to support the contention that the routine emissions are trivial. However, Linus Pauling has pointed out that the effects of even natural radioactivity are not trivial, since they may account for some 10 percent of all cancer deaths. Therefore any additional exposures of comparable magnitude will bring about unacceptable human harm.

THE DISPOSAL OF RADIOACTIVE WASTES

Refer back to Figure 10.7 (page 294) to note the items labeled "radioactive waste storage." These wastes produce various generations of "daughters" before a final stable isotope ends the sequence. Figure 10.9 shows the shape of the decay curve of a hypothetical isotope with a half-life of one month. During the early and intermediate stages of decay, the wastes are held in temporary storage (Fig. 10.10). Eventually they will be concentrated until they are solid. Then they will be put in a "permanent" resting place. Where? Disposal at sea was formerly used extensively, until it was banned by international convention. The most appealing location has been considered to be an abandoned salt mine or salt cavern (Fig. 10.11). These geologic formations are attractive because some of them have been stable for millions of years and have not been in contact with subsurface waters. The U.S. Atomic Energy Commission seriously considered a salt mine under the town of Lyons, Kansas (population about 5000). Lyons sits over two salt mines, one abandoned, the other in daily operation. Their main tunnels come to within 550 meters of each other. Some of the previously mined cavities were left filled with water. These circumstances raised such serious doubts about the ability of the abandoned mine to maintain its dry integrity for "millions of years" that, in 1972, the Atomic

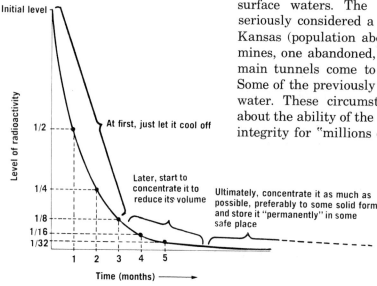

Figure 10.9 Disposal of a hypothetical radioactive waste product with a one-month half-life.

302

Figure 10.10 A "tank farm" under construction, showing large tanks to hold high-level liquid wastes. (From Inglis, op. cit.)

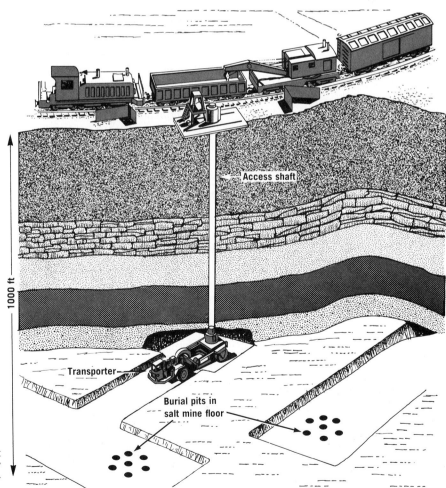

Figure 10.11 Final disposal of radioactive wastes in a salt mine. (© 1971 by The New York Times Company. Reprinted by permission.)

Access shaft

1000 ft

Transporter

Burial pits in salt mine floor

Energy Commission abandoned the Lyons site. The general approach, however, is still considered reasonable, and the search for suitable sites continues.

One suggestion has been to bury the wastes under the Antarctic ice cap. The method is appealing. If hard radioactive masses were placed on the surface of the ice, they would melt their own way down to bedrock. The hole would then rapidly reseal itself by freezing. It is estimated that the average Antarctic temperature has remained below freezing for over one million years, so that the radioactive wastes may be expected to be securely isolated from the biosphere for a similar period into the future. Critics point out, however, that once the canisters reach bedrock, they will continue to give off heat and will melt the ice around them. The resulting water, contaminated with radioactive matter, may flow out from beneath the ice sheet into the open ocean. Moreover, the entire ice sheet may eventually be able to slide over its warm, wet bed (pulverizing the canisters along the way) and eventually raise the sea level by five meters or so. Proponents of the method believe these dangers are remote, but all agree that a very careful environmental impact study is required.

More recently, the sea has been reconsidered, but rather than resting on the bottom, the wastes would be buried deep in the underlying sediment (Fig. 10.12). The sites would be selected from a region of ocean floor that is far from any geologic activity and that is expected to be undisturbed for hundreds of thousands of years. The deep clay layers would seal off the wastes and separate them from any contact with the biosphere.

Meanwhile the wastes are being held in carefully watched vaults (Fig. 10.13), awaiting a final decision.

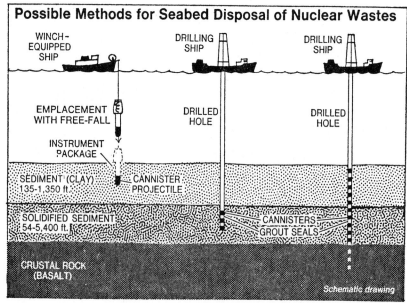

Possible Methods for Seabed Disposal of Nuclear Wastes

WINCH-EQUIPPED SHIP

DRILLING SHIP

DRILLING SHIP

EMPLACEMENT WITH FREE-FALL

INSTRUMENT PACKAGE

DRILLED HOLE

DRILLED HOLE

SEDIMENT (CLAY) 135-1,350 ft.

CANNISTER PROJECTILE

SOLIDIFIED SEDIMENT 54-5,400 ft.

CANNISTERS GROUT SEALS

CRUSTAL ROCK (BASALT)

Schematic drawing

The New York Times/March 3, 1977

Figure 10.12 Engineering scheme, left, would involve free-falling canister with monitor attached. Canister would embed itself in soft, deep clay, and monitor could then be recovered. Other methods involve multiple-canister deposition.

304

Figure 10.13 "Fission's graveyard."

TERRORISM AND SABOTAGE

Uranium and plutonium inventories are not always fully accounted for, although the discrepancies are more likely to be in the books than in the atoms. Nevertheless, the question remains, "Can they be stolen, and, if so, can amateurs convert them into bombs?" A related question is, "Can saboteurs damage nuclear plants so that radioactive matter is released to the environment?" Some writers have speculated on this matter (see bibliography); some physics majors have written term papers on how to make a bomb (but they didn't make one); and some terrorist acts have actually been threatened and attempted. All these circumstances have led to increased security measures around nuclear installations. Since our past experience is (fortunately) so limited, we can only guess at the possibilities.

On August 4, 1977, the federal government reported that nuclear facilities in the United States were unable to trace more than 8000 pounds of highly enriched uranium and plutonium as of September, 1976, but that there was no evidence that the material had been stolen.

The two agencies responsible for assuring the safekeeping of the highly dangerous materials, ERDA and NRC, contended that the uranium and plutonium had simply become trapped in machinery, wiping cloths, and scrap and then lost in the crude statistical and measuring systems.

10.10   NUCLEAR FUSION

Energy can be obtained by the combination, or **fusion,** of nuclei of certain light elements, particularly certain of the isotopes of hydrogen. The fusion reaction does not create any environmental problems at this time (except for explosions of the "hydrogen" bomb in warfare or in the testing of weapons), because no useful fusion reactors have yet been developed. However, practically everyone hopes that they will be developed, because nuclear fusion promises "inexhaustible" energy (see Chapter 8).

There are three isotopes of hydrogen: **protium, deuterium,** and **tritium.** Some information about them is given in Table 10.2. Fusion reactions can occur between protium and deuterium, between deuterium and deuterium, and between deuterium and tritium.

Unlike the fission reaction, however, fusion cannot be triggered by neutrons; what is required instead is that the fusing nuclei be fired at each other at very high speeds so as to overcome their initial repulsions. In other words, the temperature must be very high; the resulting fusion is therefore called a **thermonuclear reaction.** If a large mass of hydrogen isotopes fuses in a very short time, the reaction cannot be contained and it goes out of control; this is the explosion of the "hydrogen bomb." On the other hand, useful energy could be extracted from fusion if it were possible to devise what is called "controlled thermonuclear reaction."

To initiate fusion, the temperature must be raised to about 40 million degrees Celsius for the deuterium-tritium reaction and 400 million degrees for the deuterium-deuterium reaction. Major problems are imposed by these requirements. Of course, the problems are much more severe at the higher temperature needed for the deuterium-deuterium reaction—so much more severe, in fact, that present efforts to develop fusion reactors are directed toward the "cooler" deuterium-tritium reaction. But then we need a source of tritium, which is not naturally available on earth, and this requirement, too, poses problems.

Thermonuclear reactions are extremely difficult to control because no materials can withstand the required temperatures. Instead, all substances decompose into their free atoms, and the atoms themselves decompose into a mixture of positive nuclei and free electrons that is called a **plasma.** No rigid container exists that can survive long enough to confine a plasma for the useful production of thermonu-

TABLE 10.2 Isotopes of Hydrogen

ISOTOPE	OTHER NAMES	RADIOACTIVE?	NATURAL ABUNDANCE (%)
Hydrogen-1	"Ordinary" hydrogen "Light" hydrogen Hydrogen Protium	No	99.985
Hydrogen-2	"Heavy" hydrogen Deuterium	No	0.015
Hydrogen-3	Tritium	Yes (12-year half-life)	None

clear energy. Instead, what is envisaged is a sort of "magnetic bottle," which does not consist of a physical substance at all, but rather is a magnetic field so designed that it will confine the charged particles of the plasma in which the thermonuclear reaction is going on. The useful energy will have to be extracted from the process in the form of the kinetic energy of the evolved neutrons. Since the neutrons carry no charge, they will penetrate the magnetic field and escape from the plasma. The energy of the speeding neutrons can then be extracted by a moderator, just as in a fission reactor. If the moderator is water, the energy will create steam which can drive a turbine.

We come now to the tritium problem. Remember that the deuterium-tritium reaction is cooler than the deuterium-deuterium reaction, and hence preferable to it, but tritium is not naturally available. However, tritium is produced by the reaction between neutrons and lithium-6, a light isotope which constitutes 7.6 percent of natural lithium. To exploit this process, the entire fusion reactor would be encased in a sheath or blanket in which molten lithium is continuously circulated. The lithium would absorb the neutrons, supply the tritium and then release its heat to water in a heat exchanger (Fig. 10.14).

It should not be surprising that all these requirements impose engineering problems so difficult that they have not yet been solved. But if they were solved, what kind of world would be created?

First, there would be abundant energy. It is estimated

Figure 10.14 Schematic drawing of a thermonuclear power plant.

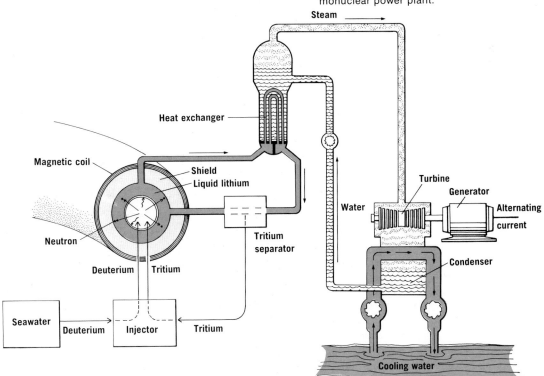

307

that if everyone used as much energy as the average North American does, the deuterium in the oceans could provide energy for about a billion years. This statement, by itself, does not describe the kind of world that such an abundance would create, but it does imply the likelihood of drastic social and environmental change.

Perhaps the foremost difficulty would be the thermal pollution that such an abundance could create. In fact, some scientists believe that this problem sets an upper limit on the rate at which any new energy sources could be developed.

Could the fusion reactor get out of control and go off like a hydrogen bomb? Nuclear scientists are entirely confident that the answer is no, an explosion could not occur. The reason is that the hydrogen isotopes are continuously fed into the reactor and are continuously consumed; they do not accumulate. The total quantity of fuel in the plasma at any one time would be very small—about two grams or so—very far below the critical mass required for a runaway reaction. If the temperature were to drop, or the plasma somehow dispersed itself, the reaction would stop; in effect, the fusion would turn itself off. The situation is rather analogous to that of a burning candle; if something goes wrong, the flame goes out, the candle does not explode.

Would there be a problem of environmental radioactivity? The answer here is yes, because both tritium and neutrons could be released. Tritium is radioactive (half-life, 12 years) and, being an isotope of hydrogen, combines with oxygen to form radioactive water. The tritiated water could conceivably enter food webs and harm living organisms. However, the energy released by tritium (in the form of fast electrons) is so weak that it would be virtually harmless if its source were outside the body. Furthermore, sound engineering practice should be able to restrict any tritium emissions to inconsequential levels.

Neutron release is another potential hazard, but we must remember that the neutrons stop when the reaction stops. However, neutrons are absorbed by atomic nuclei, and the new atoms that are thereby produced are radioactive. As a result, there could well be substantial quantities of radioactive matter to be disposed, but in general, the problem should be much less difficult than that of wastes from fission reactors.

10.11 THE NUCLEAR CONTROVERSY AND THE OKLO PHENOMENON

A rather recent development in the nuclear controversy has arisen as the result of an unexpected discovery. Recall first the disputes about the "ultimate" storage of radioactive wastes in salt mines or other stable repositories.

PLATE 1

The most favorable life conditions for different species of plants and animals are clearly displayed by the zonation on rocks along the Atlantic coast. (Courtesy of William H. Amos.)

PLATE 2

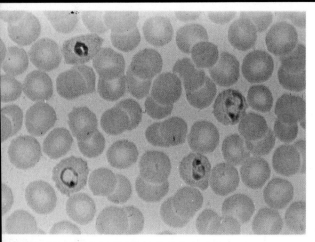

(A) Human red blood cells, four of which are infected with *Plasmodium*, the parasite that causes malaria. (Courtesy of Dr. Herman Zaiman.)

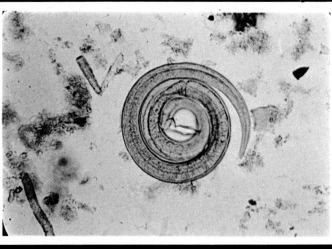

(B) Larva of the pork roundworm, *Trichinella spiralis*. (Courtesy of Dr. Herman Zaiman.)

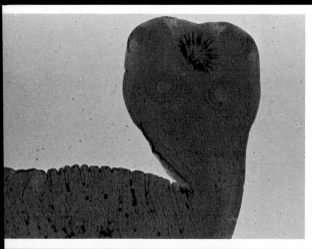

(C) The anterior end of the pork tapeworm, *Taenia solium*, as it appears in the intestine. (Courtesy of Dr. Herman Zaiman.)

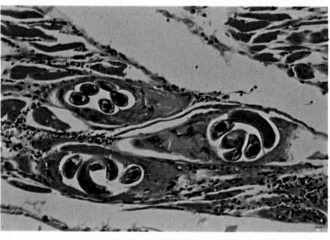

(D) Young *Trichinella* roundworms in muscle. (Courtesy of Dr. Herman Zaiman.)

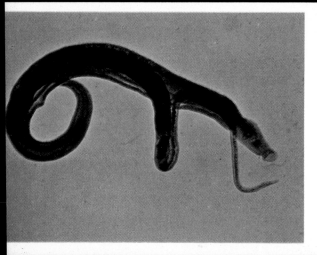

(E) *Schistosoma japonicum*, male and female flatworm parasites which infect the blood vessels of millions of people. The female lies within a canal of the larger male. (Courtesy of Dr. Herman Zaiman.)

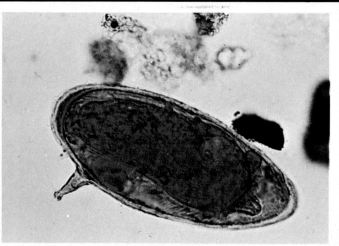

(F) An egg of the flatworm parasite, *Schistosoma mansoni*, from human stool. Note the characteristic spine on the shell. (Courtesy of Dr. Herman Zaiman.)

PLATE 3

Two examples of species that have established niches despite severe environmental resistance. (Left) California cholla cactus. (Below) California bristlecone pine. Some of these trees have reached an age of 7000 years and are the oldest living organisms. (Ed Cooper Photos.)

Island of Hawaii ferns, lichens on lava. These represent the first successional stage on fresh lava flows. (Ed Cooper Photo.)

PLATE 6

Industrial air pollution (West Virginia).

Direct discharge of human wastes into the ocean (an outhouse in the San Blas islands, Panama)

Direct use of solar energy (drying coffee beans in Guatemala).

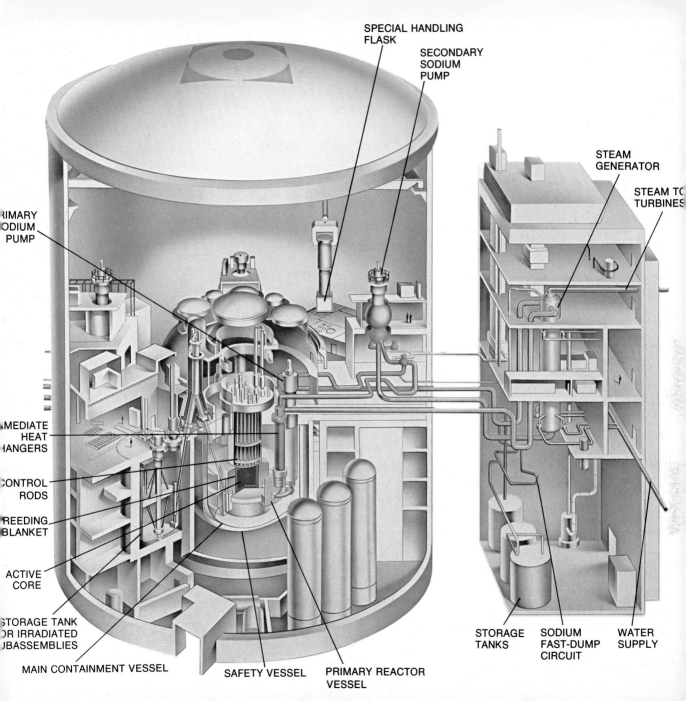

SPECIAL HANDLING
FLASK

SECONDARY
SODIUM
PUMP

STEAM
GENERATOR

STEAM TO
TURBINES

PRIMARY
SODIUM
PUMP

MEDIATE
HEAT
HANGERS

CONTROL
RODS

REEDING
BLANKET

ACTIVE
CORE

STORAGE TANK
OR IRRADIATED
UBASSEMBLIES

MAIN CONTAINMENT VESSEL

SAFETY VESSEL

PRIMARY REACTOR
VESSEL

STORAGE
TANKS

SODIUM
FAST-DUMP
CIRCUIT

WATER
SUPPLY

PLATE 8

Agricultural monocultures. (Left) California strawberries, Salinas Valley, California. (Below) Apple orchard in bloom, Bethel, Connecticut. (Ed Cooper Photos.)

PLATE 9

Patterns of smoke plumes. (Courtesy of R. S. Scorer.)

(A) Steam plume, consisting largely of water droplets.

(B) Combustion smoke, consisting largely of carbon particles.

PLATE 10
SMOKE TYPES.

(C) Flare, in which a waste gas burns freely in air. The visible flame contains incandescent particles of carbon, showing that the combustion is incomplete. As the particles cool, they stop glowing and appear as a black smoke, like (B).

(D) Smoke containing oxidized metals, especially iron. (B, C, and D from Connecticut Lung Association.)

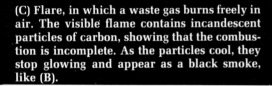

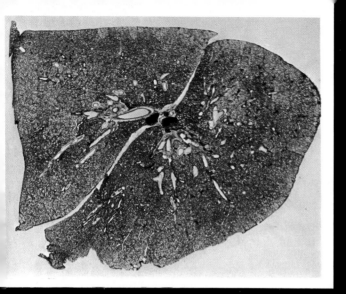

(Left) A normal lung.

(Below) A lung from a person with emphysema. Note the destruction of the normal alveolar structure with the formation of large air spaces that are very inefficient in the process of gas exchange between the breathed air and the bloodstream.

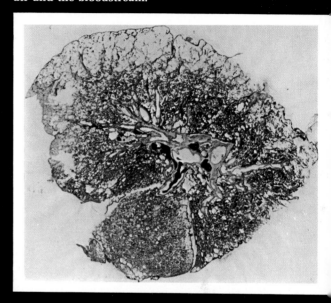

(Above and Right) Damage of monuments by air pollution. (Courtesy of R. S. Scorer.)

PLATE 12

(A) White Cascade petunia leaves show injury from ozone. White-flowered varieties tend to be more susceptible than those with colored flowers. (Courtesy of the United States Department of Agriculture.)

(B) Tomato leaflet grown in laboratory in unfiltered air containing oxidants. (Courtesy of the United States Department of Agriculture.)

(C) Peony displaying symptoms of fluoride damage. (Courtesy of the United States Department of Agriculture.)

(D) Typical sulfur dioxide injury to white birch leaf at right following two hours of exposure to two parts per million. (Courtesy of the United States Department of Agriculture.)

The arguments have been partly about geological factors (how "permanent" is the stability of a salt mine?) and partly about political factors (how reliable are future governments that we expect to monitor the storages for thousands of years?). One "fact," at least, seemed certain—no such storage of wastes from nuclear fission had ever occurred on Earth before. But recent evidence has shown that at least three natural fission reactions functioned about 2 billion years ago at what is now the Oklo uranium mine in the African country of Gabon. The evidence includes the finding that uranium in this mine contains as little as about 0.3 percent of the uranium-235 isotope, instead of the 0.7 percent that is normally found in uranium everywhere on Earth. If this reduction resulted from nuclear fission, then some plutonium-239 was formed, just as in a modern reactor, and must have produced fission products of the type that now concern us.

Chemical analysis has confirmed the presence of such products *and the fact that many have remained in place.* These observations, in addition to being scientifically interesting, have been used by proponents of nuclear energy to argue that the migration of similar radioactive wastes from carefully projected repositories is not a matter of serious risk to future human generations. Of course, this argument, too, has been countered. Not *all* of the wastes have remained in place. And the Oklo phenomenon may not be unique. In fact, some scientists conjecture that there may have been many natural nuclear reactors when the uranium-235 abundance was much greater than it is now. Depending on local geological conditions, the wastes from some of them may have remained in place, while those from others may have spread into the environment. The point is that our experience with nuclear power is, after all, of quite recent origin and that new aspects of the issue are constantly emerging. These include both the development of new methods for safe operation and the discovery of new things that can go wrong. The continued expansion of nuclear power depends on the successful control of environmental hazards over very long periods of time.

10.12 CASE HISTORY: *The Incident at Browns Ferry*

The Browns Ferry Nuclear Power Plant was constructed by the Tennessee Valley Authority (TVA) as a prototype for future U.S. power production. Its size, design, and generating capacity were to serve as a model for perhaps 1000 such plants, which would make nuclear energy the primary source of the nation's electricity by the end of the century.

Browns Ferry is a complex of three nuclear reactors, located on the red clay banks of the Tennessee River near Decatur, Alabama. It went into operation on August 1, 1974, and was eventually to supply electricity for the needs of about two million people.

At noon on March 22, 1975, Reactors 1 and 2 were operating at full power, delivering 2.2 million kilowatts of electricity. Just below the control room an electrician and his engineering

Work crew at the Browns Ferry nuclear power plant. (Courtesy of the Union of Concerned Scientists, Cambridge, Mass., from "Browns Ferry: A Regulatory Failure.")

aide were engaged in some post-construction modifications of the plant. They were in a space called the "cable spreading room," which is where electrical cables come together from various parts of the plant and are then separated, or spread, and routed through different tunnels to the reactors. The mission of the men was simple enough – to plug air leaks where the cables passed through a wall into the reactor building. It is not very difficult to detect air movements; ventilating and air conditioning workers do it all the time. Since air is invisible, the method is simply to introduce something visible into the air stream, such as a puff of smoke. At Browns Ferry, the two men were using candles! One of the cable pass-throughs was a bit too high for the electrician to reach, even while standing on an air duct. His 20-year-old aide, Larry Hargett, who had long arms, volunteered. This was a technical violation, because Hargett was new and untrained, and his official function was merely to check that the electrician tested each pass-through. What happened next is described in Hargett's own words:

We found a 2 × 4 inch opening in a penetration window in a tray with three or four cables going through it. The candle flame was pulled out horizontal, showing a strong draft. [The electrician] tore off two pieces of foam sheet for packing into the hole. I rechecked the hole with the candle. The draft sucked the flame into the hole and ignited the foam, which started to smolder and glow. [The electrician] handed me his flashlight with which I tried to knock out the fire. This did not work, and then I tried to smother the fire with rags stuffed in the hole. This also did not work, and we removed the rags. Someone passed me a CO_2 extinguisher with a horn which blew right through the hole without putting out the fire, which had gotten back into the wall. I then used a dry chemical extinguisher, and then another, neither of which put out the fire.

Then –

- 12:35 PM (about 15 minutes after the fire started). A fire alarm was turned in. Part of the delay was caused by confusion over which telephone number to call. In spite of the alarm, the reactors were not shut down.
- 12:40 PM. The operator of Reactor 1 in the control room noted that the pumps in the emergency core cooling system (ECCS) had started. Some indication lights on the control board were glowing brightly, others were dimming and going out; alarms were ringing, and smoke was coming out of one of the control panels.

310

The starting point of the Browns Ferry fire: densely packed cables leaving the cable spreading room into the Unit 1 reactor building on the right. The workman, standing on the air duct at the bottom, reached up into the cable wall with a lighted candle, igniting the plastic foam. (Courtesy of the Union of Concerned Scientists, *ibid.*)

- 12:50 PM. The power level in Reactor 1 began to behave erratically.
- 12:51 PM. The pump quit. The operator shut down Reactor 1 by inserting the control rods.
- 12:55 to 1:00 PM. Electrical supply needed to control and to power the ECCS for Reactor 1 was lost. So were the normal feedwater system, the reactor core spray system, the reactor core isolation cooling system, and most of the instrumentation that tells the control room what is going on. In addition, warning lights for Reactor 2 started to blink.
- 1:00 PM. Reactor 2 was shut down.
- 1:45 PM. The high-pressure ECCS for Reactor 2 was lost.

The next hours saw frantic attempts to prevent the water in Reactor 1 from boiling off to an extent that would expose the core and start a meltdown. The water level normally stands 200 inches (508 cm) *above* the top of the core. Pumps were switched around and pressures adjusted in accordance with the capabilities of the makeshift pumps. The water level dropped to 48 inches (122 cm), but then held.

- 2:43 PM. One of the plant's four standby diesel generators failed.

Meanwhile, the firefighting efforts had not been going well. The control room was filling with smoke. The neoprene* cable covers were burning and giving off sickening fumes. Much of the firefighting equipment in the plant was hard to get at, cumbersome to use, or even unserviceable. The cables continued to burn for six hours. Professional firemen from the Athens, Alabama, fire department had been on

*Neoprene is a chlorinated synthetic rubber, and it emits chlorinated fumes when it burns.

the scene since 1:30, and the fire chief recommended water but was overruled by the plant superintendent. Around six o'clock the superintendent finally deferred to the fire chief, whose men then put out the fire in about 15 minutes.

There were further difficulties, which were gradually worked out during the night.
- 4:30 AM (next morning). Normal shutdown of Reactor 1 was finally established.

The Browns Ferry incident was significant because it illustrated several kinds of failures that can occur in nuclear plants:

(a) *Failure of redundancy.* Recall that redundancy involves the duplication of safety devices. The doctrine assumes that the devices are independent of each other, so that if the chance that device #1 will fail is 1/100, and the chance that device #2 will fail is the same, then the chance that *both* will fail is 1/(100 × 100), or 1 in 10,000. But if a *single event* knocks out both devices, this type of calculation is meaningless. Such an event is called a **common-mode failure.** The fire at Browns Ferry was a single event that caused multiple failures.

(b) *Failure of prediction.* The Rasmussen study is based on the reasoning that accidents can be predicted from past experience and that *new* kinds of accidents are unlikely because if they were likely, they would have happened already. Such reasoning is both fallacious in theory and contrary to experience. It is fallacious because it assumes that the improbability of any *one* new accident applies to the improbability of *all* new accidents taken together. It is contrary to experience because we all know that "crazy" events happen all the time.

(c) *Failure of regulations.* The investigations of the Browns Ferry fire show a repeated history of inspections that were inadequate, rules that were broken, obvious dangers that went unheeded. Regulations are words written on paper. By themselves, they do not prevent accidents.

(d) *Human failure.* No combinations of safety equipment, safety warning devices, and safety rules have yet been devised that can keep up with the complexities of human behavior. If carelessness, ambition, fatigue, impatience. poor communication, embarrassment at admitting past failures, or other human frailties militate against safe practice, the reliability of inanimate systems is at risk. At Browns Ferry the plant superintendent admitted to the federal inspectors, "I was aware that polyurethane was flammable, but it never occurred to me that these penetrations were being tested using candles." Other senior managers at the plant, however, knew that candles were being used, but thought the sealant materials were not flammable. The candles had, in fact, been used for more than two years; there had been fires before that had been put out successfully; it had been concluded that the practice should be stopped; yet nothing was done.

TAKE-HOME EXPERIMENTS

1. **Radioactive decay.** You are not going to attempt any home-made bombs, but you can illustrate nuclear processes by other means. Shuffle a deck of cards. The exact number doesn't matter, but the more cards the better, so you may even combine two decks. However, count your cards before you start. Most card shuffling is far from thorough, so mix your cards at least a dozen times. Now turn the deck so that you see the faces of the cards. Fan them out, and discard any card *that follows a card of the same color.* If you do this correctly, you will be left with a set of cards of alternating colors—red, black, red, black, etc. Count your remaining cards and record the number. Now shuffle your hand again (even *more* thoroughly than before) and repeat the process. Do this twice more, for a total of four shuffles. Now plot your results on a graph of the type shown in the sketch. Repeat the entire process as many times as you have patience for, so that there will be several points on the graph corresponding to each shuffle. Now draw a line that fits most of the points. This curve is the pattern of *first order decay,* showing a fall-off by "half-lives." Can you explain why the card procedure generated this type of curve?

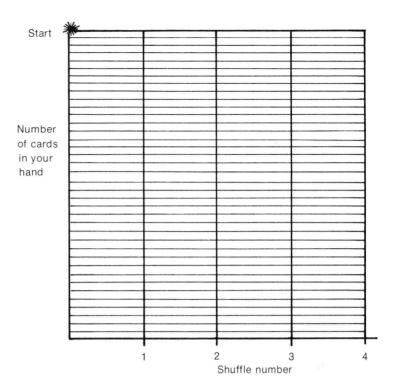

Start

Number
of cards
in your
hand

1   2   3   4

Shuffle number

2. **Branching chain reaction.** This experiment won't set off an explosion either, but it will illustrate the principle of one. You will need about three dozen wooden matches, some glue, a little aluminum foil, and a pencil. Lay the matches down on the foil in the pattern shown in the sketch. Run a layer of glue over each row of matches, and set a pencil in the top layer of glue. Do not let any glue touch the match heads. Let the assembly dry overnight, then pick it up carefully and remove any foil that might be sticking to it. Now, using the pencil as a handle and holding it at the far end, suspend the entire assembly over a sink or wash basin partly filled with water, and light the bottom end of the match first. Hold the assembly vertically, so that it burns uniformly upward. Describe the results. What factors make this process a branching chain reaction? In what ways could the rate of branching be speeded up?

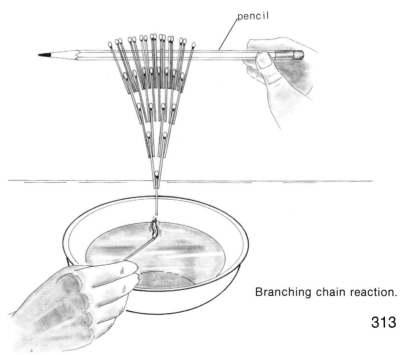

pencil

Branching chain reaction.

PROBLEMS

1. **Nuclear initials.** Identify the following sets of letters: (a) ECCS; (b) LOCA; (c) AEC; (d) ERDA; (e) NRC.

2. **Definitions.** Define radioactivity; isotope; radioisotope.

3. **Chain reaction.** What is a chain reaction? Explain chain propagation (lengthening); chain branching; chain termination; critical condition. Can the spread of a rumor among a large group of people function as a chain reaction? If so, illustrate how the chain could branch or terminate. Define the critical condition of such a system.

4. **Half-life.** A Geiger counter registers 512 counts per second near a sample of radioactive substance; 48 hours later the rate is 256 cps (counts per second). What is the half-life of the radioactive substance? What rate will the counter register after an additional 144 hours?

5. **Half-life.** Cesium-137 is a radioactive waste product whose half-life is 30 years. It is chemically similar to potassium, which is an essential element in plants and animals. Its compounds are readily soluble in water.
   (a) How long will it take for 1000 mg of cesium-137 to decay to 125 mg?
   (b) Since cesium compounds are soluble, would it be wise to dump this isotope into an open holding pond and let it dissolve and decay until only negligible quantities remain? Defend your answer.

6. **Health.** Outline the types of damage to the body that can result from exposure to high-energy radiation. Can there ever be any benefits? Explain.

7. **Health.** Give two reasons why a 70-year-old woman does not face so serious a problem regarding the health effects of radiation as a 17-year-old.

8. **Health.** The formula for thyroxin, an essential chemical growth regulator, is $C_{15}H_{11}I_4NO_4$. Which of the following radioactive waste products of nuclear reactors pose a particular threat to the thyroid gland: CH_3I containing iodine-131; or radon? Justify your choice.

9. **Health.** Let us suppose that the establishment of a nuclear reactor system in the United States were to result in the deaths of a certain number of people each year from cancers that they would not have contracted if the radiation levels were lower. What effects of nuclear power plants can you imagine that might *increase* the general level of health?

10. **Health.** Relate the type of cells most easily affected by radiation to the symptoms of early somatic radiation sickness.

11. **Reactors.** What are essential features of a nuclear fission reactor? Explain the function of each.

12. **Neutrons.** List the possible fates of neutrons in a fission reactor. Which of these events should be favored, and which should be inhibited, in order to (a) shut down a reactor, (b) breed new fissionable fuel, and (c) produce more energy?

13. **Natural nuclear reactor.** The "Oklo phenomenon" (p. 308) has been assumed to be the result of a naturally occurring chain reaction. (a) How could a natural rock formation approximate a nuclear reactor? Be specific in your answer: What factors might have conserved neutrons? What substance could have served as the neutron moderator? (b) What do you think is the chance of a naturally occurring nuclear explosion? Explain.

14. **Breeder reactors.** Explain how breeder reactors differ from non-breeders in fuel, moderator, coolant, and any other features.

15. **The fuel cycles.** (a) Outline the steps in the "once-through" uranium fuel cycle. Do you think it is appropriate to use the word "cycle" in this context? Defend your answer. (b) Outline the steps in the plutonium fuel cycle. What are its major differences from the uranium fuel cycle? What new environmental hazards would the plutonium cycle produce?

16. **Nuclear safety.** Describe any specific series of events that would cause all of the safety features of a nuclear plant to fail

and a radioactive cloud to be released to the atmosphere.

17. **Nuclear safety.** Do you think it would be reasonable to set safety limits in nuclear power plants that would prohibit *any* release of radioactive matter? Defend your answer. If your answer is no, what criteria would you use to set the limits?

18. **Nuclear safety.** Compare or contrast the control systems in a nuclear fission plant with the homeostatic mechanisms of a natural ecosystem, giving special attention to the following points: (a) Are the two sets of mechanisms comparable in that they both respond in a manner that tends to alleviate stress, or are they conceptually different and therefore not to be compared? (b) Is the effectiveness of a typical natural homeostasis a reasonable goal for the nuclear industry, or must the industry do better? Give some example of perturbations in natural systems and state whether variations from the norm of similar magnitude would be acceptable in a nuclear power plant.

19. **Radioactive wastes.** Sand-like radioactive leftovers from uranium ore processing mills, called "mill tailings," have been used to make cement for the construction of houses in Colorado, Arizona, New Mexico, Utah, Wyoming, Texas, South Dakota, and Washington. These tailings contain radium (half-life 1620 years) and its daughter radon (a gas, half-life 3.8 days), as well as radioactive forms of polonium, bismuth, and lead. Radon gas seeps through concrete, but is chemically inert. Are the following statements true or false? Defend your answer in each case:

(a) Since radon has such a short half-life, the hazard will disappear quickly; old tailings, therefore, do not pose any health problems.

(b) Even if the radon gas is present, it cannot be a health problem because it is inert and does not enter into any chemical reactions in the body.

(c) Continuous ventilation that would blow the radon gas outdoors would decrease the health hazard inside such a house.

20. **Nuclear explosion.** Explain why a critical mass of pure fissionable material must be exceeded if an explosion is to occur. Why is it thought that a nuclear reactor could not explode in the manner of a bomb?

21. **Safety.** (a) What are the five principles of industrial safe practice? (b) How are they applied to the construction and use of automobiles? of the place where you live? of a nuclear plant?

22. **Fusion.** Suppose that someone claims to have found a material that can serve as a rigid container for a thermonuclear reactor. Would such a claim merit examination, or should it be ignored as a "crackpot" idea not worth the time to investigate? Defend your answer.

23. **Fusion.** Outline the reasons why fusion reactors are expected to be far less serious sources of radioactive pollutants than fission reactors.

24. **Browns Ferry.** What is meant by a "common-mode" failure? What do such failures imply about the principle of redundancy?

BIBLIOGRAPHY

There are many books on atomic energy and nuclear engineering; many of them assume previous training in physics and chemistry. The following two texts, however, present somewhat more elementary introductions:

Alvin Glassner: *Introduction to Nuclear Science.* New York, Litton Educational Publisher, Van Nostrand-Reinhold Books, 1961.

Samuel Glasstone: *Sourcebook on Atomic Energy.* New York, Litton Educational Publisher, Van Nostrand-Reinhold Books, 1958.

(Glassner's book is based on a short course given at Argonne National Laboratory since 1957, and presupposes only one year of college physics. Glasstone is more comprehensive and offers more introductory matter.)

The pro-nuclear viewpoint is very clearly presented in:

Bernard L. Cohen: *Nuclear Science and Society.* Garden City, N.Y., Anchor Press/Doubleday, 1974. 268 pp. Paperback.

Anti-nuclear viewpoints are given in:

John J. Berger: *Nuclear Power—The Unviable Option.* Palo Alto, Calif., Ramparts Press, 1976, 384 pp. Paperback.

Union of Concerned Scientists: *The Nuclear Fuel Cycle.* Revised Ed. Cambridge, Mass., Massachusetts Institute of Technology, 1974. 291 pp. Paperback.

For more advanced books, refer to the following:
George I. Bell: *Nuclear Reactor Theory.* New York, Van Nostrand-Reinhold Books, 1970. 619 pp.

Peter John Grant: *Elementary Reactor Physics.* New York, Pergamon Press, 1966. 190 pp.

M. M. El-Wakil: *Nuclear Energy Conversion.* New York, Intext Educational Publishers, 1971. 666 pp.

Two excellent books that integrate social and technical aspects of the problems of nuclear energy are:
David Rittenhouse Inglis: *Nuclear Energy: Its Physics and Its Social Challenge.* Reading, Mass., Addison-Wesley Publishing Co., 1973. 395 pp.

Henry Foreman, ed. *Nuclear Power and the Public.* Minneapolis: Univ. of Minnesota Press, 1970. 272 pp.

Specific discussion of nuclear hazards may be found in:
Geoffrey G. Eichholz: *Environmental Aspects of Nuclear Power.* Ann Arbor, Mich., Ann Arbor Science Publishers, 1976. 681 pp.

Problems of sabotage and terrorism are considered in:
Mason Willrich and Theodore B. Taylor: *Nuclear Theft: Risks and Safeguards.* Cambridge, Mass., Ballinger, 1974. 252 pp.

The Rasmussen Study is published under the following title:
Reactor Safety Study, U.S. Nuclear Regulatory Commission, October, 1975. 198 pp. Copies can be purchased from the National Technical Information Service, Springfield, Virginia. A separate *Executive Summary* (12 pp.) is also available.

The Browns Ferry accident is reported by:
Daniel F. Ford, Henry W. Kendall, and Lawrence S. Tye: *Browns Ferry: The Regulatory Failure.* Cambridge, Mass., Union of Concerned Scientists, 1976. 72 pp. Paperback.

Finally, the excellent periodical *Bulletin of the Atomic Scientists* carries many articles on the problem.

Unit V

Rural Land Use

11

AGRICULTURAL SYSTEMS

11.1 INTRODUCTION

When people first evolved from their apelike ancestors, their diet depended on what they could collect from day to day. Hunters dragged food back to their dens or were hunted and dragged back to some other predator's lair. They competed with other herbivores for plant foods. Life was very hard during drought, flood, or pestilence. Because technology was so limited and human population so small, early cultures did not appreciably alter the environment. Stone and wooden tools for digging and hunting were competitive with the tusk of the mammoth and the claw of the tiger but certainly were not overwhelmingly superior.

Even when farmers first began to cultivate the land, their activities had little impact on global ecosystems. People lived close to their food supplies, and their wastes were returned to the farmlands directly (Fig. 11.1). Thus,

Figure 11.1 A primitive agricultural field in New Guinea. (From Roy A. Rappaport: The Flow of Energy in an Agricultural Society. *Scientific American*, Sept., 1971, p. 116. Copyright by Scientific American, Inc. All rights reserved.)

the consumption of nutrients was balanced by a return of nutrients. Of course, even a simple agricultural system is potentially disrupting, for it promotes the growth of a few species where many species once lived, but the extent of the environmental alteration was small in early cultures. In fact, humans are not alone in favoring those plants and animals which benefit them most; bees selectively pollinate the most succulent flowers, ants protect their herds of aphids, and sage grouse scratch the soil around the roots of sagebrush.

When agricultural yields first enabled some individuals to pursue goals other than cultivation of the soil and to live far from sources of food, the cycle was broken, for nutrients were not all returned to the farmland from which they came. The result was an ecological imbalance. As food-growing technology became more efficient, the ecological disruptions became more severe.

Of course, farmers understood for many years that their choices were either to refertilize the land or, eventually, to move elsewhere. In some areas, they have been quite successful in keeping farmlands fertile. For example, some regions of China, Japan, and Europe have been farmed successfully for thousands of years. During that period of time the soil has even been enriched by fertilization with human and animal manure and various other materials of biological origin. In other regions, however, agriculture still continues to destroy huge areas of previously fertile land.

The history of agriculture includes both successes and failures. While the total harvest of food has increased steadily for thousands of years, more than half of humanity is malnourished, and famines due to crop failures continue to occur. Although recent technological accomplishments, both in agriculture and in industry, have enabled people to rise to their present level of biological ascendancy, present farming practices are drastically altering the Earth's ecosystems. In fact, many soil scientists fear that the planet's future ability to produce high yields of food is being endangered.

11.2 AGRICULTURAL DISRUPTIONS

Farmers in many early civilizations carelessly destroyed the land that once supported them. One story of land destruction comes from the area of the Fertile Crescent, the "cradle of civilization." The Tigris-Euphrates valley gave birth to several great civilizations. We know that highly sophisticated systems of letters, mathematics, law, and astronomy originated in this area. Obviously, then, people had the time and energy to educate themselves and to philosophize. We can deduce that the food supply must have been adequate. Today much of this

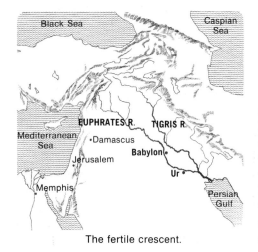

The fertile crescent.

region is barren, semi-desert, badly eroded, and desolate. Archeologists dig up ancient irrigation canals, old hoes, and grinding stones in the middle of the desert. What must have happened?

Part of the story starts at the source of the great rivers in the Armenian highlands. The forests were cleared to make way for pastures, vineyards, and wheat fields. But croplands, especially if poorly managed, cannot hold the soil and the moisture year after year as well as natural forests or grasslands. As a result, large water runoffs such as those from the spring rains or melting mountain snows tended to flow down the hillsides rather than soak into the ground. These uncontrolled waters became spring floods. As a protective measure, canals were dug in the valleys to drain the fields in the spring and to irrigate them in the summer and fall. Later, devastating wars resulted in the abandonment of the canals and croplands. The neglected canals turned to marsh, the formation of marshes shrunk the rivers, and the diminished rivers could no longer be used to irrigate other areas of land. The water table sank, so that the yields per acre today are less than the yields 4000 years ago.

Land has been destroyed with monotonous regularity. Many people know that the agriculture of India cannot quite keep up with the needs of its people and that, consequently, famine is never more than a few dry seasons away. It is less well known that poor farming practices have led to erosion and soil depletion, which have completely or partially destroyed approximately two thirds of India's cropland.

OZYMANDIAS

I met a traveller from an antique land
Who said: Two vast and trunkless legs of stone
Stand in the desert . . . Near them, on the sand,
Half sunk, a shattered visage lies, whose frown,
And wrinkled lip, and sneer of cold command,
Tell that its sculptor well those passions read
Which yet survive, stamped on these lifeless things,
The hand that mocked them, and the heart that fed:
And on the pedestal these words appear:
"My name is Ozymandias, king of kings:
Look on my works, ye Mighty, and despair!"
Nothing beside remains. Round the decay
Of that colossal wreck, boundless and bare
The lone and level sands stretch far away.

P. B. Shelley (1817)

The desert to which Shelley referred was the Sahara.

Aerial photograph of old irrigation canals along the Tigris River. (Photo courtesy of Iraq Tourist Authority.)

Another example is ancient Carthage, which was founded on the shores of the Mediterranean in North Africa amid dry but fertile grasslands. Grain was grown in abundance. Today much of this area has become part of the Sahara Desert. In fact, a large part of the Sahara is a by-product of overplowing and overfarming of fertile land, which led to a dramatic reduction of the amount of available moisture. This process is illustrated in greater detail by a more recent agricultural disaster—the American Dust Bowl.

The early European settlers found millions of acres of virgin land in America. The eastern coast, where they first arrived, was so heavily forested that even by the mid-eighteenth century, a mariner approaching the shore could detect the fragrance of the pine trees over 300 kilometers from land. The task of clearing land, pulling stumps, and planting crops was arduous. Especially in New England, long winters and rocky hillsides contributed to the difficulty of farming. It was natural that pioneers should be lured by the West, for here, beyond the Mississippi, lay expanses of prairie farther than the eye could see. Deep, rich topsoil and rockless, treeless expanses promised easy plowing, sowing, and reaping. In 1889 the Oklahoma Territory was opened for homesteading. A few weeks later the non-Indian population there rose from almost nil to close to 60,000. By 1900, the population was 390,000—a people living off the wealth of the soil. In 1924 a thick cloud of dust blew over the East Coast and into the Atlantic Ocean. This dust had been the topsoil of Oklahoma (Fig. 11.2).

In each of the earlier examples one might contend that

Figure 11.2 Dust storm. (© Arthur Rothstein, New Rochelle, N.Y.)

the destruction of the land was really caused by changes in climate rather than by mismanagement of the land. Indirect evidence, however, strongly implicates farmers. For instance, in the areas between the Tigris and Euphrates rivers where the canals were not destroyed, the land remains fertile. Similarly, some areas of North Africa near the Sahara still support trees believed planted by the Romans, and geological evidence indicates relatively constant weather patterns in these areas. However, in the case of the Oklahoma Dust Bowl we *know* that agricultural practices, and not climate, destroyed the land.

Recall from Chapter 3 that a natural prairie is a diversified ecosystem with homeostatic mechanisms to protect itself from spring floods and summer droughts. The settlers' contribution to the prairie was not particularly farsighted. They planted large fields of single crops, thus destroying the naturally balanced system. They killed bison to make room for their cattle, then killed the wolves and coyotes to prevent predation of these herds. Moreover, they often permitted their cattle to overgraze. In overgrazed land, the plants, especially the annuals, become so sparse that they cannot reseed themselves. The land itself, therefore, becomes very susceptible to soil erosion during heavy rains. In addition, the water runs off the land instead of seeping in, resulting in a lower water table. Because the perennial plants depend upon the underground water levels, depletion of the water table means death for all prairie grasses. The whole process is further accelerated as the grazing cattle pack the earth down with their hooves and block the natural seepage of air and water through the soil.

The introduction of the plow to the prairie had an even more severe effect, because the first step in turning a

Before the dust storms; grain growing in Southern Colorado in 1916. (Courtesy of the State Historical Society of Colorado.)

prairie into a farm is to plow the soil in preparation for seeding. At this point, of course, the soil is vulnerable, since the perennial grasses, which normally hold the soil during drought, have already been killed.

If the spring rains fail to arrive, the new seeds won't grow and the soil will dry up and blow away. On the other hand, if, after plowing, the spring rains are too heavy, the soil may easily wash away before the seeds have an opportunity to grow. Even after seeds have sprouted, the practice of pulling weeds between the rows loosens some soil and makes it susceptible to erosion by heavy rains.

Over a period of 20 to 35 years, the fertility of the soil in Oklahoma slowly decreased. Incomplete refertilization and loss of soil from erosion by wind and water took their toll. Crops of wheat and other grains were planted on dry, barren ground. But the Oklahoma prairies have always been subject to periodic droughts. When a particularly severe drought struck during the 1930's, the seeds failed to sprout, and a summer wind blew the topsoil eastward into the Atlantic Ocean.

The droughts that killed the Oklahoma farms had no lasting effect on those prairies left untouched by agriculture. In fact, these virgin lands are still fertile. In a few thousand years, perhaps, the wind-scarred Dust Bowl will regain its original fertility. We say "perhaps" because similar destruction of the North African prairie left the land so barren that nothing remained to hold the rain that did fall. Two thousand years after the farms failed, one can stand in the center of the ruins of a wealthy country estate and watch the sands of the great Sahara blow by.

The previous examples discuss some of the agricultural failures that have led to the loss of large areas of fertile cropland. Even now, farming is destroying productive soil faster than natural processes are renewing it. Despite this loss of total fertile area, greatly improved farming techniques have steadily increased worldwide agricultural yields.

11.3 THE FLOW OF ENERGY IN INDUSTRIAL AGRICULTURE

A wild oat plant in an unfarmed prairie and a domestic oat plant in a farmer's field live in very different environments. To survive, the wild oat must compete successfully with its neighbors for sunlight and moisture. A plant that is tall, that sprouts early, or that has an effective root system, has a competitive advantage. The energy a wild plant needs to grow a tall stalk or a deep root comes only from the Sun. On the other hand, a cultivated oat needs external aid to help it survive. The farmer waters it when necessary, removes competitive plants (weeds), and loosens

Figure 11.3 Plants compete for space, light, water, and nutrients in a natural grassland.

the soil to stimulate the growth of root systems. Since all the seeds in the field are planted at the same time and are of the same variety, competition is minimal and the plant does not need a tall stalk or a unique and fast-growing root system. In other words, a wild plant must use some of its incident solar radiation for survival and some for the production of seeds or bulbs. A farmer adds auxiliary sources of energy to encourage the growth of varieties of plants that provide abundant food for human consumption.

As previously mentioned, individuals within a given species differ significantly. Particularly important among individual plants are the differences in total weight of edible matter. Thus, by replanting seeds only from those plants that produce the most grain, regardless of the viability of that plant in an *untended* field, a farmer can breed high-yielding plants over the course of time. In today's laboratories, scientists often crossbreed different varieties of plants, hoping that a new combination of existing traits will produce new plants with desirable characteristics (Fig. 11.4). Alternatively, the existing characteristics may themselves be changed by inducing mutations artificially. These varied and sophisticated manipulations lead to the same kinds of choices: the agronomist must decide whether or not to select a plant for further breeding, and ultimately whether to adopt it for commercial planting. Traditionally, agronomists have chosen the plant varieties that yield the most desirable crop—usually with emphasis on quantity. As a result of such selections, however, crops have become increasingly dependent on cultivation for survival. For instance, a variety of Oriental rice developed in the late 1960's produces more grain than wild Oriental rice, but the new breed is also shorter, less resistant to disease, and more dependent on irrigation than its predecessor. In addi-

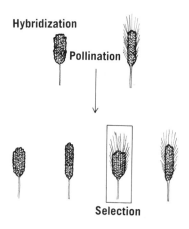

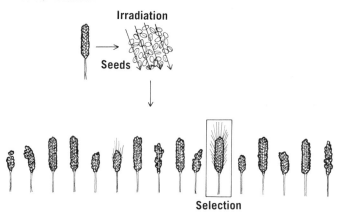

Figure 11.4 Scheme illustrating the general procedure for crossbreeding of plants.

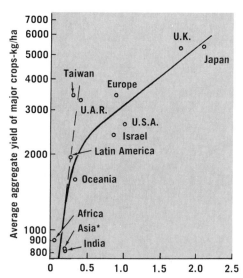

Horsepower per hectare (1 hectare = 2.471 acres)

*Excluding Mainland China

Figure 11.5 The relationship between crop yields and energy.

tion, the new seeds have lost their biological clocks and can sprout any time they are planted, whereas native rices are geared to the seasons, and if left uncultivated, new seedlings will germinate only in the proper growing season. The new seeds require care, and care requires energy—energy to plant, to weed, to control pests and disease, and to irrigate.

Similarly, animal breeders have produced varieties of livestock that are more efficient than prairie grazers in converting grain to meat and dairy products. But the farmer must invest energy to realize these high yields. A prairie chicken lays 10 to 20 eggs per year, all of them in the spring, then broods them and cares for the chicks. A domestic hen can produce over 200 eggs in a year, but most hens will not sit on a fertilized egg long enough for it to hatch. Therefore, most breeds of domestic chickens require artificial incubation for propagation.

Farmers in industrially developed countries have large supplies of fossil fuels available to them. They use these resources to supplement the energy that the plants receive

326

Cattle in a feedlot.

from solar radiation (Fig. 11.6). Let us examine the overall energetics of a typical modern agriculture system—the mechanized production of beef. Remember that in a natural system, for every 12 calories of energy a plant receives from the Sun, only about 1/2 calorie is available for the production of animal tissue by a herbivore that consumes the plant. Some of the "lost" 11½ calories are required by the plant for its metabolism, some by the animal for its metabolism, and some of the energy is used by the animal to search for more food. In a feedlot, a steer need not move, but may stand in front of his feedbin, eating hay or grain that the farmer, with the help of a tractor, baled or threshed and brought in from the field. In addition, food additives and growth hormones, synthesized in factories pow-

Cattle on the range.

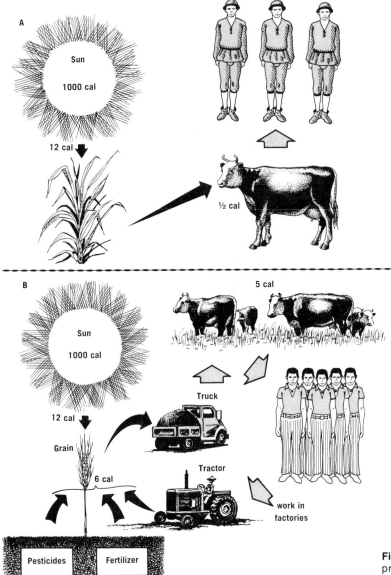

Figure 11.6 Comparison between beef production on the range (*A*) and in a feedlot (*B*).

ered by coal or oil, help to increase growth. In a feedlot, about 6 calories of fossil fuel energy are used for every 12 calories that the plant matter receives from the Sun. The total weight gain of the animal under these conditions corresponds to about 5 calories.

Thus, although a feedlot system is 10 times more efficient in converting sunlight to meat than a natural system is, the feedlot requires large quantities of fossil fuels.

Production of beef in feedlots is based on a technological cycle. High food production is needed for the urban workers who provide the technology required to maintain the high food production. The individual components of the cycle are interdependent. Large populations of human beings depend on high agricultural yields, while at

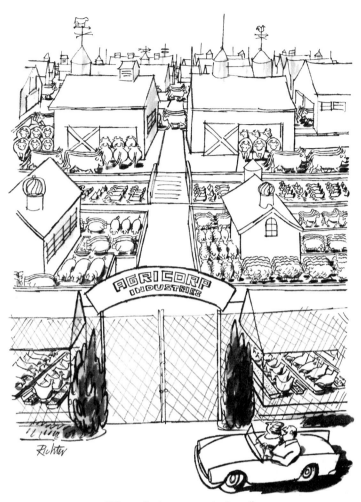

"Now, *that's* a corporate farm."

Drawing by Richter; © 1976 The New Yorker Magazine, Inc.

the same time these high yields depend on high industrial outputs. If the cycle is broken, farmers will be forced to revert to more primitive forms of agriculture. This will lead to lower yields, a lower carrying capacity for humans, and ultimately, mass starvation.

Modern agriculture requires so much energy that the price and even the availability of food is closely linked to the price of gasoline. When the cost of petroleum rose abruptly in 1973, many farmers in the less developed nations could afford neither fertilizers nor fuels. As a result, many people starved. In the industrialized nations, the fuel shortage led to higher food prices. Because increasingly severe fuel crises are projected for the future, it is important to understand the role of energy in modern agriculture. Modern farming depends on four fundamental techniques: (a) chemical fertilization, (b) mechanization, (c) irrigation, and (d) the chemical control of disease, insects, and weeds. All four techniques require large inputs of energy. The use of energy in agriculture in the United States is summarized in Table 11.1. Fertilization, mechani-

TABLE 11.1 Use of Energy from Farm to Table in the United States, 1940 and 1970* (Multiply All Values by 10^{12} Kcal)

	1940	1970
On Farm		
Fuel for tractors	70.0	232.0
Electricity	0.7	63.8
Energy to manufacture fertilizers	12.4	94.0
Energy to produce agricultural steel	1.6	2.0
Energy to manufacture farm machinery	9.0	80.0
Energy to manufacture tractors	12.8	19.3
Energy to irrigation	18.0	35.0
Subtotal	124.5	526.1
Processing Industry		
Food processing industry	147.0	308.0
Energy to manufacture food processing machinery	0.7	6.0
Energy for paper packaging	8.5	38.0
Energy to manufacture glass containers	14.0	47.0
Energy to manufacture aluminum and steel cans	38.0	122.0
Gasoline for food transport	49.6	246.9
Truck and trailer manufacture	28.0	74.0
Subtotal	285.8	841.9
Commercial and Home		
Commercial refrigeration and cooking	121.0	263.0
Energy to manufacture refrigeration machinery	10.0	61.0
Home refrigeration and cooking	144.2	480.0
Subtotal	275.2	804.0
Grand total	685.5	2172.0

*Note some interesting facts shown in the table. By 1970, almost 1½ times as much energy was used to manufacture cans as to manufacture fertilizer, and more energy was consumed by the food processing industry than was used to fuel tractors on the farm. (Modified from John S. Steinhart and Carol E. Steinhart: *Energy: Sources, Use, and Role in Human Affairs,* North Scituate, Mass., Duxbury Press, 1974. Used with permission.)

zation, and irrigation will be considered in the following sections, and the control of diseases, insects, and weeds will be discussed in the next chapter.

11.4 CHEMICAL FERTILIZATION

Farmers fertilized their crops with manure, straw, or dead fish long before they understood the chemistry of fertilization. Today, mined and manufactured fertilizers are used so extensively (Fig. 11.7) that the total energy requirement of the fertilizer industry constitutes a major portion of agriculture's large demand for fossil fuel. The chemistry of fertilization is essentially the same as the chemistry of the nutrient cycles discussed in Chapter 3. Oxygen and carbon are readily available to all plants that are exposed to air, and are of no concern to the fertilizer industry. Chemists have learned to convert atmospheric nitrogen to synthetic plant fertilizers by producing ammonia:

$$N_2 + 3 H_2 \rightarrow 2 NH_3$$
$$\text{ammonia}$$

Nitrogen makes up about 78 percent of the atmosphere by volume, and the supply is therefore practically unlimited. We cannot run out of hydrogen either, because it is obtained from water; however, the synthesis of ammonia requires large expenditures of energy. Therefore, the price of nitrogen fertilizers reflects the cost of crude oil.

Mineral fertilizing is qualitatively different from nitrogen fertilizing. If the minerals cannot be recycled by the reuse of dead plant matter or by such material as bone meal, miners must dig the needed minerals from some geological deposit. Thus, phosphorus, potassium, calcium, magnesium, and sulfur, all important constituents of a well-balanced fertilizer, are extracted from the earth. Sometimes these minerals are simply pulverized and dusted onto the soil; or they may be chemically treated in order to enhance the speed or efficiency of uptake by plant roots. Of these five minerals, all but phosphorus are plentiful in the Earth's crust. Known geological deposits of phosphate rock (an oxidized form of phosphorus) are limited (see Chapter 8). Without an adequate supply of inexpensive phosphorus, highly productive agriculture would be impossible.

In spite of the fact that manufactured fertilizers have raised agricultural yields across the world, and in spite of the general health and vitality of peoples raised on fertilized crops, many agronomists question the use of inorganic fertilizers and advocate "organic" gardening and farming.

What is meant by the term "organic farming"? Organic farmers use only natural fertilizers and pesticides and do

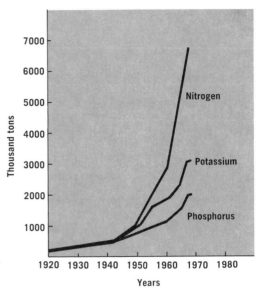

Figure 11.7 Fertilizers in U.S. agriculture. Use of plant nutrients, 1920 to 1968.

A truck is loaded and ready to spread liquid fertilizer on a field in the Sacramento Valley, California.

*The word comes from the Greek *chele,* meaning "claw." Chelates bind atoms from two or more different directions, much as a lobster grabs on to its prey.

not rely on manufactured or highly processed chemicals. Plants need soil with properly regulated nutrients, density, moisture, salinity, and acidity. Soil conditions have traditionally been regulated by the **humus,** a very complex mixture of compounds resulting from the decomposition of living tissue. A given piece of tissue, such as a leaf or a stalk of grass, is considered to be humus rather than debris when it has decomposed sufficiently in the soil system so that its origin becomes obscure. Compared to nonorganic soil, humic soil is physically lighter, it holds moisture better, and it is more effectively buffered against rapid fluctuations of acidity. Additionally, certain chemicals present in humus aid the transfer and retention of nutrients. For example, calcium ions, Ca^{2+}, can exist in water solutions from which they are usable by plants. However, atmospheric carbon dioxide reacts with water to form carbonate ion, CO_3^{2-}, which reacts with calcium in water to form the sparingly soluble compound, calcium carbonate, $CaCO_3$. A striking consequence of its meager solubility has been the deposition of large masses of $CaCO_3$ on earth in the form of limestone and marble. But it is not *completely* insoluble; moreover, its solubility depends largely on the soil acidity, and under certain conditions, the calcium may be liberated. Under other conditions, it is very insoluble and the calcium is not readily available to plants, even though it is present in the soil. Moreover, if dissolved calcium is not used immediately by plants, it may travel with water droplets down below the root zones, where it becomes unavailable. This movement of free ions into the subsoil and underground reservoirs is called **leaching.**

Richly humic soils provide chemical reaction pathways for metal ions that are not available in simple inorganic solutions. Certain chemicals in the humus, which are known as **chelating* agents,** react with inorganic ions such as Ca^{2+} to form a special class of compounds known as chelation complexes. Ions bonded in chelation complexes are held tightly under some conditions, but are easily released under others. A calcium ion chelated by humus will not react readily with carbonate to form $CaCO_3$, nor will it leach easily. Rather it will tend to remain bonded to the chelating agent and thus be retained in the humic matter.

Plants and soil microorganisms have evolved mechanisms whereby they can release chelated ions readily and incorporate them into living tissue, as shown schematically in Figure 11.8. Naturally, the specific chelation chemistry is different for each soil nutrient, but in general, humus maintains a nutrient reservoir and enhances the ease with which nutrients are used by the soil organisms.

When soil humus is not maintained, the efficiency of converting fertilizer to plant tissue is low and the ability of soils to store reserves of nutrients is poor. As a result, large

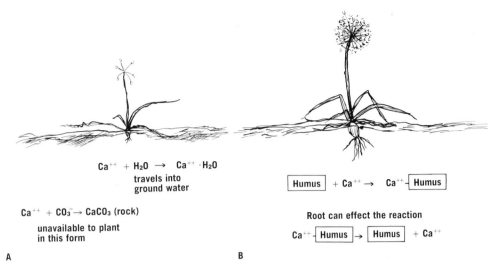

$$Ca^{++} + H_2O \rightarrow Ca^{++} \cdot H_2O$$
travels into
ground water

$$Ca^{++} + CO_3^= \rightarrow CaCO_3 \text{ (rock)}$$
unavailable to plant
in this form

A

$$\boxed{Humus} + Ca^{++} \rightarrow Ca^{++} \boxed{Humus}$$

Root can effect the reaction

$$Ca^{++} \boxed{Humus} \rightarrow \boxed{Humus} + Ca^{++}$$

B

Figure 11.8 Schematic illustration of pathways for transfer of inorganic nutrients. *A,* Without humus. *B,* With humus.

quantities of nitrogen or mineral fertilizers are leached into surface and ground waters or tied into the soils in chemically unusable forms. This loss of nutrients represents a waste of energy and ore. In addition, lost fertilizers collect in the waterways and add significantly to water pollution in agricultural areas. Compounds that are nutrients on land are also nutrients to aquatic life, and serve to feed algae and plankton, thereby upsetting the food webs in rivers and lakes. Nitrogen fertilizers are also potentially harmful in that the nitrate ion, NO_3^-, may be converted to cancer-producing compounds.

Another argument against the practice of chemical farming is related to the nutritional value of the crops. Several studies have shown that the food value of many grains and vegetables decreases when high concentrations of fertilizers are used. In one study, alfalfa yields were increased dramatically with high applications of chemical fertilizers; however, cattle fed on that alfalfa became ill from cobalt and copper deficiencies. These trace minerals, necessary to the animals' health, were not present in the fertilizer, and were therefore lacking in the alfalfa.

Nothing in the preceding discussion implies that the use of synthetic "nonorganic" fertilizer, *per se,* results in nutritionally poor crops. Plants cannot tell whether a given molecule of nutrient came from a bag of "nonorganic" fertilizer or from a bag of compost. The major deficiency of commercial "nonorganic" fertilizers is that many contain no more than three nutrients: nitrogen, potassium, and phosphorus. Trace elements could be incorporated into manufactured fertilizers, but the expense is high and is not compensated for by a large increase in yield. As a result, the continued use of commercial "nonorganic" fertilizers may lead to nutritional imbalances. On the other hand, one doesn't render a soil nutritionally perfect simply by tossing in a little manure. Only by using the proper mixture of fortifying materials such as manure, wood ashes, bone, and blood meal does the "organic" farmer grow prime foods.

The lowering of nutritional values caused by improper fertilization does not present a serious problem to most inhabitants of the developed nations. Indeed, many foods do not appear to be harmed by large applications of fertilizer, and the decreased food value of many others is insignificant to people who can afford a rich and varied diet. Secondly, transportation and refrigeration enable shoppers to purchase foods from many different geographical areas. Because some soils in different regions are naturally endowed with an overabundance of certain trace minerals, a person eating foods raised in many different soils is likely to ingest an adequate supply of minerals. The problem of the nutritional value of individual foods is most serious in areas where families eat a subsistence diet grown in one field.

Although commercial fertilization has been quite successful in the past, its problems are now demanding increased attention. Perhaps most foreboding is the fact that the humus content of soils is continuing to drop. One solution would be to formulate more sophisticated artificial fertilizers to include synthetic chelating agents. A second solution would be to initiate a massive program of naturally constituted fertilization through recycling. There is no doubt that the soils in the United States, for example, could be improved if the two billion metric tons of animal feces, slaughterhouse offal, and other organic residues produced each year were composted and returned to the land.

Only some of this material is now used as fertilizer. Bulldozers push great quantities of valuable organic matter into landfills. The rationale for this waste has traditionally been that it is often cheaper to discard manure and buy "nonorganic" fertilizers than to transport and spread heavy, bulky "organic" matter. However, the economic picture is rapidly changing. When the price of fuel increased dramatically in 1973–1974, the cost of "nonorganic" fertilizers also rose dramatically. As a result, in 1975 large-scale "organic" farming was no more expensive than "nonorganic" farming. When one considers not only the yield per dollar in a single growing season but also the long-term health of the soil, the direct economic and ecological advantages of "organic" fertilization become significant. The use of organic wastes as fertilizers not only aids the farmer but benefits the community as well, for if manure and sludge are discarded, some of the wastes wash into streams and ground water and pollute them. Although the cost of economic externalities is not reflected in the farmer's account book, it is paid for by our society.

11.5 MECHANIZATION IN AGRICULTURE

High agricultural yields require specific varieties of seeds, an active fertilizer program, and pest control. The use of tractors does not improve the yield per hectare. For example, most Japanese farms vary in size from about 0.4 to 2 hectares (about 1 to 5 acres). Many farmers own small, two-wheeled rotary tillers, but very few own four-wheeled tractors, and heavy farm machinery is virtually unknown. Yet the yields per hectare in Japanese agriculture are the highest in the world. However, these yields require great quantities of fertilizer and much labor. Therefore, the total auxiliary energy demand of Japanese farming is high even though the use of heavy equipment is minimal.

The fact that a tractor can outperform any animal in pulling a plow is not the only advantage of mechanization. Another advantage is the fact that modern equipment can perform many intricate operations simultaneously. A single device can plow, fertilize, and plant in one pass through a field, while another can pick the fruit, sort it according to size, bag it, weigh the bags, and release the labeled sacks for delivery to market. As a result of such extensive mechanization of agriculture, farms in the United States now use more petroleum than does any other single industry, and more than one calorie of fossil fuel energy is needed to produce one calorie of food value.

As mechanization becomes more and more intricate, and as the individual devices get larger, agricultural yields per acre actually *decrease*. A machine that processes 10 rows of crops is necessarily less precise than one that handles only two or three rows, because of unevenness of ground levels and other natural variations. The result is that the larger machine causes more damage and hence more waste. A mechanical tomato picker requires a variety of tough-skinned tomatoes that all ripen at the same time, because the machine has no eyes to differentiate between red and green.* Such varieties have in fact been developed, but their simultaneity of ripening is not perfect, and hence, again, the machines make waste. There are machines that shake apples from trees into collecting frames, but they bruise some apples. Furthermore, all of these problems are aggravated by the crowding of plants that is made possible by the greater productivity of fertilized soils.

According to the data in Table 11.1, gasoline for tractors accounted for almost 45 percent of the energy used on farms in the United States in 1970. What would happen if American farmers were to revert to more manual labor and fewer machines? In other words, suppose that the factory workers who build mechanized tomato pickers were to move to the country and pick tomatoes. This type of system would be similar to Japanese farming, in which high inputs of fertilizer, pesticides, and labor give rise to high

Mechanization in agriculture. *Above,* Combine. *Below,* Tomato picker. (Photos courtesy of Rorer-Amchem, Inc., Ambler, Pennsylvania.)

---

*More truthfully, between different shades of green. Most tomatoes are reddened with ethylene gas after picking.

yields. A labor-intensive agricultural system could operate in the United States and other industrial nations only if major changes were made in social structures and in living standards. A major apple grower in Connecticut has said, "If you had to hire out-of-work people to pick apples, as I do, you would be appalled at how many have lost the ability to adapt to new manual tasks. This is a human loss, not just an economic one."[*]

11.6 IRRIGATION

Irrigation is almost as old as agriculture itself. The ancient Egyptians, Babylonians, Chinese, and Incas all brought water from nearby rivers to increase yields of crops. These practices have continued so that today much of the world's crops of vegetables and some grains depend on irrigation. With imported water, marginal farmland has become more productive, and even former deserts are being farmed. Despite these successes, irrigation leads to some environmental problems.

When rain water falls on mountain sides, it collects in small streams above and below ground, and it filters over, under, and through rock formations. In the process of flowing into a large river, the water dissolves various mineral salts present in the rock and soil. Usually these salts are concentrated in the oceans. However, in using river water for irrigation, farmers are bringing slightly salty water to

---

[*]Robert Josephy: *Farming in the Urban Northeast.* Address delivered at the annual meeting of the American Association for the Advancement of Science, Boston, February 22, 1976.

Flood-water irrigation in the valley of Teotihuacán, Mexico, May 1955. (From Eric R. Wolf: *Sons of the Shaking Earth.* Phoenix Books, University of Chicago Press, 1959, p. 77. William. T. Sanders, photographer.)

Irrigation system in the United States.

their fields. When water evaporates, the salt remains. Thus, over the years, the salt content of the soil increases. Because most plants cannot grow in salty soil, the fertility of the land decreases. In Pakistan an increase in salinity decreased soil fertility alarmingly after a hundred years of irrigation. In parts of what is now the Syrian desert, archeologists have uncovered ruins of rich farming cultures. However, the land adjacent to the ancient irrigation canals is now too salty to support plant growth.

Irrigation is threatening agricultural production in parts of California, which produces about 40 percent of the vegetables consumed in the United States. The richest vegetable-growing area in California is the San Joaquin valley, where virtually all commercial operations rely heavily on irrigation. As the irrigation intensifies, so does the threat of destructive salination of the soil and the ground water. Most of the valley is now serviced by a shallow underground drainage system to divert the brackish waters, but this system now appears inadequate, and new drainage projects are being proposed.

Can desalinated sea water be used to irrigate vast desert regions? Advocates of such measures point to the fact that if all the deserts lying within 500 kilometers of the ocean were to be developed and planted with grain, the total grain-producing acreage of the world would be multiplied four-fold, from the present 650 million hectares to a phenomenal 2.6 billion hectares. Of course, any method of desalinating sea water requires so much energy that any such system is impractical at the present time.

11.7 AGRICULTURE AND RURAL LAND USE

People need food to survive, and therefore large expanses of rural land must be devoted to farming. But peo-

The South Platte River basin is potentially one of the richest agricultural regions in Colorado, yet much of the farmland has been preempted by commercial interests.

ple also need fuel, transportation systems, places to live, and manufacturing and distribution centers where they can produce and trade the goods necessary for survival. Mines, roads, pipelines, factories, homes, stores, and warehouses all require space. What happens when a prime site for a new harbor is also a valuable estuary system, when a coal mine lies under a wheat field, or when the land containing a vegetable farm becomes so valuable that it is profitable to convert it to a shopping center? When such conflicts arise, who decides how the land is to be used?

In most nations today, economic factors control these decisions. In the last example cited above, if a developer offers the vegetable farmer enough money, the farmer may sell, and the developer will bulldoze the topsoil away and cover the land with concrete and asphalt. Bit by bit, many prime agricultural areas have been converted to nonagricultural uses. In the 55 year period between 1920 and 1975, cropland in the United States decreased by a phenomenal 26 million hectares. Some of this land has been abandoned to forests because it was only marginally fertile in the first place, and some has been lost to erosion, but large fertile areas have been converted to residential or commercial use. During this period, production per hectare on land that has remained agricultural has risen so rapidly that the net food production in the nation has increased substantially despite the losses of farm area. What will happen in the future? Will continued conversion of agricultural land to other functions eventually lead to worldwide famine? If so, is there any way to reverse current trends and stabilize the system?

Legal and economic structures vary from nation to nation. In most regions of the United States, land is taxed according to its market value; the more valuable the land, the higher the taxes. Imagine, then, what happens in some agricultural communities. One farmer sells fields to a manufacturing corporation, which builds a factory. The presence of the factory boosts land prices in the surrounding area. Rising land prices leads to rising taxes, which become a great burden to neighboring farmers. As profits from farming decline, and the market value of farmland skyrockets, many more farmers sell their land. The spiral continues until an agricultural region is converted into an industrial one. If agricultural land were taxed differently from land used for other purposes—as in fact it is in a few localities—such a spiral might be prevented. Alternatively, zoning regulations could be enacted that prevent such conversion.

In other situations (discussed in Chapter 8), valuable fuels or mineral deposits lie under fertile farmlands. If the ore is mined and the farmland destroyed, the world community may be trading a short-term gain for a long-term loss. In the near future we must ask ourselves whether our current socioeconomic structures are facing these issues satisfactorily or whether significant changes are necessary.

11.8 WORLD FOOD SUPPLY

Worldwide food production and population growth have been in a nip-and-tuck race in recent decades. Widespread use of pesticides and fertilizers caused food production in the 1950's to increase faster than population. During most of the following decade, however, the situation was reversed; high birth rates and some poor crop years in many areas caused population growth to outstrip gains in agricultural yields, and a great many people starved. During the latter part of the 1960's, significant agricultural development (see Section 11.9) led to increased crop yields and a reduction of famine. Gains were slight, however, and when a serious drought struck India, Pakistan, and the grazing lands just south of the Sahara Desert between 1972 and 1974, millions again became famished. In 1975 and 1976, when the rains returned to India and Africa, the situation improved. A bumper crop of grain in North America in 1976 further alleviated the world food problem. Despite these gains, serious famines still threaten the world, and a flood or drought in any major food-producing region could again cause widespread starvation.

How can one even think about, let alone describe, the human pain and misery that result from hunger and famine? So we ask ourselves, why, in modern times, must people die for want of enough to eat? It is not enough to say that the world food problem is the result of an interacting set of social, political, economic, natural, and technical difficulties. We must try to examine some of the separate components of the problem and search for solutions to them. These components may be grouped into four categories:

(a) **Population**

As discussed in Chapters 6 and 7, the developing world, where food is already scarce, is faced with imminent large increases in population. On the other hand, in industrial nations, where food is plentiful, the growth of human populations is much slower.

(b) **Overconsumption**

A person utilizes the energy in a grain of wheat most efficiently if the wheat is eaten directly. If a cow eats the wheat first, and the person then eats the cow, most of the original energy is dissipated to the environment. However, even though meat represents an inefficient conversion of agricultural yields to food energy, animal protein is a valuable source of needed nutrients. Therefore, a healthy and efficient diet would consist mostly of vegetable matter with modest quantities of meat and fish. In many developed nations, people eat meat as a

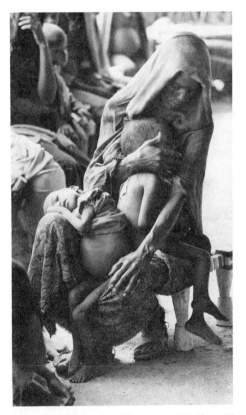

Starving mother and children in India. If world resources were apportioned more evenly, such misery could be avoided. (Photo by Henri Bureau – Sygma.)

339

major component of their diet. For example, the diet of the average person in North America utilizes a total of 900 kg of grain per year. Only 90 kg, however, is consumed directly. The other 810 kg is used as animal feed to produce meat, milk, and eggs. In India, by contrast, the average person utilizes only 180 kg of grain per year. Many people in affluent societies are simply overweight and would be healthier if they ate less. Even more food could be conserved if people whose diet consisted largely of meat would eat more grain instead. Not only would the nutritional quantity of such a modified diet be ample but, according to most nutritionists, its quality would actually be improved.

(c) **Political and Economic Barriers to Food Production**
During the 1970's, total world resources have been more than adequate to enrich farmlands and to supply ample diets for everyone. Given enough fertilizer, fuel, and machinery, the crop yields in the Ganges River delta in India could be increased sixfold, the impoverished Sudan in southeast Africa could be a bread and cereal basket for the world, and the rice paddies of the Philippines could feed the hungry throughout the South Pacific. But great quantities of fuel and machin-

Successful farming in southern Africa. (Courtesy of South African Consulate General.)

ery are being used for armaments, and the remaining resources are not apportioned evenly.

The average farmer in Japan or the United States has ample supplies of fertilizers available. The soils are rich, and many crops are grown under nearly ideal conditions. Therefore, any additional fertilizers yield only small increases in food production. On the other hand, the soil in India and Pakistan is so impoverished that any additional fertilizer has a large positive effect. For example, one kilogram of fertilizer applied on the average farm in India yields 10 additional kilograms of grain, whereas the same kilogram of fertilizer applied to the average farm in the United States yields only 5 additional kilograms. However, the average Indian or Pakistani farmer cannot afford to buy more fertilizer. Clearly, world food problems would be alleviated more quickly if industrial nations shipped fertilizer and not just food to such nations. Unfortunately, in 1973, when fuel shortages led to fertilizer shortages, industrial nations responded to internal political and economic pressures and *restricted* fertilizer exports. As a result, many people starved.

It is not easy to change present economic and political patterns, but such changes would alleviate considerable suffering.

(d) **Climate**

Farmers have always been at the mercy of the weather. If the temperature and rainfall are favorable, bumper crops can usually be expected, but frost, flood, or drought often lead to famine. In recent years there has been grave concern that world climates are changing, for cool temperatures, monsoon failures, and frequent drought conditions have been occurring.

If world climate continues to undergo disruptions, the human race is indeed in a grave situation. During the period between 1900 and 1950, human development accelerated rapidly. This era witnessed an abnormally warm global climate that led to high agricultural productivity. In addition, significant technological and medical advances supported the rapid growth of the human population. More land than ever was used for agriculture, and more water than ever was used for irrigation. Development continued in the 1960's even as the climate began to deteriorate, until at the present we are indeed in a precarious position. Dryland farming extends right to the edge of most of the world's deserts, mountainside pastures creep up the slopes of the major mountain ranges, grains are grown in northern Canada and Russia nearly into the Arctic, and farmers are already pushing into the great jungle basins. There is only limited room for further expansion. Yet, even with all the agricultural development, nearly half of the human population is undernourished and millions are literally starving. The

The effects of drought in the Sahel region of north Africa. (Courtesy of Alain Nogues— Sygma.)

problem that causes great concern is this: Food production and population have been pushed to the limit during an unusual period of particularly mild weather. If the climate should become less favorable, even if it should return to the "average" conditions of the past two thousand years, farms in marginal areas will continue to fail, and many more people will starve. No one is willing to predict that unfavorable climate change *will* occur, but most scientists say that such climate change is likely. An international group of prominent meteorologists, under the auspices of the International Federation of Institutes for Advanced Studies (IFIAF), issued the following statement on October 3, 1974:

The nature of climate change is such that even the most optimistic experts assign a substantial probability of major crop failure within a decade. If national and international policies do not take such failures into account they may result in mass death by starvation and perhaps in anarchy and violence that would exact a still more terrible toll.... We cannot safely assume that the cooling trend that we now face is a temporary aberration in a normally benign climate. It is at least as probable that the climate of our immediate future will be "worse" than the present one as it is that we will see a return to the "better" conditions of the immediate past.

As the world food situation continues to deteriorate, many people are looking for various technical solutions to alleviate the problem, at least temporarily. Some of the developments and problems will be discussed in the following sections.

11.9 INCREASES IN CROP YIELDS AND THE GREEN REVOLUTION

Next time you are walking in the country, look closely at the wild grasses growing in meadows and fields. Notice that each plant has a slender stalk, a few leaves, and a small cluster of seeds at the top. These seeds are rich in starch, protein, and vitamins, but they are so small that they are not harvested and processed for human consumption. Ancient farmers living in the Stone Age probably started cultivating grains that were only slightly more productive than these modern "weeds." As various farmers replanted only the seeds of the largest, healthiest plants, modern strains of wheat, rice, and other grains were developed gradually over the centuries. These grains were generally well adapted to local growing conditions. They were genetically adjusted to peculiarities in soil conditions, water supply, length of growing season, and seasonal temperatures, and they were at least partially resistant to local diseases and pest infestations. However, as population increased in the nineteenth and twentieth centuries, it became apparent that traditional farming practices were not adequate to feed the world's people. Therefore agriculturists searched for ways to increase crop yields. It was ob-

The wheat stalk on the right supports much larger quantities of grain than the native grass shown on the left.

vious that heavy doses of fertilizer could augment food production, at least to some extent. But there appeared to be a limit to the quantity of fertilizer that could be utilized. When a native grain plant is fertilized heavily, the leaves grow broader and larger, thereby shading nearby plants, and the stalk grows to be long and thin. The heavy grain causes the elongated stalk to break and bend, and the grain falls to the ground and rots.

In the mid-1960's an interdisciplinary team of scientists working in Mexico and the Philippines developed new varieties of wheat and rice that were adaptable to tropical climates and were capable of producing higher yields than any native grains. These varieties have short, upright leaves so plants can be grown close together without shading each other. In addition, their stalks are short and thick so that they will not bend and break when the plant is heavily fertilized. The potential yields of these new varieties are so spectacular that many people have heralded their introduction as the **Green Revolution.** For example, in Mexico in the 1940's wheat fields averaged 750 kg/hectare, whereas farmers planting the new strains of seeds in the 1970's average 3200 kg/hectare. In India and Pakistan, massive shipments of Green Revolution wheat in the late 1960's raised the wheat harvest between 50 and 60 percent during a period of two growing seasons. In Colombia, rice production increased by a factor of $2\frac{1}{2}$ despite the fact that the area of cultivation remained nearly constant (Fig. 11.9). Such increased food production across the globe has reduced famine and increased the well-being of millions.

On the other hand, figures of national grain production do not always reflect the fate of all people within that nation. To understand the total impact of the Green Revolution we must understand that a "wonder seed," by itself, does not produce large quantities of food. The seed only car-

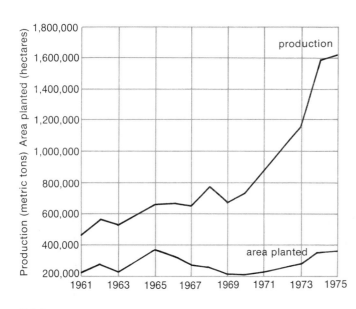

Figure 11.9 Rice production in Colombia following introduction of improved "Green Revolution" seeds in 1969.

Workers planting rice at the International Rice Research Institute, Los Baños, Laguna, Philippines.

ries the genetic potential for high grade production. If this production is to be realized, the plant must be fertilized heavily, watered, and protected from disease and insects. The new grain varieties planted in an impoverished soil and dependent on variable rainfall for growth produce equal or smaller yields than the native grains, which have been cultivated in such poor areas for centuries. In addition, the new grain varieties are less resistant to insect pests and fungal diseases than the traditional plants. As a result, farmers who invest in the seed, the fertilizer, and the advanced irrigation systems must also invest in pesticides.

Reviewing the situation, we see that if a farmer has the necessary capital and knowledge to buy the new seeds and care for them properly, high yields and an eventual profit can be realized, but if a modern integrated farming program is not practiced or if the money is not available for the needed fuel, fertilizer, pesticides, and water, the Green Revolution may provide no help.

The use of the new seeds has led to many other problems, such as:

(a) The taste and texture of hybrid rice IR-8, for example, differ from those of traditional rice, and it commands a lower price. In Pakistan, many wholesalers refuse to handle this variety.

345

Old way of harvesting wheat in India by cutting off stalks at the base by hand. (© Food and Agricultural Organization of the United Nations. Via delle Terme di Caraculla 00100, Rome, Italy.)

If food surpluses from agriculturally productive countries are to be used in other parts of the world, they must find their way into hungry stomachs. Will a newly introduced food be accepted by the residents of the poor countries? This question poses what Professor Paul Rozin of the University of Pennsylvania has called the "omnivore's paradox." Being omnivorous offers both opportunity and risk. An omnivore may choose foods from a wide selection of nutrients from several trophic levels. On the other hand, there is always a danger that untried materials may be unpleasant or toxic. The omnivore must therefore decide between limiting its choice of foods or risking harm by trying new ones—hence the "paradox." Humans, alone among omnivores, employ particular groups of flavors (for example, chili in Mexico or curry in India) to identify their foods culturally with what had been considered "safe" in their environment for generations. However, this exclusively human strategy often leads to rejection of new foods that are not culturally labeled with traditionally accepted flavors. Hence the distribution of food among the world's hungry is a demanding and complex social issue.

(b) In many rice cultures, paddy farmers have raised fish in irrigation canals. Now fertilizer and pesticide pollution of the canals have killed fish. Thus, increased total caloric content of many people's diets has been associated with decreased protein content.

(c) The problem of protein deficiencies has been aggravated by the curtailment of production of soy beans and other protein-rich legumes, which are now less profitable for some farmers than grains. Attempts to develop new high-yield varieties of these crops have thus far been unsuccessful.

(d) The new varieties of grain mature earlier than the old ones. While this enables the prosperous farmer to grow two crops a year instead of one, many other farmers are finding that since harvest time now coincides with rainy weather, the grain can no longer be dried in the sun. Instead, mechanical driers are often needed, and mechanical dryers must be powered by fossil fuels.

(e) National economic problems have arisen because of frequently insufficient foreign exchange to purchase imported pesticides and chemical fertilizers. Even if the developing nations were to build their own chemical factories, many would have to buy fuels, phosphate rock, and other raw materials.

The consequences of the Green Revolution have been mixed. Food production in the developing nations has increased. Starvation has decreased, and millions are liv-

346

ing with a new-found freedom from hunger and want. On the other hand, many rural people have sunk into deeper poverty, and homeless and untrained rural poor are migrating in large numbers to the city slums.

11.10 OTHER APPROACHES TO INCREASING FOOD PRODUCTION

INCREASES IN CROP ACREAGE

Most of the prime fertile land in the world is currently being farmed, so expansion of crop acreage will have to be in areas that have been uneconomical to cultivate in the past, such as semi-arid prairies, deserts, swamps, forests, and jungles. The farming of these areas will be feasible only if massive quantities of energy and capital become available. Thus, the southern Sahara could be farmed if nuclear power were available to desalinate water; huge regions of southeast Africa could be arable if their swamps were drained; and food production could be made more reliable in Bangladesh if rivers were dammed to increase irrigation during droughts and reduce flooding during periods of heavy rainfall. Unfortunately, the nations that need such improvements most can afford them least.

FOOD FROM THE SEA

People have often regarded the sea as a boundless source of food, especially protein. In truth, most of the central oceans are biological deserts (see page 54), and the more productive continental shelf areas are already heavily fished. In 1970, approximately 7 million metric tons of fish were harvested from the seas. This yield supplied 10 percent of the world's edible protein. During the period between 1970 and 1974, world fish production dropped steadily as the populations of many commercially important aquatic species became depleted. Marine ecologists believe that the oceans could support a larger sustained catch if maritime nations were careful and cooperative. But international maritime cooperation has not been particularly successful (see Section 5.7), and stocks of sardines, salmon, tuna, flatfish, whales, and other sources of marine protein are being overexploited. Even if the fish harvest were maximized, most marine ecologists agree that the sea is capable of supplying no more than 15 percent of the world's needed protein.

Other serious threats to the world fisheries are the pollution of the oceans, the urbanization of coastlines, and the destruction of estuaries. As increasing quantities of

This gladioli plantation is being grown on land reclaimed from the Negev desert in Israel. (Courtesy of Israel Government Tourist Office.)

347

Fish from the sea will provide a continuous source of protein only if people harvest wisely.

sewage, oil, pesticides, and other chemicals enter the ocean, plankton and fish are being poisoned. In particular, disasters such as the Nantucket oil spill in December, 1976, are having a devastating effect on world fish populations. (This subject will be discussed further in Chapter 14.) In many nations, it is often cheaper to purchase a salt water marsh and convert it to filled land than to purchase prime industrial sites. Ecologically, the destruction of an estuarine marsh is disastrous for, as was mentioned in Chapter 3, estuaries are nursery grounds for many fish that spend their adult lives in deeper waters. It has been estimated that one hectare of estuary produces enough young per day to grow into 270 kg of marketable fish.

THE CULTIVATION OF ALGAE

Recall from Chapter 3 (Fig. 3.7) that terrestrial plants, which operate at about one percent efficiency, are poor converters of solar energy into the nutrient energy of leaf tissue. Underwater algae do the job much more efficiently. If the underwater efficiencies could be maintained in concentrated cultures at full sunlight, an area of half a square meter could feed a person on a sustained basis. Even so, it does not follow that algae farming will solve food problems in the future, for algae do not grow as well in concentrated surface cultures as they do underwater. As light intensity is increased, the conversion rate decreases, so that at full sunlight algae are only about four times as efficient as wheat or rice. Furthermore, the advantage of this effi-

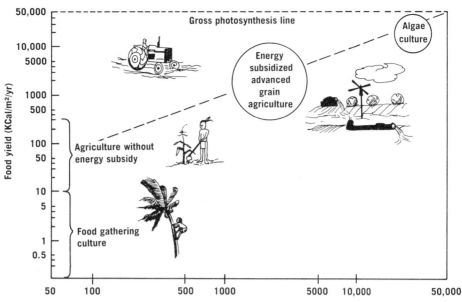

The relation between food yields and auxiliary energy.

Food yield (KCal/m²/yr)

Auxiliary fossil fuel energy in work flows (KCal/m²/yr)

ciency is offset by the fact that algae are not well adapted to concentrated growing conditions, and tremendous inputs of work are required to maintain high yields.

PROCESSED AND MANUFACTURED FOODS

Food chemists have pursued three different concepts in developing new food products: (a) alteration of good natural foods so as to make them more acceptable or more marketable, (b) conversion of agricultural wastes to edible food, and (c) the chemical synthesis of food. Let us discuss each in turn.

Alteration of Natural Foods

Perhaps the most familiar example is the conversion of vegetable oils into the butter substitute, oleomargarine. Food conversion is becoming increasingly important in both the developed and the developing nations. In some areas such high-quality protein sources as soybeans, lentils, and other legumes, though readily available, are unpopular. One solution is to use concentrated vegetable protein as a base for manufactured foods that taste like something else. Thus, soy protein "hamburger," soy "milk," soy "bacon," and other simulated products are readily available in retail stores in many parts of the world. In some areas, undernourished people spend their money on soft drinks rather than on protein-rich foods. Recently, soy protein has been used as an ingredient in soft drinks, and

Simulated ground meat made from soybean flour. (Courtesy of Archer Daniels–Midland Company.)

349

various sweetened, carbonated, soy beverages are now sold in Asia, South America, and Africa.

Conversion of Agricultural Wastes

When cooking oils are extracted from peanuts, soybeans, coconuts, cottonseeds, corn, or any other vegetable seed, the residue is a waste mash high in vegetable protein. At present, most of these residues are fed to cattle or discarded. It is relatively easy to incorporate the proteins into synthetic foods. Cottonseed mash contains a natural insect repellent which is toxic to humans. Research is under way to find an economical process to remove such poisons.

Sawdust, straw, and other inedible plant parts also represent a major source of waste materials. The major chemical component of sawdust and straw is cellulose. Cellulose consists of very large molecules which, when heated with dilute acid and water, decompose into a mixture of various sugars. Yeasts can thrive on this sugar mixture if other nutrients such as urea (a source of nitrogen and mineral salts are added to the culture medium. In turn, yeasts are sources of high-quality protein and can be eaten directly. Alternatively, yeast protein extract can be used as an additive in manufactured or processed foods. For example, whole wheat is deficient in the amino acid lysine, which is necessary for protein synthesis in the human body. Lysine can be extracted from yeast cultures; in India, it is added to dough in government-owned bakeries for the production of enriched bread.

Yeast culture, like algae culture, requires large quantities of auxiliary energy to synthesize the fertilizers and to maintain the growth chambers. Yeasts are heterotrophs, while algae are autotrophs and can derive energy from the Sun. Thus, yeasts do not represent a primary source of food but rather are organisms that can convert low-quality organic compounds into high-quality foods if sufficient additional energy is supplied. The acids and microorganisms present in the digestive system of cattle are also capable of converting cellulose and urea into protein, and future livestock feed may contain only sawdust, urea, and inorganic chemicals.

Chemical Synthesis

The high cost of maintaining yeast organisms or feed-lot cattle is prompting chemists to bypass the heterotrophic organisms completely and synthesize amino acids directly from coal and petroleum. Synthetic lysine still costs about as much as lysine extracted from yeast cultures and requires about as much energy to produce. Other synthetic foods include vitamins used for food additives and artificial pie fillers derived from petrochemicals.

It is encouraging to know that bread in New Delhi is

more nutritious than ever before, but we must recognize that such dietary advances are concomitant with the trend from the range cow to the feedlot cow to synthetic amino acids and that our existence is thus becoming increasingly dependent on technology and fossil fuel reserves.

11.11 FOUR COMPARATIVE CASE HISTORIES:
Family Farming in India, Japan, China, and the United States

INDIA

A woman follows a cow down the path, collects the dung, dries it, and uses it as fuel for cooking. She has no alternative sources for heat because there are virtually no trees in the area, and coal or oil is too expensive. Nor can her family afford to purchase chemical fertilizers. Therefore, when manure is used as a fuel to cook food from this year's crop, the yield of next year's harvest is being jeopardized. Loans to purchase fuel or fertilizer would help, but even that is unavailable, so there appears to be no immediate relief. A continuous downward spiral develops, and her family sinks into deeper poverty. With adequate capital and technology, one family's fields could be made to yield food for five families, but in her case there is barely enough food for anyone.

JAPAN

A poor farming family in Japan lives on 0.8 hectare (2 acres) of hilly land. Twenty percent of the land (0.16 hectare, or 0.4 acre) is devoted to growing the paddy rice needed to feed the family. Another small plot of 0.16 hectare is

Farming in India. (Photo by J. P. Laffont—Sygma.)

Farming in Japan. (Courtesy of Japan National Tourist Organization.)

used for growing upland rice, which is made into New Year's cakes and sold at a profit. The remainder of the land is given over to mulberry leaves, which are used to feed thousands of silkworms. The family's capital assets include a single-room mud hut, which they share with the silkworms; a two-wheeled power tiller, which they use to prepare the soil; and a small motor scooter with which they transport goods to market. The equipment is mortgaged. The cash from the silk and rice cake production is used to buy fish meal and soybeans and to repay loans on the mortgage. In their spare time the adults weave baskets of bamboo strips to supplement their income. Though they are poor, they maintain high crop yields by intense fertilization, so there is always enough to eat.

CHINA*

A mainland Chinese family has two children, whose grandparents remember periodic famine

---

*Since this picture of China is taken from reports of observers who visited the country on a government-sponsored tour, and since free travel through mainland China is not allowed, we are not sure whether this picture reflects the situation throughout most of the country.

and mass starvation. The family lives in association with a government-controlled agricultural community called a **commune**. The commune administration works in concert with the national government to assure that adequate supplies of fertilizer and fuel are available every year, that excess grain is sold to urban markets, and that the most recent agricultural technology is made available at the local level. New strains of short-stalked rice are planted, fertilized, and irrigated. When possible, two or three crops are grown successively in a single field every year so that the ground never lies idle. In addition, in some areas it has been shown that increased efficiency can be realized if corn and wheat are grown in alternate rows in the same field at the same time, and the family takes advantage of this finding. However, such planting makes machine harvesting virtually impossible, so the crops are cut and stacked by hand. The people work hard, and there are few frills in life, but there is enough to eat.

UNITED STATES

An unemployed sawmill worker is puttering around his home in the Maine countryside. It is a sunny day in August, but the dark clouds in the western sky presage rain. By United States standards, central Maine is a marginal agricultural area; individual holdings are small (an average farm is 100 to 200 hectares), and farm income is low. A new four-wheel drive pickup pulls into the driveway and a woman in her early sixties steps out.

"Hi, neighbor. Do you think you could help us get the hay in before the storm? If it rains on the cut hay, it'll rot. We can't pay much—two dollars an hour—but we need the help."

"Well, I won't work for two dollars an hour, but if you want to lend me a tractor to pull stumps after hay season's over, I'll work."

The deal is made. When they arrive at the farm, two generations of family are already on the job. All pitch in and work hard. The hay trailer was homemade in 1932, and one wheel threatens to collapse, so it must be driven slowly. The tractors, too, are old, and perform erratically. The neighbor asks why they don't invest in some automated hayloading equipment such as the ones used on large ranches in the West.

"Not enough capital. We borrowed heavily for new milking equipment and repairs on the barn [built in 1915], and as long as the price of milk stays low, we can't afford anymore machinery for a few years at least."

Baling hay in the United States.

As the storm approaches, they all work faster. Each time the wagon is loaded, it is driven slowly to the barn, unloaded, and returned to the fields. The conveyor belt that carries the bales to the interior of the barn jams repeatedly, and the women are constantly running back and forth with pitchforks to release hay caught in the belt, while the men load and stack the hay.

After four loads, the job is done, minutes before the rain falls. The neighbor is invited to stay for dinner in a new prefabricated mobile home. They all watch color TV and then sit down in the modern all-electric kitchen to a plentiful supply of steak, instant "heat'n eat" french fried potatoes, and a large salad from the garden.

TAKE-HOME EXPERIMENTS

1. **Sand and soil.** For this experiment you will need a small shovelful of *dry* sand or sandy gravel. Mix one portion of the sand with an approximately equal weight of *dry* rotten leaves, coffee grounds, grass clippings, or other organic matter. Place about 100 g of the unmixed portion of sand in one small dish and 100 g of the sand–organic mixture in the other. Add about 20 ml of water to each, and then weigh both dishes accurately. Place them side by side on a table and weigh them every half-hour for several hours. Does the weight change? Why? Which sample loses water faster? Why? Discuss the implications of this experiment for agricultural practice. (Note: 1 fluid ounce of water = about 28 g.)

2. **Sand and soil.** Fold two pieces of filter paper as shown in the accompanying sketch, and place one in each of two funnels. Fill one funnel with dry sand and another with dry rich topsoil, but be sure that the level of the dirt is below the level of the filter paper. Carefully weigh two small glasses or beakers and place one under each funnel. Now, using another

container, dissolve 3 g of table salt in 100 ml of water. Pour 50 ml of this solution into each funnel and collect the water that drips out in the two preweighed beakers. Carefully heat each container until all the water has evaporated and a residue of salt remains. Weigh the beakers. Which beaker contains more salt? Explain.

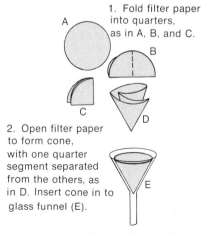

1. Fold filter paper into quarters, as in A, B, and C.

2. Open filter paper to form cone, with one quarter segment separated from the others, as in D. Insert cone in to glass funnel (E).

How to Make a Filter

353

3. **Erosion.** Find a place near your school where you can observe the effects of erosion. Describe what is happening, and, if possible, photograph the site. What agent (wind, ice, water, or other) is causing the breakdown and removal of material? Collect samples of the rock or soil as it appears before being perturbed, and then collect samples of the sediment that is being carried away. Compare the textures of the two materials. In your example is the erosion directly harmful to humans and their activities? Discuss.

PROBLEMS

1. **Agricultural ecosystems.** Explain why agriculture is more disruptive to natural ecosystems than are hunting and gathering of food.

2. **Destruction of ecosystems.** A common logging practice in the United States today is **clearcutting,** the removal of all trees and bushes to make room for logging roads across hillsides. Conservation groups attack clearcutting because it destroys wilderness recreation areas. Will clearcutting affect people other than those who enjoy wilderness? Explain.

3. **Agricultural disruptions.** In 1973 a series of devastating floods ravaged sections of Bangladesh. Many have attributed these floods to unsound logging practices in the mountains of Nepal, at the headwaters of the rivers that flow through Bangladesh. Explain how logging in Nepal could affect farming in Bangladesh.

4. **Dust bowl.** Drought recurs in North America approximately every 20 to 22 years. Discuss the relationship between these periodic droughts and the Dust Bowl disaster of the 1930's. Did the drought cause the land destruction? Defend your answer.

5. **Longevity of agriculture.** If a given land area is farmed for many years, does its productivity necessarily decrease? Justify your answer.

6. **Natural vs. agricultural ecosystems.** Briefly discuss the relative importance of each of the following characteristics to a plant species existing (a) in a natural prairie; (b) in a primitive agricultural system; (c) in an industrial agricultural system: (i) resistance to insects, (ii) a tall stalk, (iii) frost resistance, (iv) winged seeds, (v) thorns, (vi) biological clocks to regulate seed sprouting, (vii) succulent flowers, (viii) large, heavy clusters of fruit at maturity, (ix) ability to withstand droughts.

7. **Energy.** List the energy inputs that are added to a wheat field in industrial agriculture.

8. **Modern agriculture.** Explain how modern beef production differs from traditional beef production on the open range. Which is more efficient in utilizing sunlight to produce meat? Why? Which is more efficient in utilizing total energy input to produce meat? Why?

9. **Industrial agriculture.** What are the major techniques of industrial agriculture? Under what circumstances of use or misuse may each of these methods engender a loss of fertility of the land?

10. **Fertilizers.** Imagine that you saw the following sign at a farm supply center, "SALE—Oxygen and carbon fertilizers. This fertilizer contains all the oxygen and carbon your plants will need! Guaranteed! Only $5.00 for a 50 kilogram bag." Would you consider the offer to be a good buy? Why or why not?

11. **Definition of "organic."** (a) Write all the definitions of the word "organic" that you can find in a good dictionary that were not given in this text. (b) Obtain definitions of the word "organic" from organic gardening publications or from the proprietors of health food stores, and compare them critically with the definitions given in this text.

12. **"Organic" foods.** "Organic" food is grown and processed without using any unnatural product. Reliable commercial sources contend that 10 times as much "organic" food is sold as is produced. (a) "Organic" food can be sold to the consumer at a relatively high price. How is that relevant to the above contention? (b) Suppose a farmer grows a crop of tomatoes "organically." He can show his field to several potential buyers and promise each he will deliver tomatoes from that field. What if the farmer promises more than the field delivers? How can the buyer be certain he has received the organically grown tomatoes and not produce from some other field? (How would you test whether a tomato has been grown organically?) (c) How can a consumer be certain that his purchased "organic" tomatoes are genuine? (d) If you wanted to buy a jar of organic honey, what would convince you that the product was unadulterated? Its appearance? taste? the reputability of the store you bought from? the label? the price? (e) "Organic" granola is a cereal made from many ingredients; the more elaborate formulations include oatmeal, wheat germ, other grains, honey and other natural sweeteners, almonds, sesame seeds, and dried fruit. At what points in the growing of the raw materials and in the processing of the granola itself might unnatural foods or contaminants enter the product? Do you feel contamination would be likely to be accidental or deliberate? Do you think that some governmental body should be charged with the duty of insuring that the label "organic" refers to an uncontaminated product? Consider the cost of policing and the relevance of natural foods to nutritional health in the United States.

13. **Humus.** Briefly discuss some of the chemical and physical properties of humus.

14. **Fertilizers.** Assess the advantages and disadvantages of using manure as a fertilizer. Do you feel manure should be used to fertilize crops? (Remember that a field fertilized with manure may stink; however, such odors are not harmful to health.)

15. **Mechanization.** One hundred years ago most farmers used horses or oxen to plow fields and harvest crops. Today tractors are commonly used. Imagine that the fossil fuel crisis became so severe that farmers were forced to rely on animal power once again. Would the total food production increase, decrease, or remain constant? Defend your answer.

16. **Mechanization.** A large combine cuts and threshes wheat efficiently. Does a combine improve the yield of grain produced per acre? Why or why not?

17. **Irrigation vs. other uses for water.** Water from the Colorado River in the southwestern part of the United States is diverted both for irrigation and for domestic use. Some is allocated locally and some is piped to California. In addition, the coal and oil shale companies in the Southwest need the water for industrial expansion. However, environmentalists feel that the level of the river should remain high to preserve natural ecosystems. A serious snow drought during the winter of 1976–1977 greatly diminished the water level in the Colorado River. Discuss the impact of the drought on the various groups that want to use the water. If water were to be rationed, which group should receive the greatest supply? Defend your selection.

18. **Farming practices.** The Old Testament commands farmers to allow their land to rest, unused, every seventh year. In what way would such a rest be beneficial to the land?

19. **Agriculture in India.** Which of the following innovations do you believe has had significant effects on agriculture in India? Which has had a moderate effect? No effect? (a) pesticides, (b) the tractor, (c) inorganic fertilizers, (d) the electric motor, (e) the high-pressure irrigation pipe. Defend your answers.

20. **Rural land use.** Outline some approaches to legislation that might prevent agricultural land from being converted to other uses.

21. **World food production.** Describe the types of problems that would have to be solved in order to be able to allocate world resources so that everyone had enough to eat.

22. **Climate.** Explain why a small shift in climate could seriously affect world food supplies.

23. **The Green Revolution.** Examine the curve for rice yields in Colombia in Figure 11.9. If there were no significant advances in agricultural sciences, how would you expect the curve to grow in the near future? (Recall the discussion of graphical extrapolation in Chapter 6.)

24. **Green Revolution.** Discuss the advantages and disadvantages of Green Revolution grains.

25. **Green Revolution.** Explain why the Green Revolution has not helped the very poorest farmers.

26. **Agriculture in marginal lands.** Discuss some problems inherent in farming deserts; jungles; temperate forest hillsides; the Arctic.

27. **Food from the sea.** Jacques Cousteau has said, "The shores of the rivers are the roots of the oceans." What do you think he meant?

28. **Aquaculture.** Aquaculture, or seaweed farming, is likely to be practiced along shorelines and in coastal bays or marshes. Outline a series of events which might lead to a situation where successful aquaculture would *decrease* the total food harvest from the sea. Do you think that this set of events is likely to occur?

29. **Synthetic foods.** It is reasonable to believe that food chemists will be able to synthesize economically a complete and balanced diet from coal in the near future. Outline some advantages and disadvantages of the chemical synthesis of foods.

30. **Synthetic foods.** How does the probable energy requirement necessary to make a soy drink compare to the direct use of soybean soups?

31. **Sources of protein.** Mushrooms are high in protein and some varieties can be grown to maturity in 10 days to two weeks. Discuss the advantages and disadvantages of mushrooms as a source of protein. *(Hint:* Mushrooms and yeasts are both varieties of fungi.)

32. **Synthetic food.** Discuss the advantages and disadvantages of the 3-step food production sequence outlined below, as compared with agriculture.

$$CO_2 + H_2O \xrightarrow[\text{using nuclear energy}]{\text{factory process}} C_6H_{12}O_6 \text{ (sugar)}$$

$$\text{sugar} + \text{urea} + \text{yeasts} \longrightarrow \text{protein}$$

$$\text{protein} + \text{sugar} \xrightarrow[\text{processes}]{\text{industrial}} \text{imitation bread, meat, string beans, etc.}$$

BIBLIOGRAPHY

Two sets of articles on food and agriculture are:
Scientific American, the entire issue of September, 1976.
Philip H. Abelson (ed.): *Food: Politics, Economics, Nutrition, and Research.* (A compendium of articles from *Science.*) Washington, D.C., American Association for the Advancement of Science, 1975. 202 pp.

Two excellent books that discuss the power base of industrial agriculture are:
William J. Jewell: *Energy, Agriculture and Waste Management.* Ann Arbor, Mich., Ann Arbor Science Publications, 1975. 540 pp.
Howard T. Odum: *Environment, Power, and Society.* New York, Wiley-Interscience, 1971. 331 pp.

The reader who wishes to investigate the Green Revolution and food in the future should refer to:

Lester R. Brown: *By Bread Alone.* New York, Praeger Publishers, 1974. 272 pp.
Lester R. Brown: *Seeds of Change: The Green Revolution and Development in the 1970's.* New York, Encyclopaedia Britannica, Praeger Publishers, 1970. 205 pp.
Francine R. Frankel: *India's Green Revolution.* Princeton, N. J., Princeton University Press, 1971. 232 pp.

Two excellent books on agriculture and food resources are:
John Harte and Robert H. Socolow: *The Patient Earth.* New York, Holt, Rinehart and Winston, 1971. 364 pp.
Kusum Nair: *The Lonely Furrow: Farming in the United States, Japan, and India.* Ann Arbor, University of Michigan Press, 1970. 336 pp.

CONTROL OF PESTS AND WEEDS

12.1 COMPETITION FOR HUMAN FOOD

Perhaps the most persistent and difficult problem in the agricultural production of large quantities of food has been competition from small herbivores. If a cow wanders into a field, the farmer can chase her away. An insect, a field mouse, the spore of a fungus, or a tiny root-eating worm (a nematode) is more difficult to deal with. Since these small organisms reproduce rapidly, their total eating capacity is very great. In addition to their voracity, these pests may be carriers of disease. The bubonic plague, carried by a flea which lives on rats, swept through medieval Europe, killing as much as one third of the total population in a single epidemic. Malaria and yellow fever, spread by mosquitos, have killed more people than have all wars.

Not all insects, rodents, fungi, and nematodes are pests. Most do not interfere with people, and many are directly helpful. Millions of nematodes live within a single square meter of healthy soil. Most are necessary to the process of decay and hence to the recycling of nutrients. Fungi, too, are essential to the process of decay in all the world's ecosystems. Certain species of fungi are used for the production of antibiotics, bread, beer, and cheese. Rodents are part of natural ecosystems; thus, squirrels help to spread pine seeds, and lemmings provide a staple food for almost all carnivores in parts of the Arctic. The role of insects in the maintenance of the biosphere is extensive and versatile. For instance, bees are essential to the life cycle of most flowering plants. In their search for food, bees inadvertently transfer pollen from flower to flower and thereby ensure fertilization. In fact, it is a common agricultural practice to move hives into crop areas at time of bloom to assist or ensure pollination. Many insects, such as species of springtails, are part of the process of decay.

Rat flea

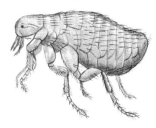

Insects are the prime food source of many animals that are vital, in turn, to the maintenance of natural balance. For example, the diet of many species of birds includes both insects and fruit. Fruit seeds transferred intact through the bird's digestive system and deposited at distant locations have been an important contributing factor in the continuing existence of certain plants. Thus, insects are essential to the survival of many species of birds that participate in the life cycle of many wild fruit trees, which, in turn, help to support wildlife. In addition, many carnivorous or parasitic insects feed on insects that eat crops. The periodic invasion of some African villages by driver ants is a fascinating illustration of insect ecology. Many disease-carrying rodents and insects live in the village houses and pose a constant threat to the human population. At periodic intervals, however, millions of large driver ants invade the villages, chase away the inhabitants, and eat everything that remains. When the people return, they find that their stored food supply is gone, but so are all the cockroaches, rats, and other pests—everything has been eaten.

Pests have lived side by side with people for millions of years. At times pest species have bloomed and brought disease and famine. But most of the time, natural balance has been maintained and humans have lived together with insects in reasonable harmony. Several factors have made this accommodation possible. First, the total human population, and hence the total food requirement, has been much smaller in past years than it is now. Also, many species of food plants produce pesticides for their own protection. For instance, the roots of some East Indian legumes contain the insecticide **rotenone,** and many wild cereal plants such as wheat, corn, rye, and oats are naturally resistant to fungal attack. Such innate protective measures are no longer sufficient to guarantee an adequate food supply for the human population. One reason is that the natural controls do not always work too well, especially in the short run. People have become less willing to accept pestilence and have strived to combat the insect blooms which bring misery and death. Perhaps even more critical is the fact that the human population is now so large that tremendous quantities of food are needed. One way to increase crop yields is to reduce competition from insects.

12.2 INSECTICIDES—INTRODUCTION AND GENERAL SURVEY

Chemical control of insects is not new. Marco Polo introduced pyrethrum to Europe after learning of its use by farmers in the Far East. However, systematic spraying of crops was not carried out until the early twentieth century. By the early 1900's, rotenone, pyrethrum, nicotine, kerosene, fish oil, and compounds of sulfur, lead, arsenic, and mercury were in common use. These formulations did not change appreciably until the 1940's. The most significant breakthrough occurred around the start of World War II, when the insecticidal properties of a chemical called DDT were discovered. DDT was far cheaper and more effective against almost all insects than the previously known control methods. The use of DDT led to dramatic early successes: it squelched a threatened typhus epidemic among the Allied army in Italy; anti-mosquito programs saved millions from death from malaria and yellow fever; and pest control, leading to increased crop yields all over the world, saved millions more from death by starvation.

Enthusiastic supporters of DDT predicted the complete destruction of all pest insects within the foreseeable future; Paul Müller, the chemist who first discovered its insecticidal properties, received a Nobel Prize. But within 30 years the promise of insect-free abundance had been broken, and the "miracle" chemical that was to have achieved it had fallen from grace. On January 1, 1973, all interstate sale and transport of DDT in the United States was banned except for use in emergency situations in which life is immediately threatened. During the next two years several other "miracle" pesticides were banned as well. At the present time many other insecticides and herbicides are being re-evaluated, and more bans can be expected in the future. Here we will review the story of this transition, a classic example of the interplay between technology and the dynamic balance of natural environments.

To focus attention only on DDT, however, would be incomplete, for many other chemical insecticides have been in use during the past 30 years. Some of the most common of these are listed in Table 12.1. Note the first three classes of compounds—**organochlorides, organophosphates,** and **carbamates.** These words designate categories of structurally related chemicals. Chemical compounds usually act on living systems or on other compounds in a manner that is roughly predictable from the composition and the structure of their molecules. It is therefore convenient to classify compounds according to molecular composition and structure. Roughly speaking, all organochloride insecticides have similar mechanisms of action in living tissues, and they are all about equally soluble (or insoluble) in various solvents such as water, alcohol, or fats. Organophosphates, as a group, also demonstrate a recognizably coherent set of properties that are quite different from those of the organochloride compounds.

The insecticides in these three groups are not produced by living organisms anywhere on the Earth but are *synthetic* products of laboratories and chemical factories. The last grouping in Table 12.1, naturally occurring or "organic" (see page 332) pesticides, is not a set of chemically related compounds. Nicotine, pyrethrum, and ro-

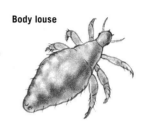

Body louse

TABLE 12.1 Common Chemical Insecticides*

COMPOUND CLASS	DESIGNATION
Organochlorides (also called chlorinated hydrocarbons)	Aldrin Chlordane DDD DDT Dieldrin Endosulfan Endrin Heptachlor Lindane Toxaphene
Organophosphates	Diazinon Malathion Parathion
Carbamates	Sevin (carbaryl)
Natural pesticides	Nicotine Pyrethrum Rotenone

*Chemical formulas of various pesticides are shown in Appendix C–12.

tenone are similar to each other in that they are all of biological origin—that is, they are extracted from the living tissue of certain naturally occurring plants—but their chemical structures are vastly different.

The names of the individual compounds in Table 12.1, with a few exceptions, are trade names. Thus the word Aldrin gives no more chemical meaning to the nature of the insecticide than the word Detroit gives a geographical meaning to the size and location of that city.

12.3 THE ACTION OF CHEMICAL PESTICIDES— BROAD-SPECTRUM POISONING

Chemical insecticides were initially received with great enthusiasm because they are inexpensive, easy to use, fast-acting, and effective against a wide range of pests. Their promise engendered an uncritical optimism. It was imagined, for example, that a tomato farmer, faced with a mid-season invasion of some insects that started to eat the green fruit, would not even have to identify the pest. He would simply call in an aerial spray company and would expect 90 percent destruction of his pests by the next day or two. A simple problem, a simple solution—or so it seemed.

Yet if we examine the problem more closely, complexities emerge. A tomato field under attack by some insect is not merely a two-species system. The tomatoes and pests are but members of a large agricultural ecosystem of thousands of species that include predator insects, bacteria, parasites, and many types of soil dwellers, as well as carnivorous, herbivorous, and omnivorous birds and other migratory animals. Therefore, despite the undeniable fact that innumerable successes of spray programs have been recorded during the past 30 years, it is important to look more closely into the intricacies of the problem.

The use of nonselective sprays has often led to the destruction of the natural controls on relative population sizes. As an example, when DDT and two other chlorinated hydrocarbons were used extensively to control pests in a valley in Peru, the initial success gave way to a delayed disaster. In only four years, cotton production rose from 490 to 730 kilograms per hectare. However, one year later the yield dropped precipitously to 390 kilograms per hectare, 100 kilograms per hectare less than before the insecticides were introduced. Studies indicated that the insecticide had destroyed predator insects and birds as well as insect pests. Then, with natural controls eliminated, the pest population thrived better than it had ever done before.

But, you may ask, how could the pests stage a comeback? Why didn't the predators stage an equal comeback? Why couldn't the farmers combat the pest resurgence with more spraying?

One of the major problems with chemical insecticides is that many insects become resistant to the poisons. In other words, a given insecticide at a given concentration often becomes less effective after some years of use. It appears as though the chemical has diminished in potency, although the composition has, of course, remained unchanged. To understand the reason for this phenomenon, recall that the chemistry of plants and animals changes from time to time as a result of random mutations of their reproductive cells. A mutant has a good chance of survival if its particular mutation protects it from a hostile environment. This mechanism of random mutation has allowed insects to adapt to their environment for millions of years, and it is this process that protects insects from pesticides. In areas where spraying is heavy, strains of insects have evolved which are genetically resistant to a particular chemical. Genetic resistance to insecticides is an extremely serious problem. By 1945 at least a dozen species had developed some resistance to DDT; by 1975 the number had increased to 200 species. About 35 of these resistant species carry disease and about 80 others are serious agricultural pests. Since resistant parents tend to pass this character on to succeeding generations, the old pesticides are rendered ineffective. This effect was directly demonstrated in an experiment in which DDT-resistant bedbugs were placed on cloth impregnated with DDT. They thrived, mated, and the females layed eggs normally. The young, born on a coating of DDT, grew up and were healthy. Attempts to change pesticides have in many cases simply produced strains of insects that are resistant to more than one chemical.

The problem of resistance of insects to poisons can be compounded if the pest becomes resistant and various predators do not. If this happens, the pests have a new biological advantage and can thrive in greatly increased numbers. This is favored by three factors:

(a) *Insect pests are often smaller and reproduce at a greater rate than their predators.* More frequent reproduction engenders more frequent mutation and thus improves the chances that resistant strains will arise. Furthermore, once a resistant population of insect pests appears, it can repopulate its ecological niche much faster than can a larger, more slowly reproducing species. In effect, the pest species develops resistance faster than the predator.

(b) *Predators generally eat a diet richer in insecticides than that of the original pests.* Because chemical poisons are not immediately excreted by herbivorous insects (or any other organism for that matter), the concentration of the chemical in their bodies becomes greater as more contact is made with the poison either directly or through the ingestion of sprayed leaf tissue. Since death may be delayed for some time after poisoning, many poisoned but living insects will be eaten by their natural enemies. In this way

Cockroaches that were highly susceptible to DDT were, of course, the first to succumb, and their generations have all but died out. Some of the originals, however, were bred in laboratories where they were carefully protected from any exposure to insecticides. Their descendants are now rare strains of "old-fashioned" cockroaches.

Many insects are predators. Here a praying mantis is eating a Satyrid butterfly. (Courtesy of Emeritus Professor Alexander B. Klots, Biology Department, the City College of the City University of New York.)

predators eat a diet that is more concentrated in poison than the diet of the herbivores, the original pests.

(c) *There are always fewer predators than herbivores (including pests) in an ecosystem.* A species with a small population runs the risk of extinction, because it is easier for a group containing only a few individuals to be wiped out by some disaster than it is for a group containing many. Therefore, the species of herbivores (the pests, which are more numerous) have a greater chance of survival than the species of predators.

We can now reconstruct what happened in Peru. Although the pesticides caused a rapid decrease in the pest population, resistant mutants soon displaced the susceptible individuals. By this time the situation had become worse than it was originally, because the natural predators did not achieve immunity so well as the pests, and therefore the pests were less under control than ever.

Another effect of broad-range pesticides is that killing one species sometimes leads to the bloom of another pest. Consider the story of the spider mite in the forests of the western United States. The spider mite feeds on the chlorophyll of leaves and evergreen needles. Because in a normal forest ecosystem predators and competition have kept the number of mites low, mites have never been a serious problem. However, when the United States Forest Service sprayed with DDT in a campaign to kill another pest, the spruce budworm, complications arose. The budworms were effectively killed, but the insecticide also poisoned such natural enemies of spider mites as ladybugs, gall midges, and various predator mites. The next year the forests were plagued with a spider mite invasion. Although spraying had temporarily controlled the spruce budworm, the new infestation of spider mites proved to be more disastrous.

These examples are two of the many cases in which broad-range insecticides have engendered more serious problems than they have solved, and where the goal of increased agricultural production has ultimately been frustrated. The response to such failures has often been to increase the number of sprayings or the amount of insecticide per spraying.

Pesticides are also poisonous to humans and a wide variety of other species of animals. There have been thousands of deaths—accidents, suicides, and homicides—attributed directly to insecticide poisoning. In addition, farm workers who are regularly in contact with high concentrations of pesticides have often been poisoned by direct exposure. Since many of these men and women are migrant workers, who move from field to field, they usually are not familiar with the spraying schedule on a particular farm and often become exposed to high pesticide concentrations unknowingly.

Wild animals and even livestock have also been affected. For example, when several communities in eastern Illinois were sprayed aerially in an effort to stop the westward movement of the Japanese beetle, many species of birds were almost completely annihilated in the sprayed area, ground squirrels were almost eradicated, 90 percent of all farm cats died, some sheep were killed, and muskrats, rabbits, and pheasants were poisoned. These unwanted side effects might have been considered a necessary price to pay for pesticidal success, but the cost did not yield the desired benefit; the Japanese beetle population continued its westward advance.

In a natural ecosystem, bees pollinate many plant species as they travel from flower to flower. But bees are also killed by pesticides. When the bee population is decimated, crop losses arise from lack of pollination. In some

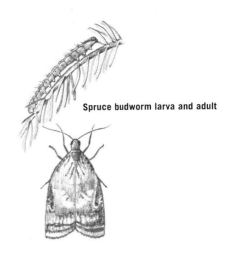
Spruce budworm larva and adult

Migrant farm workers near Salinas, California. These people are not told of past spray applications and are sometimes exposed to high concentrations of pesticides.

cases these losses have been more damaging than the insect attack itself.

12.4 THE ACTION OF CHEMICAL PESTICIDES—PERSISTENCE IN THE ENVIRONMENT

Most naturally occurring organic compounds are **biodegradable,** that is, capable of being broken down by some organisms. Biodegradability is a phenomenon that has developed with the evolution of species; organisms evolve to consume compounds that provide energy or raw material. However, most of the chemical insecticides are synthetic and have only recently appeared in the environment. Some of these are similar enough to naturally occurring compounds to be easily biodegradable. The others decompose more or less rapidly through the chemical action of water, environmental acids and bases, atmospheric oxygen (especially in the presence of sunlight), and perhaps other agents. The organochlorides are persistent chemicals; they do not degrade rapidly in the environment. For example, in some soils DDT, or Aldrin, has been detected in appreciable quantities 15 years after a single spray application. Such persistence is a serious ecological problem, for not only does a single spray application subject non-target organisms to continuous exposure to poisons for many years, but with continued annual spraying, the insecticide accumulates in the environment.

The persistence of organophosphates and carbamates is more difficult to assess accurately. If Parathion, a representative organophosphate pesticide, is dispersed in loamy soil, 70 percent of the original quantity will decompose after one month, and over 98 percent will decompose after four months. This evidence suggests that Parathion is nonpersistent. But what happens when Parathion decomposes? The atoms that build a Parathion molecule do not disappear; the molecule breaks apart into smaller fragments, and these fragments react with chemicals in the soil to produce new compounds. Recently, researchers have asked, "Are the decomposition products of Parathion poisonous to soil organisms?" "Are they harmful to natural ecosystems?" "Are they biodegradable?" At the present time, no one knows the answers to these questions, and more research must be done before we know whether or not the organophosphate pesticides decompose to produce compounds that cause long-term harmful effects to natural ecosystems.

12.5 PHYSICAL DISPERSAL OF INSECTICIDES

A chemical pollutant that is nonbiodegradable not only persists in the immediate environment but also lasts while it is being mechanically and biologically transported throughout the biosphere. The physical dispersal of organochlorides has been particularly well studied. To understand the mobility of insecticides, we must first review some of their physical characteristics. Organochlorides are relatively insoluble in water, evaporate slowly, have a strong tendency to adhere to tiny particles of dust, dirt, or salts, and are soluble in the fatty tissues of living organisms.

Now consider the fate of an application of DDT or one of its sister compounds sprayed from an airplane onto a crop. Typically, the insecticide is mixed with some inert ingredient so that at a calculated airplane speed and spray rate, a dose of between $1/2$ and 6 kilograms of insecticide is applied per hectare. The exact amount depends on the type of insecticide, the species of pest, and the nature of the crop. Not all the spray will hit the exact target. The accuracy is dependent on such factors as the skill of the pilot, the number and position of electrical wires and trees that must be dodged, and the wind direction, velocity, and turbulence. Some of the insecticide that misses the desired target lands on nearby houses, roads, streams, lakes, and woodlots, while some is carried into the air either as a gas or as a "hitchhiker" on particles of dust or on water droplets. Therefore, a great deal of insecticide travels long

Crop duster dusting sulfur to retard mildew on grapevines 20 miles south of Fresno, California. (Courtesy of EPA-DOCUMERICA, photographer, Gene Daniels.)

"There I was, coming in low at a hundred and seventy-five m.p.h. with forty-two acres of broccoli to the left of me, eighteen acres of asparagus to the right of me, eighty acres of carrots straight ahead! Power lines all around! My target: seven acres of badly infested garlic smack in the center...." (Drawing by Dedini; © 1972, *The New Yorker Magazine, Inc.*)

365

Figure 12.1 Physical dispersal of insecticides with some average values for DDT levels.

Labels within figure:
Trade winds .1-.3 ppb
Rain .1-.3 ppb
Fat of man 6-12 ppm
Rivers and lakes .001-.2 ppb
Fat of cows .5 ppm
Soil 2-10 ppm
Ground water .001-.2 ppb
Ocean depths .001 ppb

distances in the air. Indeed, all organisms on Earth are subject daily to some exposure to these chemicals. Before most organochlorides were banned, almost all rain water contained a measurable concentration of pesticides. Additionally, some of these poisons enter the tradewind currents and circulate around the globe. This dispersal has been so complete that even penguins in Antarctica carry measurable quantities of pesticides in their fatty tissues.

Patterns of dispersal of organochlorides and some average DDT concentrations in 1970 are shown in Figure 12.1. The effects on non-target species (including humans) are taken up in the following two sections.

12.6 BIOLOGICAL CONCENTRATION OF ORGANOCHLORIDE INSECTICIDES—EFFECTS ON NON-TARGET SPECIES

The danger of high concentrations of poisons in the soil arises from the fact that fertile soil contains much living matter. One kilogram of rich farm earth contains up to 2 trillion bacteria, 400 million fungi, 50 million algae, and 30 million protozoa, as well as worms, insects, and mites. These organisms are vital for continued fertility of the soil. They fix nitrogen, they break down rock and thus make minerals available to the plants, they retain moisture, they aerate the soil, and they bring about the essential process of decay. Without these organisms, the plants above ground usually die. The effect on these organisms of an increasing concentration of poison in the soil is largely unknown. In many heavily sprayed areas of the world, farmers are harvesting more food per acre than ever before. Yet some facts are coming to light which may presage future disaster. Studies in Florida have shown that some chlorinated pesticides seriously inhibit nitrification by soil bacteria. Termites have not been able to survive in soils that were sprayed with Toxaphene 10 years previously. Similarly, Endrin present in as low a concentration as one part per million (ppm) caused significant changes in the population of soil organisms and, consequently, in the relative concentrations and availability of important soil minerals. As a result, beans and corn grown in soils treated with Endrin contained different nutrient values than the same crops grown in untreated soils—some minerals were increased, some decreased. The net result, however, was a decrease in the total plant growth. In another instance, Aldrin routinely sprayed on a golf course depleted the number of earthworms.

As is the case for many types of ecological disruptions, the long-term effects of insecticides in soils are not known. Perhaps many soil organisms are or will become resistant to the pesticide accumulations. But the stakes in the gamble are high, for if life in the soil dies, plants will soon succumb.

Not all the pesticides in the soil remain fixed there. Some fraction of the total seeps into groundwater reservoirs and hence into drinking water supplies. A government study of private wells in Illinois in 1971 showed that all surveyed contained water contaminated with insecticide. A large fraction of soil-based insecticides gets carried into the world's surface water supply along with the tons of sediment eroded from agricultural systems. As a consequence, even though several organochloride pesticides have already been banned, all major rivers in the United States still contain measurable insecticide concentrations in the parts per billion (ppb) range. In river water at 7.2°C (45°F), which is normal temperature for trout, 1.4 ppb of Endrin will kill half of a population of rainbow trout in three days. Similarly, many other species of fish cannot survive insecticides in concentrations greater than about one to ten parts per billion. Trout in the United States today live in the more mountainous regions, which are not so polluted, while those that were once native to the Great Lakes water systems or the lower Missouri River can no longer be found. Under severe conditions, such as occurred during 1950 in parts of the South, when heavy rains followed heavy spray applications, agricultural runoff was so high that the residual pesticide concentrations were raised a thousandfold. Nearly all of the fish in many watersheds were killed.

Inevitably, if insecticides are present in major rivers, they must also be present in the ocean, and the concentrations must be highest in estuary systems, that is, at river mouths and in coastal bays. One of the most serious problems in estuary systems is that insecticides reduce photosynthesis carried out by plankton. DDT at a concentration of 1 ppb can reduce phytoplankton activity by 10 percent and at 100 ppb by 40 percent, as compared with that of plankton grown in unpolluted waters.

Recall that the effect of a given spray application is more severe on carnivorous insects than on the original pest. In light of the fact that persistent pesticides are present throughout the biosphere—in air, soil, water, and on plant tissue—it is not surprising that biological magnification of DDT and its sister compounds occurs throughout numerous food webs, with carnivores of all kinds receiving the highest concentration in their tissues. In fact, the chemistry of organochloride compounds is particularly amenable to this biological transport. Their persistence makes them survive long enough to travel. Moreover, they are fat-soluble, and, therefore, when they enter an organism, they dissolve and are stored in fatty tissue.

The normal cleansing mechanisms of the body depend on the transport of unwanted substances through aqueous media such as blood and urine. However, because organochloride compounds dissolve in body fats, they are not easily carried away by water solutions. Therefore they accumulate in the bodies of most animals.

Consider the widely publicized case of the spraying of Clear Lake in California in the 1950's with DDD for the control of biting pests (Fig. 12.2). After the project was

Clear Lake, California.

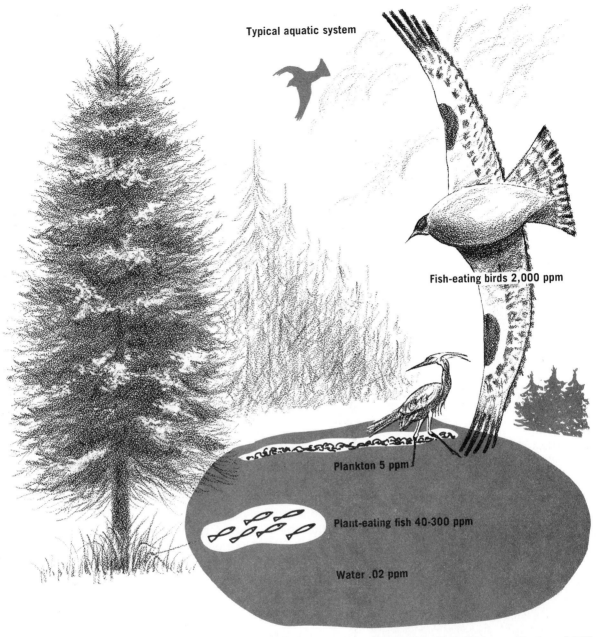

Typical aquatic system

Fish-eating birds 2,000 ppm

Plankton 5 ppm

Plant-eating fish 40-300 ppm

Water .02 ppm

Figure 12.2 Biological magnification of DDD (Clear Lake, California).

completed, the water contained 0.02 ppm of DDD. However, plankton living on the surface of the water contained 5 ppm of the insecticide in their tissues. Plant-eating fish had magnified this concentration to 40–300 ppm (depending on the species); some carnivorous fish and birds at the top of the food chain had as much as 2000 ppm of DDD in their tissues. Thus, it is not surprising that fish and birds can be poisoned by an insecticide that originates in waters in which its concentration seems innocently low. In fact, in Clear Lake as well as in other aquatic systems, it has been observed that a year or so after introduction of a pesticide, the chemical, though undetectable in the water, is still present in the food web.

369

Figure 12.3 Biological magnification of DDT after spray application for Dutch elm disease.

444 ppm in Robin

141 ppm in earthworm

Soil 9.9 ppm

When peregrine falcons are subjected to DDT poisoning, many eggs crack before hatching, and chicks often die prematurely. The two dead chicks and cracked egg in this nest all resulted from such poisoning. (Courtesy of the Peregrine Fund.)

The effects of high levels of insecticide on carnivorous fish and birds have been particularly severe. In many cases the animals die soon after application of the insecticide. In others, sublethal doses impair the normal activity or reproduction of the animals, resulting in either a delayed death of part of the population or a decreased population growth. For example, in many sprayed areas, insect-eating fish, such as salmon fry, are subjected not only to the poison itself but also to starvation when their food supply is destroyed. Trout and salmon with insecticide in their systems avoid cold waters and tend to lay their eggs in warmer areas, where the young have a smaller chance of survival. Pesticides can be fatal to animals that store food energy in fat for use during winter months. Trout build up a layer of fat during the summer months, when food is plentiful. In areas where the land has been sprayed, this fat contains high concentrations of DDT. During winter, the fat is used as a source of energy. The DDT released into the bloodstream upon fat breakdown has been known to kill the animal. The eggs of fish also contain a considerable amount of fat, which is used as food by the unborn fish. In one case, 700,000 hatching salmon were poisoned by the DDT in their own eggs. Finally, the evidence seems undeniable that DDT poisoning is responsible for the sharp decline in populations of many birds, especially those that are secondary or tertiary consumers (Fig. 12.3). One manifestation of the early stages of DDT poisoning is the inability to metabolize calcium properly. In birds, this has led to the production of thin-shelled eggs. Often these weakened eggs crack or break in the nest, with resulting prenatal death. Laboratory birds that are fed low concentrations of DDT in their diet produce eggs with thin and weakened shells. The populations of several species of birds—among them, the peregrine falcon, the pelican, and some eagles—are declining so rapidly that many conservationists fear that they will become extinct in the near future.

12.7 EFFECTS OF INSECTICIDES ON HUMAN HEALTH

Most chemical pesticides are poisonous to humans as well as to insects. The organophosphates, which have been used extensively in North America since 1973, are much more poisonous than the banned DDT that they replaced.

Since the mid-1940's, the incidence of acute fatal poisonings due to pesticides has been about constant at one per million people per year. At present, more than half these cases are children who are exposed to the toxic chemicals through carelessness in packing or storage. The balance of fatal poisonings has occurred largely among those who work with the chemicals either in production or in agriculture. Although it is relatively easy for most people to avoid exposure to large doses of insecticides, it is nearly impossible to avoid exposure to trace contaminants in food, in the air, and in drinking water supplies. We all carry measurable quantities of insecticides in our bodies. What are the chronic, long-term effects of these chemicals? Unfortunately, it is impossible to answer this question with certainty. If there were two large groups of people and one group ate contaminated food while the second group ate pesticide-free food, and we studied these groups for a few decades, then perhaps we could measure the long-term harmful effects of pesticides. However, not only would it be morally unjustifiable to perform such an experiment, but also it would be impossible to control the diets of large groups of people. Instead we must rely on studies of small groups of people and of laboratory animals. Such studies are always ambiguous. In addition, most experimental subjects are examined for only a few years, although it is known that some diseases, most notably cancer, may appear 10 to 20 years after exposure to a foreign toxin.

It can safely be said that after a decade and a half of study, proof for or against the carcinogenicity (cancer-producing qualities) of pesticides in humans is still lacking. The following describes some of the types of study that have been carried out.

(1) **Direct feeding of DDT to human subjects.** When DDT was ingested by healthy volunteers at doses as high as 35 mg per day for nearly 22 months and the subjects were then observed, some for about five years following the end of the treatment period, no adverse clinical or chemical effects were seen.

(2) **Examination of men who have had long-term occupational exposures to DDT.** Several studies of a population of workers who were estimated to have daily intakes of DDT nearly 400 times the "normal" have shown absolutely no evidence of any toxic effect. These studies have not, of course, gone unchallenged. The major (and completely justified) criticisms are that the numbers of people observed were too small to draw any sort of valid conclusion about anything but the grossest kind of effect. In addition, the length of time that most of the subjects have been observed is too short to warrant confident conclusions about an effect such as carcinogenesis, for cancer can take many years to develop after exposure to a cancer-causing agent. Other studies have suggested that exposure to insecticides may be related to increased abnormalities of pulmonary function, but here, too, it is difficult to establish whether or not there are long-term effects.

(3) **Experiments with laboratory animals.** Within the past decade, several groups have shown that laboratory rodents fed DDT develop tumors more frequently than those not fed DDT. These results sound simple enough, but

proper interpretation is difficult for several reasons. (a) Questions have been raised about faulty experimental design in the most important of these experiments. (b) The doses of DDT fed to the animals were huge in comparison to those to which human beings are exposed. (c) The strains of laboratory mice used were highly inbred and have a very high *spontaneous* incidence of tumors; hence they may not react to chemicals as would "normal" outbred strains. (d) Most of the induced tumors were a certain type of liver tumor. Pathologists disagree about whether these tumors are typical cancers. In at least a few cases, however, the tumors did seem to behave like malignant tumors in human beings. (e) It is always very difficult to extrapolate any kind of results across species lines, for things that are true of mice are not necessarily true of humans.

This is the strongest available evidence concerning the chronic health effects of DDT. From a purely scientific standpoint, we simply do not know whether DDT has any chronic effects in humans.

12.8 PESTICIDE LEGISLATION

In 1962 Rachel Carson wrote *Silent Spring,* a very popular book which suggested that exposure to DDT might increase the occurrence of cancer in human beings. The book dramatically increased public awareness of the possible harmful effects of environmental chemicals.

Federal regulations that require the registration of pesticides before they can be moved in interstate commerce are the principal means of controlling the harmful effects of pesticide usage in the United States. Registration consists of a determination that the compound, as formulated, will control the pest under the conditions of use prescribed and that the conditions of use will be safe for the applicator, for people, and for beneficial plants and animals. If the use will result in a residue on a food product, then a residue tolerance must be obtained. This tolerance must be sufficiently low to be deemed safe for human consumption and must not be higher than the amount required to obtain effective control of the particular pest involved.

Regulation of the use of pesticides at the point of their application is the responsibility of state and local authorities. States vary substantially in the degree of restriction which they impose on pesticide use. Generally, they require federal registration or its equivalent. Some of them have greatly restricted the use of some persistent pesticides.

Research on the acute and chronic effects of pesticides on humans and other living organisms receives considerable attention. Monitoring of air, water, soil, fish and wildlife, and humans is done on a nationwide scale. The Environmental Protection Agency has been assigned responsibility to unify the regulatory functions involving pesticides.

During the late 1960's pressure by environmental groups began steadily mounting. Finally the Environmental Protection Agency forbade most uses of DDT and other chlorinated hydrocarbons in the United States. This decision was made not only because of the suspicion of carcinogenicity but also because of all the other effects on various aspects of the biosphere that we have mentioned previously, and because many experts felt that the organophosphates were satisfactory alternatives to the use of DDT. On occasion, the EPA has granted exemptions to the law. For example, in 1974, the federal government sanctioned the use of DDT to control tussock moths in the Northwest. An emergency section of the law banning DDT permits its use in public health emergencies. Thus, DDT has been used to control rabid bats in several parts of the country and to kill fleas carrying bubonic plague germs on rodents in California in 1976.

The ban on DDT and other pesticides for use in the United States has not prevented U. S. government agencies from helping other countries to procure the very pesticides prohibited in the United States. The Agency for International Development (AID), for instance, has promoted the use of DDT, Aldrin, and Dieldren, all of which are banned in the U. S., for use in Africa, Asia, and Latin America, where their application is permitted. Because pesticides spread so readily throughout the biosphere, their regulation should be subject to international agreements. Little progress has been made in multinational cooperation on the use of pesticides.

The bans of the various pesticides have been both praised as farsighted and damned as shortsighted. Perhaps the most vociferous critic has been Dr. Norman E. Borlaug, the winner of the 1970 Nobel Peace Prize for his work in developing high-yield wheat strains. He has emphatically denied that such chemicals as DDT are significant contributors to the deterioration of the environment and fears that all other pesticides will soon be banned, leading to failures in agriculture and possible mass starvation. These predictions seem overly gloomy, but they do illustrate the levels of passion that this whole question has aroused. Other critics argue that the shift away from organochloride insecticides has led to increased reliance on organophosphates, which are *more* poisonous to non-target species than DDT and its sister compounds. Moreover, since biochemists do not fully understand the action of the organophosphate decomposition products, we run the risk of outlawing one class of compounds and substituting an ultimately more harmful one.

On the other hand, those who support the ban on DDT

Norman E. Borlaug (1914–   ) is an American plant scientist best known for his work in plant pathology, pesticides, and wheat breeding. His results led to methods for increasing agricultural yields. He was awarded the Nobel Peace Prize in 1970.

373

as well as restrictions on the use of other pesticides take the viewpoint that any substances as biologically active, persistent, and foreign to natural food webs as the organochloride insecticides can be *presumed* to be harmful to human health. They cite examples such as cigarette smoke and asbestos dust, whose injurious effects often emerge only several decades after exposure, to illustrate how insidious these dangers can be. They prefer to be slightly cautious than to risk a large-scale proliferation of cancer in the next decade. Supporters of the ban also point out that so many insect species had developed genetic resistance to DDT by the early 1970's that the pesticide was losing its usefulness anyway, and a shift to other spray compounds was due. Finally, they point out that a permissive attitude toward chemical pesticides encourages their proliferation, whereas a more restrictive stance serves as a drive to develop other, more environmentally sound methods of pest control, such as those that are described in the following section.

When experts disagree, it is often very discouraging to the nonspecialist, who develops a feeling of hopelessness about ever being able to determine which decision is the "right" one. Probably even more important than having a definite opinion, however, is having a reasonable grasp of the foundations on which decisions are based.

12.9   OTHER METHODS OF PEST CONTROL

USE OF NATURAL ENEMIES: PREDATORS, PARASITES, AND PATHOGENS

We have shown how destruction of predator populations by indiscriminate spraying may cause upsurges in pest populations. It is reasonable to assume that the opposite treatment — importation of predators — may be an effective control measure.

The new imports may be more individuals from a native species, or predators foreign to the area. Recall that DDT spraying against the Japanese beetle in Illinois caused widespread havoc among other species. The Japanese beetle, a native of the Orient, was inadvertently imported with a shipment of some Asiatic plants. In the absence of any effective natural control, the beetles thrived on the eastern seaboard of the United States and gradually became a major pest. Scientists then searched for natural predators and imported several likely species. One of these, an Oriental wasp, provides food for its young by paralyzing the Japanese beetle grub and attaching an egg to it. When the young wasp hatches, it eats the grub as its first food. The life cycle of the wasp is dependent upon the grub of the Japanese beetle; it does not naturally breed on the grubs of other insects. This type of species-specific control does not seriously affect the rest of the ecosystem. Importation of a species of bacteria that infects Japanese beetles in their native environment has also helped to control the pest. The spores of the bacteria, known as milky disease, are commercially available.

Corn earworm

Insects can be controlled through the use of certain strains of viruses. A virus effective against the cotton bollworm and the corn earworm has been approved by federal regulatory agencies in the United States and is now ready for mass production. Viral strains that combat several other species of pests should be commercially available in the near future.

There are many advantages to importing enemies of pests. Because these agents are living organisms, they reproduce naturally, and one application can last for many years. Most insect parasites and pathogens are very specific and do not interfere with the health of vertebrates. No harmful or questionable chemicals are introduced into the environment.

With these encouraging advantages, one may wonder why insect enemies are not used more frequently. The major objections to the use of predators, parasites, and pathogens are often based on social and economic considerations (see Section 12.11).

Another difficulty with the use of natural enemies to control insect pests is that the technique is neither fast-acting nor simple to apply. The farmer must know something about the life cycles of the pest and of the control organism to choose the proper time to release the predators, pathogens, or parasites. Although ultimately a high level of control can be reached, results cannot be expected immediately, for it takes time for organisms to act. In the interim between the onslaught of a pest infestation and the growth of the enemy population, produce might be blemished and the crop size reduced. Certainly it is a common human trait to choose the chance of avoiding an immediate, visible loss, even if it involves the risk of later harm by some invisible poison. But the prevalence of such behavior does not necessarily make it the best strategy in the long run.

STERILIZATION TECHNIQUES

Pests can be controlled without killing them directly if the adults of one generation are sterilized to prevent production of viable offspring. The screwworm (the parasitic larvae of the screwworm fly) is a serious cattle pest that has been responsible for large financial losses to many ranches. Some years ago, the United States Department of Agriculture initiated a program in the southeastern states to raise male screwworm flies, sterilize them by irradiation, and release them in their natural breeding grounds

Screwworm fly

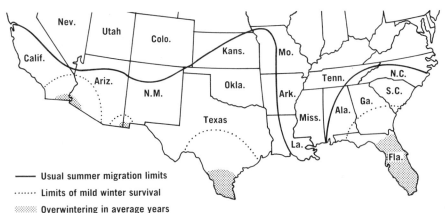

Figure 12.4 Screwworm distribution in the United States.

— Usual summer migration limits
······ Limits of mild winter survival
▓▓ Overwintering in average years

(Fig. 12.4). If a sterile male mates with a normal female, she will lay eggs but they will not hatch.

For several years, millions of sterilized males were released annually to mate with healthy females living in the area. Initially, the program was a spectacular success, and screwworm infestations were completely eliminated from 1962 to 1971. But in 1972 the screwworm population increased dramatically and infested nearly 100,000 cattle. During the following three years, screwworms continued to cause serious losses of cattle. What had happened?

Researchers reasoned that perhaps the screwworms raised in captivity evolved into a domestic strain that was no longer well adapted to natural conditions. For some reason these domestic irradiated flies no longer mated suc-

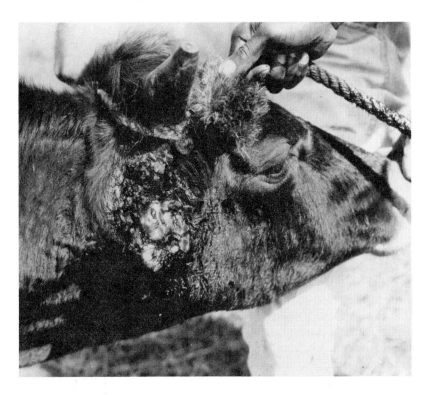

Screwworm larvae infestation in ear of a steer. An untreated, grown animal may be killed in 10 days by thousands of maggots feeding in a single wound. (Courtesy of USDA.)

TABLE 12.2 Mathematical Population Decline When a Constant Number of Sterile Males are Released Among an Indigenous Population of One Million Females and One Million Males

GENERATION	NUMBER OF VIRGIN FEMALES	NUMBER OF STERILE MALES RELEASED	RATIO OF STERILE TO FERTILE MALES	NUMBER OF FERTILE FEMALES IN THE NEXT GENERATION
1	1,000,000	2,000,000	2:1	333,333
2	333,333	2,000,000	6:1	47,619
3	47,619	2,000,000	42:1	1107
4	1107	2,000,000	1807:1	less than 1

cessfully with wild females. Perhaps the males could not fly so far or fast, perhaps they lost resistance to natural predators or disease organisms; no one knows for sure.

Sterilization techniques applied to other species have sometimes led to even more serious problems. The initial success of the screwworm fly program depended on the fact that the sperm of a wild adult male can be killed without harming the adult. In most insect species, however, any sterilization process debilitates the male so that he cannot compete successfully with untreated, naturally occurring insects.

Even if the sterilization problem can be solved, the situation must be adjusted so that those fertile males that were part of the naturally occurring population have a small probability of finding and impregnating a female. In the case of female insects who copulate many times with different partners, it is virtually impossible to reduce the chance of fertile matings to levels that are low enough to effect control. The case of the screwworm is particularly favorable because a virgin female fly copulates only once in her lifetime, and if the ratio of sterilized to nonsterilized males is sufficiently high, then control is possible (see Table 12.2). In practice, the numbers of sterilized males required to achieve this flooding is often prohibitively great. The screwworm program was successful because the flies migrate into small, well-defined southern areas during the cold months, and the program was initiated during a severe winter which had already decimated the native population. When sterilized males were released against the codling moth, a serious pest of apple and pear orchards, the initial control was successful, but migration of fertile males from nearby areas spoiled the early gains.

Codling moth larva and adult

CONTROL BY HORMONES

Many insects begin their life in some larval stage and later metamorphose into a mature adult. As an example, a caterpillar is a larva that later matures to become a moth or butterfly. When an insect is in the larval state, it continuously produces a chemical called the **juvenile hormone.**

377

Figure 12.5 The role of hormones in insect metamorphosis. The juvenile hormone, secreted mainly during the larva stage, keeps the caterpillar in this immature state until it is ready to metamorphose into a pupa and adult. The hormones must be secreted in the right amounts at the right time. (Redrawn from *Fortune*, July, 1968.)

As long as sufficient quantities of juvenile hormone are in the animal's system, it remains a larva. It is only when the flow of that biochemical agent stops that the animal metamorphoses. If an insect larva is artificially sprayed with the juvenile hormone specific to its species, it will never mature into its adult form. Because insects can neither mate

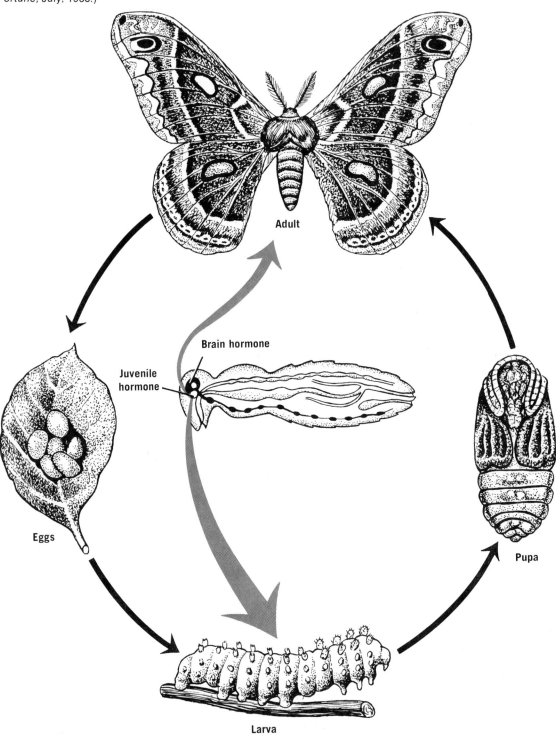

Adult

Brain hormone

Juvenile hormone

Eggs

Pupa

Larva

nor survive long as larvae, such a spray application is eventually lethal. Therefore, juvenile hormones can be used as insecticides. It is likely that widespread use of these chemicals or their analogues would produce minimum environmental insult, because they are biodegradable and active only against specific insects. With the aid of careful timing, pests can be destroyed without killing their predators.

The first hormone control chemical to be marketed commercially is an agent that combats three species of floodwater mosquitoes, including *Anopheles albimanus,* the malaria vector in South and Central America. Spray doses of 150 grams per hectare (about 2 ounces per acre) have resulted in complete control of this mosquito.

Despite this success, some technical difficulties with hormone control have arisen. One is that although natural juvenile hormone is stable in the body of a caterpillar, it is not stable in the environment and often breaks down chemically before it can act. This problem has been circumvented by the discovery of organic chemicals that are structurally similar to natural juvenile hormone and biologically active, but more stable. Another difficulty is in choosing the best time for the spray application. Juvenile hormone is an effective insecticide only during the relatively brief period of an insect's life cycle when the *absence* of juvenile hormone is essential to normal insect development. If the chemical is sprayed before the larva is ready to metamorphose, it has no effect. One final problem arises because juvenile hormones do not kill the larvae directly. In a test study of the effectiveness of this technique in Colorado against a potato beetle pest, the hormone prevented metamorphosis effectively, but the caterpillars didn't die soon enough; they grew to gigantic size and ate the entire crop.

Since the timing of spray applications is so crucial, considerable expertise and precision are necessary for effective control. In recent years, biochemists have been studying a new class of compounds, **hormone inhibitors.** These chemicals block the action of juvenile hormone while the animal is still a larva and disrupts its life cycle so severely that the insects are killed directly. Active research on hormone inhibitors is under way, but commercial distribution is not expected before 1980.

SEX ATTRACTANTS

In many species of insect, a virgin female signals her readiness to mate by emitting a small amount of a species-specific chemical sex attractant, called a **pheromone.** The males detect (smell) very minute quantities of pheromone and follow the odor to its source. Attempts have been made to bait traps either with the chemical or, more simply, with live virgin females. In this program, only minute quantities of natural chemicals are released into the air by evaporation, and environmental perturbations are nonexistent. However, as in the case of the sterile male technique, there is a statistical problem in areas of heavy infestation. Very simply, if there are a million males in the area, and 1000 traps are set, and if each trap catches 100 males before they find a female, the control project is only 10 percent effective.* Further, the traps and the labor of setting them up are expensive. In parts of Europe, the cost of this program has been reduced by having schoolchildren set the traps as a part of an educational program. Still, the trapping technique is probably best suited for special applications, such as for the control of migration where the migratory populations are small, or for control after a single effective spray application.

Some successes have been registered by the confusion technique. In one application millions of cardboard squares were impregnated with pheromone and dropped from an airplane. With a high ratio of cardboard squares to females, males become confused and were observed to try to copulate with the cardboards. Another approach has been to spray the area with pheromone so that the entire atmosphere smells of virgin females (to the male insect, that is, not to the human nose). Then the males are unable to track down a mate.

USE OF RESISTANT STRAINS OF CROPS

It has already been mentioned that some plants are naturally resistant to pests because they synthesize their own insecticides. For some time, plant breeders have been actively developing more resistant strains and have succeeded in a number of instances, such as in the production of a variety of alfalfa that is resistant to the alfalfa weevil, and of various strains of cereal crops resistant to rust infections.

The successes achieved have been encouraging, and further research deserves support, but it should be remembered that the technical problems are difficult. The new plant variety must produce high yields as well as maintain the natural resistance to other diseases. Moreover, genetic adaptation is not stagnant, for throughout biological time genetic defense in one species traditionally has been met by genetic changes in the attacking organism to neutralize the defense. Thus, many resistant crops lose their immunity after several years, and new resistant varieties must be developed.

CONTROL BY CULTIVATION

Uniformity is not typical of virgin land masses. The systems created by modern agriculture differ from natural

---

*$\dfrac{1000 \text{ traps} \times 100 \text{ males/trap}}{1,000,000 \text{ males}} \times 100\% = 10\%$

systems, for farms tend to specialize in a very few species of plants, whereas areas untouched by people do not. For instance, in Kansas and Nebraska thousands of acres of land are covered almost exclusively with wheat fields. The result of such specialization is that a mold, fungus, or insect that consumes wheat has a vast food supply and an extremely hospitable environment. No barriers to spreading exist, so the pest can grow quickly in uncontrolled proportions. The advantage of diversity in nature was discussed in relation to the ability of an ecosystem to survive perturbations (see Chapter 3). One such perturbation is attack by specific consumers. A fungus that attacks wheat will spread more slowly if half the plants in a field are something other than wheat, because the spores have a reduced chance of landing on a wheat plant. Moreover, if the disease spreads slowly, there is more time for the development of natural enemies of the fungus or of naturally resistant strains of wheat. Therefore, one solution to the problem of pests is to grow plants in small fields, with different species grown in adjacent fields.

Unfortunately, the mechanization that is essential to modern agriculture is best suited to large fields of single crops.

Simply planting small fields of different crops is effective in itself, but by judiciously choosing the companion plants, additional success can be realized. For example, the grape leafhopper, a pest of vineyards, can be controlled by a species of egg parasite that winters in blackberry bushes. Knowledgeable grape-farmers therefore maintain blackberry thickets. Conversely, one variety of stem rust that attacks North American cereal crops must live part of its life

Large, homogeneous fields provide rich environments for insect pests. (See also Color Plate 8.)

cycle on barberry bushes, and selective destruction of these plants will reduce rust infestations.

Other methods of cultivation that have been successful in controlling pests include (a) crop rotation so that a given pest species cannot establish a permanent home in one field, (b) specific plowing and planting schedules to favor predators over pests, and (c) planting certain weeds that some omnivorous insects, which prey on pests during part of their life cycle, need for food during other stages.

MISCELLANEOUS CONTROL TECHNIQUES

The Egyptian Nile sparrow is a grain-eating bird that appears at harvest time in huge and destructive flocks. Modern control practices have been frustrated by the fact that one cannot poison the grain just prior to harvesting. However, the Chinese have successfully used an interesting technique against their own sparrows since the beginning of recorded history. They have discovered that the birds take to the air when bothered by noise, and furthermore, that they die of exhaustion if forced to fly for more than 15 minutes at a time. When a sparrow infestation threatens, farmers over a wide area race through fields yelling, beating gongs, or generally making noise in a coordinated effort. The technique is often successful in destroying entire flocks.

Some more modern, but no more successful, pest control practices are: (a) use of sound or light to attract insects to traps; (b) use of sound as a repellent; (c) use of proper sanitation such as swamp drainage for mosquito control; (d) introduction of innocuous insects which are effective competitors of pest species. For example, some populations of biting insects such as gnats can be controlled by introducing species which don't bite people but which otherwise occupy niches similar to those of the pests.

INTEGRATED CONTROL

If we have learned any lesson during the DDT era, it is that insects are not passive to control measures, and that any approach to pest management is subject to resistant reactions from the insect world. Therefore, many scientists now believe that we should never again rely on unilateral attack, but rather we should use as many available control measures as possible in an integrated and well-planned manner. Integrated control will never be so conceptually or technically simple as straightforward chemical control can be. It will be slower-acting and more expensive initially, although experts hope it will be more effective and cheaper in the long run. Despite its complexities, integrated control has been successful in many areas.

Recall from Section 12.3 how the yields of cotton in a valley in Peru were reduced after several years of insecticide use. Following the crop failure, a new program was initiated. First, predacious and parasitic insects were imported from nearby valleys. Second, all cultivation of marginal soils was banned in an effort to cull out weak plants, which are often breeding centers for disease. Third, the cycles of planting and irrigation were adjusted so as to interrupt most effectively the life cycles of the pests. Last, all spraying with synthetic organic insecticides was banned except under the permission and supervision of a panel of scientists. The result was that, despite the fact that some land was purposely not planted and very little insecticide was used, cotton yields were higher than had ever been recorded in the area.

12.10 HERBICIDES

During the past 10 years, the use of herbicides for the chemical control of unwanted plants of all types has been so increasingly popular that it has resulted in a multimillion-dollar industry. In many respects, the use of herbicides has closely paralleled insecticide uses. The discovery of potent chemicals for application against noxious weeds has led to opportunities for selling them in large quantities. As a result, there has been a great deal of indiscriminate use, which in turn has produced chemical pollution of the environment despite only partial success of the control programs.

There are about 75 different herbicidal chemicals in common use, but the wide variety of structural formulas makes classification into a few neat chemical groupings impossible. Most herbicides are toxic to mammals but are not nearly so persistent in the environment as organochloride insecticides, for they decompose in a few months. Although herbicides, like organochloride insecticides, become concentrated in the food web, the degradability of herbicides reduces the severity of the problem.

Scientists have investigated the carcinogenicity and teratogenicity (tendency to cause birth defects) of many herbicides. One compound, known as 2,4,5-T, has been shown to be a particularly active teratogenetic agent. Although the relationship between commercial preparations of 2,4,5-T and birth defects is undisputed, there is considerable argument as to whether the teratogen is actually 2,4,5-T or some impurity produced in the manufacturing process. Meanwhile, as the argument continues, the 2,4,5-T, impurity and all, can be purchased at your local hardware store.

USES OF HERBICIDES IN AGRICULTURE

Weed control, like pest control, is necessary for efficient agriculture. In labor-intensive agricultural communities such as the ones found in mainland China, men and women often simply uproot weeds with hoes or shovels. In North America, farmers cannot afford to hire small armies of fieldhands to hoe fields, weed irrigation ditches, and cut brush along fence lines, so it is inevitable that chemical control should be in common use. As we have learned by now, widespread use of chemical poisons often disrupts natural ecosystems. As one example, herbicides were used extensively to kill sagebrush in semi-arid land of the western United States in an effort to encourage larger yields of grasses. The sagebrush ecosystems, however, have evolved to cope effectively with this harsh climate of dry summers and cold winters. Many ecologists believe that the sagebrush–natural grass system may provide more year-round food for cattle than the farm-hay system. Perhaps in such a delicate situation the probability of chemical pollution should swing the decision in favor of the undisturbed range.

In another study, scientists learned that corn fields sprayed with herbicides were more prone to insect and disease attack than fields that were weeded manually. When the insect populations increase, farmers are likely to use more pesticides, resulting in a spiraling pollution problem.

Herbicide use in the United States increased steadily during the 1970's so that by 1974, 220 million kilograms were used annually. Many of these compounds are directly poisonous to humans, many are suspected of causing cancer or mutations, and most of the decomposition products of these chemicals are unknown.

Agricultural scientists are faced with a dilemma. If herbicides were banned directly, the price of food would rise sharply, and even the availability of some staples might be threatened. Some agronomists are studying integrated control techniques, but development of such programs is not likely to occur for a decade or more. Meanwhile, the use of herbicides continues, chemical pollution of the environment is increasing, and no one knows how this pollution is affecting human bodies and world ecosystems.

CLEARING OF RIGHTS-OF-WAY

Herbicides are used extensively by industry and by all levels of government to restrain the natural advance of brush and forest on various rights-of-way. Transmission lines, railroads, highways, county roads, fire control roads in national forests, and hikers' paths in wilderness areas have all been subjected to chemical weed control. Despite

Herbicides are used to clear power-line rights of way.

obvious agreement that brush and trees must be kept back from roadways to allow drivers to see traffic patterns around turns, and despite the fact that other rights-of-way must be maintained, the present system of maintaining these openings by repeated mowing and spraying might not be the cheapest and most effective technique available. Shrub and tree seedlings sprout easily in cut grass, so the objective of preserving lawns perpetuates the need for further maintenance. On the other hand, if the grasses and shrubs are allowed to grow and once a year, or once every several years, the tall shrubs and trees are selectively killed, the rights-of-way can be maintained for a fraction of the cost and a fraction of the pollution. The question of aesthetics is subjective; some feel that a field of natural grasses and shrubs, green in the spring and yellowed by late August, is beautiful, while others feel it looks unkempt and ugly.

HOME GARDENING

Prodigious quantities of herbicides are sold yearly to homeowners for control of crabgrass and other lawn and garden weeds. Consider two neighbors, one who believes that it is his right to sit in his backyard on a Sunday afternoon and breathe clean air and the other who believes that it is his right to spray against crabgrass. Whose right takes precedence? If the spray drifts downwind from one person's garden to another's hammock, does the latter have any recourse to action? The question is not simple either legally or morally, but the problem is typical of many environmental controversies.

12.11 ECONOMIC FACTORS IN PEST CONTROL

Section 12.9 described various environmentally attractive approaches to controlling pests. The reader may wonder, then, why such advantageous methods are used so infrequently. One possible answer is that the environmentally more desirable control methods are often the least profitable to produce and sell. For example, a natural predator, parasite, or pathogen, once deployed against a target insect pest, tends to maintain its own population by reproduction. Therefore, a private company that sells them to a farmer will not enjoy the same rate of reorders as will a manufacturer of a chemical pesticide, which does not reproduce itself. Furthermore, a new chemical can be protected by a patent, whereas a microbe or a ladybug cannot.

Even among chemical pesticides, a compound that is effective against a single species, and is therefore less generally disrupting to the environment, will enjoy a smaller sales market than will a broad-spectrum insecticide that can be used for a variety of crops. The most highly specific agents, such as pheromones, will be used in the smallest quantities.

The commercial development of any new product is expensive. In the 1970's, the cost of research, testing, and obtaining approval for use of either a new chemical insecticide or a new biological control measure is about $5 to 10 million and requires several years' time. Understandably, the economic incentive favors the product that has the greater sales potential as well as the protection of a patent. This incompatibility of sound economic and ecological practices presents a serious problem of environmental management. Some of the general aspects of such contradictions were discussed in Chapter 2. What is needed in the area of pest control, at the very least, is a greatly increased level of government support of research and development that private companies lack the incentive to underwrite.

12.12 OTHER CHEMICAL POLLUTANTS: PCB'S

Pesticides and herbicides represent only a small portion of our widespread chemical pollution. A great many other types of compounds that are used routinely in manufacturing and in chemical and industrial processes are eventually released into the environment. Polychlorinated biphenyls (or PCB's for short), a class of organochloride chemicals structurally related to DDT, are extremely valuable in many industrial applications.* Their unique electrical insulating properties make them useful in the manufacture of transformers and other electrical components. They are also used in the production of plastic food containers, epoxy resins, caulking compounds, various types of wall and upholstery coverings, and as ingredients in soap, cosmetic creams, paint, glue, self-duplicating ("no-carbon") paper, waxes, brake linings, and many other products. In the late 1960's scientists began to realize that PCB's were also serious environmental poisons. In 1968 the accidental contamination by PCB's of some cooking oil in Japan caused several thousand people to suffer from enlarged livers, disorders of the intestinal and lymphatic systems, and loss of hair. In New York City, workers in an electrical factory that used PCB's complained of similar ailments. Laboratory studies have shown that PCB's interfere with reproduction in rodents, fish, and many species of birds and monkeys. They are also suspected of being carcinogenic, but conclusive evidence is lacking. Moreover, PCB's, like DDT, are environmentally persistent, easily transported through the world's ecosystems, and soluble in the fat of living animals. Therefore, quantities of PCB's that have been accidentally spilled into waterways or that have leaked onto the ground have been dispersed throughout the

---

*Chemical formulas of various pesticides are shown in Appendix C–12.

environment. In recent years, PCB's have been found in cow's milk, many inland and deep sea fish, most meats, and in people's bodies as well. This persistence means that the contamination will continue far into the future. Even when PCB's do decompose, problems will continue, for evidence indicates that the decomposition products are even more poisonous than the original material.

As evidence mounted regarding the harmful effects of PCB's, chemical manufacturers voluntarily curtailed production and use so that by 1972 applications were restricted to the manufacture of electrical components. In September, 1976, Monsanto Chemical Corporation, the last manufacturer of PCB's in the United States, announced that it would eliminate all production no later than October 31, 1977.

Before a new pesticide, drug, herbicide, or food additive is released for marketing in the United States, it must be submitted to the Environmental Protection Agency (EPA) for testing. The immediate toxicity and long-term carcinogenicity and teratogenicity are studied. This research costs millions of dollars and requires several years to complete. Yet, despite all the care, the test procedures are often questionable, and the results are never absolute. Many "safe" products have later been recalled as being "unsafe," as new data become available. Industrial chemicals are screened much less carefully, for they are not sprayed on food or used in its preparation. Nevertheless, many, like the PCB's, may find their way into our bodies.

What can be done to stop this contamination? It is possible to screen *all* chemicals entering the environment for their short-term effects, but such a program would be enormous. Long-range effects can be determined only with studies extending at least decades.

12.13 CASE HISTORY: *The Boll Weevil*

Many of the most serious agricultural pests are immigrants that have moved to a new and hospitable environment relatively free of the predators and pathogens native to their original home. The boll weevil is a small insect—about 0.65 cm (¼ in) long—which first migrated to the United States from Mexico in about 1880. Weevils feed on the buds and bolls of young cotton plants, thereby destroying the valuable fiber. Since there are few natural enemies of the weevil in the United States, population blooms of these insects are common. If left uncontrolled, a light infestation of 10 weevils per hectare can increase to 10,000 weevils per hectare within a single growing season and

> One time I seen a Boll Weevil, he was settin' on a square.
> Next time I seen the Boll Weevil, he had his whole darn family there;
> He was lookin' for a home
> He was lookin' for a home.
>
> The farmer take the Boll Weevil, put him on the ice;
> Boll Weevil said to the farmer, You's treatin' me mighty nice;
> And I'll have a home,
> And I'll have a home.
>
> Folk ballad

Boll weevil on a cotton plant. (Courtesy of USDA.)

Cotton flower. (Courtesy of USDA.)

completely consume an entire crop. The weevils' reproductive and eating capacity is so great that they are responsible for 200 to 300 million dollars' worth of crop losses in the United States annually.

In the early 1900's farmers used cultural controls to combat the weevils. Cotton was planted and harvested as early in the year as possible, and the plant stalks and leaves were burned immediately after harvest. If a large portion of the weevil population could be destroyed by fire, they reasoned, perhaps an early crop could be harvested the following year before the weevils could repopulate.

This nip and tuck race with growing weevil populations was often unsuccessful, and many crops were completely destroyed. In the 1930's a new pesticide, calcium arsenate, was introduced, but it was so highly toxic to non-target species (including soil microorganisms) that it was largely replaced by DDT in the 1940's. By 1960, many DDT-resistant strains of boll weevil had evolved. Therefore, when DDT was banned in 1973, many farmers had already been using organophosphates for several years.

Broad-spectrum pesticides were effective in partially controlling the population of weevils, but many secondary problems were raised. Cotton is not the only crop grown in the South. Tobacco, corn, peanuts, soybeans, and vegetables are also commercially important. Boll weevils have few natural enemies in the United States, but the insect pests of other major crops have traditionally been controlled by natural enemies. When cotton fields were sprayed heavily with broad-spectrum chemicals, adjacent regions were contaminated also, and many natural predators were killed. As a result, epidemics of tobacco and corn pests became a problem for the first time.

Obviously, a new strategy was necessary. A

Open cotton boll. (Courtesy of USDA.)

Special Committee on Boll Weevil Eradication was formed with government support and a five-pronged attack on this insect pest was outlined. Farmers were asked to (1) spray lightly with chemical pesticides during the growing season; (2) spray again in the fall just before the weevils hibernate for the winter; (3) destroy all plants, leaves, and stalks just after harvest to kill any remaining insects; (4) set traps in the early spring baited with a sex pheromone to catch and destroy females emerging from hibernation; and (5) release sterile male weevils to prevent fertile matings of any remaining females.

A pilot project was initiated in 1971 to see if such a program could eradicate the boll weevil forever. The results of the project were encouraging, and weevil infestations were generally reduced in the target area, but even with such a careful and complete program, total annihilation was not realized. Research teams found that 95 percent control was feasible, but it is extremely difficult to destroy every weevil in a region. Despite an intense campaign, small pest populations survived in several safe havens. One farmer who was cheating on his cotton allotment and income taxes did not tell authorities about a hidden field. Weevils survived there and later migrated outward. Another farmer claimed that pesticides were killing his chickens and also refused to cooperate. Many owners of roadside stands and restaurants kept small plots of cotton to attract Yankee tourists, and for the most part these smalltime "growers" did not participate in the program, so more weevil breeding populations survived. Finally, wild cotton plants grow throughout the South, and many weevils bred in small cotton patches in the woods and swamps, far from agricultural areas.

The United States Department of Agriculture (USDA) seriously contemplated a billion-dollar program to annihilate the boll weevil once and for all throughout the United States. But in view of the problems inherent in the complete destruction of the species, the program was abandoned, and the weevil war continues to this day.

TAKE-HOME EXPERIMENT

Insect pests. Identify one species of insect pest that inhabits your area. Does this pest carry disease, cause economic loss, or is it merely an annoyance? Try to control the pest in a small, well-defined area without using chemical sprays. You can use baited traps (available at many hardware stores), screens to isolate the target area, fly swatters, imported predators, or any other non-spray techniques. Comment on the cost and effectiveness of your control program.

PROBLEMS

1. **Insects and people.** What are the harmful effects and the benefits that insects bring to people? By what mechanisms did people accommodate themselves to insects before the production of modern pesticides?

2. **Chemistry of insecticides.** Consider the following table:

Insecticide	Compound Class	Chemical Formula	Melting Point, °C
DDT	organochloride	$C_{14}H_9Cl_5$	108.5
Parathion	organophosphate	$C_{10}H_{14}NO_3PS$	6.1

(a) A compound has the formula $C_2H_8Cl_6$ and the melting point 104°C. What compound class would you guess that it belongs to?

(b) Another compound is a common insecticide and has the melting point 2.8°C. With no further information, could you make a guess as to whether it is an organochloride or an organophosphate chemical? Would you be sure of your answer?

3. **Resistance to pesticides.** Explain why it often becomes necessary as time goes on to use larger quantities of a given pesticide to achieve the same results.

4. **Pesticides and predators.** Pesticides have been known to be more harmful to predators than to the pests they are designed to control. What factors could account for this selectivity?

5. **Chemical control.** Explain why it is ecologically unsound to replace complex natural insect controls completely with chemical ones. Explain why it has worked nevertheless in many instances.

6. **Biodegradability.** The first sentence in Section 12.4 states that most natural organic compounds are biodegradable. Is this statement correct or should it have said *all*? Does the existence of coal and oil deposits on Earth have any bearing on this question? Defend your answer.

7. **Biodegradability.** Synthetic compounds are less likely to be biodegradable than naturally-occurring ones. Account for this difference.

8. **Biodegradability.** Explain how a pesticide may sometimes cause damage to an ecosystem even after the original material has decomposed.

9. **Wrong targets.** Explain how soil runoff from agricultural systems can be deadly to fish in nearby streams.

10. **Pesticide solubility.** What advantages and disadvantages would result from the use of a pesticide that was soluble in water? Do you think it likely that it would be practical to use a water-soluble pesticide? Defend your answer.

11. **DDT solubility.** Imagine that you are going to eat a fish that is contaminated with DDT. Which of the following methods of preparation would be most effective in reducing the insecticide concentration: (a) broiling on a charcoal fire; (b) deep-fat frying; (c) boiling in water; (d) steaming; (e) baking in an oven; (f) none of the above, that is, just eating it raw. Defend your choice.

12. **Pesticides in aquatic systems.** Explain why carnivorous fish are generally more susceptible to low levels of pesticides in the water than are herbivorous fish.

13. **Pesticides in aquatic systems.** If no mea- surable quantities of DDT were found in the water of a pond, would that necessarily mean that the aquatic ecosystem was unpolluted by DDT? Explain.

14. **Sublethal doses.** An amateur ecologist studying wildlife populations before and after a heavy spray application determined that since no animals were directly killed by the spray, no harm had resulted. Would you agree? Explain.

15. **Birds.** Do you think that peregrine falcons might become resistant to DDT? Would you think that resistance might save the birds from extinction? Defend your answers.

16. **Health effects.** Explain why it is difficult to determine whether there is a relationship between DDT and cancer.

17. **DDT ban.** Outline the arguments for and against the DDT ban.

18. **Alternatives.** What methods of pest control are available as alternatives to the use of chemical sprays?

19. **Resistance.** Would you think that the development of resistance to insect parasites, predators, and pathogens would be a serious problem if widespread use of these natural enemies were initiated?

20. **Insect predators.** When insect predators are used to control pests, it is undesirable to annihilate the pest population completely in a given area. Explain why.

21. **Sterile males.** Give three reasons why the sterilized male approach was initially successful for the control of screwworms. Why did the success slacken? Explain how these unique conditions might not be met with other pests.

22. **Sterile females.** What would you think of a program to release sterilized females instead of sterilized males?

23. **Control by hormones.** Explain why it is important to spray with juvenile hormone at a precise time. If a spray application were successful on May 1 of one year, would it be safe to assume that farmers could spray successfully on that date every year?

24. **Control by hormones.** Describe the advantages of hormone inhibitors as compared with juvenile hormone sprays.

25. **Integrated control.** Explain why one indiscriminate spraying with DDT could destroy the effectiveness of an integrated control program.

26. **Integrated control.** Birds often become major pests in vineyards because they eat the grapes. (a) Which of the following control programs would you recommend for bird control: (i) spreading poison; (ii) broadcasting noise from a loudspeaker system; (iii) shooting; (iv) covering the vineyard with some fencing material? Discuss. (b) Do you think that it might be wise to initiate research directed toward: (i) developing a sterilization program against the birds, or (ii) developing new strains of grape which would be unpalatable or poisonous to birds? Discuss.

27. **Integrated control.** Sometimes an integrated control program may not become effective until the second season of its application. Moreover, crop losses during the first season may actually be greater than average. Do you feel that it would be good policy for the government to subsidize such losses in an effort to improve environmental quality? Defend your answer.

28. **Pesticides and social attitudes.** A worm that thrives in the core of an apple does not usually eat a large proportion of the apple, but people do not like wormy apples anyway. Discuss the social and economic factors influencing the choices between wormy apples and apples that might contain pesticide residues.

29. **Herbicides.** Ragweed is considered to be a major plant pest because its pollen causes misery to hayfever victims. In natural systems ragweed is characterized as an early successional plant. A few years ago, the State of New Jersey initiated a program to eliminate ragweed. Thousands of acres of roadways and old fields were sprayed with herbicides. Why do you think this program failed?

30. **Herbicides.** Outline briefly some of the parallel problems that exist between herbicide use and insecticide use.

31. **Nematocides.** Another category of broad-spectrum pesticides, known as **nematocides,** is used against soil pests such as cutworms. What types of ecological problems would you guess would develop from use of these chemicals?

32. **PCB's.** Explain why pollution from PCB's will remain a problem long after the manufacture and sale of these compounds has been curtailed.

33. **Boll weevils.** The five-point program initiated for weevil control called for spraying with chemical insecticides. Discuss seom of the effects of this spray campaign on agriculture in the South. Suggest a five-point integrated control program for boll weevils that does not involve chemical poisons.

BIBLIOGRAPHY

Several recent and authoritative books on pesticides are:
W. W. Fletcher: *The Pest War.* New York, John Wiley and Sons, 1974. 218 pp.
Rizwanul Haque and V. H. Freed (eds.): *Environmental Dynamics of Pesticides.* New York, Plenum Press, 1975. 387 pp.
C. B. Huffaker (ed.): *Biological Control.* New York, Plenum Press, 1971. 511 pp.
David Irvine and Brian Knights (eds.): *Pollution and the Use of Chemicals in Agriculture.* Ann Arbor, Mich., Ann Arbor Science Publishers, 1974. 136 pp.
Fumio Matsumura: *Toxicology of Insecticides.* New York, Plenum Press, 1975. 503 pp.
Robert L. Metcalf and William H. Luckman (eds.): *Introduction to Insect Pest Management.* New York, John Wiley and Sons, 1975. 587 pp.

A few older general references include:
W. W. Kilgore and R. L. Doutt (eds.): *Pest Control.* New York, Academic Press, 1967. 477 pp.
David Pimental: *Ecological Effects of Pesticides on Non-target Species.* Washington, D.C., U. S. Government Printing Office, 1971. 219 pp.
U. S. Department of Health, Education, and Welfare: *Report of the Secretary's Commission on Pesticides and Their Relationship to Environmental Health.* Washington, D.C., 1969. 677 pp.
The book that started much of our current concern about pesticides, and a more recent sequel to it, are the following:
Rachel Carson: *Silent Spring.* Boston, Houghton Mifflin Co., 1962. 368 pp.
Frank Graham, Jr.: *Since Silent Spring.* Boston, Houghton Mifflin Co., 1970. 333 pp.

Unit VI

Pollution

13

AIR POLLUTION

13.1 SENSATIONS

A medical treatise on malaria, written in 1827, stated, "It has long been familiar to physicians that there was produced by . . . marshes and swamps, a poisonous . . . substance, the cause . . . of fevers, . . . and to this unknown agent of disease the term marsh miasma has been applied." The word miasma comes from the Greek word for pollution, and malaria is derived from the Italian expression for bad air, *mala aria*. It is now understood that malaria is carried by mosquitoes that breed in swamps and not by the gases that are produced there. But it is easy to imagine how swampy air came to be associated with disease. After all, rotting food stinks and is dangerous to your health, so it is tempting to blame the hazard on the odor. Furthermore, some malodorous gases are very toxic, and it is therefore wise, as well as natural, to seek fresh air.

After the first use of coal, around the beginning of the fourteenth century, foul air also came to be widely associated with black smoke, which leaves its evident marks on darkened buildings, soiled clothing, and grimy hands and faces. Smokes of other colors, from various industrial operations, produce direct irritation of the eyes, nose, and throat. When smoke and foul odors occur together, the effect can be truly repulsive, sometimes even terrifying.

We now know that not all unpleasant odors are harmful, and that some white smokes consist only of water droplets. But conversely, some air pollutants are toxic and even deadly, and air that looks clear and smells clean is not always wholesome.

Air pollutants and their effects are now studied with the aid of sophisticated chemical, medical, and statistical methods. However, the major driving force for preventing or controlling air pollution results from the political, social, and economic actions of people. And what moves most people to action is what they can see or smell.

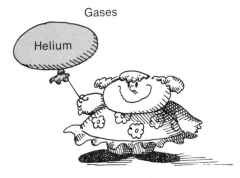

Gases

Helium

OXYGEN

CITY GAS

13.2 GASES AND SMOKES

GASES

Think about gases from your everyday experience to recognize what properties they have in common and in what ways they may differ from one another. Air is a gas; so is the "natural gas" that we burn in a kitchen stove. (The "gas" used as automobile fuel is not in this category, but refers to gasoline, which is a liquid.) The aroma of a flower and the stench of a rotten egg are gases. When water or "dry ice" evaporates, it turns into a gas. Some gases, such as carbon monoxide, are poisonous; others, such as oxygen, are necessary for the maintenance of life; still others, such as nitrogen, are chemically inert in the human body.

Gases disperse readily in space, and tend to occupy uniformly whatever volume is available to them. Thus, if some ammonia solution is spilled on the floor and evaporates, the resulting ammonia gas will move into the entire space of the room. Once the gas is dispersed, it never settles out. The ammonia gas will never return to the dish from which it evaporated.

Gases are easily compressed; for example, it is easy to squeeze an inflated balloon, thus reducing the volume that the gas occupies. When the pressure is reduced, the gas expands back to its original volume.

SMOKES

Gray smoke rises from a chimney; does it therefore disperse like a gas? The answer is surely no, for it eventually settles and collects on smooth surfaces, such as the windshields of automobiles. Unlike a gas, smoke is opaque. It is readily visible by day, and even at night it can be seen in the reflection of a flashlight beam.

Or as another example, think of water boiling in a teapot (Fig. 13.1). What emerges at the spout is invisible; it is steam, a true gas. At a short distance from the spout, however, a visible mist appears, which consists of tiny water droplets. Still farther, as these droplets evaporate, they become gaseous again and are visible no longer.

MOLECULES AND LARGER PARTICLES

The properties of gases and smokes arise in large measure from the sizes of the particles that make them up. Gases consist of molecules; each molecule is a combination of a small number of atoms—often only two or three, and, in some cases, only one. The molecules are in constant motion; they move in straight lines between collisions with

'...And from the top there'll be a spectacular view.'

The Gateway to the West arch, in St. Louis, Missouri. *Left,* Before its completion, a cartoonist speculates. *Right,* The finished product. (Cartoon by Engelhardt, in the *St. Louis Post-Dispatch,* from U.S. Public Health Service Publication No. 1561: *No Laughing Matter.* Photo from O'Sullivan: Air Pollution. *Chemical and Engineerings News,* June 8, 1970.)

each other or with the walls, floor, or ceiling. The molecules are so small, so speedy, so numerous, and their collisions so frequent, that nothing in our experience (not even, say, a swarm of gnats) can be called upon to help us visualize them in a quantitative way. Most of the space in a volume of gas is "empty." At ordinary pressures, the molecules themselves occupy less than one percent of the total volume. Thus "one liter of gas" consists of many molecules darting about in almost one liter of empty space.

Smokes consist of particles. A particle is a very small portion of matter, and therefore a molecule is a particle, and so is an atom or even an electron. But in air pollution

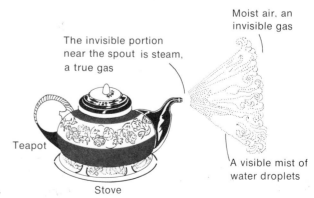

Figure 13.1 Steam and mist.

usage the word particle has come to have a more restricted meaning: it refers to portions of matter that, although small, are much larger than molecules, large enough, in fact, to settle out, or at least to reflect or scatter a beam of light that shines on them. (Hence smoke is visible.) Grains of sand are particles, and so are droplets of mist and tiny organisms like protozoa or bacteria. But the classification of viruses is not so definite; are they small particles or large molecules? The boundary line is not always sharp.

13.3 THE ATMOSPHERE

The predominantly gaseous envelope that surrounds the Earth is its **atmosphere**, and the stuff of which it consists is **air**. Air is not constant in its composition in various parts of the globe, especially when it becomes polluted. However, the concentrations of certain of its *gaseous* components at sea level vary very little, and it is therefore reasonable to consider them to be a constant portion of the Earth's atmosphere. The concentrations of atmospheric gases are typically expressed in terms of their proportions by number of molecules. Such expressions are reasonable because gases do consist of separate, individual molecules, and because, at any given temperature and pressure, the volume of any gas is directly proportional to the *number* of molecules it contains, and is independent of the *kinds* of molecules present.* Thus if there is 1 molecule of argon in 100 molecules of air, we say that the concentration of argon is "one percent by volume."

The concentrations of gaseous pollutants are more frequently expressed as "parts per million" (ppm) and sometimes "parts per billion" (ppb), by volume. The meanings of these expressions are:

percent by volume	= volume of pollutant per 100 volumes of air	= number of molecules of pollutant per 100 molecules of air
ppm by volume	= volume of pollutant per 1,000,000 volumes of air	= number of molecules of pollutant per 1,000,000 molecules of air
ppb by volume	= volume of pollutant per 1,000,000,000 volumes of air	= number of molecules of pollutant per 1,000,000,000 molecules of air

To change percent to ppm, multiply by 10,000. To change ppm to ppb, multiply by 1000.

Try to get some feeling for the sizes of these numbers. Is 1 part per billion a large concentration or a small one? One billion pennies laid out rim to rim in a straight line would extend almost halfway around the Earth's equator.

*This statement is known as Avogadro's Law, after Amadeo Avogadro, who suggested it in 1811.

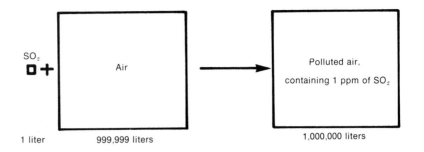

SO$_2$	Air		Polluted air, containing 1 ppm of SO$_2$
	1 liter	999,999 liters	1,000,000 liters

If one of these pennies were a bad one (a "pollutant") it would indeed be hard to find. This example makes 1 ppb sound small. But if air were contaminated with 1 ppb of SO$_2$ (much less than in city atmospheres), there would be 27,000,000,000 (27 billion) molecules of SO$_2$ per cubic centimeter. This example makes 1 ppb sound large. Don't let either of these examples fool you. Concentration values by themselves do not provide information on the effects of pollutants. For some substances, 1 ppb is inconsequential; for others it is highly significant.

With the arithmetic in hand, let us now look at the gaseous composition of "natural" air, that is, of the gaseous components that are not the result of human intervention. The most variable component of air is water vapor, or moisture, whose concentration may range from a negligibly small value in a desert to about five percent in a steaming jungle. If we neglect the moisture, the composition by volume of dry air is roughly 78 percent nitrogen, 21 percent oxygen, and one percent of other gases (Fig. 13.2). A more detailed breakdown is given in Table 13.1.

Note that the title of Table 13.1 refers to "natural" air, not to "pure" air. When is air "pure" and when is it "polluted," and does "natural" mean "pure"? These questions are difficult to answer. A "pollutant" is a harmful or

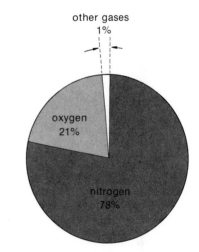

Figure 13.2 Approximate gaseous composition of natural dry air.

TABLE 13.1 Gaseous Composition of Natural Dry Air

	GAS	CONCENTRATION (BY VOLUME) ppm	percent
"Pure" Air	Nitrogen, N$_2$	780,900	78.09
	Oxygen, O$_2$	209,400	20.94
	Inert gases, mostly argon, (9300 ppm) with much smaller concentrations of neon (18 ppm), helium (5 ppm), krypton and xenon (1 ppm each)	9,325	0.93
	Carbon dioxide, CO$_2$	315	0.03
	Methane, CH$_4$, a natural part of the carbon cycle of the biosphere; therefore, not a pollutant although sometimes confused with other hydrocarbons in estimating total pollution	1	
	Hydrogen, H$_2$	0.5	
Natural Pollutants	Oxides of nitrogen, mostly N$_2$O (0.5 ppm) and NO$_2$ (0.02 ppm), both produced by solar radiation and by lightning	0.52	
	Carbon monoxide, CO, from oxidation of methane and other natural sources	0.3	
	Ozone, O$_3$, produced by solar radiation and by lightning	0.02	

undesirable impurity. Certainly sulfur dioxide, SO_2, is an air pollutant. Some SO_2 is discharged into the atmosphere from smelters, oil burners, and coal-fired power plants. Some SO_2 is also emitted by natural sources such as volcanoes and hot springs. Since these sources are found only in certain parts of the globe, SO_2 is not present in a uniform concentration throughout the atmosphere, and it is not considered to be a "normal" component of the air. On the other hand, carbon monoxide, also an air pollutant, is generated all over the globe as part of the Earth's carbon cycle, so it is both "natural" and, in the absence of human sources, fairly constant in its concentration. Its natural concentration, however, is well below the toxic range.

Shall we then consider air pollutants to be only those impurities which are present in sufficient concentration to produce some measurable adverse effects on either living organisms or on materials? But our knowledge of adverse effects changes as we study them. Instead, it is convenient simply to define a substance called "pure air," and to classify any other component as a pollutant. This is the basis for assuming that pure air is a gaseous mixture of the first six components of Table 13.1: nitrogen, oxygen, inert gases, carbon dioxide, methane, and hydrogen in the concentrations shown in the table, or close to them, plus any additional moisture that may be present. Of course, any significant variation in these compositions could be harmful; for example, air containing 10 percent CO_2 would be poisonous, and air containing 10 percent H_2 or 10 percent CH_4 would be explosive. Thus, CO_2 in high concentrations is a pollutant. All other gases, regardless of concentration, whether of human or nonhuman origin, as well as all particulate matter, are then considered to be pollutants.

Table 13.1 does not include nongaseous or "particulate," components. The "natural" concentrations of particulate matter in the air vary much more than those of gaseous matter. Thus, if we analyzed air in various parts of the Earth away from human activities, the composition of the gases would be very close to the values in Table 13.1. But the particulate matter would vary widely from place to place. It would include nonviable (not capable of living) particles such as airborne soil granules, volcanic dust, and salts from evaporation of sea spray. It would also include viable particles such as plant and insect matter.

It is not meaningful to express the concentrations of airborne particulate matter in "composition by volume," because particles tend to settle out and do not distribute themselves throughout a given volume. Instead, the concentration of particles may be represented simply in terms of their total mass per unit volume of air, typically with units such as micrograms per cubic meter (μg/m^3). Figure 13.3 illustrates the relative sizes involved. Study the figure carefully, noting the following:

(a) The size scale, shown in micrometers (μm) is not

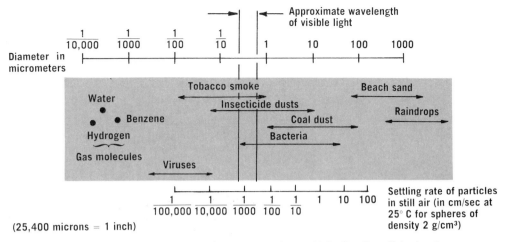

Figure 13.3 Small particles in air.

linear, but progresses by factors of 10. The largest diameter shown, 1000 μm, is equal to 1 mm; it is the size of large grains of sand or small raindrops. If you have a metric ruler, look at the millimeter division to see what size this is. Now imagine you divide your millimeter into 100 parts; each part is one hundredth of a millimeter, or 10 μm, and is too small for you to see. This is the size of a particle of coal dust suspended in the air in a mine tunnel, or the size of a rather large bacterium. The large bacterium (10 μm) must in turn be cut into 1000 parts for each to be the size of a virus (0.01 μm), but still the virus is many times larger than a water molecule.

(b) Turn your attention now to the lower scale, which tells how rapidly particles settle in still air. If you are fortunate enough to be sunning yourself at a sandy beach while reading this book, toss a handful of sand into the air. (Otherwise, just imagine it.) Most of the sand falls rapidly, but if there are some very fine grains they will linger, and you may see them glisten in the sunlight as they descend in a leisurely manner. If it were windy, the small grains could be kept aloft indefinitely and carried over long distances. In fact, as we well know, strong winds will lift sand grains from the ground and blow them into the air. Now consider particles of tobacco smoke, which have settling rates that are much slower than even the gentlest stray currents of air in the quietest spaces we know. The conclusion is that such tiny particles do not settle at all! Do they then remain aloft forever? The answer is no, they do not, for they are removed by processes other than gravitational settling. Air currents push them around until they hit a solid surface, where they may stick. Furthermore, they usually have or acquire some electrical charge, and the consequent electrostatic forces are much greater than gravitational ones, so they are attracted to surfaces of opposite charge, or to other particles.

The names used to describe airborne particles are often confused and inconsistent, referring sometimes to size,

401

TABLE 13.2 Classification of Airborne Particles

	DIAMETER LESS THAN 1 MICROMETER*	DIAMETER GREATER THAN 1 MICROMETER*
Aerosols Smokes Fumes	May be solid or liquid, depending on their origin	Dusts (solid particles) Mists (liquid droplets)

*25,400 micrometers = 1 inch. A micrometer is also called a micron, but this usage is old and is avoided by the SI system.

sometimes to source, and sometimes to a solid or liquid state. Table 13.2 is a rough guide to current usage. The word **aerosol** is used very commonly and should be remembered; it refers generally to any small particle in air. The 1-micrometer distinction between aerosol and dust or mist is not at all precise; many workers in air pollution refer, for example, to 10-micrometer aerosol particles.

13.4 SOURCES OF AIR POLLUTANTS

Some harmful substances are introduced to the atmosphere by natural processes. Volcanoes erupt and discharge great masses of dust and sulfurous gases. Forest fires rage and release quantities of materials in a few hours that the slower processes of decay would take many years to produce. And so there were pollutants on Earth even before the advent of technology. However, these have generally been isolated and sporadic events.

Of course, massive pollution from industrial activities is a recent development. Furthermore, the resulting contaminants, by the nature of the activities that produce them, are likely to be emitted to the air in regions where many people live. Therefore the effects, even if small on a global scale, may be locally very severe.

What are the human sources of air pollution? The following paragraphs will summarize the major categories.

STATIONARY COMBUSTION SOURCES

Certainly the burning of fuel for heat is the oldest form of "civilized" air pollution. Since the Industrial Revolution, combustion has also been used extensively to generate power. The major fuels have been coal and petroleum (including natural gas).

Coal is largely carbon, which, when it burns completely, produces carbon dioxide, CO_2. Petroleum consists largely of hydrocarbons (compounds of hydrogen and carbon), which, when they burn completely, form CO_2 and water, H_2O. If CO_2 and H_2O were the only products, the burn-

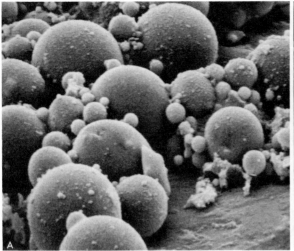

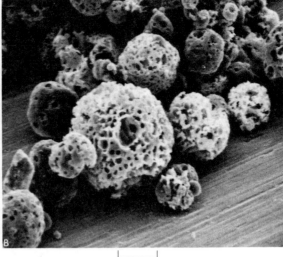

|_____|
10 μm

Particulate matter sampled from the effluent of (A) coal-fired and (B) oil-fired power plants. Illustrations obtained with scanning electron microscope, courtesy of Atmospheric Sciences Research Center of the State University of New York at Albany.

ing of fossil fuels would not pollute the air. But the circumstances are not so ideal, because fossil fuels contain other chemicals and combustion produces other products. Coal is always mixed with some of the incombustible mineral matter of the Earth's crust; when the coal burns, some of this mineral ash flies out the chimney. The smoke it produces is called, aptly enough, **fly ash**.

Sulfur is an essential element of life, and since coal and oil are derived from living organisms, these fuels always contain some sulfur. When the fuels burn, so does the sulfur, to produce a mixture of oxides, mainly **sulfur dioxide**, SO_2, and **sulfur trioxide**, SO_3. From the viewpoint of its harmful effects and the difficulties involved in preventing its discharge into the atmosphere, SO_2 is probably the most significant single air pollutant. High SO_2 concentrations have been associated with major air pollution disasters of the type that have occurred in large cities, such as London, and that were responsible for numerous deaths. The trioxide, SO_3, reacts rapidly with atmospheric moisture to produce sulfuric acid,

$$SO_3 + H_2O \longrightarrow H_2SO_4$$
sulfuric acid

in the form of a strongly acidic mist that corrodes metals, destroys living tissue and deteriorates buildings.

Nitrogen, like sulfur, is common to living tissue, and therefore is found in all fossil fuels. This nitrogen, together with a small amount of atmospheric nitrogen, also oxidizes when coal or oil is burned. The products are mostly NO and NO_2:

$$N_2 + O_2 \rightarrow 2NO$$
nitrogen oxide

$$2NO + O_2 \rightarrow 2NO_2$$
nitrogen dioxide

403

For purposes of convenience in reporting analytical results, chemists frequently group NO and NO_2 together under the general formula NO_x. NO_2 is a reddish-brown gas whose pungent odor can be detected at concentrations above about 0.1 ppm, and it therefore contributes to the "browning" and the smell of some polluted urban atmospheres.

Finally, the incomplete combustion of carbon and of hydrocarbons produces a variety of pollutants. Carbon generally yields considerable quantities of carbon monoxide,

$$2C + O_2 \rightarrow 2CO$$
carbon monoxide

a gas that is colorless, odorless, and nonirritating, yet very toxic. Accidental deaths occur, for example, from CO escaping into a room from a faulty gas heater, or from charcoal cooking grills used indoors. Hydrocarbons generate a wide variety of products of incomplete combustion, some gaseous, others particulate. Particles that consist mostly of carbon are called, collectively, **soot**. Many of these substances are known carcinogens.

MOBILE COMBUSTION SOURCES

With the demise of the steam locomotive, the prime sources of energy for mobile engines are gasoline, diesel fuel, and jet fuel. The internal energy of these materials, like that of coal and oil, lies in the ability of their carbon and hydrogen atoms to oxidize, respectively, to CO_2 and H_2O. The resulting air pollutants, therefore, do have features in common with those from stationary combustion sources, but there are important differences.

Mobile power plants, such as automobile engines, must be more compact than stationary ones, and it is not feasible for them to be fitted with massive air pollution control equipment. The fuels they use are therefore more highly refined than those in large stationary power plants, and so automobiles are not a major source of fly ash. However, as a result of chemical reaction conditions that are characteristic of mobile combustion sources, some of their emissions are different from those of the larger stationary plants. For example, leaded gasoline generates various particulate lead compounds. In addition, many of the products of incomplete combustion of gasoline undergo reactions with oxides of nitrogen in the presence of sunlight to produce a complex mixture of pollutants called **oxidants**, and the atmospheric effect they generate is called **photochemical smog**. These pollutants are typical of the atmospheres of sunny urban areas with considerable automobile traffic, such as Los Angeles. They exhibit various properties in common, including certain toxic and irritating effects on people, various patterns of damage to vegetation, and the ability to produce cracks in natural rubber (Fig. 13.4).

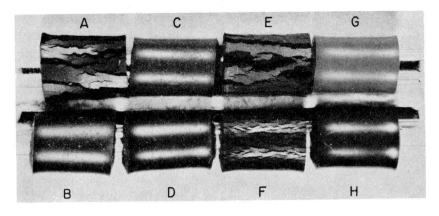

Figure 13.4 Effect of ozone exposure on samples of various kinds of rubber. *A*, GR-S; *B*, Butyl; *C* and *D*, Neoprene; *E*, "Buna-N"; *F*, Natural rubber; *G*, Silicone; *H*, "Hypalon." (Photo courtesy of F. H. Winslow, Bell Telephone Laboratories. From Stern: *Air Pollution*, 2nd ed. New York, Academic Press, 1968.)

Chemical formulas of various pollutants are shown in Appendix C-13.

CHEMICAL MANUFACTURING INDUSTRIES

This source is so varied and complex that it will be possible only to suggest some generalizations and to highlight a few examples. As a rule, gases are produced when large molecules are broken down into smaller ones. Such transformations occur in practically all industrial operations that process organic chemicals at high temperatures. Airborne particulate matter is generated in many mechanical operations, such as blasting, drilling, crushing, grinding, mixing, and drying. (See Color Plates 6 and 10.)

Metallurgical processes, especially those involved in the production of iron and steel, copper, lead, zinc, and aluminum, are very significant sources of air pollution.

In some instances, new findings cast suspicion on some chemicals that were once thought to be environmentally harmless. An example is the class of compounds that contain both chlorine and fluorine, especially the chlorofluoromethanes,* $CFCl_3$ and CCl_2F_2, which have been widely used in recent years. $CFCl_3$ is a propellant for aerosol cans, and all the material manufactured for such purpose must therefore be dispersed into the atmosphere. CCl_2F_2 is a refrigerant, and it too goes into the air, ultimately, when old refrigerators rust away. These compounds are all rather stable in the lower atmosphere. However, they may disperse to the stratosphere, where environmentally damaging reactions are possible. Such questions will be taken up in Section 13.5.

13.5 THE METEOROLOGY OF AIR POLLUTION

To assess the extent to which your health is affected by, say, sulfur dioxide, you must be concerned with the quantity or concentration that you inhale. The total quan-

*These compounds are often referred to as Freons, which are trade names of the Dupont Company. $CFCl_3$ is called Freon-11, and CCl_2F_2 is Freon-12.

tity of SO_2 in the Earth's atmosphere, or the concentration in the exhaust gas of some particular copper smelter, does not affect you directly, because you do not breathe all the world's air, nor do you stick your head into the smokestack. Therefore, you must consider how pollutants are transported in the atmosphere, and how atmospheric conditions affect their concentrations. The science that deals with these and other atmospheric phenomena is **meteorology.** The atmosphere is heated from above by direct sunlight and from below by radiation from the Earth's surface. If you were to ascend vertically from the Earth in a balloon, under average meteorological conditions,* you would find that the air temperature drops steadily with altitude. This cooling, which is shown in Figure 13.5A, occurs because the heat from the Earth's surface is dissipated at high altitudes. Thus mountain tops are generally colder than valley floors, and pilots of airplanes with open cockpits must wear heavy clothing.

---

*The "average" meteorological conditions refer to an equilibrium in which gains and losses of energy in the atmosphere are in balance. The rate of cooling is then about 1°C for every rise of 1000 meters.

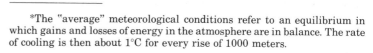

A Temperature ——▶

theoretical conditions

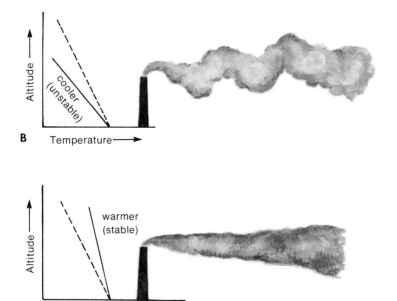

B Temperature ——▶

cooler (unstable)

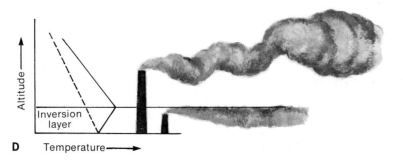

C Temperature ——▶

warmer (stable)

D Temperature ——▶

Inversion layer

Figure 13.5 Air temperature versus altitude. *A,* Theoretical condition when gains and losses of energy are in balance. *B,* Actual temperatures (solid line) are cooler than theoretical conditions. *C,* Actual temperatures (solid line) are warmer than theoretical conditions. *D,* Atmospheric inversion layer.

406

Atmospheric conditions do not always conform to this theoretical model. At any particular time, for example, the *actual* temperatures may be cooler than the equilibrium conditions. If polluted air from a chimney entered such an atmosphere, it would become warmer than its surroundings and would tend to rise. Such a situation favors turbulent mixing, which helps to dilute the pollutants. The smoke pattern would resemble that shown in Figure 13.5*B*.

Now consider the situation in which the air at some given altitude is *warmer* than the equilibrium conditions (Fig. 13.5*C*). Such a warm layer rests in a stable pattern over the cooler, denser air beneath it. This condition is called an **atmospheric inversion**. Such a situation could start to develop an hour or two before sunset after a sunny day, when the ground starts to lose heat and the air near the ground also begins to cool rapidly. This inverted condition often continues through the cool night and reaches its maximum intensity and height just before sunrise. When the morning sun warms the ground, the air near the ground also warms up, and it rises. Within an hour or two the inversion may be broken up by turbulent mixing. Sometimes, however, atmospheric stagnation allows the inversion to persist all day or even for several days. Consider now what would happen in an atmosphere in which such an inversion layer existed between the ground and, say, 200 meters, but not at higher altitudes. Any polluted air discharged into the atmosphere at elevations below 200 meters would not be able to rise; the 200-meter level would act as a ceiling, or lid, that traps the pollutants below it. In fact, even if the lower air were turbulently stirred by warm currents, the total volume available for mixing would be limited by the 200-meter ceiling, and any pollutants discharged at lower levels would concentrate beneath the ceiling. The longer the inversion lasted, the greater the build-up. However, a chimney tall enough to penetrate the 200-meter barrier would discharge its effluents into the upper atmospheric layers where they would be diluted in a much larger volume. Such a condition is shown in Figure 13.5*D*, and photographs of smoke plumes are shown in Color Plate 9.

Effluents from chimneys may contain internal energy in the form of heat and upward motion. Figure 13.6 shows such a plume which, although released below the inversion barrier, was energetic enough to pierce it.

The world's tallest chimney (as of 1977), built for the Copper Cliff smelter in the Sudbury District of Ontario, Canada, is as tall as the Empire State Building. It is illustrated in Figure 13.7 together with three smaller ones. Such heroic structures can be quite effective in reducing ground-level concentrations of pollutants. For example, during a 10-year period studied by the Central Electric Generating Board in Great Britain, the SO_2 emissions from power stations increased by 35 percent, but, as a result of

Three views of downtown Los Angeles. *Top:* a clear day. *Middle:* pollution trapped beneath an inversion layer at 75 meters. *Bottom:* pollution distribution under an inversion layer at 450 meters. (Photos from Los Angeles Air Pollution Control District.)

Figure 13.6 The photograph shows an atmospheric inversion over Hartford, Connecticut, viewed from the Talcott Mountain Science Center in Avon, about 4 km west of the city. Automobile and industrial emissions have become trapped beneath the overlying warmer air, and the murky layer containing high concentrations of sulfates, nitrates, and hydrocarbons persists until a new front restores atmospheric mixing. The plume of steam, issuing from an East Hartford industrial plant, was energetic enough to pierce the inversion but dissipated in 3–4 minutes. The photograph was taken at about 8 AM on a cold winter day; by noon, the city was no longer visible through the smog. (Photo courtesy of G. C. Atamian, Talcott Mountain Science Center, Avon, CT.)

Figure 13.7 World's tallest chimney. Built in Ontario, Canada, at a cost of $5.5 million, this chimney stands 380 meters high. (Photo courtesy of M. W. Kellogg Company, division of Pullman Inc.)

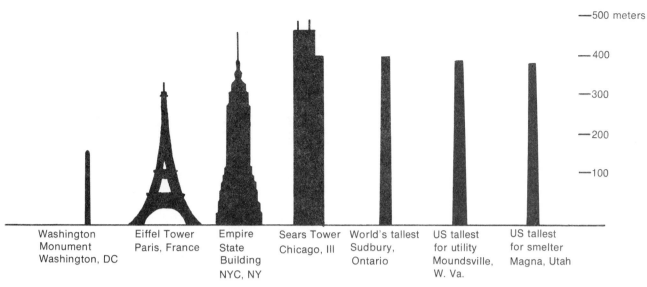

| | | | | | | | —500 meters |
| Washington Monument Washington, DC | Eiffel Tower Paris, France | Empire State Building NYC, NY | Sears Tower Chicago, Ill | World's tallest Sudbury, Ontario | US tallest for utility Moundsville, W. Va. | US tallest for smelter Magna, Utah | |

The three structures on the right are chimneys, whose heights rival the tallest buildings.

the construction of tall stacks, the ground level concentrations decreased by as much as 30 percent. However, tall chimneys do not collect or destroy anything, and, therefore, they do not reduce the total quantity of pollutants in the Earth's atmosphere.

13.6 THE EFFECTS OF AIR POLLUTION—INTRODUCTION

It was pointed out at the beginning of this chapter that in most instances the first noticeable effects of air pollution are odor and visible smoke. Of course, there are exceptions: carbon monoxide can kill quickly without warning, and many other pollutants cause harm at a much slower rate while they, too, escape detection.

Our methods of appraising the effects of air pollution are now much more sophisticated than they used to be. No longer do we blame malaria on swampy odors, nor are we unaware of the slow harm produced by cigarette smoke or coal dust. But there is much we do not know. Furthermore, much is suspected but is not easily proved—witness the controversy about the proposition that lead compounds from automobile exhaust are harmful to people. These uncertainties, taken together with the differences in interests among various sectors of the population, lead to a variety of recommendations for public policy.

It is important, therefore, to learn about air pollution effects and to be careful to distinguish among what is "known" with a high degree of confidence, what is suspected, and what is merely possible (that is, not yet proved to be untrue).

Air pollution effects may be classified into six ca-

409

tegories: (a) modification of climate, (b) harm to human health, (c) damage to vegetation, (d) injury to animals, (e) deterioration of materials, and (f) aesthetic insults. The following sections will discuss each in turn.

13.7 EFFECTS OF AIR POLLUTION ON THE ATMOSPHERE AND ON CLIMATE

The possibility that the Earth's climate, particularly its temperature, may be drifting away from its previous range of conditions is alarming. A warming portends a melting of glaciers and a flooding of the populous coastal plains. A cooling presages a new ice age. Changes in precipitation may not be quite so cataclysmic, but, nonetheless, a greater or lesser rainful, or its redistribution, can have profound effects on agriculture, and so on people.

Can the pollution of the Earth's atmosphere, then, affect its climate? To approach this question, we must first consider what happens to the solar radiation that the Earth receives. As was pointed out in Chapter 3, all of this energy is eventually radiated back into space; the fact that the Earth, in spite of various temperature fluctuations, is

Figure 13.8 Energy balance of the Earth. The sets of numbers in the dashed areas total 100 percent.

410

not suffering an unchecked heating or cooling attests to this complete exchange of radiant energy.

Figure 13.8 shows how the balance is distributed. Only 21 percent of the incident solar radiation strikes the Earth directly. The other 79 percent is intercepted by the atmosphere—the clouds, gases, and aerosol particles. Some of this intercepted radiation is reflected back to space, some is absorbed as heat, and some is rescattered down to Earth. On a global average, just about half of the heat energy received from the Sun reaches the surface of the Earth. Some of this heat is reflected back into the atmosphere, while some is absorbed, thereby warming the Earth. Heat energy is then carried back into the atmosphere by several processes, including evaporation and recondensation of water, conduction, and reradiation. The warm atmosphere radiates some heat back to Earth and some to outer space. Thus we see that a complex set of interactions occurs, in which radiation bounces back and forth between the surface and the atmosphere until it is ultimately lost to space.

REFLECTION OF RADIATION

Energy is scattered into space by reflection. Dust reflects light. Pollution makes dust. Therefore, one would expect that the effect of dust pollution is to cool the Earth. What evidence do we have to bear on the accuracy of this conclusion? So far, major volcanic eruptions far outstrip human activities in producing dust (but man is catching up). The most spectacular eruption in modern times was that of Krakatoa (near Java) in 1883; its dust particles stayed in the atmosphere for five years. Summers seemed to be cooler in the Northern Hemisphere during this period, although the extent of temperature fluctuations makes even this observation somewhat doubtful. Volcanic eruptions occur continually here and there on the Earth, and they do inject large quantities of small particles into the upper atmosphere. The best evidence seems to be that such particles do account for some back-scattering of solar radiation.

What, then, of the dust injected into the atmosphere by people? Of course, much of the larger particulate matter falls to the ground or is washed down by rain. There is some persistent introduction of small particles into the upper atmosphere, where they join the volcanic dust in reflecting sunlight. However, this effect is as yet puny compared with that of volcanic dusts. The effect of pollutant particles in the lower atmosphere, where most of them are concentrated, is more complicated. Particles can absorb as well as reflect radiation, and the net heat effect of man-made dusts near the ground is not at all easy to interpret. There is therefore no convincing evidence that pollutant

Cerro Negro Volcano (Nicaragua) blanketed the countryside and the city of Leon (approximately 27 km away) with ash from October 23 to December 7, 1968. Reprinted by permission of *Science & Public Affairs,* the Bulletin of the Atomic Scientists.

411

dusts in the lower atmosphere have any important effect on the Earth's temperature. However, the dust that settles to the ground is another matter. Most of us have seen how snow in the city can become dirty after a few days. Dust that falls on snow and ice in mountainous and polar regions depresses their reflectivities. Snow and ice that retain heat may melt, and if this occurs extensively, it may produce a rise in temperatures, with significant global consequences.

ABSORPTION OF RADIATION: THE GREENHOUSE EFFECT

If there were no atmosphere, the view from the Earth would be much like that which the astronauts see from the Moon—a terrain where starkly bright surfaces contrast with deep shadows, and a black sky from which the Sun glares and the stars shine but do not twinkle. The atmosphere protects us by serving as a light-scattering and heat-mediating blanket. As shown in Figure 13.4, about half of the incident radiation from the Sun passes through the atmosphere to the Earth; the rest is reflected or absorbed in the atmosphere. Of the heat emitted from the Earth, a large portion (represented by the arrow that curves down) is reabsorbed by the atmosphere and is, in effect, conserved, with the result that the surface of the Earth is warmer than it would otherwise be. This warming is called the **greenhouse effect,** by analogy with the ability of a greenhouse to keep its inside warmer than the outside during the daytime. The energy emitted from the Earth is, of course, invisible (the Earth does not shine); it is largely infrared radiation, sometimes called heat rays.

Some molecules in the atmosphere absorb infrared radiation, and others do not. Oxygen and nitrogen, which together compose almost 99 percent of the total composition of dry air at ground level, do not absorb infrared. On the other hand, molecules of water, carbon dioxide, and ozone do absorb infrared. Water plays the major role in absorbing infrared because it is so abundant. Ozone is the least important because there is so little of it. Carbon dioxide is also important, particularly because our combustion of fossil fuels produces more and more of it. Taking all factors into account, our best estimate is that the worldwide carbon dioxide concentration has increased from about 290 ppm in 1870 to 322 ppm in 1970.

Some scientists estimate that the carbon dioxide concentration will increase enough in the next 40 years to warm the Earth significantly, perhaps by as much as $0.5°$ C. It is impossible to predict the effects of such a temperature change on world climate. At present the Earth appears to be cooling, not warming. But we cannot predict what will happen in the future.

PRECIPITATION

The aerosol particles in cities whose atmospheres are polluted serve as nuclei for the condensation of moisture. Fogginess, cloudiness, and perhaps rainfall are usually increased considerably in contrast with the less polluted countrysides. However, the rise of warm air from cities, and the circulation of cooler surrounding airs that is thereby induced, can sometimes reverse this situation.

The worldwide effects of pollutants on precipitation are even more difficult to assess. The concentration of small particles in the lower atmosphere is increasing over wide areas. The lead compounds from automobile exhaust are of particular concern because some of them, particularly lead chloride and lead bromide are crystallographically similar to cloud-seeding compounds. That is to say, they serve particularly well as nuclei for the condensation of water droplets or ice crystals. This action could lead to more precipitation, if the droplets coalesce, or less precipitation, if the nuclei are so small that the droplets stabilize and coalescence is inhibited.

Ice fog from emissions from industrial processes, Edmonton, Alberta. (Photo by Dr. James W. Smith, Toronto, Canada.)

JET AIRCRAFT

There has been concern that jet aircraft, and particularly the supersonic transport (SST), whose flight patterns are in the 18,000- to 20,000-meter range, could engender stratospheric air pollution with consequent changes in climate. Jet exhaust contains water, CO_2, oxides of nitrogen, and particulate matter. The question is, can the effects of these pollutants be harmful? The answer, at this time, is speculative. For example, it is estimated that a fleet of 500 SST's over a period of years could increase the water content of the stratosphere by 50 to 100 percent, which could result in a rise in average temperature of the surface of the Earth of perhaps 0.2° C and could cause destruction of some of the stratospheric ozone that protects the Earth from ultraviolet radiation. On the other hand, the particulate matter would nucleate clouds of ice crystals and would increase stratospheric reflection to some extent. This could lead to a cooling of the Earth's temperature, probably of only a small magnitude. One visible effect of such continued high-altitude pollution would be that the skies would gradually become hazier and lose some of their blueness. Since the water vapor is predicted to warm the atmosphere and the particulate matter may cool it, we don't know what the net effect of SST aircraft will be.

Jet exhaust. (Photo by David F. Hall; courtesy of the Connecticut Lung Association.)

413

Depletion of the ozone layer by chlorofluoro-methanes.

DEPLETION OF THE OZONE LAYER

The Sun emits light over a wide range of wavelengths, including infrared, visible, and ultraviolet (UV). It is the UV radiation which, having high energy, tans our skin. Heavier doses of UV can cause burns and can increase the chances of skin cancer. If more of the UV light that reaches our upper atmosphere were to penetrate to the surface of the Earth, the risks of such damages would be increased.

Plants, too, might be adversely affected, and preliminary data suggest that the growth of some food crops, such as tomatoes and peas, is retarded by high doses of ultraviolet light. Fortunately, the high-energy UV radiation is removed in the upper atmosphere by a series of photochemical reactions involving molecular oxygen and ozone. In the first step, sunlight striking an oxygen molecule breaks it apart (dissociates it) into two oxygen atoms:

$$O_2 + UV \text{ radiation} \rightarrow O + O$$

The radiation is thus absorbed and is converted to chemical

energy. Atomic oxygen is highly reactive and combines with O_2 to form ozone.

$$O + O_2 \rightarrow O_3$$

This reaction releases energy, but in the form of infrared rather than ultraviolet radiation. Thus the biologically harmful high-energy ultraviolet is converted to benign infrared radiation. Ozone then absorbs additional UV radiation to dissociate according to the reaction.

$$O_3 + \text{UV radiation} \rightarrow O_2 + O$$

Thus UV is removed by both the formation and the destruction of ozone.

Ozone also can be removed by other, nonphotochemical processes. One such reaction occurs normally in the upper atmosphere:

$$O_3 + O \rightarrow 2O_2$$

There has been recent concern that some air pollutants are thinning out the protective ozone barrier. Of particular interest are the chlorofluoromethanes (Freons), $CFCl_3$ and CF_2Cl_2 (see page 405). We shall use the former to represent the pair. In the lower atmosphere, $CFCl_3$ is inert and does not affect living systems. Partly for this reason it has been used as a propellant in aerosol cans. Its function is simply to provide the pressure that propels the liquid out as a fine mist. However, quantities of gas have been moving to the ozone layer of the upper atmosphere. There it readily dissociates into two particles:

$$CFCl_3 + \text{radiation} \rightarrow CFCl_2 + Cl$$

The atomic Cl can react to remove ozone:

$$2O_3 \xrightarrow[\text{of Cl atoms}]{\text{in the presence}} 3O_2$$

The result is that $CFCl_3$ ultimately removes ozone, and since ozone absorbs ultraviolet radiation, any heavy concentration of $CFCl_3$ in the upper atmosphere could conceivably allow large amounts of harmful ultraviolet light to reach the life-supporting layers of the Earth. It has been calculated[*] that if the projected increases in the use of these compounds materialize, and if they are not destroyed in the lower atmosphere, the total global abundance of ozone may be reduced by more than 20 percent over the next 50 years.

It is impossible to predict accurately what is going to happen if current trends continue. More and more chlorofluoromethanes are being released into the upper atmo-

[*]P. Turco and R. C. Whitten: *Atmospheric Environment*, 9:1045, 1975.

sphere, and there is some evidence that the ozone level is being reduced.

Those who favor new developments such as spray deodorants and SST's assume that the human inventiveness that produced them will also find ways to prevent or undo any adverse effects. (In scientific jargon, such a cure is called a "technological fix.") Environmentalists, on the other hand, take the position that global effects are so complex that we do not understand them well enough to dare to tamper with them and that, in any case, each "technological fix" will only lead to a new set of problems.

Perhaps human activities will bring about a deterioration of global climate, or perhaps we will learn how to regulate temperature and rainfall patterns for human benefit. But it may just be that global forces are so great and complex that we will primarily be observers of the weather that affects our very existence.

13.8 EFFECTS OF AIR POLLUTION ON HUMAN HEALTH

As mentioned previously, air pollutants may be either particulate or gaseous. The fate of a particle once it is inhaled depends largely on its diameter. If it is greater than about two micrometers (microns), it is usually trapped in the nasal passage or in the mucus of the bronchi (Fig. 13.9) and is generally coughed up and perhaps swallowed. If the diameter is less than two micrometers, the particle may be carried all the way through the air passages into the air sacs (alveoli) of the lung, where it may be trapped by specialized cells, or alternatively it may be absorbed into the bloodstream.

For gases the determining factor is largely the solubility of the gas in water. Since biological tissues are rich in water, a water-soluble gas like SO_2 will rapidly dissolve in the soft tissues of the mouth, nose, throat, bronchi, and eyes where it produces the characteristic dry mouth, scratchy throat, and smarting eyes that most city dwellers have sometimes experienced. By contrast, NO_2, which is relatively insoluble, may bypass this part of the respiratory tract and be carried to the alveoli where in very high doses it may cause gross accumulation of fluid in the air spaces, and thus make effective lung function impossible.

It must be emphasized, however, that the net toxic effect when various pollutants are inhaled together may be different from the sum of the effects of these same pollutants if they are inhaled separately. An interaction that produces *more* than a merely additive effect is called **synergism.** It is known, for example, that SO_2 may be adsorbed onto particles; if these particles are smaller than two micrometers, molecules of SO_2 may thus gain access to

"Synergism" (from the Greek for "working together") originally referred to the doctrine that the human will cooperates with God's grace in the redemption of sinners. Later, the word was used in medicine to describe the cooperation between a drug and a bodily organ, or between two bodily organs, in carrying out some organic function. In modern scientific usage, synergism refers to any combination of actions in which the result is more than that which would be attained if the actions were entirely independent of each other. In other words, in a synergistic process the whole is greater than the sum of its parts.

416

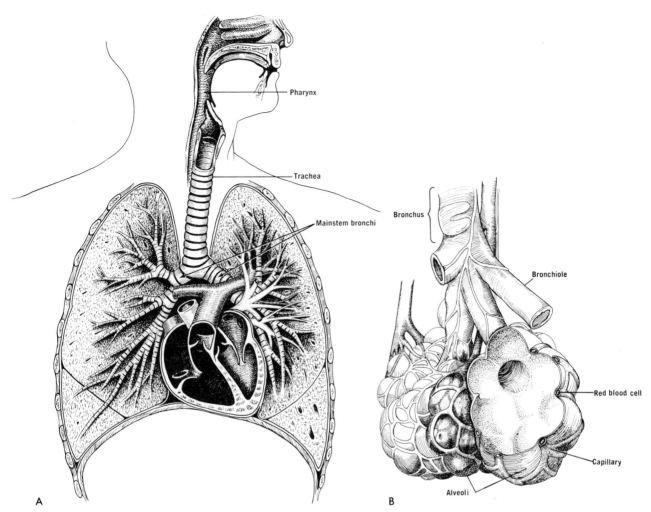

A

B

Figure 13.9 A diagram of the human respiratory tract. Air is breathed into the pharynx and from there travels through the *trachea*, or windpipe, which divides into the two *mainstem bronchi*. Each of these subdivides into airways which in turn subdivide into conduits of increasingly smaller caliber until they terminate in the *alveoli*, or air sacs. It is in the air sacs that gases in the inspired air exchange with the gases dissolved in the blood. This gas exchange is the function of the lung. Lining the walls of the airways are cells, some of which form glands that secrete mucus. The walls are also covered with hairlike projections called *cilia*, which wave in synchrony with one another and serve to propel the mucus and any impurities from the air that may be suspended or dissolved in it up toward the *pharynx*, where it is usually swallowed.

the alveoli in greater concentrations than they would otherwise. The retention of carcinogenic hydrocarbons in the human body has been shown to be greatly enhanced if they are first adsorbed onto soot particles. Also, oxygen and water can react with SO_2 to form sulfuric acid and with NO_2 to form nitric acid. It is clear, then, that the interaction of the various pollutants with each other may have great significance in determining the types of toxicities that result when a human being breathes polluted air.

As previously noted, most city residents have frequent personal experiences with what might be called the nuisance effects of dirty air. Indeed, if one defines ill health simply as the absence of a feeling of physical well-being, then city residents are almost universally the victims of air pollution. Whether the dry mouth, smarting eyes, and offended nose constitute grounds in themselves for limiting pollution of the air is not a trivial question. However, there are many instances in which increased air pollution levels have had very serious consequences on the level of health of the exposed population.

417

"Thank you for not smoking."

Drawing by Booth; © 1977 The New Yorker Magazine, Inc.

ACUTE EFFECTS

Much of the public attention to air pollution has been focused on acute episodes in which deaths occurred during periods of pollution and were clearly caused by its toxic effects. The first such dramatic instance in the United States occurred in Donora, Pennsylvania, in 1948, and is described in the Case History at the end of this chapter.

In 1952 the population of London experienced for four days a similar build-up of the usual air contaminants that affected the British Isles. During that period, and continuing for the next two to three weeks, there were nearly 4000 more deaths in London than would have been expected for that time of year. As in Donora, those most affected were in the older age groups and generally had disease of the heart or lungs prior to the pollution episode. A much larger number of individuals became ill, chiefly with exacerbations of chronic respiratory and cardiac ailments.

Such disasters are fortunately infrequent. They all have certain common patterns. First, those who become severely ill usually are elderly or have chronic heart or lung disease. Secondly, as the London disaster showed, the increase in illness and mortality following the acute episode actually lasted for weeks. What probably happened was that the pollutants interfered with the normal mechanisms of the respiratory tract for dealing with invading pathogenic organisms, and that this lowered resistance continued for a time after subsidence of the intensified air pollution.

PERSONAL AIR POLLUTION—CIGARETTES

The air whose quality is important to your health is the air *you* breathe, not the entire global atmosphere. If the outdoor air in any community resembled the self-polluted air that a smoker inhales, it would be considered a national disaster. How "dangerous to your health" are cigarettes?

In 1964 the Office of the Surgeon General of the United States Public Health Service published a volume entitled *Smoking and Health.* This represented the attempt of a distinguished panel of medical scientists, chemists, and statisticians to evaluate all the existing evidence concerning the effects of smoking on health. The aim of the panel was not simply to define whether there was an *association* between smoking and various diseases, for such associations had been known for years. Rather, its purpose was to establish whether smoking *causes* disease. The evidence that supports such a conclusion is derived from many studies. Some typical findings have been:

- Average smokers have an approximately tenfold risk of developing and dying of lung cancer, a sixfold risk of death from pulmonary disease, and a nearly twofold risk of dying from coronary heart disease, as compared with a nonsmoking population. There are also a host of other illnesses which are significantly associated with smoking.

- The more you smoke, the greater are your chances of getting lung cancer.
- Chemical analysis of tobacco and cigarette smoke has revealed the presence of at least seven distinct polycyclic hydrocarbons which have been shown to be able to produce cancer in animals. Cigarette smoke also contains polonium-210, a radioactive substance that may be carcinogenic.

CHRONIC EFFECTS OF COMMUNITY AIR POLLUTION

Another question of concern to medical scientists for many years has been the issue of whether exposure to mildly polluted air results in a higher rate of illness and death than is experienced by those who breathe relatively clean air. Many illnesses have been examined, but the ones that have prompted more research than any others have been diseases of the respiratory tract: lung cancer, chronic bronchitis, and emphysema. Bronchitis is an inflammatory condition of the bronchi. When the bronchi become inflamed, the cell layers that line them become thickened and the mucus content of the air passages increases markedly. The small airways usually become constricted, and segments of the lung may become infected with bacteria or viruses. Emphysema, by contrast, refers to a breakdown in the walls of the alveoli themselves. (See Color Plate 11.)

Various studies have shown that lung disease is up to four times as prevalent in cities as in rural areas. Before incriminating air pollution alone, however, one must first consider in what other respects cities differ from rural areas. Clearly, there are many differences. Population density, occupations of the inhabitants, socioeconomic levels, ethnic composition, and even possibly smoking habits, age distribution of the population, and other factors may vary from city to country. If they do differ, one cannot assume that differences in the incidence of any disease from city to country may not be due to one of these, or other, unidentified factors. Many of the better studies of this problem take one or more of these variables into account. The preponderance of the evidence indicates that, after most of these factors have been taken into consideration, there still remains an excess incidence of lung disease in urban environments. The most likely element of the "urban factor" seems to be air pollution, though conclusive data is lacking.

Moreover, cigarettes and the urban environment seem to have a more than additive effect on the incidence of lung cancer. That is, if one compares urban non-smokers with rural non-smokers, the difference in incidence of lung disease is small; but urban smokers have a much greater incidence of the disease than rural smokers. These studies indicate that the combination of smoking and living in large cities is particularly pernicious.

Thus far, the only effects of air pollution we have considered are those on mortality from lung disease. It is quite clear, though, that much of the illness brought about by breathing polluted air may not cause death, but rather just increased suffering and debility. Moreover, studies from Japan and the United States both indicate that effects on the respiratory system may in fact begin in childhood. Children living in highly polluted areas in both countries have been shown to have a much increased incidence of acute respiratory infections, and the Japanese studies have also documented increased airway constriction during periods of high pollution.

Those with pre-existing cardiac disease, particularly the type caused by narrowing of the blood vessels to the heart (so-called coronary heart disease) are, as we have mentioned, at high risk of serious illness when subjected to very high levels of air pollution. Recently, scientists studied chronic heart patients as they were driven in automobiles along a freeway in the Los Angeles area during heavy morning traffic. After a 90-minute drive, the subjects showed significantly less ability to do a controlled series of exercises without experiencing pain in the chest (this is a standard test to assess the severity of coronary heart disease); their pulmonary function also showed deterioration, and their blood levels of carbon monoxide had risen significantly. In order to show that these effects were related to the breathing of polluted air rather than, say, the harrowing experience of simply driving along a freeway in southern California during rush hour, the experiment was repeated the next day, but this time the subjects breathed purified compressed air during the trip. After this part of the experiment, their exercise testing and lung function test showed no deterioration at all. These studies show clearly that air polluted by ordinary automobile exhaust at a level that is probably attained most days during the year can have deleterious effects on the cardiovascular function of people whose heart is already compromised by other disease. One wonders how many people have died (or killed others) on crowded highways from heart attacks triggered by dirty air.

13.9 DAMAGE TO VEGETATION

Air pollution has caused widespread damage to trees, fruits, vegetables, and ornamental flowers (see Color Plate 12). In fact, the total annual cost of plant damage in the United States has been estimated at close to one billion dollars. The most dramatic early instances of such effects were seen in the total destruction of vegetation by sulfur dioxide in the areas surrounding smelters (Fig. 13.10), where this gas is produced by the "roasting" of sulfide ores.

We now know that there is a wide variety of patterns of plant damage by air pollutants. For example, all fluorides appear to act as cumulative poisons to plants, causing collapse of the leaf tissue. Photochemical (oxidant) smog bleaches and glazes spinach, lettuce, chard, alfalfa, tobacco, and other leafy plants. Ethylene, a hydrocarbon that occurs in automobile and diesel exhaust, makes carnation petals curl inward and ruins orchids by drying and discoloring their sepals.

13.10 INJURY TO ANIMALS

Countless numbers of North American livestock have been poisoned by fluorides and by arsenic. The fluoride effect, which has been the more important, arises from the fallout of various fluorine compounds on forage. The ingestion of these pollutants by cattle causes an abnormal calcification of bones and teeth called **fluorosis,** resulting in loss of weight and lameness (see Fig. 13.11). Arsenic poisoning, which is less common, has been transmitted by contaminated gases near smelters.

13.11 DETERIORATION OF MATERIALS

Acidic pollutants are responsible for many damaging effects, such as the corrosion of metals and the weakening or disintegration of textiles, paper, and marble. Hydrogen sulfide, H_2S, tarnishes silver and blackens leaded house paints. Ozone, as previously mentioned, produces cracks in rubber.

Particulate pollutants driven at high speeds by the wind cause destructive erosion of building surfaces. And the deposition of dirt on an office building, as on a piece of apparel, leads to the expense of cleaning and to the wear that results from the cleaning action (see Fig. 13.12). The total annual cost in the United States of these effects is very difficult to assess but has been estimated at several billion dollars. See Color Plate 11.

13.12 AESTHETIC INSULTS

A view of distant mountains through clear, fresh air is aesthetically satisfying, and an interfering acrid haze is therefore a detriment. However, there is no direct way to

Figure 13.10 The Copper Basin at Copperhill, Tennessee. A luxuriant forest once covered this area until fumes from smelters killed all of the vegetation. (U.S. Forest Service photo. From Odum: *Fundamentals of Ecology,* 3rd ed. Philadelphia, W. B. Saunders Co., 1971.)

Figure 13.11 A cow afflicted with fluorosis.

Figure 13.12 Old post office building being cleaned in St. Louis, Missouri, 1963. (Photo by H. Neff Jenkins. From Stern: *Air Pollution*, 2nd ed. New York, Academic Press, 1968.)

measure a subjective experience. Suppose a factory emits unpleasant odors that blanket a community. How can people be compensated for such a loss? If someone is physically injured, one guideline for judging proper compensation is to estimate the degree to which the victim suffers loss of income. But how can one assign a dollar penalty to a stink? Various attempts have, in fact, been made. In one study, people in a community were asked how much they would be willing to pay to get rid of the odors that annoyed them. While theoretically valid, this approach runs into the practical difficulty of separating what people say they would pay from what they actually would pay if they had to comply with their own responses.

A more fruitful approach lies in determining what people actually have paid to obtain an odor-free environment. Economic theory implies that if odors are bothersome, people should be willing to pay more to live in an odor-free area. Thus two similar properties in similar neighborhoods should sell for different prices if one is affected by odors and the other is not.

Another approach makes use of the fact that homeowners often purchase air-cleaning devices to get rid of odors, and the dollar cost of these devices is taken as a measure of the aesthetic insult. How well do such measures work? Poorly. In our complex society, we have not yet learned how to isolate aesthetic effects from the many other driving forces that mediate our economic behavior. To complicate matters still further, unpleasant aesthetic effects cannot be neatly separated from the other disruptions caused by air pollution. The acrid haze referred to above is sensed not only as an annoyance, but also as a harbinger of more direct harm, somewhat as the smell of leaking gas forebodes an explosion. Thus the evident pollution engenders anxiety, and anxiety may depress our appetites, or rob us of sleep, and these effects, in turn, can be directly harmful.

13.13 AIR POLLUTION STANDARDS AND INDICES

The wind dies down, and the air is still, but traffic continues to move, homes are heated, and industry operates. The air becomes hazy, murky, then uncomfortable. How dangerous is it? Shall we close down the factories, stop the traffic? Clearly, to answer such questions we must measure something. But what? SO_2? Particulate matter? Carbon monoxide? All of them? After the measurements have been made, what is the most reasonable basis for action? The usual approach is to establish a set of air pollution standards that can serve as guidelines for governmental policies or regulations. Since any action will depend on the concentrations of pollutants, the conditions under which the analyses are carried out must be specified.

An air pollutant can be measured at the point where it is discharged to the atmosphere, such as at the chimney top, or in the surrounding (ambient) atmosphere where people live. The concentration at the source is not *directly* related to effects on human health, for we do not expect to find people in chimneys. Nonetheless, such measurements are valuable aids to the enforcement of air pollution control regulations. When permissible limits are established for such sources, they are called **emission standards**. The ambient concentrations are those in the air that people breathe, and the recommended limits are called **ambient air quality standards**.

Air quality standards that are established to guard human health are called **primary standards**. However, it is also important to avoid secondary effects such as damage to vegetation, deterioration of materials, and reduction of visibility. These objectives are not as important as the protection of human health, but they often require even lower concentrations of air pollutants. These maximum permissible levels are called **secondary standards**.

Of course, a given concentration of pollutant does more damage over an entire day than in a single hour. In setting standards, therefore, it is important to specify the time span over which the measurement is averaged.

Once the standards are established and the measurements made, what is to be done with them? A chemical analysis of a sample of air does not protect anyone's lungs any more than a fire alarm puts out a fire. There are two directions for action. One is to alert people so that they may protect themselves; the other is to stop the trouble at the source. Unfortunately, hazardous levels of air pollution are not always so evident to the senses as a raging fire. It is therefore very important to be able to establish criteria for warning people when air quality standards are exceeded.

Look at Table 13.3, which displays air quality standards and criteria for action, as set up by the U. S. Environmental Protection Agency (EPA). The table contains important information, and the trends it shows are more interesting than the absolute values of the numbers. Note the following:

- For any one pollutant and averaging time, say eight hours' exposure to carbon monoxide, the higher the concentration, the greater the hazard (obviously).
- For any one pollutant, the longer the averaging time, the lower the permissible concentration. Thus, for example, the standard allows exposure to 35 ppm of carbon monoxide for one hour, but if the exposure lasts for eight hours, the concentration must be reduced to 9 ppm.
- The hydrocarbon entry is interesting; note that it applies

TABLE 13.3 Federal Ambient Air Quality Standards and Episode Criteria (Adopted April 30, 1971)

AIR POLLUTANT	AVERAGING TIME	AIR QUALITY STANDARDS		CRITERIA FOR DECLARING AN AIR POLLUTION EPISODE			
		Secondary (for plants, materials, etc.)	Primary (for human health)	Alert	Warning	Emergency	Significant harm
Sulfur dioxide, ppm	1 yr		0.03				
	24 hr		0.14	0.3	0.6	0.8	1.0
	3 hr	0.5					
Particulate matter, $\mu g/m^3$	1 yr	60	75				
	24 hr	150	260	375	625	875	1000
Carbon monoxide, ppm	8 hr	9	9	15	30	40	50
	1 hr	35	35				125
Oxidants, ppm	1 hr	0.08	0.08	0.1	0.4	0.6	0.7
Nitrogen dioxide, ppm	1 yr	0.05	0.05				
	24 hr			0.15	0.3	0.4	0.5
	1 hr			0.6	1.2	1.6	2.0
Hydrocarbons, ppm	3 hr (morning)	0.24	0.24				

only to morning hours. The reason is that the standard recognizes only that hydrocarbons are converted to smog by the action of sunlight. Therefore the hydrocarbons exhausted from automobiles during the morning traffic rush can be irradiated throughout the day—for this reason they are called **smog precursors.** Hydrocarbons from evening traffic, on the other hand, will usually be dissipated during the night.

All this may be very interesting, but when you read your newspaper or listen to a broadcast for your city's daily air quality report, you are not presented with a set of numbers like those in Table 13.3. More likely you are given a one-word summary, such as "acceptable" or "unhealthy." This brief description is the result of an arbitrary assignment of numbers to certain ranges of air pollutant concentrations, followed by some arithmetical procedure for combining these numbers to get a single air pollution index. The index may be expressed in the form of a number or an adjective such as cited above. By 1975, some 40 such indices had been initiated by air pollution agencies in the United States and Canada. Unfortunately, no two were alike. In fact, inconsistencies were so great that an atmosphere that would be characterized as "unhealthy" in New York could be called "normal" in Miami and "excellent" in Cincinnati. A study of the problem, sponsored by the Council on Environmental Quality and the EPA, recommended that a single index be adopted everywhere. Table 13.4 shows such a model index. Information that can help individuals who want specific suggestions related to their own physical conditions is given in Table 13.5.

Drawing by Sidney Harris.

TABLE 13.4 **Model Air Quality Index for Cities**

RANGE OF CONCENTRATION OF AIR POLLUTANTS*	DESCRIPTIVE TERM
Up to secondary air quality standard	Good
Higher than secondary standard but not above primary standard	Satisfactory
Higher than primary standard but not above "alert" level	Unhealthful
Above "alert" level	Hazardous

*Averaging times: 24 hr for SO_2 and particulate matter; 8 hr for carbon monoxide; 1 hr for oxidants and nitrogen dioxide. Hydrocarbons are not included. Concentration ranges are based on the Federal Air Quality Standards and episode criteria.

TABLE 13.5 Possible Adverse Health Effects When Concentrations Exceed Primary Air Quality Standards

POLLUTANT	EFFECT
Carbon monoxide	Impaired exercise tolerance in persons with cardio-vascular disease Decreased physical performance in normal adults
Sulfur dioxide	Increased hospital admissions for respiratory illness Aggravation of asthma and cardiorespiratory symptoms in elderly patients with related illness
Particulate matter	Aggravation of chronic lung disease and asthma Aggravation of cardiorespiratory disease symptoms in elderly patients with heart or chronic lung disease Increased cough, chest discomfort, and restricted activity
Oxidants	Aggravation of chronic lung disease and asthma Irritation of the respiratory tract in healthy adults Decreased visual acuity; eye irritation Decreased cardiopulmonary reserve in healthy subjects

13.14 CONTROL OF AIR POLLUTION EMISSIONS

It is easy to think of air pollution control as something one does to the smoke stack, or to the tail pipe, to stop the emissions. This is, of course, a valid concept, but it is not the only one. Instead, you could put the control device on your own face; such equipment is called a gas mask. Could you put a gas mask on your house? In a way, yes. Houses are leaky, and the air filters in and out in patterns that depend on the wind and on the construction of the walls, windows, and doors. But you could blow enough air into your house through some air-cleaning device at a sufficiently high rate that a slight pressure is built up inside, and all leaks are directed outward.

One could also make cultural choices of many kinds that serve to reduce air pollution. Many of these were discussed in the sections of Chapter 9 that deal with energy sources; for example, solar energy is nonpolluting and, of course, one could simply use *less* energy. The role of public policy in establishing alternative strategies for air pollution control will be discussed in Section 13.15. Here we will deal only with technical approaches.

There are two general categories of methods for controlling air pollution at the source: (a) The pollutants are separated from the harmless gases and disposed of in some way other than by discharge into the atmosphere; or (b) the pollutants are somehow converted to innocuous products which may then be released to the atmosphere.

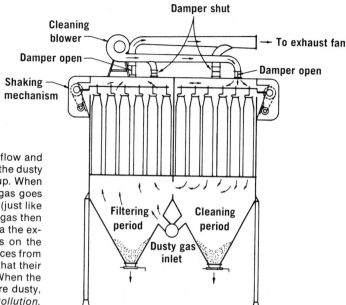

Figure 13.13 Typical bag filter employing reverse flow and mechanical shaking for cleaning. The figure shows the dusty gas being blown from the inlet toward the left and up. When the dusty gas reaches the six bags on the left, the gas goes into the bags while the dust remains on the outside (just like a vacuum cleaner running backward). The cleaned gas then comes out the tops of the bags and is discharged via the exhaust fan to the atmosphere. Meanwhile, the bags on the right, which have collected dust on their outer surfaces from the previous cycle, are being shaken and blown so that their dust falls to the bottom, where it can be removed. When the bags on the right are clean and those on the left are dusty, the air flow pattern is reversed. (From Stern: *Air Pollution.* 2nd ed. New York, Academic Press, 1968.)

CONTROL OF POLLUTANTS BY SEPARATION

Particulate matter may be retained on porous media (filters) which allow the gas to flow through. Such separations are possible because particles are much larger than gas molecules. For handling large gas streams, the filters are often in the form of cylindrical bags, somewhat like giant vacuum cleaner bags, from which the collected particulate matter is periodically shaken out (Fig. 13.13).

There are various mechanical collection devices that depend on the fact that particles are *heavier* than gas molecules. As a result, particles will settle faster and can be collected in a chamber that allows enough time for them to settle out. However, as evidenced by the data on settling rates shown in Figure 13.3, such methods are practical only for very large particles. More important than their settling rate is the fact that heavier particles have more *inertia*. As a result, if a gas stream that contains particulate pollutants is whirled around in a vortex, the particles may be spun out to locations from which they may be conveniently removed. A device of this sort, called a **cyclone**, is shown in Figure 13.14.

Particles may also be removed from a gas stream by **electrostatic precipitation**. Refer to Figure 13.15 for a view of the equipment, which operates as follows: A dusty gas is blown upward through a tube that contains a fine wire (called the discharge electrode) running through the center. A high voltage on the wire transfers electrical charge to nearby dust particles. This charge forces the dust particles to drift to the outer tube, where they stick to the wall and to each other. The accumulated dusty layer can be

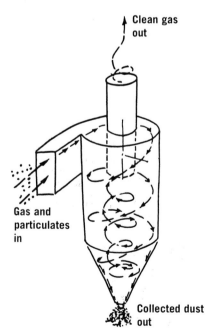

Figure 13.14 Basic cyclone collector. (From Walker: *Operating Principles of Air Pollution Control Equipment.* Bound Brook, N.J., Research-Cottrell, Inc., 1968.)

427

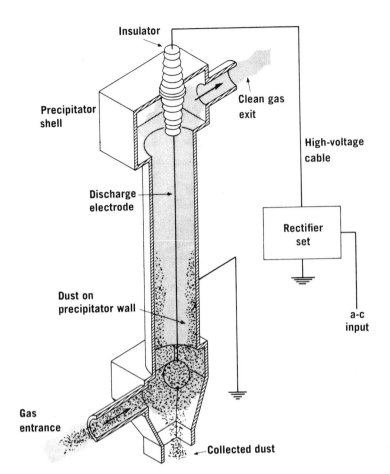

Figure 13.15 Basic elements of an electrostatic precipitator. (From White: *Industrial Electrostatic Precipitation.* Reading, Mass., Addison-Wesley Publishing Company, 1963.)

Labels in figure: Insulator; Clean gas exit; Precipitator shell; High-voltage cable; Discharge electrode; Rectifier set; Dust on precipitator wall; a-c input; Gas entrance; Collected dust

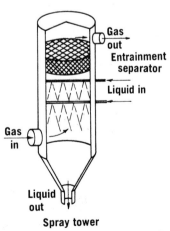

Spray tower

Figure 13.16 Schematic drawing of a spray collector, or scrubber. (From Stern: *Air Pollution,* 2nd ed. New York, Academic Press, 1968.)

Labels in figure: Gas out; Entrainment separator; Liquid in; Gas in; Liquid out

removed by rapping the wall, so that the deposit falls to the bottom where it can be collected and disposed. True, the rapping returns the dust to the air stream from which it came, but the particles are no longer in their original finely divided state. Instead, they are large agglomerates, which fall readily. Electrostatic precipitators operate at efficiencies of 99 percent or higher and can convert a smoky exhaust to a visually clear one. Most electrostatic precipitators are used in power plants that burn fossil fuels.

Pollutant gases cannot feasibly be collected by mechanical means, because their molecules are not sufficiently larger or heavier than those of air. However, some pollutant gases may be more soluble in a particular liquid (usually water) than air is; they may therefore be collected by a process that brings them into intimate contact with the liquid. Devices that effect such separation are called **scrubbers** (Fig. 13.16). The methods of making contact between gas and liquid include spraying the liquid into the gas and bubbling the gas through the liquid. Ammonia, NH_3, is an example of a gas that is soluble in water and can be scrubbed out of an air stream.

Gas molecules adhere to solid surfaces. Even an apparently clean surface, such as that of a bright piece of silver, is covered with a layer of molecules of any gas with

428

which it is in contact. The gas is said to be **adsorbed** on the solid. "Adsorbed" means "held on the surface of a substance," and is different from "absorbed," which means "held in the interior of a substance." The quantity of gas that can be adsorbed on an ordinary piece of non-porous solid matter, such as a coin, is too small to be of any consequence as a means of collecting pollutants. However, if a solid is perforated with a network of fine pores, its total surface area (which includes the inner surfaces of the pores) may be increased so much that its capacity for gas collection becomes significant. Such a solid is **activated carbon,** which can have many thousands of square meters of surface area per kilogram.

Activated carbon is made from natural carbon-containing sources, preferably hard ones, such as coconut shells, peach pits, dense woods, or coal, by charring them and then causing them to react with steam at very high temperatures. The resulting material can retain about 10 percent of its weight of adsorbed matter in many air purification applications. Furthermore, the adsorbed matter can be recovered from the carbon and, if it is valuable, recycled into the process or product from which it had escaped.

CONTROL OF POLLUTANTS BY CONVERSION

By far the most important conversion of pollutants is by oxidation in air. Oxidation is applied most often to pollutant organic gases and vapors, rarely to particulate matter. When organic substances containing only carbon, hydrogen, and oxygen are completely oxidized, the sole products are carbon dioxide and water, both innocuous. However, the process is often very expensive because considerable energy must be used to keep the entire gas stream hot enough (about 700°C) for complete oxidation to occur. If the pollutant is sufficiently concentrated, its own fuel value may contribute a large part of this energy. Also, the required combustion temperature may be reduced by using a catalyst.

There are a number of possible chemical conversions of pollutants other than combustion in air. These include the chemical neutralization of an acid or a base and the oxidation of pollutants by agents other than air, such as chlorine or ozone.

13.15 LEGISLATION, PUBLIC POLICY, AND STRATEGIES IN AIR POLLUTION CONTROL

In Western society, the significant beginnings of legislative approaches to air pollution control occurred during the early use of coal in Great Britain, starting in the fourteenth century. During the reign of Edward II (1307–1327), for example, a man was put to torture for filling the air with a "pestilential odor" through the burning of coal. Later, less brutal and (one hopes) more effective methods of regulation took the form of taxation, restriction of the movement of coal into congested areas such as the City of London, and application of the common law of nuisance. A "nuisance," in legal tradition, is anything that is obnoxious to the community, or to individuals in the community, for which some legal remedy may be found.* The law of nuisance embraces two separate concepts: First, nuisance is considered to be a minor crime, to be regulated with "flexibility" in such a way as to guard the public peace. Second, nuisance relates particularly to the use and enjoyment of land or property. Ordinarily a person charged with creating a nuisance will be ordered to pay for damages. In cases where the nuisance is both severe and continuous, an injunction may be granted. (An injunction is a legal order to stop a wrongful act.)

This brief historical background foreshadows some of our most pressing current issues of public policy in air pollution control. True, we no longer torture polluters, although it is possible under United States federal law for responsible company executives to be sentenced to prison for violation of air pollution regulations. Many communities establish regulations that are based on the nature of the localities whose air is polluted. For example, higher concentrations are allowed in areas zoned for industry than in residential areas. Such discrimination may be regarded as a form of regulation of land use. Although the legislative protection of pure air no longer depends entirely on statutes declaring air pollution to be a public nuisance, much of the tradition that pollution is only a minor crime remains. In particular, it is still customary to be "flexible" in administering air pollution regulations, particularly when an injunction would close down an otherwise lawful business where the immediate economic well-being of the community is seen to be involved.†

In the United States, the first federal legislation exclusively concerned with air pollution was enacted in July, 1955. A very modest beginning, it authorized the Public Health Service to perform research, gather data, and provide technical assistance to state and local governments. The major air pollution legislation in the United States is now the Clean Air Act of 1963, together with a series of amendments added during the 1960's and 1970's. This legislation recognized, at the start, that polluted air crosses state boundaries, and that in such instances, if individual states did not act to correct the problems, the federal government could do so. The Act further called for the publication of documents on air quality criteria and control techniques (see the bibliography to this chapter). The States

---

*From W. Selwyn, *Law Nisi Prius*, 1817, "If A build an house so as to hang over the land of B, whereby the rain falls over B's land, and injures it, B may maintain an action against A for this nuisance."

†Remember, such a view may require a degree of blindness to economic externalities (see Chapter 1).

were then to develop ambient air quality standards and plans for implementing them.

It turned out that these measures were not sufficient to promote rapid progress in air clean-up, and the 1970 amendments to the Clean Air Act called for the development of national ambient air quality standards. These are the standards discussed in Section 13.13 and displayed in Table 13.3. The law also limits the quantities of air contaminants that may be emitted by any new factory. These standards are distinct from the ambient air quality standards, since they are intended to control directly the pollution given off by a specific source, such as a power plant or foundry.

Furthermore, the legislation is not static. Debates on environmental policy occur in every session of Congress, and new amendments are added from time to time.

In view of all these legislative safeguards, how is it that you can still see and smell air pollution if you travel through industrial areas of the United States? This deficiency may be blamed, in part, on the usual problems of enforcement—administrative complexities, inadequate staffing, judicial delays. But there is more to it than that. Recall first the traditional reluctance to close down an otherwise lawful business for the "minor" crime of committing an air pollution nuisance. Currently, this reluctance takes the form of legal or administrative extensions of compliance dates for any of a variety of reasons. For example, a given company or even an entire industry may claim that it will take more time to develop, test, and install the control systems needed to reduce their air pollution emissions than the law allows. Representatives of the government may or may not agree. Lengthy negotiations usually follow, and extensions are often granted, especially if a plant is experimenting with "innovative" technology that promises to be more efficient than existing control methods. In some cases there may be a penalty for delayed compliance, such as a monthly payment, which may be refunded to the company when compliance is finally achieved. Temporary measures, such as using cleaner fuels or shutting down the entire plant during temperature inversions, may be accepted as a substitute during the extension period. During such negotiations, a common industry viewpoint is that "we are reaching the limits of the law of diminishing returns.... The time has come for carefully weighing the full cost and impact of marginal improvements in pollution abatement here and elsewhere against the economic benefits which would result from investment of such dollars in productive, job-making tools."*

The usual countervailing argument takes three forms: First, the improvements in air pollution control that are being sought are not marginal at all, but are essential if we are to reduce chronic impairments to human health, which are not so readily evident. Second, the economic cost of

*U. S. Steel Chairman Edgar B. Speer, in reference to negotiations during 1976 regarding air pollution controls at its Clairton (Pennsylvania) Works.

431

pollution control must be weighed against the actual economic difficulties that the pollution produces. Third, the unstated assumption that the control measures will be costly is itself open to question. It often turns out that the extensive modifications needed to abate pollution require a redesign of obsolete equipment. Although costly, such changes eventually result in a more efficient operation that compensates for the original investment and even yields a savings.

More basic to the objective of air quality than the problem of delay in applying control methods is the fundamental question of how standards should be set. For example, should there be different standards for different kinds of communities, or for different varieties of weather? After all, delay implies only a temporary setback, but the adoption of a standard, far from being a mere tactic, reflects an established public policy. It may come as a shock to the reader that some environmentalists* are altogether against air pollution standards. They argue that if a standard is established, industrialists will feel free to pollute up to the allowable level and that such a policy should be opposed because *any* air pollution is undesirable.

Most policymakers disagree with this position. They reason that if standards are too permissive they should be made more stringent, and that progress in changing our way of life is so slow that we had better make use of the tools available to us.

However, whether one agrees or not, it is important to understand the basis for the viewpoint against air pollution standards. Remember that the traditional aim of air pollution legislation has been the prevention of acute damage to health, especially in disastrous episodes. Such legislation led to the adoption of standards, and it is assumed that if the standards are not violated, disasters will not occur. But this concept leads to an additional assumption; namely, that it is not only legal but also reasonable and *advantageous* to pollute the air to a level that is close to the standards, so long as those barriers are not breached. In fact, this position was taken by industry from the very beginning of modern efforts to develop legislation for regulating air pollution. The following quotation is representative of that period:†

The objective of air pollution control should be to limit the amount of foreign material in the air so that there will not be "too much," but at the same time allow the atmosphere to function usefully to its fullest capacity. This principle . . . provides for "reasonable or natural use" of the air. It is logical to regard the atmosphere as one of many natural resources which are being and should be used to technical and economic advantage.

The strategy derived from this principle is known as **air resource management.** If this appears to you as a rather upside-down view of air pollution standards, per-

*See, for example, R. S. Scorer: "The Crime of Pollution Standards." *Atmospheric Environment* 5:819, 1971.

†From a paper entitled *A Rational Approach to Air Pollution Legislation,* presented by the Subcommittee on Legislation Principles of the Manufacturing Chemists' Association, at a meeting of the Air Pollution Control Association, June 12, 1952.

"I mean, you can have the cleanest air in the world but if you can't manufacture anything what the hell good is it?"

haps it is. To help you to think about it, consider the following analogy. Imagine that some citizens complain about the "visual pollution" of advertising billboards along public highways. To satisfy the complainants, a law is passed that requires a separation of at least a mile between any two billboards. A victory for the citizens? Perhaps, if the signs along some cluttered roadways are thinned out. But perhaps also there is a 100-mile stretch of unspoiled parkway somewhere that may now be defiled with 100 billboards. Thus, a legal limit may be viewed either as a means of preventing a disaster or as a license to pollute, depending on which way you look at it.

The analogy is hardly far-fetched, for there are many areas which, like the unspoiled parkway, enjoy a much greater freedom from pollution than is required by national standards. There have been attempts in Congress to legislate against any "significant deterioration" of such areas. Such attempts are in direct opposition to the principle of air resource management, which views a pure atmosphere as a container into which pollutants may be discharged, up to a limit. Industrial lobbyists view regions whose air has never been polluted as being "undeveloped" and, by implication, "poor." Such an outlook calls for an air resource management strategy that would allow the introduction of heavy industry. To fail to do so, it is argued, "would mandate undeveloped areas into eternal poverty."[*]

The relationships among poverty, economic development, and air quality, however, are far more complex than such a simple analysis suggests. First, it is not always easy to identify who is poor. A family in a nonindustrialized

---

[*]The quoted portion is attributed to Gary Knight, environmental lobbyist for the U. S. Chamber of Commerce, as cited in *Science* **192**:534, 1976.

area living in a modest dwelling that is heated by wood and whose diet includes vegetables from their own garden and meat from an occasional hunt or fish from a stream may not regard themselves as poor even though their cash income may be low. Furthermore, the new factories that would utilize their clean air as a "resource" might import management and engineering personnel and offer the lower-paying jobs to local residents. True rural poverty, where amenities for a comfortable life are not available, often exists in regions that are not attractive to industry either. A final irony is the finding* that in developed urban areas it is usually the poor who are clustered in localities where the concentrations of air pollutants are highest.

Thus the various controversies that surround the concepts of air pollution standards, air resource management, and the preservation of pristine atmospheres become inseparable from questions of public policy regarding land use, for the regulation of what may be done to the air determines, in large measure, what may be done on the land.

*Julian McCaull: "Discriminatory Air Pollution." *Environment* 18:26, 1976.

13.16 CASE HISTORY: *The Donora Smog Episode*

Donora is an industrial town of about 14,000 people, located on the inside of a sharp horseshoe bend of the Monongahela River, about 48 km (30 miles) south of Pittsburgh, Pennsylvania. The area along the river bank (Fig. 13.17) is occupied by a steel-and-wire plant and by a zinc-and-sulfuric-acid plant. Taken together, the plants extend for some 5 km (3 miles) along the river bank. Across the river (east, which is to the right in the figure), the hills rise sharply to a height of some 105 meters (350 ft) above the river bank; on the Donora side the elevation is a bit higher, though the rise is more gradual. These hills tower far above the factory chimneys, and as a result, any atmospheric inversion threatens to convert Donora into a basin that collects pollutants from the stacks along the river bank. Anyone familiar with the hill country of Pennsylvania and West Virginia knows that foggy days are common there, especially in the fall. The blackened houses and treeless hills of Donora show vividly that the atmospheric stagnations which shelter these fogs do in fact trap the pollutants that come from the chimneys.

One such polluted fog settled over the area during the last week of October, 1948. It started Tuesday morning, October 26, and seemed somewhat heavy and motionless, but the townspeople had seen many such fogs before. By the next day, however, there was such a dead calm

The actual operations in a chemical manufacturing plant are much more complex than are revealed by the equations in textbooks of introductory chemistry. In the Donora zinc plant, for example, the objective of the operation is to convert the raw material, zinc sulfide, ZnS, into two valuable products: zinc, Zn, and sulfuric acid, H_2SO_4. In practice, pollutants escape during various stages of the manufacturing process. For example, not all of the sulfur becomes sulfuric acid; some is emitted from the stacks in the form of sulfur dioxide, SO_2, or sulfur trioxide, SO_3, or as a fine zinc sulfide dust. Furthermore, the raw material is by no means a pure zinc compound. Zinc is in the same chemical family as cadmium, and the two metals occur together in ores. As a result, the plant produces cadmium as one of its products, and also releases compounds of cadmium, together with those of zinc and other metals, among its particulate pollutants. All in all, the pollution chemistry is often more complex than the manufacturing chemistry and this complexity makes it difficult to establish simple relationships between what a plant produces for sale and what pollutants it releases to the environment.

that it was considered unusual, even for Donora. Streams of sooty gas from locomotives did not even rise, but hung motionless in the air. The visibility was so poor that even natives of the area became lost. Such dense fog is much worse than mere darkness, which can be pierced by a flashlight beam. During such episodes, a driver cannot see the side of the road, nor even the

434

Figure 13.17 Region of Donora, Pennsylvania.

white line that marks the center. If he leaves his car to explore the immediate area, he may become disoriented and lost and be unable to find his way back unless he can hear the sound of the idling engine.

The smog continued through Thursday, when it seemed to become thicker with what may be called chemical "sensations," for such perceptions go beyond the sense of smell: they involve various receptors in the nose and mouth and give rise to sensations of tickling, scratchiness, irritation, and even taste. Sulfur dioxide, particularly, is a gas that imparts a bittersweet taste at the back of the tongue. The fog had piled up in height as well, and the mills, except for the tops of their stacks, had become invisible.

Some illnesses had begun, rather gradually, on Tuesday, and their number had increased on Wednesday and Thursday, but they were yet too scattered to alarm the community. On Friday, however, the rate of illness soared, so that altogether about 40 percent of the exposed population developed some symptoms of illness before the episode was over. The symptoms included smarting of eyes, nasal discharge, constriction in the throat, nausea and vomiting, tightness in the chest, cough, and shortness of breath.

The first death came to a retired steelworker named Ivan Ceh at 1:30 Saturday morning; the second occurred an hour later. By 10 A.M. nine bodies lay at one undertaker's, one at a second undertaker's, and another, the eleventh death so far, at a third. Knowledge of the extent of the disaster gradually spread through the town. It was a long time before the citizens of Donora, accustomed as they were to smog, began to realize that a real calamity was at hand. Help then began to come in from neighboring towns and hospitals (there was none in Donora) and an emergency-aid station was in operation by Saturday night. The death toll that day reached 17. Two more people died on Sunday, and at six o'clock on Sunday morning the factories finally began to shut down. By afternoon it started to drizzle. It really rained Sunday night and Monday, and then it was all over, except for the lingering illnesses and one more death, which occurred a week later.

There are two sequels to the Donora episode. One involved Donora itself and its people. The other involves all of us, everywhere.

435

Donora, as the fog thickens.

436

Donora Calendar, 1948

SUN	MON	TUES	WED	THURS	FRI	SAT
Oct. 24	Oct. 25	Oct. 26 Fog settles in	Oct. 27 Fog thickens	Oct. 28 Strong chemical irritations in fog	Oct. 29 Fog very thick 4700 illnesses	Oct. 30 Fog very thick 400 illnesses 17 deaths
			810 illnesses			
Oct. 31 Rain starts in the afternoon 2 deaths	Nov. 1 Rains all day. Smog clears	Nov. 2	Nov. 3	Nov. 4	Nov. 5	Nov. 6
Nov. 7	Nov. 8 1 more death					

To Donora, after the killing smog lifted, came teams of investigators from the U.S. Public Health Service and the Pennsylvania Department of Health. They studied the clinical evidence, the sources of the pollutants, and the local weather patterns.* They found that of the nearly 6000 people who became ill during the air pollution episode, most did so because of severe irritation of the respiratory tract; those who became sickest tended to be older than 60 and to have pre-existing disease of the heart and/or lungs which rendered them more sensitive to the irritant effects of the pollutants. The study of pollutant sources identified the emissions from the various manufacturing operations and made specific recommendations for their reduction. The meteorological study confirmed what everyone knew, namely that "a definite relationship . . . existed between the concentration of contaminants and atmospheric stability."

In spite of these studies, however, the *specific* cause of the disaster was not identified. It is important to understand what this statement means and what its implications are. It tells us that the study did not reveal enough for a chemist to be able to duplicate the killer smog in a laboratory flask by mixing together a col-lection of known ingredients. In kitchen parlance, one may say that the study did not furnish the recipe for the smog. Why would one want to have such a recipe? The answer is: to know what must be *avoided* in the atmosphere.

This brings us to the second, broader sequel to Donora, because it presents a general question. If it is not known exactly what it is in air pollution that harms people, how does one know what standards to set? The clinical study of the Donora episode led to the following conclusion, as stated in the government report:

It does not appear probable from the evidence obtained in the investigation that any one of these substances (irritant or nonirritant) *by itself* was capable of producing the syndrome observed. However, a combination of two or more of these substances may have contributed to that syndrome.

There have been deaths caused by air pollution before and since, but in the United States, Donora was a landmark in public attitudes toward air pollution. The incident shocked people into the realization that air pollution can be deadly, that simplified experiments with laboratory animals exposed to individual pollutants do not predict how humans will be affected in polluted outdoor atmospheres, that factories should shut down during air pollution episodes before deaths occur, and that the establishment of air quality standards must take into account, to the extent that good judgment can serve as a guide, the possibilities of unknown and subtle effects.

---

*Reported in *Air Pollution in Donora, Pa. Epidemiology of the Unusual Smog Episode of October 1948.* Public Health Bulletin No. 306, U.S. Public Health Service, Washington D.C., 1949. The description refers to 1948, not to the present. The mills have since been shut down.

TAKE-HOME EXPERIMENTS

1. **Dustfall measurement of particulate air pollution.** Get about three half-gallon wide-mouthed jars, such as restaurant-size mayonnaise jars, and wash them out. If you live in a hot, dry climate, fill the jars about one-fourth full of distilled water. In winter or rainy season, fill them only to a height of about $\frac{1}{2}$ to 1 inch. If you expect freezing weather, use a 50-50 mixture of water and rubbing alcohol. (Distilled water may be available from your school, automobile service station, or drugstore. "Deionized" water will also do. You may use rain water if you filter it first through a coffee filter or a paper towel.) Set the jars in *open* areas where you wish to measure dustfall, not under a tree or any part of a building. The jars should be elevated, preferably at least six feet, to avoid contamination from coarse windblown material, such as soil, that does not reflect general air pollution levels.

   Leave the jars exposed for 30 days. Visit them at least once a week to replace any evaporated water by refilling the jars to the $\frac{1}{4}$-level mark. (If the water dries out completely, the test is invalid, because the wind may blow fine dusts out of the jar.) In rainy weather, check that the jar does not overflow. If the level approaches the top, stop the test.

   Your next job is to evaporate all the water and collect and weigh the residue. There are several possible procedures. One is to heat the jar directly on a hot plate or electric stove, turned on to the *lowest* setting, and using a wire gauze or other convenient spacer between the jar and the hot surface. Even so, the jar may break. A better procedure is to pour the liquid, in stages, into a smaller open dish for evaporation. In the laboratory, you would use a previously weighed evaporating dish. In the kitchen, a clean frying pan can be used. Again, evaporate *slowly*. Make sure to remove *all* the solid matter from the jar, using a small rubber kitchen scraper. Do not overheat the solid residue at the end. Finally, weigh the residue, expressing the result in milligrams (mg). If you do not have a balance, you can make a rough estimate from the fact that a drop of water from an ordinary medicine dropper weighs about 50 mg. Use your ingenuity to set up a little homemade balance on which you can counterweight your dust with drops of water. If this sounds too complicated, even an "eyeball" estimate will be informative. Now, to calculate the dustfall pollution:

   (a) Measure the diameter of each jar opening in centimeters.
   (b) Calculate the area of each opening:

   $$\text{area} = \frac{\pi \times (\text{diameter})^2}{4}$$

   (c) Calculate the dustfall:

   $$\text{Dustfall} = \frac{\text{wt of dust (mg)}}{\text{area (cm}^2) \times \text{time (days)}} \times 30 \, \frac{\text{days}}{\text{month}}$$

   This result expresses dustfall in mg/cm^2 per month. To convert to tons of dust per square mile per month, multiply your answer by 28.

   Derivation:

   $$\frac{1 \text{ mg/cm}^2 \times 2.2 \times 10^{-6} \text{ lb/mg} \times 5 \times 10^{-4} \text{ ton/lb}}{0.0011 \text{ ft}^2/\text{cm}^2 \times 3.6 \times 10^{-8} \text{ mi}^2/\text{ft}^2}$$
   $$= 28 \text{ tons/mi}^2$$

   Compare your results with reports from your local health or environmental agencies.

2. **Observation of smoke shade.** When fuel burns inefficiently, some of the unburned carbon particles are visible as a black or gray smoke. The density of such smoke coming from the stack of a factory or power plant can be estimated visually by comparing it with a series of printed grids. These grids, first suggested by Maximilian Ringelmann in 1898, are formed from squares of black lines on a white background, as follows:

Ringelmann No.	Black (%)	White (%)
1	20	80
2	40	60
3	60	40
4	80	20

Select a suitable location for observing a smoke plume from a stack. Hold the chart

No. 1   No. 2   No. 3   No. 4

in front of you and view the smoke while comparing it to the chart. The light shining on the chart should be the same as that shining on the smoke. For best results, the sun should be behind you.

Match the smoke with the corresponding Ringelmann smoke grid (1, 2, 3 or 4). Record your results and the time of your observation.

Calculate "observed smoke density" as follows:

Observed smoke density (percent) for a single observation

$$= \text{Ringelmann number} \times 20$$

Observed smoke density (average percent) for a number of observations

$$= \frac{\text{sum of all Ringelmann numbers} \times 20}{\text{number of observations}}$$

Check the air pollution regulations in

your community and compare them with your findings.

3. **Dispersal of gases in air.** Ask someone to pour a little fragrant liquid, such as cologne water or shaving lotion, into a saucer or other shallow open dish. The dish should be covered with plastic or aluminum foil and set in one corner of a quiet room. Now sit down in some other part of the room where you can keep busy with a quiet activity, such as reading a book. Ask your friend to remove the foil from the dish, while you note the time. Now read your book until you become aware of the smell, and note the time again. Repeat the experiment under various conditions, in different locations, perhaps with a fan blowing, and with different sources of odor. Note your findings with regard to the various factors that influence the dispersal of gases in air.

PROBLEMS

1. **Sensations of air pollution.** (a) Which two human senses are most often involved in the direct perception of air pollution? (b) Are other senses ever involved? If so, suggest instances in which they are. (c) Direct perceptions are not always reliable in detecting air pollution. Suggest an instance in which an air pollutant may escape human detection, and an instance in which something that is perceived as an air pollutant is really harmless.

2. **Gases.** The discussion at the beginning of Section 13.2, together with the sketches on page 396; makes reference to 11 gases. List them. How many of them would be considered to be pollutants if they were found in the atmosphere?

3. **Definitions.** Define aerosol; dust; mist.

4. **Gases and particulate matter.** Identify each of the following substances as either a gas, smoke, dust, or mist. (a) A gray material stays suspended in the air without settling. A flashlight beam that shines through it is clearly visible, even when viewed from the side. (b) A brown transparent material is in a closed container. When the container is opened, the brown color becomes lighter, first near the top, then throughout the container. Finally, the material disappears entirely from the container. (c) Black particles slowly settle from the air to the ground. The settled material feels gritty. (d) Transparent particles slowly settle from the air to the ground. The settled material feels wet.

5. **Natural air.** Table 13.1 refers to 14 gases. List them in descending order of their concentrations in natural dry air. (Careful—you must separate some gases that are grouped together in the table.) What is the sum of all their concentrations, expressed in ppm? Divide this number by 10,000 to get the sum expressed in percent. Is the total 100 percent? If not, how can you account for the discrepancy?

6. **Concentrations.** A concentration of 1 ppb of ethylene (C_2H_4) gas in the air cannot be smelled and has no demonstrable effect on people, animals, or materials. However, it does produce dried sepal injury in growing orchids—to such a degree that they become unfit for sale. Furthermore, ethylene at a concentration of 0.1 percent is used by food wholesalers to ripen bananas. Care must be exercised in this process, for if the concentration rises to 2.7 percent, the ethylene-air mixture become explosive.

   Some ethylene is produced by fruit as part of the natural ripening process during its growth, and some ethylene is also discharged from the exhausts of automobiles and trucks, especially from those with diesel engines.

   (a) Taking all these facts into consideration, would you classify ethylene as an air pollutant? Would your answer depend on its concentration? On its source? On your occupation?

   (b) Would you invest your savings in a greenhouse for growing orchids that was located near a banana-ripening building? Near a busy highway?

7. **Concentrations.** The concentrations of particulate matter are never expressed in "percent by volume." Explain why they are not. What is a better expression?

8. **Sources of air pollutants.** State which of the following processes are likely to be sources of gaseous air pollutants, particulate air pollutants, both, or neither: (a) Gravel is screened to separate sand, small stones, and large stones into different piles. (b) A factory stores drums of liquid chemicals outdoors. Some of the drums are not tightly closed, and others have rusted and are leaking. The exposed liquids evaporate. (c) A waterfall drives a turbine, which makes electricity. (d) Coal is heated in a large oven at 1000°C to drive off volatile matter, which is piped away to be refined into various chemicals. After 18 hours, the oven door is opened and a great ram ejects the glowing, sizzling residue, which is **coke,** onto a railcar, where it is quenched with water.

9. **Air pollutants.** List five gaseous and three particulate air pollutants, and identify their possible sources.

10. **Definitions.** What is a hydrocarbon? a chlorofluoromethane? an oxidant?

11. **Carbon monoxide.** Carbon-14 is a radioactive form of carbon, with a half-life of 5700 years, that is continuously produced in the Earth's atmosphere. As a result of this production, atmospheric CO_2 has a slight but constant level of radioactivity. The carbon-14 content of coal and petroleum, on the other hand, is negligible.

    (a) Using these facts, outline a method by which you could determine what proportion of the Earth's CO is natural and what is man-made. State the assumptions implicit in your method.

    (b) If it were established that the carbon-14 content of atmospheric CO is steadily decreasing, what would this finding imply as to the source of the CO?

12. **Particles.** Is the speed of settling of particles in air directly proportional to their diameters? (If the diameter is multiplied by 10, is the settling speed 10 times faster?) Justify your answer with data from Figure 13.3. Is a settling chamber a good general method of air pollution control? Explain.

13. **Inversion.** Under some conditions, two inversion layers may exist at the same time in the same vertical atmospheric structure. Draw a diagram of temperature vs. height that shows one inversion layer between the ground and 200 meters, and another aloft, between 1000 and 1200 meters, while the temperature variations between them approximate equilibrium conditions.

14. **Chimneys.** "Since tall chimneys do not collect or destroy anything, all they do is protect the nearby areas at the expense of more distant places, which will eventually get all the pollutants anyway." Argue for or against this statement.

15. **Meteorology.** Describe the meteorological conditions most conducive to the rapid dispersal of pollutants. Describe those that are least conducive.

16. **Ozone layer.** (a) Which two chemical elements are common to both DDT and the Freons? (b) Which one of these elements constitutes a potential danger to the ozone layer if it is released in the upper atmosphere? By what means can the ozone layer be weakened? (c) Which of the substances cited in part (a) is most likely to be transported to the upper atmosphere? Why?

17. **Climate and air pollution.** Imagine that a given factory, with a fixed rate of emission of gaseous air pollutants, could be located either in the tropics or in the Arctic. Assuming that the population densities in the two areas were equal, which location would you choose? Defend your answer.

18. **Climate.** Refer to Figure 13.8. (a) What percent of the incident solar energy is received by the Earth? (b) Does the Earth's surface receive any additional energy? If so, from what source? (c) Is the amount of energy emitted by the Earth greater, less, or the same as that which it receives from incident solar radiation? Explain.

19. **Climate.** Imagine that you must determine whether some particular climatic effect, such as increased fog or rainfall in a given area, is caused by human activity. Which of the following experimental method(s) would you rely on? Defend your choices. (a) Compare current data with that of previous years, when population and industrial activity were less. (b) Compare the effects during weekdays, when in-

dustrial activity is higher, with those on weekends, when it is low. (c) Compare effects during different seasons of the year. (d) Compare effects just before and after the switch to or from Daylight Saving Time, to see whether there is a sharp one-hour shift in the data. (e) Compare effects in different areas where populations and industrial activities differ.

20. **Air pollution effects.** Of the six categories of air pollution effects cited in Section 13.6, how many do you think could be produced by sulfur dioxide, SO_2, or by air pollutants derived from SO_2? Justify your answer for each category.

21. **Health.** The United States government has banned the domestic use of DDT. Since cigarettes have been implicated in a rather frightening extent of chronic disease causation, discuss the proposition that they too should be banned.

22. **Cigarettes and air pollution.** Which of the following groups of subjects would you study to learn the separate effects of cigarettes and air pollution on human health? Explain. (a) Urban smokers vs. urban nonsmokers; (b) urban smokers vs. rural smokers; (c) urban smokers vs. rural nonsmokers; (d) urban nonsmokers vs. rural smokers; (e) urban nonsmokers vs. rural nonsmokers; (f) rural smokers vs. rural nonsmokers.

23. **Control methods.** Suppose that you keep some animals in a cage in your room and you are disturbed by their odor. Comment on each of the following possible remedies, or some combination of them, for controlling the odor: (a) Spray a disinfectant into the air to kill germs. (b) Install a device that recirculates the room air through a bed of activated carbon. (c) Clean the cage every day. (d) Install an exhaust fan in the window to blow the bad air out. (e) Install a window air conditioning unit that recirculates and cools the room air. (f) Install an ozone-producing device. (g) Spray a pleasant scent into the room to make it smell better. (h) Light a gas burner in the room to incinerate the odors. (i) Keep an open tub of water in the room so that the odors will dissolve in the water.

24. **Control methods.** Distinguish between separation methods and conversion methods for source control of air pollution. What is the general principle of each type of method?

25. **Control equipment.** Explain the air pollution control action of a cyclone; a settling chamber; a scrubber; activated carbon; an electrostatic precipitator; an incinerator.

26. **Air quality standards.** A report of air pollutant concentrations shows 24-hour average values of 1.0 ppm for sulfur dioxide, 1000 $\mu g/m^3$ of particulate matter, and 0.5 ppm of nitrogen dioxide. However, this analysis was carried out at the top of a 200-meter stack of a power plant, the only factory in town that is a potential source of air pollution. If you were the health officer or mayor, what action, if any, would you recommend? Would you require any additional information? If so, describe the data you would request.

27. **Smog.** What is a smog precursor? Which pollutant shown in Table 13.3 falls into this category?

28. **Air pollution index.** If your morning news broadcast announced that the air pollution index for the day was "unhealthful," what effect would that have on your day's activities? What other information would you want to help you decide what to do?

29. **Air quality standards.** Using the data of Table 13.3 and the model air quality index of Table 13.4, describe each of the following atmospheric conditions as good, satisfactory, unhealthful, or hazardous: (a) sulfur dioxide, 0.25 ppm for 24 hours; (b) carbon monoxide, none detected all day long, except between 10 and 11 AM, when a sudden wind shift blew in air from a large oil fire and brought the average concentration during that hour to 125 ppm; (c) particulate matter 100 $\mu g/m^3$, together with oxidants, 0.01 ppm, and sulfur dioxide, 0.04 ppm, each averaged over 24 hours; (d) particulate matter, 200 $\mu g/m^3$ over 24 hours.

30. **Public policy.** Explain the legal concept of a nuisance. What implications does this concept have with regard to air pollution regulations?

31. **Legislation.** Describe some of the important features of the Federal Clean Air Act of 1963 and its amendments. (If you or the entire class wish to study this in more detail, get a copy of the Act from your U.S. representative or senator.)

32. **Public policy.** Summarize arguments that have been advanced against the adoption of ambient air quality standards. How have these arguments been answered?

33. **Public policy.** Explain the strategy of air resource management and summarize the arguments in favor and in opposition to it.

34. **Public policy.** How is air pollution policy related to strategy of land use?

BIBLIOGRAPHY

The basic text on air pollution is a five-volume work:
Arthur C. Stern: *Air Pollution.* 3rd Ed. New York, Academic Press, 1976–1977. Volume I, 715 pp.; Volume II, 656 pp.; Volume III, 799 pp.; Volume IV, 946 pp.; Volume V, 672 pp.

Three good one-volume texts are:
Arthur C. Stern, Henry C. Wohlers, Richard W. Boubel and William P. Lowry: *Fundamentals of Air Pollution.* New York, Academic Press, 1973, 492 pp.
Samuel J. Williamson: *Fundamentals of Air Pollution.* Reading, Mass., Addison-Wesley Publishing Co., 1973. 473 pp.
Samuel S. Butcher and Robert J. Charlson: *An Introduction to Air Chemistry.* New York, Academic Press, 1972. 241 pp.

Among the air pollution technical publications of the U.S. Environmental Protection Agency are two important series of documents on specific air pollutants. One series deals with "air quality criteria," which are established from our knowledge of the effects of air pollutants. The other series outlines "control techniques." The titles of the first series are:
Air Quality Criteria for Particulate Matter
Air Quality Criteria for Sulfur Oxides
Air Quality Criteria for Carbon Monoxide
Air Quality Criteria for Photochemical Oxidants
Air Quality Criteria for Hydrocarbons
Air Quality Criteria for Nitrogen Oxides

Two of the control documents are:
Control Techniques for Particulate Air Pollutants
Control Techniques for Sulfur Oxide Air Pollutants

For a very brief introductory text, refer to:
National Tuberculosis and Respiratory Disease Association: *Air Pollution Primer.* New York, 1969. 104 pp.

Various popular books that warn of the dangers of air pollution appeared in the 1960's. Some representative ones are:
L. J. Battan: *The Unclean Sky.* New York, Doubleday and Co., 1966. 141 pp.
D. E. Carr: *The Breath of Life.* New York, W. W. Norton & Co., 1965. 175 pp.
Howard R. Lewis: *With Every Breath You Take.* New York, Crown Publishers, 1965. 322 pp.

A moving human account of the Donora smog episode is given by:
Berton Roueché: *Eleven Blue Men.* Boston, Little, Brown, 1953; chapter entitled "The Fog."

14

WATER POLLUTION

14.1 THE NATURE OF WATER POLLUTION

Water is more widely distributed over the surface of the Earth and in the tissues of its living organisms than any other substance. From the Earth's rivers, oceans, clouds, and icecaps, to our own body fluids, water is the prime medium in which physical and chemical transformations, particularly those of biological significance, take place.

The water droplets or ice crystals in high clouds are quite pure, and so is the snow that is driven down from these regions. But at the Earth's surface, liquid water comes in contact with many other chemicals and mixes more or less intimately with them. What, then, is "pure" water, and in what form does it become "polluted"? The chemist thinks of a "pure" substance as one whose molecules are all alike. But although water gushing out of a rock fissure on a high mountain contains some mineral matter, it is hardly polluted. The pollution of water, then, is the addition of undesirable foreign matter that deteriorates the *quality* of the water. Water quality may be defined as its fitness for the beneficial uses which it has provided in the past—for drinking by people and animals, for the support of a wholesome marine life, for irrigation of the land, and for recreation.

It will be helpful to consider the differences in polluting a gas, a liquid, or a solid. You have already learned from the previous chapter that the molecules of a gas such as air move about rather independently of one another. Therefore, the molecules of a gaseous pollutant can simply enter some of the unoccupied space. Solids, on the other hand, behave very differently. Suppose you were to try to pollute an iron bar. Immerse it in water or oil, expose it to bacteria or viruses, and its impurities remain on the surface while its internal composition is unchanged. The reason for this resistance is that the atoms of iron are strongly and closely bonded to each other, and are very dif-

Crater Lake in Oregon was originally formed when a volcanic mountain was destroyed by a violent explosion. (Photo by John Paul, Ph.D., University of California, Berkeley.)

ficult to displace. This strong tendency for the atoms to retain their respective positions and not to be dislocated is what makes iron a solid. Nor is it easy to pollute ice, except by the freezing of polluted water.

Liquids are intermediate between gases and solids in their readiness to accept contamination. The attractive forces between molecules in the liquid state are strong enough that a sample of liquid (for example, a raindrop) holds itself together. However, the attractive forces are not so strong as they are in solids; they are not strong enough to prevent the molecules from sliding past one another. Such molecular relocations manifest themselves in the familiar phenomenon of liquid flow. Now, when a molecule of a liquid relocates, it leaves behind a vacant site, or a "hole." This vacancy can be occupied by another molecule of the same type, or by a molecule of a foreign substance.

What determines whether a given liquid will accept or reject a given type of foreign matter? We drop a lump of sugar into a glass of water and note that it dissolves. But a piece of lead does not. Since sugar is a solid, the sugar molecules must be attracted to each other—otherwise they would fly apart and sugar would be a gas. But the sugar molecules are also strongly attracted by water molecules, and can be pulled away from other sugar molecules to occupy sites surrounded by molecules of water. Thus, sugar dissolves in water. Atoms of lead are attracted to each other much more than they are to water molecules; thus, lead is insoluble. Therefore, the ease with which a liquid can become contaminated by dissolved foreign matter depends on the chemical relationships between the molecules of the liquid and foreign molecules.

A contaminant may be harbored by a liquid without being dissolved in it. If we grind our piece of lead to a fine powder and stir it into the water, the suspended lead is a pollutant. However, the ease with which foreign matter can be suspended in a liquid also depends to some extent on the mutual attraction between the foreign particles and the liquid molecules.

Water is not a typical liquid. Corn oil is more like cottonseed oil, kerosene is more like gasoline, and grain alcohol is more like wood alcohol than *anything* is like water. One of the consequences of the unique physical and chemical properties of water is that it invites or accepts pollution readily, sometimes through mechanisms that are quite unexpected. Of course, water is the universal liquid medium for living matter; it is therefore uniquely prone to pollution by living organisms, including those that carry disease to humans. Contamination pathways that involve suspension, solution, and biochemical change are not necessarily separate and distinct from each other, and many of these complex processes can occur *only* in water. Therefore, to understand water pollution, it is essential first to consider the nature of water itself.

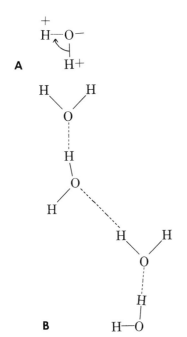

A

B

Figure 14.1 *A,* Structural formula for water, showing partial electrical charges. *B,* Water molecules bonded to each other (dashed lines).

A drop of water hanging from a leaf. The forces that bond water molecules to each other pull the drop into a spherical shape.

14.2 WATER

Water can be decomposed to hydrogen and oxygen; hence it consists of these two elements. The water molecule, represented by the formula H_2O, consists of one oxygen atom bonded to two hydrogen atoms. The molecule has a bent shape with a 105 degree angle between the bonds.

The negative charges (electrons) are crowded somewhat closer to the oxygen atom than the positive charges (protons) are. The effect of this is a separation of charges with the negatively charged part of the molecule nearer the oxygen atom and positively charged parts nearer the hydrogens. These electrical charges attract their opposites in other water molecules, with the result that liquid water consists of aggregates of H_2O molecules bonded to each other as indicated in Figure 14.1. This strong aggregation accounts for the fact that water remains a liquid up to 100°C at normal pressure, in sharp contrast to the behavior of other substances of similar or even high molecular weight.

The electrical forces that bind water molecules to each other can also serve to bind water molecules to those of foreign substances. Therefore, water is an unusually good solvent, especially for substances which have separated centers of positive and negative electrical charge. Such substances are, typically, inorganic compounds, such as the compounds of the metallic elements. Water is a poor solvent for substances whose molecules do not have separated centers of positive and negative charge; examples are the hydrocarbon substances derived from petroleum, such as gasoline, oil, and grease.

It is energetically costly to heat water, or to boil it, or to melt ice, because work must be done to break the bonds that hold the molecules together. The energies required are greater for water than for practically any other liquid. This is equivalent to saying that water stores heat energy more effectively than other liquids do. Thus, water is a good heat sink, or, looking at it the other way, water is a good cooling agent. These properties of water are important in economic aspects of water purification, because any operations that involve heating, and especially boiling, of water require large amounts of energy.

14.3 THE HYDROLOGICAL CYCLE (WATER CYCLE)

Most of the water that moves through the biosphere does so in response to physical forces — the movement of air and ocean currents, the flow of rivers, the fall of rain, the creep of glaciers, evaporation from surfaces, and transpiration through the porous outer barriers of plants and animals. Human technology has contributed other drives, such as the thrust of pumps and the flush of toilets. Some water is cycled by chemical changes, notably by photosynthesis, which rearranges the atoms of water molecules so that they become incorporated into the structure of plant matter, and by oxidation, which produces water again and releases it back to the hydrosphere.

Now, look at Figure 14.2. The total amount of water on Earth is about 1.35 billion cubic kilometers. Only about one percent of this vast amount, however, is in the form of inland waters such as lakes, streams, and underground storages. Figure 14.2 shows these percentages, as well as the quantities of water that are cycled through different portions of the Earth. Note that water leaves the atmosphere as rain and snow, and returns by evaporation. It also moves by flow of liquid water and ice on the Earth's surface and through the ground. Vast amounts of water pass through the atmosphere, although the quantity of water held there at any one time is only a very small fraction of the total amount on Earth. This fact implies that if human activities might ever trigger serious disruptions of the hydrological cycle, the most likely means would be by manipulation of the water content of the atmosphere.

As rain or snow falls to earth, it intercepts various at-

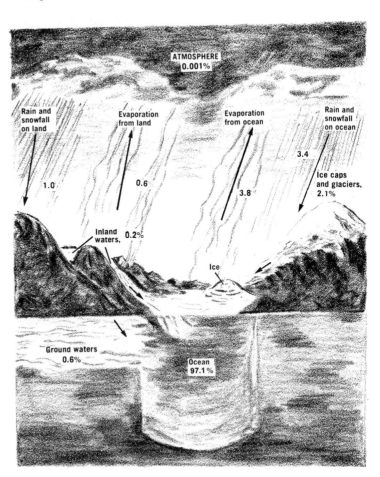

Figure 14.2 Hydrological cycle. Numbers near arrows are in geograms (10^{20} grams) of water transferred per year. One geogram of water is about 100,000 km³. The percentages refer to the water stored in different portions of the Earth.

447

mospheric contaminants. By the time it reaches the ground, therefore, it contains foreign matter from natural sources such as dust particles, carbon dioxide, microorganisms, and pollen. It may also contain traces of industrial pollutants such as sulfur dioxide, sulfuric acid, or pesticides. As the water runs over or percolates through the land, its load of impurities continuously increases. These substances are fed into the ocean, where they accumulate. The ocean is thus the Earth's ultimate sink for water-borne impurities. The concentration of dissolved matter in the ocean rises very slowly, however, because its volume is so large.

14.4 TYPES OF IMPURITIES IN WATER

It is useful to classify foreign substances in water according to the size of their particles, because this size often determines the effectiveness of various methods of purification. Figure 14.3 shows a spectrum of particles arbitrarily divided into three classes: suspended, colloidal, and dissolved. Let us consider each in turn as we refer to the figure.

Suspended particles, which have diameters above about one micrometer, are the largest. They are large enough to settle out at reasonable rates and to be retained by many common filters. They are also large enough to absorb light and thus make the water that they contaminate look cloudy or murky.

Colloidal particles are so small that their settling rate is insignificant, and they pass through the holes of most filter media; therefore, they cannot be removed from water by settling or ordinary filtration. Water that contains

Figure 14.3 Small particles in water.

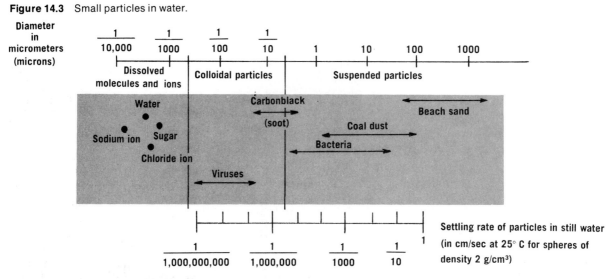

(25,400 microns = 1 inch)

448

colloidal particles appears cloudy when observed at right angles to a light beam. (The same phenomenon occurs in air—colloidal dust particles can best be seen when observed at right angles to a sharply focused light beam in an otherwise dark room.) The colors of natural waters, such as the blues, greens, and reds of lakes or seas, are caused largely by colloidal particles.

Dissolved matter does not settle out, is not retained on filters, and does not make water cloudy, even when viewed at right angles to a beam of light. The particles of which such matter consists are no larger than about 1/1000 micrometer in diameter. If they are electrically neutral, they are called molecules. If they bear an electric charge, they are called ions. Cane sugar (sucrose), grain alcohol (ethanol), and "permanent" antifreeze (ethylene glycol) are substances that dissolve in water as electrically neutral molecules. Table salt (sodium chloride), on the other hand, dissolves as positive sodium and negative chloride ions.

Natural waters contain substances in all three of the categories outlined above, as shown in Table 14.1. They range in quality from tastily potable to poisonous; in saltiness, from fresh rainwater (non-salty) to brackish (partly salty, as where river water starts to mix with sea water), to

TABLE 14.1 Impurities in Natural Waters

SOURCE	PARTICLE SIZE CLASSIFICATION				
	Suspended	Colloidal	Dissolved		

Atmosphere	←Dusts→		Molecules
			Carbon
			    dioxide, CO_2
			Sulfur
			    dioxide, SO_2
			Oxygen, O_2
			Nitrogen, N_2
				Positive ions
				Hydrogen, H^+
					Negative ions
					Bicarbonate,
					    HCO_3^-
					Sulfate, SO_4^{2-}

Mineral soil and rock	←Sand→
	←Clays→
	←Mineral soil particles→
			Carbon
			    dioxide, CO_2
				Sodium, Na^+
				Potassium, K^+
				Calcium, Ca^{2+}
				Magnesium,
				    Mg^{2+}
				Iron, Fe^{2+}
				Manganese,
				    Mn^{2+}
					Chloride, Cl^-
					Fluoride, F^-
					Sulfate, SO_4^{2-}
					Carbonate, CO_3^{2-}
					Bicarbonate,
					    HCO_3^-
					Nitrate, NO_3^-
					Various
					    phosphates

Living organisms and their decomposition products	Algae
	Diatoms
	Bacteria
	←Organic soil (topsoil)→
	Fish and
	    other
	    organisms
		Viruses
		Organic
		    coloring
		    matter
			Carbon
			    dioxide, CO_2
			Oxygen, O_2
			Nitrogen, N_2
			Hydrogen
			    sulfide, H_2S
			Methane, CH_4
			Various organic
			    wastes, some
			    of which produce odor
			    and color
				Hydrogen, H^+
				Sodium, Na^+
				Ammonium,
				    NH_4^+
					Chloride, Cl^-
					Bicarbonate,
					    HCO_3^-
					Nitrate, NO_3^-

ocean water, to the heavy concentrations of a landlocked evaporation sink such as the Dead Sea or the Great Salt Lake.

14.5 NUTRIENTS, MICROORGANISMS, AND OXYGEN

All animals and most plants require oxygen for metabolism of food. Aquatic animals utilize the oxygen dissolved in the waters they occupy. The solubility of oxygen in water is low: one liter of air at 25°C contains 0.27 gram of oxygen; one liter of water in contact with air at 25°C contains 0.0084 gram of oxygen. Therefore, one liter of aerated water contains only about one-thirtieth as much available oxygen as one liter of air. Furthermore, when oxygen has been removed from water, it is not rapidly replaced unless there is turbulent mixing with air, as in the "white water" of shallow rapids.

You don't have to read a book to be convinced that clear running water in which trout and salmon thrive is better for people than stagnant cloudy water that is slimy with plant growth and hospitable to sludge worms. Of the two ecosystems, the flowing trout stream is more delicately balanced; its homeostatic mechanisms are the less effective, and, therefore, it is the more susceptible to change. As a result, a relatively small shift in environmental conditions may readily dislocate the ecosystem of such a stream in a way that adversely affects its water quality. We must understand the nature of the aquatic energy systems to be able to protect our waters.

Aquatic food webs are similar to land-based ones in that they are powered by solar energy (Fig. 14.4). Green plants absorb sunlight and use the energy to convert carbon dioxide, CO_2, to the organically bound carbon in plant tissue and to oxygen gas. Most of the photosynthesis in aquatic systems is carried out by the smallest organisms, the phytoplankton. The cycle is completed by the consumers, mainly the zooplankton, which convert the organic carbon to CO_2 and the other nutrients to their oxidized forms. However, do not the larger organisms, such as fish and whales, also play a part? Of course they do, but the amounts of energy that flow through their bodies, or the masses of oxygen and carbon dioxide that they cycle, are small compared with those of the tiny organisms. There are several reasons for these differences. First, the total biomass of large consumers is small, because they are more likely to be higher on the food pyramid (see page 45). Furthermore, much of the mass of small organisms is near their surfaces, and, therefore, they have closer access to dissolved oxygen and can utilize it more rapidly. Thus, a pound of one-celled primary consumers uses more oxygen

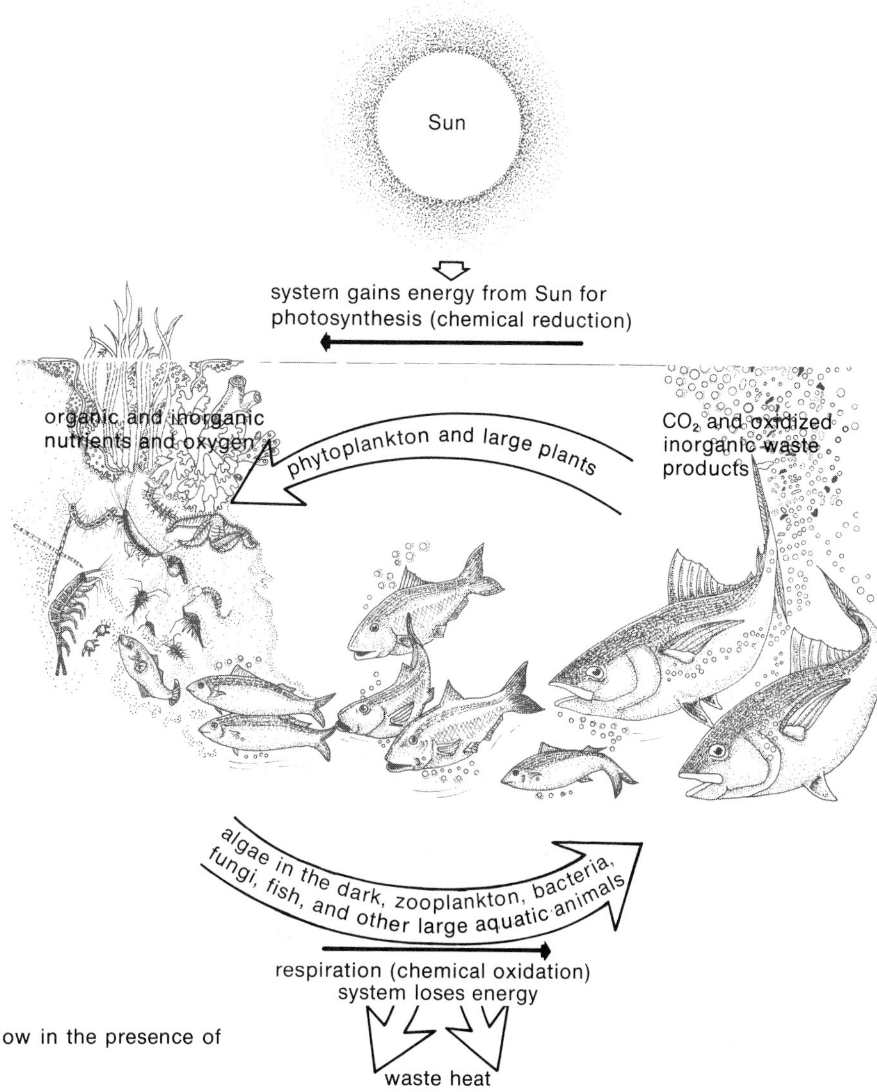
Sun
system gains energy from Sun for
photosynthesis (chemical reduction)
organic and inorganic
nutrients and oxygen
phytoplankton and large plants
CO_2 and oxidized
inorganic waste
products
algae in the dark, zooplankton, bacteria,
fungi, fish, and other large aquatic animals
respiration (chemical oxidation)
system loses energy
waste heat

Figure 14.4 Aquatic energy flow in the presence of oxygen.

in an hour than a one-pound trout, and there are more pounds of one-celled organisms in a pond than of large organisms. It is for these reasons that gas exchange among large aquatic animals is a relatively insignificant portion of the total.

Additionally, microorganisms can survive a much greater variety of environmental conditions than can more complex organisms, and, therefore, the cycle shown in Figure 14.4 is almost always operative. The adaptability of microorganisms is largely due to the fact that a great many different species are present in a sample of aquatic life. Therefore, collectively they provide a great variety of chemical reaction pathways—some can survive in the presence of substances, such as heavy metals, that would be fatally toxic to larger organisms. Finally, if all goes wrong, some microorganisms can suspend their life processes by becoming spores and thus survive by awaiting a return to a more hospitable environment.

Bacterial decomposition in the presence of air is called **aerobiosis;** it is this process that yields the most energy from a given weight of nutrients. For example, the complete aerobiosis of glucose (a sugar, $C_6H_{12}O_6$) may be represented by the equation:

$$C_6H_{12}O_6 + 6O_2 \longrightarrow 6CO_2 + 6H_2O + 4100 \text{ calories per gram of glucose}$$

Proteins contain sulfur and nitrogen as well as carbon, hydrogen, and oxygen. The aerobic decomposition of a protein is represented by the following equation:

$$\text{protein} + O_2 \rightarrow CO_2 + H_2O + \underset{\substack{\text{ammonium} \\ \text{ion}}}{NH_4^+} + \underset{\substack{\text{sulfate} \\ \text{ion}}}{SO_4^{2-}} + 4100 \text{ calories per gram of protein}$$

The reactions represented above are typical of the first stage of bacterial action in the deoxygenation of polluted waters. When organic nutrient is exhausted, the environment becomes favorable for other bacteria that can facilitate the oxidation of ammonium salts to yield additional energy:

$$\underset{\substack{\text{ammonium} \\ \text{ion}}}{NH_4^+} + 2O_2 \rightarrow \underset{\substack{\text{hydrogen} \\ \text{ion}}}{2H^+} + H_2O + \underset{\substack{\text{nitrate} \\ \text{ion}}}{NO_3^-} + 4500 \text{ calories per gram of ammonium ion}$$

This process is called **nitrification.**

All of these reactions require oxygen, and the organisms that carry them out are said to be **aerobic.** They also require nutrients, either organic or inorganic. These reactions, therefore, must stop when any one of the essentials is depleted. Whichever one is depleted fastest becomes the limiting factor. Oxygen in its molecular form, O_2, is far more abundant in air than it is in water—therefore, insects, birds, and mammals may compete for grain, but oxygen is available to all. The oxygen dissolved in waters, however, can be depleted faster than it is replaced from the atmosphere, and, therefore, oxygen becomes the limiting factor when organic nutrients are plentiful. Small organisms are more efficient energy converters than fish are, in part because of their ability to reproduce extremely rapidly in response to a sudden influx of nutrients. When the runoff of sewage or agricultural fertilizers makes waters so rich in nutrients that oxygen becomes the limiting factor, then the denitrification bacteria, the sludge worms, and some protozoa become more plentiful, while the fish decline in population. The quality of the water must be regarded as deteriorated, and the added nutrients are therefore pollutants.

Suppose, instead, that there is plenty of oxygen and it is the nutrients that become the limiting factor. Again, the primary consumers get the bulk of the nutrient and accumulate the greatest biomass. But the consumers of higher orders now get their share of the oxygen, and they can eat some of the worms and the bugs. And the trout jumping in the sparkling waters do look good to us.

Let us now return to the nutrient-rich, or polluted, conditions, and consider what happens to aquatic life when the oxygen content of waters is depleted. Does it all go dead? No, bacterial action does not stop when the molecular oxygen is gone. Instead, a new series of decompositions called **anaerobiosis** occurs. The anaerobic decomposition of sugars and other carbohydrates is called **fermentation,** and that of proteins is called **putrefaction.**

These processes are represented by the following simplified equations:

$$C_6H_{12}O_6 \rightarrow 2C_2H_6O + 2CO_2 + 100 \text{ calories per gram of sugar}$$
glucose, a   ethyl alcohol
  sugar

$$\text{protein} + H_2O \rightarrow NH_4^+ + CO_2 + CH_4 + H_2S + 370 \text{ calories per gram of protein}$$
methane   hydrogen
sulfide

Some anaerobic bacteria convert carbohydrate matter to methane:

$$C_6H_{12}O_6 \rightarrow 3CH_4 + 3CO_2 + 220 \text{ calories per gram of sugar}$$

Note that all of these reactions yield much less energy (fewer calories) than oxidation, but are still energetically profitable. Methane is very insoluble in water, and practically all of it is evolved as a gas. Hydrogen sulfide, a highly odorous gas, smells like rotten eggs. Putrefaction, therefore, makes water bubble with foul smells and makes it unlivable for fish or other oxygen-breathing animals. It may be regarded as the worst condition of bacterial pollution. The production of methane is not in itself an undesirable process (except when it accumulates in confined spaces, like sewers, to such a concentration that it becomes explosive). Its evolution means that the aquatic system has rid itself of a quantity of organic matter that would otherwise have demanded oxygen for its conversion to CO_2 and H_2O. And methane is not a toxic substance in its natural concentrations in air.

BIOCHEMICAL OXYGEN DEMAND

We have seen that nutrient matter pollutes water because it serves as food for microorganisms. Microorganisms, including any pathogens that may be among them, multiply; the oxygen is exhausted and thus becomes unavailable for forms of life (such as fish) that people prefer;

and finally, the stinks of putrefaction set in. What is the measure of such pollution? One might think that the analysis of a sample of water to determine the total amount of organic matter it contained would provide such an index. But not all organic matter is equally digestible by bacteria. In fact, some organic matter that is manufactured by industrial processes and is foreign to natural food chains may not be able to function as a nutrient at all. Such matter is said to be **non-biodegradable.** Some material, such as petroleum oil, may decompose only very slowly, so that it cannot be considered equivalent to, say, sugar as a nutrient. An appropriate measure of pollution of water by organic nutrients therefore must somehow recognize the rate at which the nutrient matter uses up oxygen, as well as the total quantity that can be consumed. Of course, the rate of biochemical oxidation depends on the temperature of the environment and on the particular kinds of microorganisms and nutrients present. If these factors are constant, then the rate of oxidation can be expressed in terms of the half-life of the nutrient. This concept is exactly the same as that applied to radioactive decay (see Chapter 10). The half-life is the time required for half of the nutrient to decompose, and the continuously decreasing rate can be represented in the form of a decay curve ("decay" here has both its biological and mathematical meanings) like that of Figure 14.5.

The chemical equation for the reaction can be considered to correspond to the "chemical oxidation" arrow of Figure 14.4:

$$\text{nutrients} + \text{dissolved oxygen} \xrightarrow{\text{microorganisms}} CO_2 + H_2O + \text{oxidized inorganics}$$

Now, imagine that, in the presence of oxygen at 20° C, the half-life of the nutrient matter is one day. (This is in fact

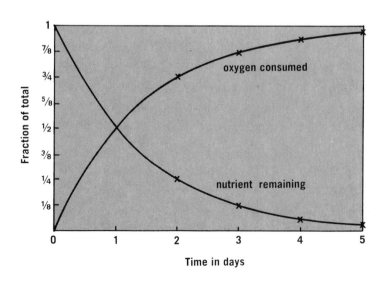

Figure 14.5 The biochemical oxygen demand (BOD) curve. The "nutrient" refers only to the maximum quantity that could be oxidized if all the oxygen were utilized. Any quantity in excess of this amount would necessarily remain unchanged.

454

the situation shown in Figure 14.5 and it is typical of experimentally observed values.) Then half of the nutrient will remain in the oxygenated water after one day; half of that, or one fourth of the original amount, after two days; one-eighth after three days; and so on. This means that one half of all the oxygen that is going to be used up will be used up in one day, another one-fourth (or a total of three-fourths) after two days, another one-eighth after three days, and so on.

The ideal shapes of the curves are shown in Figure 14.5. Note that at the end of five days almost all the nutrient is gone and almost all the oxygen that is going to be used has already been used. This length of time is considered to be a good compromise between completion of the oxidation and not having to wait forever. Therefore, a standard test is carried out by saturating a sample of the polluted water with oxygen at 20° C and determining how much oxygen has been used up after five days. The amount of oxygen thus consumed per liter of contaminated water is called the **biochemical oxygen demand**, or BOD.

14.6 THE POLLUTION OF INLAND WATERS BY NUTRIENTS

STREAMS AND RIVERS

The preceding discussion showed that unpolluted water is characteristically rich in dissolved oxygen and low in oxygen-demanding nutrients (BOD). Typical desirable levels are 8 mg of dissolved oxygen per liter and no more than about 2 mg per liter of BOD. When sewage is discharged into an unpolluted stream, the organic nutrient in the sewage creates an instantaneous increase in BOD; that is to say, the stream is polluted the moment the sewage enters it (see Fig. 14.6). But, as the shape of the decay curve shows, the demand on the oxygen is not instantaneous. Therefore, the dissolved oxygen falls off gradually, not instantaneously. Organic sewage that consists of human and animal wastes does not, of itself, kill fish; in fact it nourishes them. What does kill fish is lack of dissolved oxygen (in concentrations less than about 4 mg per liter, depending on the species); therefore, when the dissolved oxygen concentration dips below this level (about 25 km from Pollutionville, Fig. 14.6), the fish begin to die. As the river continues to flow, it recovers oxygen from the atmosphere and from photosynthesis by its vegetation, and thus it repurifies itself. Of course, if additional sewage is discharged before recovery is complete, as by closely spaced cities, the pollution becomes continuous. A river in such a condition, which unfortunately can be found near densely populated areas all over the world, supports no fish, is high

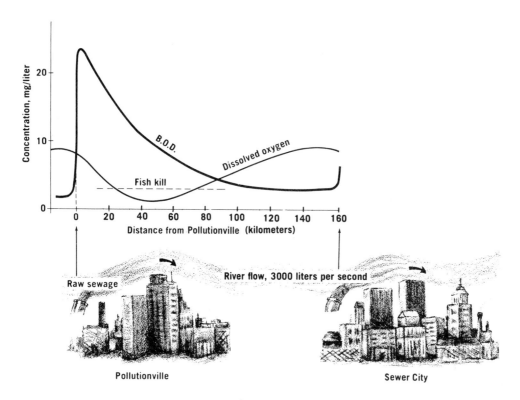

Figure 14.6 River pollution from hypothetical cities.

in bacterial content, usually including pathogenic organisms, appears muddily blue-green from its choking algae, and, in extreme cases, stinks from putrefaction and fermentation.

LAKES

Study Figure 14.7 to see the typical summer and winter conditions in a small lake in a temperate climate, such as New England. In the summer, the upper waters, called the **epilimnion** (the "surface lake"), are warmed by the sun. These warmer waters, being lighter than the colder ones below, remain on top and maintain their own circulation and oxygen-rich conditions. The lower lake waters (the **hypolimnion**) are cold and relatively airless. Between the two lies a transition layer, the **thermocline,** in which both temperature and oxygen content fall off rapidly with depth. As winter comes on, the surface layers cool and become denser. When they become as dense as the lower layers, the entire lake water circulates as a unit and becomes oxygenated. This enrichment is, in fact, enhanced by the greater solubility of oxygen in colder waters. Furthermore, the reduced metabolic rates of all organisms at lower temperatures result in a lesser demand for oxygen. When the lake freezes, then, the waters below support the aquatic life through the winter.

Limnology is the study of the physical phenomena of lakes. The prefix *epi-* (Greek) means in addition to, or resting upon. *Hypo-* (also Greek) means under, or beneath.

456

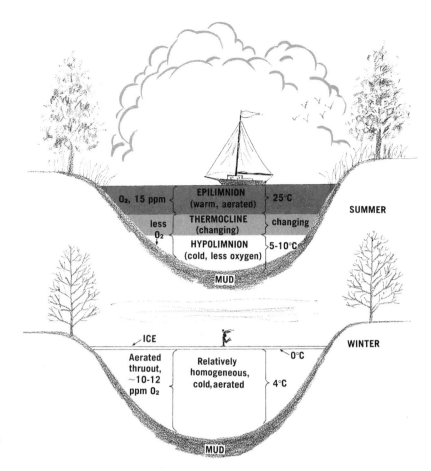

Figure 14.7 Thermal stratification in a small, temperate lake.

With the spring warmth, the ice melts, the surface water becomes heavier,* and again the lake "turns over," replenishing its oxygen supply.

Now, what are the effects of oxygen-demanding pollutants on these processes? During the summer, increased supplies of organic matter serve as nutrients in the oxygenated upper waters; the oxygen is replaced as needed by physical contact with the air and from photosynthesis by algae and other water plants. But some organic debris rains down to the lower depths which are reached neither by air nor by sunlight. Therefore, in an organically rich or "polluted" lake, the bottom suffers first. Fish that live best at low temperatures are therefore the first to disappear from lakes as the cold depths they seek become depleted of oxygen by the increased inflow of nutrients. These fish are frequently the ones most attractive to human diets, such as trout, bass, and sturgeon.

If the process we have just described continues, it leads eventually to a condition called **eutrophication,** which may be defined as the enrichment of a body of water with nu-

Eutrophic originally meant "tending to promote nutrition"; in this sense, a vitamin supplement would be a eutrophic medicine. The application of the term to describe the nourishment of natural waters as a contributor to the process of succession (and hence to pollution) was introduced into the literature of limnology about 1910.

---

*Recall that water reaches its maximum density at 4°C, so any approach to this temperature, from above or below, is accompanied by an increasing density.

trients, with the consequent deterioration of its quality for human purposes. This process occurs naturally in any lake whose nutrient inflow exceeds its outflow. Such **natural eutrophication,** which is closely associated with natural succession, is a slow process from a human point of view, frequently taking place over periods of thousands of years. In contrast, the discharge of untreated sewage and agricultural or industrial wastes into a lake hastens the process greatly, often shrinking millennia into decades. This accelerated process is called **cultural eutrophication,** in recognition of its civilized origins. Lakes in which the nutrient level is particularly high, which are characterized by abundant littoral (shore-dwelling) vegetation, frequent summer stagnation with algal blooms, and absence of cold-water fish species, are said to be **eutrophic.** Such a situation has been considered to be irreversible, and a lake that has reached it has been characterized as "dead." The justification for such a pessimistic outlook lies in the nutrients that accumulate in the muds of eutrophic lakes. These reservoirs of organic matter can supply many years of algal blooms and high oxygen demands. However, the situation is not always hopeless; more recent experience, for example, at Lake Washington (in Seattle) in the 1960's, has shown that cultural eutrophication can be reversed if the inflow of nutrients is greatly reduced.

14.7 ALGAE AND DETERGENTS—AND HOW TO WASH YOUR CLOTHES

In recent years there has been considerable public discussion of laundry detergents and their possible role in promoting the growth of algae and in deteriorating the quality of water. This is a matter of some complexity, so it is essential first to consider several basic facts and principles.

Algae are aquatic plants; they are sometimes visible as a blue-green slime on the surface of still water. As plants, algae derive energy from photosynthesis. Therefore, they consume carbon dioxide, CO_2, in the presence of sunlight and release oxygen. Like other plants, algae also need various inorganic nutrients, such as compounds of nitrogen, potassium, phosphorus, sulfur, and iron. In natural systems, growth of algae is usually limited by the small quantities of inorganic nutrients dissolved in surface waters. If the nutrient supply becomes sufficiently lavish, algae grow rapidly and can cover the surface of the water as thick, slimy mats. As some of the algae die, either through exhaustion of some essential nutrients or for other reasons, they become, in turn, food for bacteria.

Bacterial decomposition consumes oxygen, with consequent polluting effects. Such a sequence of processes results in a condition in which the water beneath the slimy surface is deficient in oxygen and therefore unable to support forms of life that people value most. Trout and bass give way to less desirable scavenger varieties such as catfish, and to leeches and worms. As we have seen, this is the process of eutrophication.

The use of modern detergents has contributed to the overfeeding of algae. To appreciate this somewhat complex situation, it will be helpful to understand something about the nature and mode of action of detergents. Recall that the ease with which a foreign substance dissolves in a liquid depends on how strongly the molecules of the two different substances attract each other, relative to the mutual attractions of like molecules. Sugar dissolves in water, and if your hands are sticky with honey or lollipops they can be washed clean by rinsing them in pure water. Vegetable oil and animal fat, however, are insoluble in water, and the pure water that rinsed away the honey will not remove grease. The ancient Romans knew that heating a mixture of animal fat and wood ashes produced a substance that could dissolve both in water and in grease and that could somehow bring these two otherwise incompatible substances together. Therefore, if you rinse your greasy hand with a mixture of water and this useful substance, which we call **soap,** the grease can be washed away. Soap functions in this manner because it is made up of long molecules that have, on one end, separated regions of positive and negative electrical charge that are strongly attracted to water molecules, and, on the other end, a hydrocarbon character that is attracted to grease molecules. This action of the soap molecule is called **detergency.**

The use of soap does not present a serious threat to the quality of inland waters unless large quantities are discharged directly into rivers and lakes. Soap is a nutrient for bacteria, not for plants, and is normally degraded quite satisfactorily by the bacterial action in sewage treatment plants. However, soap has not been a satisfactory detergent in all respects. The mineral matter in ground water contains metal ions that make soap insoluble, and thereby rob it of its detergency. This insoluble soap manifests itself as a "ring around the bathtub" or a "tattletale gray" on an otherwise white textile fabric. Water containing such mineral matter is called hard water. Water from which it is absent, such as rain, is said to be soft. The years since World War II have witnessed increasing development and use of synthetic detergents that are effective in hard water and that have various other properties with advantages over soap.

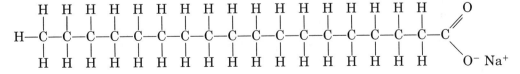

sodium stearate, a soap

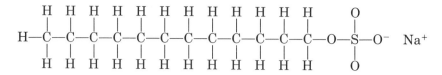

sodium lauryl sulfate, a synthetic detergent

Phosphates are ions that contain the elements phosphorus and oxygen, arranged in linear or cyclic sequences such as:

Some of the synthetic detergents manufactured during the 1960's were not biodegradable, that is, they did not serve as nutrients for bacteria.* As a result, they persisted in all their sudsiness through the sewage works, along the rivers, and occasionally right into the drinking water. This foamy problem was solved by the development of the straight-carbon-chain synthetic detergents used in recent years. These modern formulations also contain various other components—some to improve the detergent action, some to obstruct the redeposition of dirt, and some to provide bleaching and brightening actions. These additives typically include phosphates which, as noted in Chapter 11, are also used as agricultural fertilizers; they are plant nutrients. More important, phosphates are frequently in short supply in natural waters, and this deficiency serves as a limit to algal growth. When phosphate detergents are discharged into waterways, they supply these needed nutrients and promote rapid growth of algae. Sometimes the results far exceed unsophisticated expectations. In many areas of the world, especially in the great rivers and lakes of the tropical and subtropical regions, aquatic weeds have multiplied explosively (Fig. 14.8). They have interfered with fishing, navigation, irrigation, and the production of

*Bacteria can degrade unbranched hydrocarbon chains, but they find branched chains much less digestible.

Persistence of soap suds in a wastewater treatment plant.

A

Figure 14.8 The choking of waters by weeds. *A,* The dam on the White Nile at Jebel Aulia near Khartoum, Sudan. The area was clean when photographed in October 1958. *B,* The same area in October 1965, showing the accumulation of water hyacinth above the dam. (From Holm: Aquatic Weeds. *Science 166*:699–709, Nov. 7, 1969. Copyright 1969 by the American Association for the Advancement of Science.)

B

hydroelectric power. They have brought disease and starvation to communities that depended on these bodies of water. Water hyacinth in the Congo, Nile, and Mississippi rivers and in other waters in India, Pakistan, Southeast Asia and the Philippines, the water fern in southern Africa, and water lettuce in Ghana are a few examples of such catastrophic infestations. People have always loved the water's edge. To destroy the quality of these limited areas of the Earth is to detract from our humanity as well as from the resources that sustain us.

How, then, should you wash your clothes? That depends on how you balance your concern for the environment with your requirements for personal hygiene and for effective cleaning. Environmental impact would be avoided

461

"Notice how bright and white Brand X gets your clothing because of the harmful chemicals and enzymes it contains. Pure-O, on the other hand, containing no harmful ingredients, leaves your clothes lacklustre gray but protects your environment." (Drawing by Dana Fradon: © 1971 The New Yorker Magazine, Inc.)

if you used only pure water, but neither personal hygiene nor aesthetics would then be adequately served. However, the use of a combination of soap and washing soda (hydrated sodium carbonate, $Na_2CO_3(H_2O)_{10}$, also called sal soda) is minimally disrupting. Low-phosphate or non-phosphate synthetic detergents are also acceptable alternatives. Another very effective action is simply to use less detergent. Manufacturers often recommend quantities appropriate (or even excessive) for extreme conditions of water hardness. Frequently, one-half of the recommended amount gives very good cleaning, and as little as one-eighth gives fairly good cleaning. As far as personal health is concerned, it is futile to try to achieve a higher degree of sterilization in home laundry operations than is obtainable by soap alone, and the use of bleaches or other antiseptic agents should therefore be regarded as an effort in aesthetics, not in hygiene. The final point is your personal criterion of "effective" cleaning. The various additives in modern detergents do contribute whitening and brightening effects, and you must therefore decide how important it is to have "dazzling" underwear.

14.8   INDUSTRIAL WASTES IN WATER

Industrial activity, especially pulp and paper production, food processing, and chemical manufacturing, generates a wide variety of waste products that may be discharged into flowing waters. Some of these wastes are known to be poisonous; the effects of others are obscure.

A fish kill caused by water pollution.

Some have been known since antiquity, many are quite recent, and new types of wastes continue to appear as new technology develops. Many industrial wastes are organic compounds degradable by bacteria, but only very slowly, so that they may carry unpleasant odors and tastes along a watercourse for considerable distances. (Even domestic sewage contains significant quantities of non-biodegradable substances of unknown origin.) To complicate matters still further, some of these wastes react with the chlorine that is used as a disinfectant for drinking water. The result of such reaction is the production of chlorinated organic compounds that smell and taste much worse than the original waste product. More serious is the fact that such chlorinated compounds have been implicated as cancer-producing agents.

Metals corrode (oxidize) in water, and the dissolved or suspended oxidation products become pollutants. One of the oldest known waterborne metallic poisons is lead. Throughout history, the most prevalent source has been the lead piping formerly used in water distribution networks. More recently, the use of lead arsenate spray as an insecticide has contaminated surface and ground waters with both lead and arsenic. Lead is a cumulative poison, and even small concentrations, if continuously present in drinking water, may lead to serious illness or death. Arsenic, which sometimes occurs in natural waters that flow through arsenic-bearing minerals, is also a cumulative

Industrial wastes in water.

poison. The "safe" limits in drinking water for both lead and arsenic are recommended to be no higher than about 0.01 ppm.

The compounds of various other metals, such as copper, cadmium, chromium, and silver, have sometimes been implicated as industrial water pollutants. In very recent years, much attention has been given to the problem of mercury poisoning, and the Case History at the end of the chapter will be devoted to this interesting element.

14.9 THE POLLUTION OF THE OCEANS

The oceans are among the stable ecosystems of the Earth. The great mass of water implies stability, by virtue of its capacity to dilute foreign matter down to inconsequential concentrations. The total oceanic area is about 360 million square kilometers and the average depth of the major oceans (Atlantic, Pacific, and Indian, total area, 335 million square kilometers) is about 4100 meters. Of course, living organisms are not distributed uniformly in all this water. The organically most productive areas are the shallow waters near the shorelines, which make up about one tenth of the total oceanic surface and much less of the total volume. Furthermore, the magnitudes of ecological disruptions are not at all proportional to the masses or concentrations of polluting substances, but rather depend on the extent to which homeostatic mechanisms are upset. It is important, therefore, to consider the various classes of pollutants now being discharged into the "world ocean," and try to assess their possible environmental impacts.

OIL

The largest source of pollution of the oceans is oil. In this context, oil means petroleum oil; either the crude oil as it comes from the ground or one of its products derived from industrial or natural processing.

Crude oil is crude indeed, in the sense that it consists of many thousands of components of widely differing molecular weights. It is usually a dark brown, smelly liquid, about as thick as engine oil. It is largely composed of hydrocarbons, but there is an appreciable proportion of sulfur, and there are trace concentrations of metals such as vanadium and nickel.

Most hydrocarbons are less dense than water, and therefore the major portion of a mixture of hydrocarbons such as crude oil floats. However, some hydrocarbons are dense enough to sink even in sea water, and these materials, together with a portion of the metallic components, may settle to the bottom, where they have the potential to disrupt generations of aquatic organisms. Furthermore, the oxidation of floating oil also yields some products that are denser than sea water.

If a typical crude oil is heated to 100° C, some 12 percent of its volume boils off; if it is heated to 200° C, an additional 13 percent boils off. The total (25 percent) may be considered to be the volatile fraction that will evaporate from the floating oil surface within a few days. (Of course, the evaporated matter will not stay in the air indefinitely, but will eventually be returned to earth by mechanisms discussed in Chapter 13. The remaining oil is slowly metabolized by bacteria, and some of it slowly evaporates. After about three months, practically all the material that can evaporate has evaporated and all that can be eaten has been eaten. The persistent remainder is an asphaltic residue, representing about 15 percent of the original oil. These leftovers occur as small tarry lumps all over the Earth's seas.

The extent of the damage produced by oil pollution in the ocean depends in large measure on the direction it is taken by wind and by currents. There are four possibilities, none of them happy ones:

(a) The oil can be driven landward, where it may ruin beaches and shorelines that are a major recreational resource, destroy marine eggs (such as lobster eggs) floating near the shore, and kill various sea creatures on which fish and birds feed.

Oil-soaked gannet (a gull-like sea bird), Jones Beach, Long Island. (© Komorowski, from National Audubon Society.)

On December 21, 1976, the tanker Argo Merchant split apart on Nantucket Shoals off Massachusetts as it was slammed by 20-foot waves. The grounded vessel spewed more than five million gallons of crude oil into fishing waters. Bow is in foreground, with part of stern visible above it. (UPI photo.)

(b) The oil may float over a portion of continental shelf that is particularly productive of marine life. Such an area is the Georges Bank east of Nantucket Island, near where the oil tanker *Argo Merchant* split apart in December, 1976. In such a region oil can destroy populations of clams, scallops, cod, flounder, haddock, and whiting, on which thousands of people depend for a living and which furnish a source of high-quality protein food for millions of consumers. In addition, the oil destroys sea birds by matting their feathers, thus interfering with their swimming, flight, and insulation.

(c) The oil may be driven out to sea. Such an eventuality seems less immediately threatening, for there is little mid-ocean aquatic life. However, many components of oil are toxic to phytoplankton, and the cumulative effect of oil on life in the sea is unknown. The question of an overall danger to photosynthesis in the ocean will be discussed later in this section.

(d) As noted above, some of the denser components of the oil or its oxidation products may sink and disturb benthic organisms. These pollutants contain carcinogenic compounds, which may then enter the aquatic food web and be concentrated as they progress to higher and higher trophic levels, often culminating in human diets.

The most spectacular sources of oil pollution are the wrecks of tankers. Even a "medium-sized" supertanker can be nearly half a kilometer long, and wider than a football field. They are vastly larger than the great luxury steamers that used to ply the oceans. For various reasons, some of which have their roots in economic and political forces, supertankers are vulnerable to shipwreck and negligent in their "routine" discharges. These mammoth ships

466

Drawing by Chas. Addams; © 1974 The New Yorker Magazine, Inc.

represent large investments of money, and the incentive for a quick financial recovery is very strong. Their earnings can be maximized by keeping them in constant service and holding down their operating expenses. One of the most effective ways of economizing has been to sail these super-tankers under so-called flags of convenience, which means that they are registered in small non-seafaring nations whose shipping regulations are substandard to start with and haphazardly enforced.

The chief flag-of-convenience nations are Liberia and Panama; others include Costa Rica, Honduras, and Cyprus. Thus, an American owner sailing under the Liberian flag pays registration fees and tonnage taxes but not income taxes on shipping revenues; he is free to hire foreign or poorly trained seamen at low wages. American flagships must be crewed by American citizens, who must be trained and certified as to their competence, and who must be paid union wages. It has been estimated that the accident rate of flag-of-convenience ships is four times that of vessels registered by traditional maritime nations. The master of such a ship need not be unduly concerned about regulations from his "home port" (which he will probably never visit, anyway) regarding the dumping of oil sludge. There are, however, responsibilities to the country whose coastline may be polluted by an offshore spill, but these are sometimes very limited. Under United States federal law, for example, a shipping company can limit its liability to the value of the vessel if it can be shown that the tanker was seaworthy at the *start* of the voyage that ended in the accident. The limitation cannot be obtained, however, if the owner knew or should have known that the tanker was not seaworthy. Such questions often constitute the key issue in legal arguments regarding coastal oil spills. If the owner succeeds in limiting the liability to the value of the vessel, virtually nothing can be collected from the company because a wrecked tanker and its spilled oil are worthless. On the other hand, if negligence is proved, the company may be liable for many millions of dollars of damage to fishermen and to the owners of shore property.

A worldwide awareness of this problem was triggered by the disaster that occurred in 1967 when the tanker *Torrey Canyon*, sailing from the Persian Gulf towards Milford Haven, England, ran aground on a reef. The impact tore open six of her 18 tanks, and within two days, about 27,000 metric tons of escaped oil had created a slick about 30 km long. The efforts to reduce the damage were quite unsatisfactory. First the wreck was bombed and the remaining oil was set afire, but most of the volatile components had already evaporated and the fires could not be maintained, even with the addition of gasoline. Later, attempts were made to disperse the oil with detergents. The object of the treatments was to break up the oil film into small droplets so that it could be more readily washed away from the shoreline and more easily attacked by the oil-consuming bacteria. The detergent itself does not destroy the oil. The treatment was, indeed, mechanically effective, but the detergent contributed significantly to the poisoning of marine life and therefore did more biological harm than good.

New technological responses to oil spills have been developed, some of them quite imaginative. For example, there are various instruments that can "smell" oil at a distance, so that clean-up activities can be brought in quickly. Perhaps more important, spilled oil can be characterized by a sort of chemical fingerprinting, so as to identify and prosecute a violator who attempts to leave the scene. The most promising clean-up methods rely on floating barriers to confine the spilled oil to a limited area, from which it can be pumped out.

However, the wrecks of tankers are not the only sources of oil pollution, even though they are among the most dramatic. The same intractable tarry residues that eventually accumulate in the open ocean also build up in the storages of the tankers, as well as in the fuel tanks of oil-burning ships. Furthermore, an appreciable quantity of ordinary crude oil adheres mechanically to the walls of tanks even after they are drained. These oily ballasts must be dealt with somehow. In spite of mounting legal pressures, the tanks are often cleaned at sea and the waste dumped into the water, usually when the dark of the night matches the color of the oil. Modern methods of detection and clean-up are less adequate when they confront these more widely dispersed sources of pollution.

Offshore drilling operations, now being conducted on continental shelves in many parts of the world, are subject to accidents that result in the direct release of oil from well to sea. A notable incident of this type occurred off the coast of Santa Barbara, California, in 1969; it is estimated that between 75,000 and 200,000 liters of oil per day were released for 11 days.

The total quantity of oil that finds its way into the sea each year is very large. It has been estimated by various investigators that about one million metric tons of oil are spilled into the ocean each year from shipping and oil drilling operations alone. But there are also myriads of "minispills"—sludges from automobile crankcases that are dumped into sewers, routine oil-handling losses at seaports, leaks from pipes, and the like. Some oil aerosols also settle into the sea from the atmosphere. The grand total from all these sources is difficult to estimate, but it could well reach 10 million metric tons per year or more.

OTHER CHEMICAL WASTES

How does one dispose of highly toxic chemical wastes, such as by-products from chemical manufacturing, chemical warfare agents, and pesticide residues? There is no easy answer. Biological treatment systems are inapplicable if the microorganisms are poisoned by the substances they are supposed to oxidize. Chemical conversions, including burning, sometimes work, but they are costly, and the facilities are not always available. It is tempting, then, to seal such material in a drum and dump it in the sea. But drums rust, and outbound freighters do not always wait to unload until they reach the waters above the sea's abyssal depths. As a result, many such drums are found in the fisheries on continental shelves or are even washed ashore. It is estimated that tens of thousands of such drums have been dropped into the sea.

Furthermore, not all liquid industrial chemicals are packed in drums. Some are carried, like oil, in tankers. Defoliants for Vietnam, for example, were transported in chemical tankers. These ships, like any others, can be wrecked by the sea, and their captains, like the captains of oil tankers, can dump their slops where no one is watching.

Of course, all the river pollutants enter the same sink: the world ocean. The organic nutrients are recycled in the aqueous food web, but the chemical wastes from factories and the seepages from mines, including the mineral matter and the stubbornly resistant organic chemicals, are all carried by the streams and rivers of the world into the ocean.

And where do the air pollutants go—airborne lead and other metals from automobile exhaust, and mercury vapor from electrolysis operations, and the fine particles of agriculture spray dusts that ride the winds? Again, to the ocean—perhaps 200,000 metric tons of lead and 5000 of mercury per year, as well as many tons of other toxic materials. In ocean regions near large cities, such as New York, waste accumulations have rendered large areas unfit for any marine life at all, and these have come to be known as "dead seas." In other areas the riverborne organisms invade shellfish and make them carriers of diseases like infectious hepatitis. Radioactive wastes, discussed in Chapter 10, are another class of ocean pollutants. Still another threat comes from the possible future course of undersea industrial operations, such as the mining, concentration, and processing of ores from the ocean floor. All of the process wastes would go directly into the waters.

IS THERE AN OVERALL THREAT TO LIFE IN THE SEA?

It is, in general, very difficult to predict the reactions of homeostatic mechanisms to environmental stresses. We are surprised at times by the extreme fragility and at other times by the apparent stability of ecosystems. Something, we are not sure just what, goes out of adjustment and a massive "red tide" (page 76) devastates a productive shoreline, or an unchecked bloom of starfish devours a coral reef. Or conversely, thousands of tons of crude oil are dumped in mid-ocean and in a few months go quietly away (or seem to, since no one is sure of the possible long-term effects). The toxins that accumulate in the ocean have no other place to go; the Earth's ocean is their ultimate sink. Complex biotic societies are typically adaptable to changing conditions, but no one can guarantee that the organisms of the ocean will continue to survive the present rate of influx of exotic chemical wastes. The Earth's oxygen is continuously replenished by photosynthesis, and a large portion of that activity is carried out by the vegetation of the oceans. Some investigators have cautioned that the destruction of the phytoplankton or the impairment of its photosynthetic activity might seriously reduce the oxygen content of the atmosphere (not to speak of the capacity of the oceans to supply food for humanity). At this point, recall Lovelock's hypothesis (page 92) that the Earth's oxygen is maintained by Gaia, its overall biotic community, not by inorganic mechanisms. If this is so, a threat to life in the sea is a threat to life on Earth.

Along the Ganges River. (Photo by Ken Heyman.)

14.10 THE EFFECTS OF WATER POLLUTION ON HUMAN HEALTH

On a worldwide scale, the pollution of water supplies is probably responsible for more human illness than any other environmental influence. The diseases so transmitted are chiefly due to microorganisms and parasites. Two examples will illustrate the dimensions of the problem. Cholera, an illness caused by ingestion of the bacterium *Vibrio cholerae,* is characterized by intense diarrhea which results rapidly in massive fluid depletion and death of a very large percentage of untreated patients. Though its distribution in the past was virtually worldwide, it has been largely restricted during the twentieth century to Asia, and particularly the area of the Ganges River in India. During the nine years from 1898 to 1907, about 370,000 people died from this disease, and thousands of Indians continue to die each year even up to the present. In 1947, a severe epidemic occurred in Egypt with about 21,000 cases, half of whom died.

Most Americans have never heard of schistosomiasis. This is actually a group of diseases caused by infection with one of three related types of worms. (Which worm you get depends on where in the world you live.) Current estimates are that over 100 million people are infected with schistosomiasis; these cases are distributed throughout the African continent, in parts of Asia, and in areas of Latin America. Estimating the amount of human suffering caused by schistosomiasis is much more difficult than for a disease like cholera, because unlike cholera it is a cause of much chronic as well as acute disease. For both these ill-

471

nesses the main mode of transmission is from water supplies contaminated with the feces of infected individuals. Other bacterial illnesses, such as the salmonelloses (of which typhoid fever is a leading example) and viral infections like poliomyelitis and hepatitis, may also be disseminated in this way. In the case of the bacterial and viral illnesses, the organisms themselves are shed in the stool, and must be ingested by others to cause disease.

In the case of schistosomal infections, however, the eggs of the organisms are shed. They then hatch into forms which must find a certain type of snail to complete their life cycle. Once safely in the snail, the worm develops into a free-living form which leaves the snail and may infect people if ingested in drinking water. Alternatively, it may penetrate human skin on contact and enter the bloodstream.

In the United States, however, the picture is very different; in fact, nowhere is the contrast between developed and underdeveloped countries starker than in the comparison between the health effects of water pollution on the respective populations. During the decade 1961–1970, there were 130 reported outbreaks of disease attributable to contaminated water supplies in the United States; of these, all but a very few were probably due to the presence of microorganisms rather than chemicals. A total of 46,000 people became ill, but only 20 died. While the existence of such outbreaks in a technological society such as ours is deplorable, one can immediately see that water pollution is a very minor source of acute fatal illness in the United States. In this age of extreme mobility made possible by international travel, the possibility always exists that a disease such as cholera could spread to the United States and attain epidemic proportions here. That this could happen on a large scale, however, seems unlikely since about three-quarters of the American population derive their water from sources which are monitored by state and federal agencies.

The usual measure of microbiologic purity of a water supply is the so-called coliform count (coliforms are the class of bacteria present in the human intestine); therefore, the concentration of coliforms in a water supply is a measure of the amount of human fecal contamination, and not a direct measure of the number of disease-causing microorganisms. Water is generally considered safe if it contains fewer than 10 coliforms per liter. Though this method generally serves to safeguard the purity of water, its major pitfall is that some steps in water purification, notably chlorination, may destroy bacteria without killing viruses; hence viral disease may be transmitted by water that satisfies rigid bacteriologic standards.

As noted earlier, water supplies may become contaminated with a wide variety of chemical substances. It is surprising, therefore, to realize that although the potential for the production of disease from this source exists, actual accounts of major illness due to chemically contaminated water are few; the Minamata Bay disaster mentioned in the Case History at the end of the chapter is a devastating, but fortunately rare, example. But the simple fact that acute illness is uncommon does not rule out the possibility of chronic illness, about which there is very little definite information. Over the years the United States Public Health Service has suggested standards for drinking water in the form of maximal allowable concentrations of various substances, particularly metals and some classes of organic pollutants. Some of these may be acutely toxic; others produce chronic illness.

Recently, nitrates have come under close scrutiny. Under certain circumstances, when relatively large amounts of nitrates are ingested, they may be reduced to form nitrites. The toxicological significance of nitrites is twofold: (a) They can interfere with the ability of hemoglobin to bind oxygen. Severe cases of this condition have produced death in infants. (b) Nitrates may react in the body to form certain compounds that are strongly suspected of being carcinogenic. However, most Americans probably ingest far more nitrates from foods than from drinking water; indeed, potassium nitrate (saltpeter) is actually added as a preservative to certain foods.

The Department of Health, Education, and Welfare is continuing its efforts to identify possible contaminants of significance to human health and to set limits on their allowable concentrations in water supplies. The difficulty in doing so, however, is made clear when one realizes that about 12,000 toxic chemicals are used today by industry, and about 500 new chemicals are developed each year. This problem will be far more difficult to resolve than that of communicable disease prevention, for which highly effective technology already exists.

> The real authority of the Public Health Service relates only to communicable diseases and not those due to chemical contamination; this law is archaic and should be reformed. In addition, the PHS exercises authority over only those water supplies that serve interstate carriers. It turns out that these supplies serve only about 80 million people, or somewhat less than 40 per cent of the population. However, most states have adopted the PHS guidelines.

14.11 WATER PURIFICATION

Water molecules have no memory, and therefore it is silly to talk about the number of times that the water you drink has been polluted and repurified, as if the molecules gradually wore out. All that is important is how pure it is when you drink it.

The purification of water has developed into an elaborate and sophisticated technology. However, the general approaches to purification should be comprehensible, and in some cases even obvious, from a general understanding of the nature of water pollution.

Illustration continued on the opposite page

Schematic representation of a complete water purification system.

In Section 14.4 impurities in water were classified as *suspended, colloidal,* or *dissolved.* These categories are also shown in Figure 14.3 on page 448. Suspended particles are large enough to settle out or to be filtered. The colloidal and dissolved impurities are more difficult to remove. One possibility is somehow to make these small particles join together to become larger ones, which can then be treated as suspended matter. Another possibility is to convert them to a gas that escapes from the water into the atmosphere. Whatever the approach, it must be remembered that energy is required to lift water or to pump it through a filter.

With these principles in mind, let us now consider the procedures used in purifying municipal waste waters. The first step is the collection system. Waterborne wastes from sources such as homes, hospitals, and schools contain food residues, human excrement, paper, soap, detergents, dirt, cloth, other miscellaneous debris, and, of course, microorganisms. This mixture is called **sanitary** or **domestic sewage.** (The adjective "sanitary" is rather inappropriate since it hardly describes the condition of the sewage; it presumably refers to that of the premises whose wastes have been carried away.) These waters, sometimes joined by wastes from commercial buildings, by industrial wastes, and by the run-off from rain, flow through a network of sewer pipes, as shown in Figure 14.9. Some systems separate sewage from rain water, others combine them. The combined piping is cheaper and is adequate in dry weather, but during a storm the total volume is apt to exceed the capacity of the treatment plant, so some is allowed to overflow and pass directly into the receiving stream or river.

Bacterial and microbial actions occur during the flow of wastes through the sewer pipes; the high-energy food chemicals are degraded to low-energy compounds, with the consumption of oxygen. The more such action occurs before the sewage is discharged to open waters, the less occurs afterward; therefore, this process must be regarded as the beginning of purification.

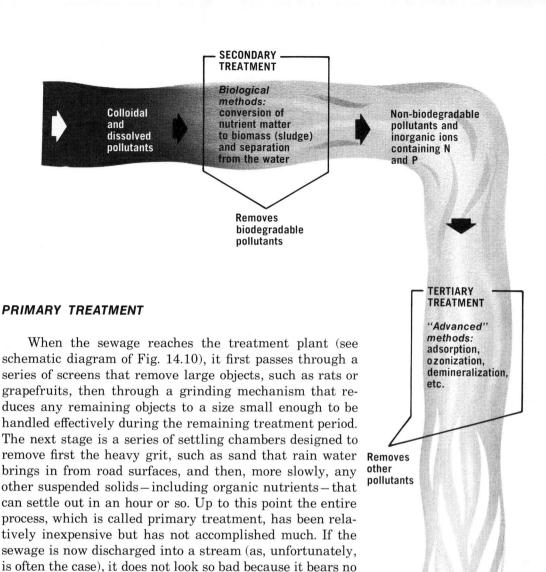

SECONDARY TREATMENT

Biological methods: conversion of nutrient matter to biomass (sludge) and separation from the water

Colloidal and dissolved pollutants

Non-biodegradable pollutants and inorganic ions containing N and P

Removes biodegradable pollutants

TERTIARY TREATMENT

"Advanced" methods: adsorption, ozonization, demineralization, etc.

Removes other pollutants

Pure water

PRIMARY TREATMENT

When the sewage reaches the treatment plant (see schematic diagram of Fig. 14.10), it first passes through a series of screens that remove large objects, such as rats or grapefruits, then through a grinding mechanism that reduces any remaining objects to a size small enough to be handled effectively during the remaining treatment period. The next stage is a series of settling chambers designed to remove first the heavy grit, such as sand that rain water brings in from road surfaces, and then, more slowly, any other suspended solids—including organic nutrients—that can settle out in an hour or so. Up to this point the entire process, which is called primary treatment, has been relatively inexpensive but has not accomplished much. If the sewage is now discharged into a stream (as, unfortunately, is often the case), it does not look so bad because it bears no visible solids, but it is still a potent pollutant carrying a heavy load of microorganisms, many of them pathogenic, and considerable quantities of organic nutrients that will demand more oxygen as their decomposition continues.

SECONDARY TREATMENT

The next series of steps is designed to reduce greatly the dissolved or finely suspended organic matter by some form of accelerated biological action. What is needed for such decomposition is oxygen and organisms and an environment in which both have ready access to the nutrients. One device for accomplishing this objective is the **trickling filter,** shown in Figure 14.11. In this device, long pipes rotate slowly over a bed of stones, distributing the polluted water in continuous sprays. As the water trickles over and around the stones, it offers its nutrients in the presence of air to an abundance of rather unappetizing forms of life. A fast-moving food chain is set in operation. Bacteria consume

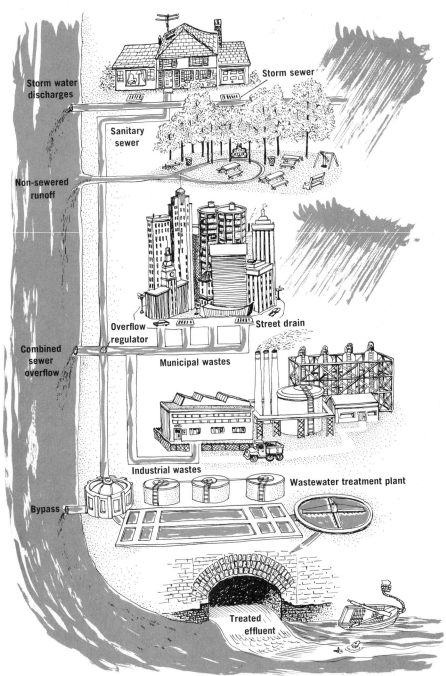

Storm water discharges

Storm sewer

Sanitary sewer

Non-sewered runoff

Combined sewer overflow

Overflow regulator

Street drain

Municipal wastes

Industrial wastes

Wastewater treatment plant

Bypass

Treated effluent

Figure 14.9 Sewer collection system.

molecules of protein, fat, and carbohydrate. Protozoa consume bacteria. Farther up the chain are worms, snails, flies, and spiders. Each form of life plays its part in converting high-energy chemicals to low-energy ones. All the oxygen consumed at this stage represents oxygen that will not be needed later when the sewage is discharged to open water. Therefore, this process constitutes a very significant purification.

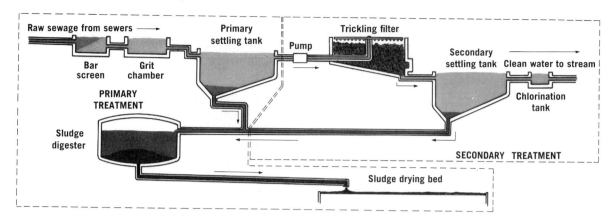

Sewage plant schematic, showing facilities for primary and secondary treatment. (From *The Living Waters*. U.S. Public Health Service Publication No. 382.)

An alternative technique is the **activated sludge** process, shown schematically in Figure 14.12. Here the sewage, after primary treatment, is pumped into an aeration tank where it is mixed for several hours with air and with bacteria-laden sludge. The biological action is similar to that which occurs in the trickling filter. The sludge bacteria metabolize the organic nutrients; the protozoa, as secondary consumers, feed on the bacteria. The treated waters then flow to a sedimentation tank where the bacteria-laden solids settle out and are returned to the aerator. Some of the sludge must be removed to maintain steady-state conditions. The activated sludge process requires less land space than the trickling filters, and, since it exposes less area to the atmosphere, it does not stink so much. Furthermore, since the food chain is largely confined to microorganisms, there are not so many insects flying around. However, the activated sludge process is a bit trickier to operate and

Figure 14.11 A trickling filter with a section removed so as to show construction details. (From Warren: *Biology and Water Pollution Control*. Philadelphia, W. B. Saunders Co., 1971. Photo courtesy of Link-Belt/FMC.)

Domed covers for two trickling filters of Mason City (Iowa) wastewater treatment plant, measuring 60 meters in diameter by 12 meters high at the apex. The enclosures help to control the release of odorous air pollutants and to maintain summer-like conditions inside, which enhances biological efficiency. Photo courtesy of Temcor, Torrance, California.

can be more easily overwhelmed and lose its effectiveness when faced with a sudden overload.

The effluent from the biological action is still laden with bacteria, and so is not fit for discharge into open waters, let alone for drinking. Since the microorganisms have done their work, they may now be killed. The final step is therefore a disinfection process, usually chlorination. Chlorine gas, injected into the effluent 15 to 30 minutes before its final discharge, can kill more than 99 percent of the harmful bacteria.

Let us now return to the sludge. Each step in the biological consumption of this waterborne waste, from sewage nutrients to bacteria to protozoa and continuing to consumers of higher orders (such as worms), represents a degradation of energy, a consumption of oxygen, and a reduction in the mass of pollutant matter. Also, and perhaps

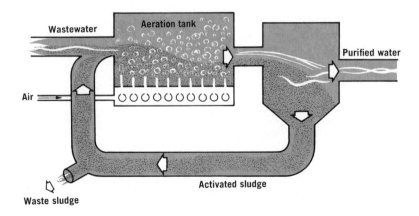

Figure 14.12 Activated sludge process.

478

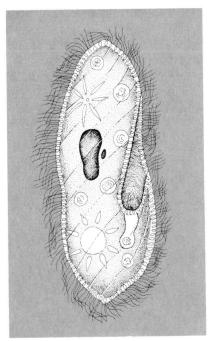

A freshwater protozoan, *Paramecium caudatum.* (From Mary Clark: *Contemporary Biology.* Philadelphia, W. B. Saunders Co., 1973.)

A

B

The Metropolitan Sanitary District of Greater Chicago transports its sewage sludge by barge and pipeline (above) to a 15,000 acre (6000 hectare) site in Fulton County, Illinois, where it is spread on agricultural areas that had been left in poor condition by strip-mining (below).

most important from a practical point of view, the process brings about an increase in the average size of the pollutant particles. Look at Figure 14.3 to see how dramatic this change can be. Sugar is dissolved in water in the form of molecules which never settle out. Partially degraded starch and protein occur as colloidal particles approximately in the same size range as viruses. Bacteria are much larger, growing up to about 10 micrometers. Protozoa are gigantic by comparison; some amoeba reach diameters of 500 micrometers and thus are comparable in size to fine grains of beach sand. Some agglomeration also occurs in the metabolic processes of the protozoa, so that their excreta are

479

Sewage sludge.

usually larger than the particles of food they ingest. Finally, when the microorganisms die, their bodies stick together to form aggregates large enough to settle out in a reasonably short time. This entire process of making big particles out of little ones is of prime importance in any system of waste water treatment. The mushy mixture of living and dead organisms and their waste products at the bottom of a treatment tank, constitutes the biologically active sludge. (See photo in margin.) Typical sewage contains about 0.6 gram of solid matter per liter, or about 0.06 percent by weight, while a liter of raw waste sludge contains about 40 to 80 grams of solid matter, corresponding to a concentration of 4 to 8 percent. Even after this magnification, however, the raw sludge is still a watery, slimy, malodorous mixture of cellular protoplasm and other offensive residues. The organic matter can undergo still further decomposition, but its high concentration engenders anaerobic conditions. Recall that anaerobiosis produces methane, CH_4, and carbon dioxide, CO_2, among other gases. Such conversions thus decrease the content of solid carbonaceous matter still further, although the process is necessarily accompanied by offensive nitrogenous and sulfidic odors. The final disposal of the sludge residue, whether by incineration, landfill, or other means, becomes a problem in the handling of solid wastes (see Chapter 15).

TERTIARY OR "ADVANCED" TREATMENTS

Although considerable purification is accomplished by the time wastewaters have passed through the primary and secondary stages, these treatments are still inadequate to deal with some complex aspects of water pollution. First of all, many pollutants in sanitary sewage are not removed. Inorganic ions, such as nitrates and phosphates, remain in the treated waters; these materials, as we have seen, serve as plant nutrients and are therefore agents of eutrophication. If chlorination is incomplete, microorganisms will remain; in any case, chlorine will remain in some form or other, frequently as chlorinated organic matter which seriously impairs the taste of the water and which can even introduce new toxins.

Additionally, many pollutants originating from sources such as factories, mines, and agricultural runoffs cannot be handled by municipal sewage treatment plants at all. Some synthetic organic chemicals from industrial wastes are foreign to natural food webs (that is, they are non-biodegradable); they not only resist the bacteria of the purification system, but may also poison them, and thereby nullify the biological oxidation which the bacteria would otherwise provide. There are also inorganic pollutants, including acids and metallic salts, as well as suspended soil particles from chemical and mining operations and from natural sources. Some of these materials occur as very fine particles from roadways, construction sites, or irrigation run-offs. These sediments are troublesome before they settle, because they reduce the penetration of sunlight, and afterwards, because they fill reservoirs, harbors, and stream channels with their silt.

The treatment methods available to cope with these troublesome wastes are necessarily specific to the type of pollutant to be removed, and they are generally expensive. A few of these techniques are described below.

Coagulation and Sedimentation

As mentioned earlier in the discussion of biological treatment, it is advantageous to change little particles into big ones which settle faster. So it is also with inorganic pollutants. Various inorganic colloidal particles are water-loving (hydrophilic) and therefore rather adhesive; in their stickiness they sweep together many other colloidal particles that would otherwise fail to settle out in a reasonable time. This process is called **flocculation.** Lime, alum, and some salts of iron are among these so-called flocculating agents.

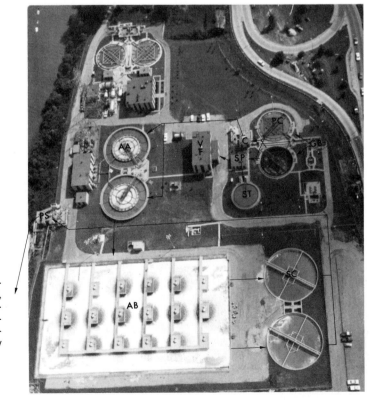

Waste treatment plant, Charleston, West Virginia. Grit basin: GB, Primary clarifier: PC, Sludge pumps: SP, Sludge thickener: ST, Vacuum filters: VF, Chlorinators: C, Aero accelerator: AA, Aeration basin: AB, Secondary clarifier: SC, Pump station: PS. (Photo courtesy Union Carbide.)

Adsorption

The process of adsorption (page 429) is by no means restricted to gases; it also takes place in liquids. As in air, the agent of choice is activated carbon, which is particularly effective in removing chemicals that produce offensive tastes and odors. These include the biologically resistant chlorinated hydrocarbons.

Other Oxidizing Agents

Potassium permanganate, $KMnO_4$, and ozone, O_3, have been used to oxidize waterborne wastes that resist oxidation by air in the presence of microorganisms. Ozone has the important advantage that its only byproduct is oxygen.

$$2O_3 \longrightarrow 3O_2$$

Reverse Osmosis

Osmosis is the process by which water passes through a membrane that is impermeable to dissolved ions. In the normal course of osmosis, as illustrated in Figure 14.13A, the system tends toward an equilibrium in which the concentrations on both sides of the membrane are equal. This means that the water flows from the pure side to the concentrated, "polluted" side. This is just what we don't want, for it increases the quantity of polluted water. However, if excess pressure is applied on the concentrated side (Fig. 14.13B), the process can be reversed, and the

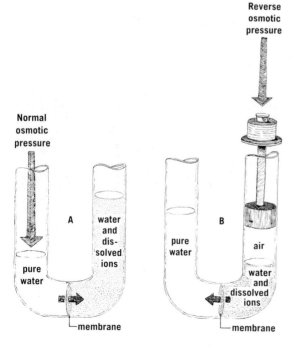

Figure 14.13 Reverse osmosis.

pure water is squeezed through the membrane and thus freed of its dissolved ionic or other soluble pollutants.

14.12 ECONOMICS, SOCIAL CHOICES, AND STRATEGY IN WATER POLLUTION CONTROL

Domestic wastes were once collected in pits called cesspools, from which they were periodically shoveled out and carted away. As cities grew denser, the task became more onerous, and toward the end of the last century it became customary to connect series of cesspools with conduits so that they could all be flushed out with water in a single operation. The next obvious step was to eliminate the cesspools, and use the piping system alone with flushing water continuously available. Thus were sewer systems born. According to some environmentalists, this was the point at which sanitary engineers and public health officers took civilization down the wrong road. What we are doing now is discharging our wastes into the public waters and then spending billions of dollars to restore the water to a quality that is fit for drinking. The wastefulness of the process is illustrated by the fact that the average toilet flush uses about 20 liters of water to carry away about ¼ liter of body wastes and that the average user of a flush toilet flushes it some seven times a day.

Few recommend a return to the outhouse or cesspool. Before examining reasonable strategies, however, let us look at the objectives we are trying to reach. According to the Federal Water Pollution Control Act of 1972, two general goals are proclaimed for the United States:

(1) Wherever attainable by July 1, 1983, water quality should be clean enough for swimming and recreational use and for the protection and propagation of fish, shellfish, and wildlife.

(2) And then, by 1985, the discharge of pollutants into navigable waters should be eliminated altogether.

In 1974, Congress passed the Safe Drinking Water Act, which was designed to assure that water supply systems serving the public meet minimum national standards for protection of public health. The Act gave the Environmental Protection Agency (EPA) responsibility for setting minimum national drinking water regulations throughout the United States. Interim regulations, which were published on December 24, 1975, and became effective June 24, 1977, set maximum levels permitted for bacteria, cloudiness, and concentrations for a number of organic and inorganic chemicals. Meanwhile, the National Academy of Sciences has conducted a study that is to form the basis of *revised* regulations to become effective in 1979.

The 1972 goals were hardly embraced by industry. An early response was, "I flatly predict zero discharge can't be

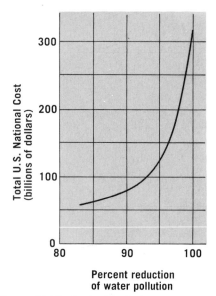

Figure 14.14 Cost of reducing water pollution. (Source: estimates from the Environmental Protection Agency, 1972.)

Chemical toilet: A new style in an old setting.

done, won't be done, and that the people of America won't want it to be done because they won't want to pay the cost."[*] The typical industrial position was that the goal of zero discharge should be replaced with water quality standards. (Compare this with the discussion on air quality standards in Chapter 13.)

The central factor that must be confronted in trying to reconcile national goals, industrial objections, and recommendations for alternate strategies is the *quantity* of water that is involved.

The average total volume of water supplied to the United States per day by rain and snow is close to 4 trillion liters. About 1½ trillion liters, more than a third of the total supply, is used daily by the manufacturing and power industries, by agriculture, and by cities and towns. If all of this water were to be purified to the highest standards of quality for drinking, the costs might well become prohibitive. Figure 14.14 shows how purification costs rise as the purity rises. The key to achieving national goals for water quality at bearable costs must therefore lie in conservation.

To return to the problem of the flush toilet as an example, it has been suggested[†] that sewage disposal systems be decentralized, with water disposed in the individual house, apartment building, or factory, rather than into the public water supply. This does not mean a return to the outhouse. Waterless toilets have been developed for boats and can be readily adapted for dwellings on land. Various companies in Sweden, the United States, Australia, New Zealand, and Japan manufacture incinerating toilets, composting toilets, enzymic-action biological toilets, and recycling oil-flushed toilets.

There have been many suggestions, over the years, to return to traditional agricultural practices by piping liquid sewage directly to farms and woodlands, where it may be sprayed as fertilizer. As an added benefit, the water, filtered by the soil, replenishes the water table. Such practices are especially beneficial in the arid areas of the Southwest and have operated in some hundreds of communities in those regions. However, they are hardly applicable to large cities, where the requisite farm areas simply do not exist. Furthermore, the use of domestic sewage as a fertilizer must not be considered as an unmixed blessing, for it is no longer a simple mixture of biodegradable organic matter. Instead, such sewage is always mixed with some industrial wastes which contain metallic compounds and other non-biodegradable chemicals that may accumulate in the soil. The agricultural use of sludge or

---

[*]John T. Connor, Chairman of Allied Chemical Company, in an address to the Synthetic Organic Chemical Manufacturers Association in New York City.

[†]Harold H. Leich: "The Sewerless Society." *Bulletin of the Atomic Scientists,* November, 1975.

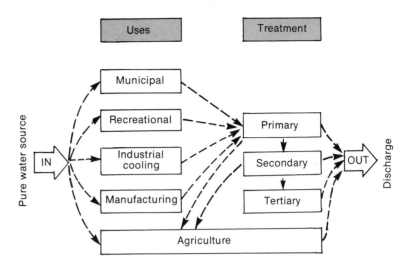

Figure 14.15 Strategies for water pollution control.

compost from household biological toilets would be a much sounder practice, especially if people could be discouraged from dumping pesticides or paints in them.

Figure 14.15 illustrates a generalized scheme of strategies for water pollution control and shows that choices must be made at various points of the system to achieve pure water at bearable costs.

14.13 CASE HISTORY: *Mercury*

Mercury has always been regarded with fascination and alarm. It is the only metal that is liquid at ordinary temperatures—hence its other name, quicksilver—and it is fun to play with. (But don't do it. Its vapor is poisonous, and at high temperatures it can vaporize rapidly enough to be deadly.) Some of its compounds, whose toxicity has been well known since the Middle Ages, have been used as agents of murder and suicide.

The mining of mercury and the extraction from its principal ore—the red sulfide, HgS, called cinnabar—has long been known to be hazardous to miners. Until the early part of the present century, mercury was used in the manufacture of felt hats. The exposed workers suffered from tremors (the "hatter's shakes"), and loss of hair and teeth. Lewis Carroll's Mad Hatter was probably inspired by this industrial syndrome.

Until very recently, however, mercury was not considered a dangerous water pollutant. Although mercury is widely distributed over the Earth, it generally occurs only in trace concentrations. Natural waters typically contain only a few parts per billion of mercury.

Metallic mercury, itself, although poisonous

Lewis Carroll's Mad Hatter, probably inspired by the syndrome of mercury poisoning in the British hatting industry.

in vapor form, is not particularly hazardous when taken by mouth as a liquid. The use of mercury as a component of dental fillings has been shown to be harmless; the mercury in the

485

teeth does not migrate to other parts of the body. Many mercury compounds are very highly insoluble; for example, it is calculated that it would require about 100 liters of water to dissolve one molecule of mercuric sulfide, HgS!

These considerations imply that mercury in water is not a potential pollutant and perhaps account for the previous lack of concern over the fact that half of the total amount of mercury mined annually is released into the environment. (About 10,000 metric tons are mined, of which 5000 metric tons are somehow "lost.") These discharges occur as waste effluents from manufacturing plants or else as the incorporation of traces of mercury into products in which it does not belong. An example is the electrochemical conversion of brine, NaCl dissolved in water, into chlorine and sodium hydroxide, as represented by the following equation:

$$2NaCl + 2H_2O \rightarrow Cl_2 + 2NaOH + H_2$$

chlorine sodium hydrogen, hydroxide which is (caustic released soda) to the atmosphere

Mercury does not appear in the equation, but it flows along the bottom of the reaction cell as an electrical boundary (electrode) at which the sodium hydroxide and hydrogen are produced. When the salt solution (brine) becomes too weak, it is discarded. This waste contains mercury, which then follows whatever watercourse is available to it. The sodium hydroxide product also is contaminated with mercury, and carries it into many products for which sodium hydroxide is a raw material. Finally, the hydrogen discharged to the atmosphere also carries some mercury vapor with it.

The complacent notion that such discharges are tolerable has been destroyed by various instances of acute mercury poisoning. Most notable was that which occurred in the 1950's in a coastal area of Japan known as Minamata Bay, where fishermen, their families, and their household cats all became stricken with a mysterious disease that weakened their muscles, impaired their vision, led to mental retardation, and sometimes resulted in paralysis and death. What the people and their cats had in common was a diet of fish, and what the fish had in their bodies was a high concentration of mercury that came from the bay waters. Minamata Bay received the mercury-containing effluent from a local plastics factory. Moreover, the mercury in the fish was present in organically bound forms that are especially hazardous to humans. These compounds are all related to methyl mercury, $H_3C—Hg—CH_3$. Such mercury compounds are sometimes used as pesticides and fungicides, and the discharge of these

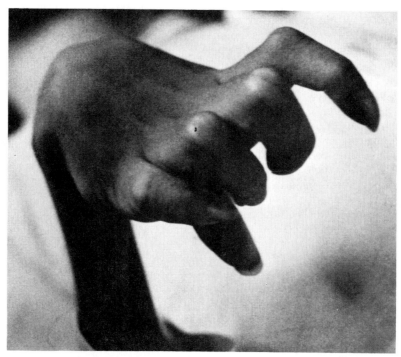

A victim of the "strange disease" at Minamata. Photo by W. Eugene Smith.

Minamata: The factory, the dump-way, the bay, and on to the sea. Photo by W. Eugene Smith.

residues into waters is therefore a serious hazard.

Following this episode, research led to the disturbing finding that metallic mercury and inorganic mercury compounds can be methylated (converted to methyl mercury) by anaerobic bacteria in the mud of lake bottoms, as well as by fish and mammals. These quantities become more concentrated as the mercury enters the food web, first in algae, then in the bodies of marine animals, and finally in fish-eating birds and land animals, including humans. The highest natural levels in meat and fish may reach several tenths of a part per million. Ultimately, decay organisms which break down animal and plant tissue serve to return the mercury to the seas. Thus, a mercury cycle operates on the Earth. The problem with industrial effluents is that their discharges create unnaturally high concentrations. The mercurial wastes that have accumulated in muddy lake bottoms therefore cannot be regarded as inert sludges; they are potential sources for biochemical conversion into forms of mercury that can enter and pass through the food chain in increasing concentrations and thus become poisons for humans.

It would be easy to draw the conclusion from these disturbing circumstances that it would be good to "abolish" mercury, if that were somehow possible. That conclusion would be incorrect. Since mercury has been present in the environment, including the food chains, throughout the history of life on Earth, we have necessarily developed a tolerance for the concentrations to which we have been exposed during evolution. "Tolerance" is not a passive detachment; it is a biochemical adjustment, and such an adjustment usually leads to dependence. It is therefore likely that we require small amounts of mercury just as we do traces of other metallic elements that would be poisonous in higher concentrations. To say that mercury is the danger is therefore only a limited truth; the more general conclusion is that the danger lies in heedless disruption of a delicately balanced natural process.

TAKE-HOME EXPERIMENTS

1. **Dissolved solids in water.** Use a clean jar to get a sample of water you wish to test. Filter the water, using a coffee filter or paper towel. Now evaporate all the water and weigh the residue by means of the procedure described on page 438. Test the following types of water samples: rain water or distilled water; tap water; water from a natural source, such as a river, stream, or pond; sea water, if available. Add to this list any other sample you consider to be of interest.

2. **Water purification with activated carbon.** Food colors are usually available in small containers from which they can be

dispensed in drops. Using such colors, prepare a set of lightly tinted solutions in ordinary drinking glasses, filled about ¾ full. Now stir a little activated carbon powder, of the type used for aquariums, into each glass. The carbon may be purchased from a pet shop, drugstore, or hobby shop. Place a saucer over each glass and allow the carbon to settle overnight. Note the effectiveness with which the colored impurities are removed.

Can you design a series of experiments to determine how much carbon is needed to remove a given amount of dye? Or to determine which dyes are easier or harder to remove?

3. **Acids and bases in water.** Acidic and basic impurities occur as pollutants in water. They can be detected conveniently by substances whose color depends on acidity or basicity; these substances are called **indicators**.

Prepare an acid solution by mixing tap water with a little vinegar or lemon juice.

Prepare a basic solution by mixing tap water with a little dishwasher soap or other strong soap.

Many juices from naturally colored foods or flowers will serve as indicators. Purple grapes, red cabbage, or blueberries are among the best choices. Grind up one of these materials and extract the juice. This juice is your indicator. Place a few drops of your indicator in the acid solution and note the color. Repeat with the basic solution. The base and acid can neutralize each other. Verify this by testing various mixtures with your indicator.

4. **Fermentation.** Mix about 30 ml (about 1 fluid ounce) of molasses with about 250 ml (about a cup) of water and about 1 gram of dry yeast (about ⅛ of a packet) in a narrow-necked bottle. Close the bottle and shake up the mixture. Now remove the cap and cover the bottle with a piece of paper secured with a rubber band, and punch a few pinholes in the paper. DO NOT SEAL THE BOTTLE; PRESSURE WILL DEVELOP. Let the bottle stand in a moderately warm room for 3 days. Do you see any bubbles in the liquid? What is the gas? Remove the paper and pour some of the contents in a cup. Smell the liquid. What substance was produced?

PROBLEMS

1. **Vocabulary.** Define water pollution; water quality.

2. **Contamination.** Discuss the relative ease of contamination of gases, liquids, and solids.

3. **Vocabulary.** Define molecule; ion; colloidal particle; suspended particle.

4. **Water pollution.** A healthy person lives in harmony with bacteria in his digestive system. Why, then, should water that contains digestive bacteria be considered to be polluted?

5. **Impurities in water.** Imagine that you had a sample of water containing all the impurities listed in Table 14.1, and that you purified it in the following successive stages: (a) Filter it through insect screening. (b) Filter it through filter paper that

removes suspended but not colloidal particles. (c) Boil it so that dissolved gases are expelled. (d) Distill it so that inorganic compounds are left behind.

List typical substances that would be removed in each step.

6. **Water pollution.** Explain how a nontoxic organic substance, such as chicken soup, can be a water pollutant.

7. **Hydrological cycle.** Why is it conceivable that despite the great mass of water on Earth, efforts to control climate by such means as cloud seeding might affect the hydrological cycle?

8. **Hydrological cycle.** (a) Do the numbers in Figure 14.2 imply that all the water on Earth is conserved? Justify your answer. Do you think that such conservation is absolutely complete? If not, how could any

net gains or losses occur? Would such differences be of any ecological importance?

9. **Vocabulary.** Define the terms aerobiosis; anaerobiosis; nitrification; fermentation; putrefaction.

10. **Oxygen.** What harmful effects on water quality result from the depletion of molecular oxygen?

11. **B.O.D.** Define biodegradability; biochemical oxygen demand. In what way is the latter a measure of water pollution?

12. **Eutrophication.** What is eutrophication? Explain how it occurs and why it is hastened by the addition of inorganic matter such as phosphates.

13. **Nutrients.** It has been suggested that the world food shortage could be alleviated if we cultivated sewage and processed the final product in the form of "algae-burgers."

   (a) Could such production be carried out on a 24-hour basis? Only during the daytime? Only at night? Explain.

   (b) If the sewage were used as the food in a "fish farm," would the product be able to feed more people or fewer people? Explain.

14. **B.O.D.** The B.O.D. curve of Figure 14.5 shows that the rise and fall occur sharply but not instantaneously. How would the curve look if both the rise and fall did start instantaneously? Which of the following is the more reasonable explanation for the non-instantaneous character of the changes: (a) some smaller discharges, such as those from individual homes or small farms, occur both before and after the main sewer effluent; (b) the sewage does not react instantaneously with oxygen. Defend your answer.

15. **Oxidation.** In the absence of molecular oxygen, O_2, some bacteria, called **facultative bacteria,** can use the oxygen content of various ions, such as sulfate and nitrate, for oxidation of organic matter. A simplified chemical equation for this process is:

$$\text{Nutrient} + NO_3^- \longrightarrow \text{Oxidized nutrient} + NO_2^-$$

Would you consider such a change more closely analogous to aerobiosis or to fermentation? Defend your answer.

16. **Energy.** (a) Show that the relative energies obtained from complete oxidation and from fermentation of sugar as shown in the relevant equations of Section 14.5 are consistent with the values given in Figure 3.1 on page 39. (*Hint:* The nutrient in an apple is mostly starch, which has about the same energy content per gram as sugar. Evaluate the ratio of oxidation energy/fermentation energy from the two sources and determine whether or not they are roughly equal.) (b) What do you think of the prospects of generating electricity by the anaerobic oxidation of sewage sludge?

17. **B.O.D.** If the half-life of nutrient matter at 20° C is one day, what percentage of it will be consumed after five days? Would it be reasonable to approximate the B.O.D. by measuring the oxygen demand created in one day, then doubling your answer?

18. **Water pollutants.** Is the speed of settling of particles in water directly proportional to their diameters? If the diameter is multiplied by 10, is the settling speed 10 times faster? (Justify your answer with data from Figure 14.3.) Is a settling pond a good general method of water pollution control? Explain.

19. **Lakes.** (a) Construct a graph in which the vertical axis is depth and the horizontal axis is temperature. Draw one curve that represents a lake in summer, another that represents winter. (b) Construct similar graphs of depth vs. oxygen content. How would the shapes of these curves change with advancing eutrophication?

20. **Home laundry.** Write up a set of specific laundry instructions that embodies your personal decisions about pollution, hygiene, and washing effectiveness.

21. **Industrial wastes.** List seven metals whose compounds may have been implicated as water pollutants.

22. **Acidic pollutants.** The contents of our stomachs are acidic, and we drink acidic

fruit juices without doing ourselves any harm. Why, then, are acids considered to be pollutants in drinking water?

23. **Oil spills.** What are four ecologically significant actions of spilled oil in the ocean? List them in what you consider to be the order of increasing environmental damage that would occur within a year after the spill. Would your order be different if you judged the consequences a decade after the spill?

24. **Oil spills.** (a) Briefly summarize the technological responses available to deal with oil spills. Which ones are preventive? Which are remedial? (b) What changes in legislation or public policy could an individual nation make to protect itself from the effects of oil spills? What factors would limit the benefits of such measures? (c) What changes of international policy would be required to protect the global ocean from the effects of oil spills? (d) Compare, in a general way, the relative effectiveness of technological responses with political responses to oil spills.

25. **Oil pollution.** The half-life of carbon-14, which is produced in the atmosphere by cosmic rays, is about 5700 years. As a result, recently produced organic matter has practically its original concentration of carbon-14, whereas "old" organic matter, such as fossil fuels, has practically none. Explain how you could differentiate between sewage and oil pollution in a stream, based on observations of carbon-14 levels.

26. **Oil spills.** Discuss some of the ecological consequences of the death of large numbers of sea birds.

27. **Oil spills.** A technique used by the French in the Torrey Canyon disaster was to dump chalk dust on the slicks to absorb the oil. The chalk-laden oil is heavier than water, and sinks. Can you suggest advantages and disadvantages of this procedure compared with the detergent method? Compared with not doing anything?

28. **Ocean pollution.** What are the various categories of ocean pollutants? List them in what you consider to be the increasing order of their threat to life in the sea.

29. **Ocean pollution.** In its article on "Sewerage," the Eleventh Edition of the Encyclopaedia Brittanica, published in 1910, states, "Nearly every town upon the coast turns its sewage into the sea. That the sea has a purifying effect is obvious. . . . It has been urged by competent authorities that this system is not wasteful, since the organic matter forms the food of lower organisms, which in turn are devoured by fish. Thus the sea is richer, if the land is the poorer, by the adoption of this cleanly method of disposal." Was this statement wrong when it was made? Defend your answer. Comment on its appropriateness today.

30. **Water quality.** What are the criteria for water that is considered fit for drinking? Is such water always safe? Is water that does not meet these criteria always harmful? Explain.

31. **Health effects.** When people infected with schistosome worms excrete the eggs via the stools, the eggs hatch into forms which must find a certain type of snail to complete its life cycle. Once safely in the snail, the worm develops into a different form which leaves the snail and is again infective for man. In thinking about ways to decrease the incidence of schistosome infection, scientists have considered two major types of measures: (a) developing compounds which kill the snails, thus preventing the worm from completing its life cycle, and (b) increasing measures of sanitation to prevent human feces from getting to the water supply. Discuss what you think would be the pros and cons of these methods specifically with reference to relative costs, chemical pollution of the environment, and effect on other forms of life besides the snails.

32. **Water purification.** An alternate method of waste water treatment is the **stabilization** or **oxidation pond,** which is a large shallow basin in which the combined action of sunlight, algae, bacteria, and oxygen purifies the water. It may be said that the stabilization pond trades time, space, esthetics, and flexibility for savings in capital and operating costs. Explain this statement.

33. **Water purification.** Biological treatment

of waste water reduces the mass of pollutant. Where does the lost matter go?

34. **Water purification.** Distinguish among primary, secondary, and tertiary types of waste water treatment.

35. **Water purification.** The following advice has been offered to tourists who wish to avoid ill effects from drinking water in areas where sanitation is uncertain or where intestinal disorders are common: Do not drink cold tap water. **Never** drink water from pitchers, carafes, or bottles which may have been reused, even if washed. On the other hand, tap water that is hot enough to burn your hand may be directly used to rinse and fill a cup and, when it has cooled, is safe to drink. Suggest a rational justification for each of these statements.

36. **Sewage.** In a combined piping system, some untreated sewage is dumped into the receiving watercourse during rainstorms. Is this procedure more acceptable than it would be in dry weather? Defend your answer.

37. **Sewage treatment.** What is flocculation, and how does it help to purify water?

38. **Sewage treatment.** List and explain four methods of "advanced" water treatment.

39. **Aquatic energy cycle.** If Figure 14.4 referred to the energy cycle in a trickling filter, should any additional organisms have been included? If so, what kinds?

40. **Sewerless toilets.** Four types of sewerless toilets are mentioned on page 484. Which two do you think require the least energy to operate? Which one requires the most?

41. **Water purification strategy.** (a) The sketch shown below is a partial copy of Figure 14.15, showing one portion of a strategy for water purification. Describe this strategy in your own words. (b) Redraw the figure shown here and, using different colors for each new sequence, add lines showing (i) recreational waters discharged after primary treatment, (ii) industrial cooling waters discharged after primary treatment, and (iii) manufacturing wastewater discharged after all three treatments. (c) Redraw the figure once again, but this time illustrate a strategy that you think might be appropriate for treatment of waters in the locality where you live.

42. **Mercury.** Do you think that the known industrial hazards of mercury should necessarily have led to the conclusion that mercury would be a water pollutant if the metal were discharged into streams or lakes? Why were the effects of mercury in water not anticipated?

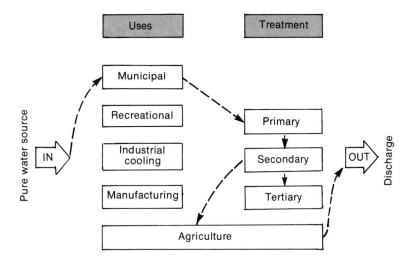

Strategy for water pollution control.

BIBLIOGRAPHY

There is a considerable volume of literature on water, on the analysis of its impurities, and on various aspects of its purification, including sewage treatment. A basic three-volume text on the properties of water itself is:

Felix Franks, ed.: *Water—A Comprehensive Treatise.* New York, Plenum Publishing Co., 1973. Vol. I, 596 pp.; Vol. II, 640 pp.; Vol. III, 430 pp.

An older but still valuable one-volume text is:

Ernest Dorsey: *Properties of Ordinary Water-Substance.* New York, Litton Educational Publisher, Van Nostrand—Reinhold Books, 1940. 704 pp.

The following texts are good sources of information on water pollution and its control:

Charles E. Warren: *Biology and Water Pollution Control.* Philadelphia, W. B. Saunders Co., 1971. 434 pp.

Metcalf & Eddy, Inc.: *Wastewater Engineering.* New York, McGraw-Hill, 1972, 782 pp.

John W. Clark, Warren Viessman, Jr., and Mark J. Hammer: *Water Supply and Pollution Control.* Scranton, Pa., International Textbook Co., 1971. 661 pp.

T. R. Camp: *Water and its Impurities.* New York, Litton Educational Publisher, Van Nostrand—Reinhold Books, 1963.

G. V. James and F. T. K. Pentelow: *Water Treatment.* 3rd Ed. London, Technical Press, 1963.

For a detailed study of eutrophication, refer to:

National Academy of Sciences: *Eutrophication: Causes, Consequences, Correctives.* Washington, D.C., National Academy of Sciences Press, 1969. 661 pp.

The following three books deal with the marine environment, although in different ways. The first is a general text. The second, on the Torrey Canyon, analyzes that disaster in detail. The third is a fascinating, well-written account of supertankers, written by an author who took a voyage on one.

National Academy of Sciences: *Beneficial Modifications of the Marine Environment.* Washington, D.C. 1972. 166 pp.

J. E. Smith, ed.: *"Torrey Canyon" Pollution and Marine Life.* London, Cambridge University Press, 1968.

Noël Mostert: *Supership.* New York, Alfred A. Knopf, 1974. 332 pp.

The Minamata poisoning is graphically described in a photo-essay book by a husband-and-wife team who were very much a part of the action:

W. Eugene Smith and Aileen M. Smith: *Minamata.* New York, Holt, Rinehart and Winston, 1975. 192 pp.

A technical book on mercury contamination is:

Rolf Hartung and Bertram D. Dinman (eds.): *Environmental Mercury Contamination.* Ann Arbor, Mich., Ann Arbor Science Publishers, 1972. 349 pp.

A number of recent chemistry texts emphasize various environmental topics, including water pollution. One such book is:

John W. Moore and Elizabeth A. Moore: *Environmental Chemistry.* New York, Academic Press, 1976. 500 pp.

The following are cited as representative of popular books that deal with the crisis of water pollution:

D. E. Carr: *Death of the Sweet Waters.* New York, W. W. Norton & Co., 1966. 257 pp.

F. E. Moss: *The Water Crisis:* New York, Encyclopaedia Britannica, Praeger Publisher, 1967. 305 pp.

G. A. Nikolaieff (ed): *The Water Crisis.* New York, H. W. Wilson Co., 1969. 192 pp.

15

SOLID WASTES

15.1 SOLID WASTE CYCLES

Today's landscape is not littered with huge mounds of dinosaur bones or ancient fern spores, for debris from living things has been traditionally re-used, and the chemicals of one organism's wastes have been incorporated into another organism's tissue. Occasionally, chemical elements are locked for long periods of time inside glaciers or geological deposits, but changing weather patterns, continental drift, upheavals of the Earth's crust, and the actions of various organisms cause some of the long-inaccessible deposits to be returned into the rapidly recycled reserves of ecosystems.

Recycling is quite inefficient in modern societies. An apple grown in an orchard in the State of Washington may be shipped to a city on the Atlantic seaboard. After someone eats it, the core is not left out to be consumed by scavengers; rather, it is stored in a garbage pail, picked up by a truck, and dumped in an area of land or ocean too polluted to support the normal scavengers of the biotic community. Similarly, the feces of a person who has eaten the apple may ultimately be discarded into a polluted area, and the farmer in Washington must purchase manufactured fertilizers from an independent source.

To cite another example, coke, which is produced from coal, is used as a raw material for manufacturing the gas acetylene, which in turn is used for making various plastics and synthetic rubber. The plastics and rubber eventually accumulate in some location such as a garbage dump; they do not return to the mine as coal. In fact, many new synthetic materials, particularly plastics and corrosion-resistant coatings for metals, were developed to be resistant to chemical changes so that they would not deteriorate during their useful lifetimes. Unfortunately, this resistance also persists after the products are discarded. The

493

movement of matter through the industrial processes, unlike the movement through the life processes, therefore generates an ever-increasing quantity of waste, mostly in the form of solid material. This does not mean that *every* industrial product eventually becomes a dead-end waste. Some products are used as raw material for other manufacturing. Other industrial products—for example, soap—can be used as food by some living organisms. As previously stated, materials that can be consumed by living organisms are biodegradable. However, the fact that a waste product is biodegradable does not necessarily mean that it is benign to the ecosystem in which it is discarded. For example, although petroleum is degraded by bacteria, the process is very slow. Tarry residues dumped along a shoreline may disrupt a particular ecosystem long before bacteria consume the tar (see Chapter 14).

This chapter will discuss the sources of solid wastes, the extent to which they are recycled, and the problems and issues involved in their disposal. Radioactive wastes were discussed in Chapter 10, and so will not be included here.

15.2 SOURCES AND QUANTITIES

Perhaps the most noteworthy characteristic of solid wastes is their variety. In our household garbage pails there are food scraps, old newspapers, discarded paper from miscellaneous sources, wood, lawn trimmings, glass, cans, furnace ashes, old appliances, tires, worn-out furniture, broken toys, and a host of other items too numerous to mention. The total quantity of solid waste is large and increasing. In the United States municipal solid wastes (which include household and commercial discards) averaged 1.2 kilograms per person per day in 1920, 2 kilograms in 1965, and about 3 kilograms in 1975. In other words, the waste disposal system of an average city must accommodate about 85 kilograms of refuse per week for every family of four, and that is a lot of trash. In the year 1975 alone, about 70 million metric tons of municipal refuse accumulated in the United States. Residents of the United States generate more solid waste than any other people. In contrast to the 3 kilograms per person per day in the United States, residents of Australia produce 0.8 kilogram per person per day, and the average resident of India produces about 0.2 kilogram per day.

Affluence is responsible for most of this trash. Packaging materials amount to about one fifth of the municipal refuse. Thus, the facts that over 50 per cent of the cost of soft drinks is in the bottles, and almost half the cost of many other items lies in their packages, do not prevent people from buying these products. But even without wasteful packaging, American citizens discard over twice as

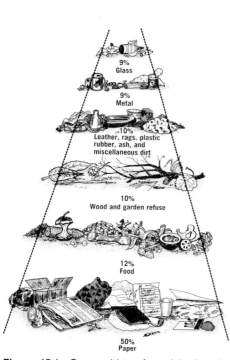

9%
Glass

9%
Metal

≃10%
Leather, rags, plastic
rubber, ash, and
miscellaneous dirt

10%
Wood and garden refuse

12%
Food

50%
Paper

Figure 15.1 Composition of municipal trash in the United States.

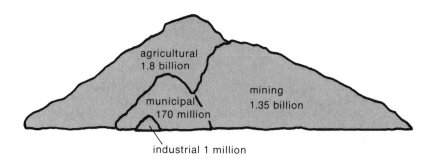

Figure 15.2 Sources and quantities of solid wastes in the United States (approximations expressed in metric tons per year.)

agricultural 1.8 billion

municipal 170 million

mining 1.35 billion

industrial 1 million

much as Australians. For example, more food is discarded, cars are junked sooner, clothes are patched less often, and tires are recapped less frequently in North America than in Oceania.

Municipal sources contribute only a fraction of the types and amounts of solid wastes discarded in the United States. Agricultural activities, for example, produce each

TABLE 15.1 Major Industrial Solid Waste Categories

Acetylene wastes	Fly ash	Precious metals
Agricultural wastes	Food processing wastes	Pulp and paper
Aluminum	Foundry wastes	Pyrite cinders and tailings
Animal-product residues	Fruit wastes	Refractory
Antimony	Furniture	Rice
Asbestos	Germanium	Rubber
Ash, cinders, and flue dust	Glass	Salt skimmings
Asphalt	Glass wool	Sand
Bauxite residue	Gypsum	Seafood
Beryllium	Hemp	Shingles
Bismuth	Hydrogen fluoride slag	Sisal
Brass	Inorganic residues	Slag
Brewing, distilling, and	Iron	Sodium
fermenting wastes	Lead	Starch
Brick plant waste	Leather fabricating and	Stone spalls (chips)
Bronze	tannery wastes	Sugar beets
Cadmium	Leaves	Sugar cane fibers
Calcium	Lime	Sulfur
Carbides	Magnesium	Tantalum
Carbonaceous shales	Manganese	Tetraethyl lead
Chemical wastes	Mica	Textiles
Chromium	Mineral wool	Tin
Cinders	Molasses	Titanium
Coal	Molybdenum	Tobacco
Cobalt	Municipal wastes	Tungsten
Coffee	Nonferrous scrap	Uranium
Coke-oven gas residues	Nuts	Vanadium
Copper	Nylon	Vegetable wastes
Cotton	Organic wastes	Waste paper
Dairy wastes	Paint	Wood wastes
Diamond grinding-wheel dust	Paper	Wool
Distilling wastes	Petroleum residues	Yttrium
Electroplating residues	Photographic paper	Zinc
Fermenting wastes	Pickle liquor	Zircaloy
Fish	Plastic	Zirconium
Flue dust	Poppy	
Fluorine wastes	Pottery wastes	

Source: Bureau of Solid Waste Management of the United States Department of Health, Education, and Welfare.

495

year over 1.8 billion metric tons of waste. About three quarters of this is manure; the balance includes forest slash from logging operations, culled fruits, slaughterhouse offal, pesticide residues, containers, and plant parts such as corncobs, leaves, and stems. Mining operations, the second major contributor, produces about 1.35 billion metric tons per year. Most of this material is rock, dirt, sand, and slag that remain behind when metals are extracted from the earth. More mineral wastes accumulate as concentrated deposits of ore are depleted and as the mineral content of the raw ores decreases. Excess rock and dirt differ from any of the other wastes mentioned in this chapter. The problems associated with strip mining, acid mine drainage, and mine reclamation were discussed in Chapter 8. Waste problems of this sort are common to other types of mining as well.

The processing of raw materials such as metals, fossil fuels, or agricultural products to manufacture airplanes, shoes, beer cans, or even balloons, always generates solid wastes. Look about the room you are in and note the number of manufactured objects. Each different kind of object was produced by a series of industrial operations, and some solid wastes were generated at each stage of production. It should not be surprising, then, that industrial solid wastes are more varied in their categories than are municipal wastes. Table 15.1 is a rather uncritical compilation of industrial sources, with some repetitive, some obscure, and some trivial entries, but it does illustrate diversity.

15.3 THE NATURE OF THE SOLID WASTE PROBLEM

Solid wastes present a many-faceted problem. The disposal of trash around the country creates litter. Its accumulation in trash cans on city streets attracts rats and flies, stimulates bacterial growth, and creates a collection problem. When large cities run out of space to dump the collected trash, a disposal problem is created. Finally, the accumulation of rusty old car bodies, cans, bottles, and other recyclable scrap hastens the depletion of non-renewable resources.

Litter is particularly vexing, because a small percentage of the population is responsible for a large nuisance. Moreover, there appears to be no solution. Advertising campaigns have been ineffective. Stiff fines, even when enacted into law, are difficult to enforce. One significant step is the development of plastic packaging materials that decompose in sunlight. For example, the plastic plate of the type shown in Figure 15.3 disappears when left outdoors. (The atoms, of course, don't disappear; they redistribute themselves to produce gases and small solid particles.) The concept of the disappearing plate is not new; food was

The ideal biodegradable package.

Nonbiodegradable containers.

served on stale bread in medieval times. In fact, in our own era, the ice cream cone may perhaps be the perfect packaging material. But what should be the ideal life span for degradable plastics? If materials are constructed to decompose rapidly, then they may become activated by the fluorescent lights in supermarkets and disappear on the shelf. On the other hand, if objects decompose too slowly, then the false sense of security that people may feel in dealing with decomposable items may result in increased carelessness and more litter.

In most cases the solid waste problem begins with collection. Usually the contents of trash cans that are placed outside near a curb are loaded into trucks which compact the refuse to increase the hauling capacity of the vehicle. This system is expensive, noisy, and disruptive of traffic. For example, one high-rise apartment building in New York City "solved" its trash collection problems by installing 400 individual garbage cans. The contents of each can are manually transferred into city sanitation trucks. If you think there must be a better system, you are right—there are several. In recent years, many owners of commercial and residential establishments have begun to use central collection boxes with a three- to ten-cubic-meter capacity. When filled, these units can be loaded into trucks hydraulically in one quick, efficient operation. In suburban or rural areas where population densities do not warrant such large containers, automatic garbage trucks, such as the one shown in Figure 15.4, can reduce collection costs and provide faster service. But the ultimate in collection efficiency is found in some Swedish cities where the garbage truck has been replaced completely by pneumatic tubes. Apartment dwellers in areas served by the pipeline simply dispose of their trash in a receiving hopper, push a button,

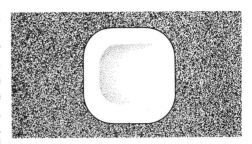
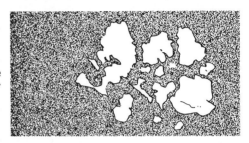

Figure 15.3 Disappearing picnic plate.

Figure 15.4 Automated collection vehicle. (From *Environmental Science & Technology* 6:415, May, 1972.)

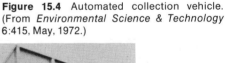

and everything is sucked into a central receiving station by a powerful vacuum system.

15.4 LAND AND OCEAN DISPOSAL

About 70 percent of the municipal refuse discarded in the United States in 1975 was deposited on land or on the ocean floor. The most primitive waste repository is the **open dump.** Waste is collected and, to save space and transportation costs, is compacted. The compacted waste is hauled to the dumping site, usually in the morning, and spread on the ground, further compaction sometimes being effected by bulldozing. Organic matter rots or is consumed by insects, by rats or, if permitted, by hogs. Various salvaging operations may go on during the day. Bottles, rags, knick-knacks, and especially metal scraps are collected by junk dealers or by individuals for their own use. In some

An open dump. (From *Sanitary Landfill Facts.* U.S. Department of Health, Education and Welfare, PHS 1970.)

communities, the accumulation is set afire in the evening (or it may ignite spontaneously) to reduce the total volume and to expose more metal scrap for possible salvage. Of course, the organic degradation, the burning, and the salvaging are recycling operations. However, there are serious detrimental features to the open dump. Its biological environment differs from those that have evolved in natural ecosystems, and is not controlled by effective regulatory mechanisms.

The result is that the organisms that multiply at the dump are not likely to be the type that are benign to people. The dump is a potential source of diseases, especially those carried by flies and rats. The fires, too, are uncontrolled and therefore always smoky and polluting. Rainfall enters the dump and removes a quantity of dissolved and suspended matter, including pathogenic microorganisms, that are water pollutants. And, of course, the dumps are ugly.

Ocean dumping is practiced by many coastal cities. Barges carrying the refuse travel some distance from the harbor and discharge their loads into a natural trench or canyon on the ocean floor. In this way most of the trash is removed from sight, though not from the biosphere. Aquatic dumping areas are almost devoid of communities of benthic animals, and thus the normal food webs in the ocean are disrupted. Although plankton and fish may survive in dump areas, they are affected by the unusual environment. For example, flounder caught in the former New York City dumping region have had off-tastes. Analysis of the stomach contents of these fish reveals that old adhesive bandages and cigarette filters constituted part of the animals' diet. Such dietary aberrations certainly cause foul flavors.

The **sanitary landfill** is far less disruptive to the environment than is uncontrolled dumping onto open land or into the ocean. A properly engineered landfill should be located on a site where rainwater, leaching through the refuse, will not pollute the groundwater. If such drainage cannot occur naturally, the site must be modified by regrading or piping to redirect the flow of water. After waste is brought to a landfill, it is further compacted with bulldozers or other heavy machinery. Large objects, such as furniture, are sometimes shredded. Each day 15 to 30 centimeters of soil is pushed over the trash to exclude air, rodents, or vermin (see Fig. 15.5). In practice, however, the distinction between the sanitary landfill and the open dump is not always sharp. For example, a thin layer of earth may be an ineffective barrier against burrowing rats, flies developing from larvae, or gases evolving from decomposition.

Sanitary landfills permit inexpensive biodegradation

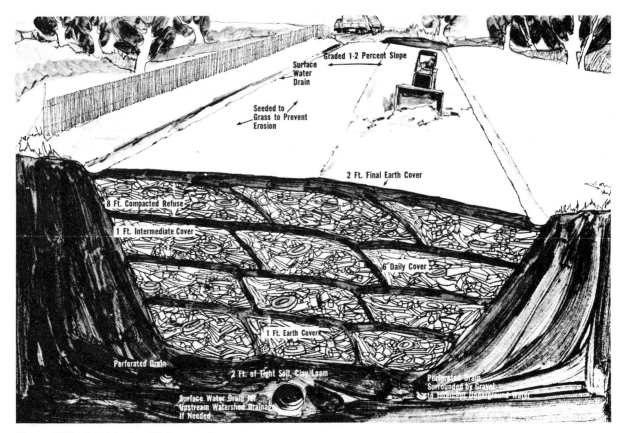

Labels within the figure:
- Graded 1-2 Percent Slope
- Surface Water Drain
- Seeded to Grass to Prevent Erosion
- 2 Ft. Final Earth Cover
- 8 Ft. Compacted Refuse
- 1 Ft. Intermediate Cover
- 6" Daily Cover
- 1 Ft. Earth Cover
- Perforated Drain
- 2 Ft. of Tight Soil, Clay Loam
- Surface Water Drain for Upstream Watershed Drainage If Needed
- Perforated Drain Surrounded by Gravel to Intercept Underground Water

Figure 15.5 Sanitary landfill in a ravine or valley. Where the ravine is deep, refuse should be placed in lifts of 6 to 10 feet deep. Cover material may be obtained from the sides of the ravine. To minimize settlement problems, it is desirable to allow the first lift to settle for about a year. This is not always necessary, however, if the refuse has been adequately compacted. Succeeding lifts are constructed by trucking refuse over the first one to the head of the ravine. Surface and groundwater pollution can be avoided by intercepting and diverting water away from the fill area through diversion trenches or pipes, or by placing a layer of highly-permeable soil beneath the refuse to intercept water before it reaches the refuse. It is important to maintain the surface of completed lifts to prevent ponding and water seepage. (Courtesy of New York State Department of Environmental Conservation.)

without much pollution, disease, or unsightliness. With proper ingenuity, landfills may reclaim spoiled land or beneficially alter the topography of an area. (Reclamation of strip mines has already been mentioned in Chapter 9.) In some cases swamps and marshes have been filled and used as building sites for apartment complexes, parks, and athletic fields. In other cases, landfill mountains have been constructed and used as ski slopes, amphitheaters, and in one case as a "soap box derby" raceway.

On the other hand, several serious problems are associated with sanitary landfills. First, land conversion is not an unmitigated gain; as we have seen in Chapter 3, the loss of marshland itself leads to ecological disruptions. Second, many large metropolitan areas are exhausting their available sites for landfill and will therefore soon be forced to transport their trash further into the countryside. In these instances, high transportation costs may well offset the low operating costs of a distant landfill. Third, and perhaps most serious, such disposal represents a depletion of resources. Food wastes and sewage sludge which could be used to enrich surface land are buried deep underground. Paper and wood scraps which could be repulped are lost, and non-renewable supplies of metals are dissipated.

15.5 ENERGY FROM REFUSE

Many metropolitan areas no longer simply dump their garbage; they incinerate it. The process, as applied to waste disposal, is more complex than simply setting fire to a mass of garbage in an open dump. A modern incinerator unit is currently used in Montreal, Canada. A crane removes refuse from a storage pit and feeds it into the furnace at a constant rate. Burning occurs on a set of three inclined grates which are agitated to insure complete combustion and a constant movement of trash into and ash out of the chamber. The ashes are ultimately removed by a conveyor belt and cooled; the metals are salvaged, and the remainder is removed to a sanitary landfill. The furnace heats the boilers and the resultant steam is sold to industry. In the Montreal incinerator an auxiliary oil burner has been installed to ensure a constant supply of steam even if a crane were to break or the sanitation workers went on strike. Finally, the furnace gases are purified using conventional air pollution control equipment.

When the Montreal plant was first built, incineration of refuse was more expensive than operation of a sanitary landfill, even crediting incineration with the value of the steam produced. The reasons behind this expense lie in the difficulties inherent in handling so heterogeneous a fuel mixture as garbage. Incinerators operate most efficiently if constant furnace temperatures are maintained, but it is difficult to control the fire when the heat content of the fuel varies significantly from load to load. Also, municipal trash contains food scraps and other wet garbage, and special problems of furnace design arise if this fraction is to be burned completely. Finally, the incineration of certain waste products produces acidic gases which corrode furnace walls and grates. Particularly notorious is polyvinyl chloride (PVC), a plastic used in the manufacture of consumer products such as rainwear, toys, containers, garden hoses, records, and credit cards. The burning of PVC produces hydrogen chloride gas which, on contact with water, becomes a solution of strongly corrosive hydrochloric acid. Even more threatening is the fact that some of the PVC decomposes before it burns, and releases vinyl chloride, a known potent carcinogen.

In spite of the difficulties of incineration, several factors have combined to alter the economic situation in recent years.

(a) Increasingly large quantities of dry paper and cardboard have appeared in refuse, thereby increasing the fuel content of trash.

(b) The price of fuel has skyrocketed since 1973, and therefore the value of steam has also skyrocketed. On the other hand, costs of handling and burning trash have increased only moderately.

A botanical garden became the final layer of this completed landfill. (From *Sanitary Landfill Facts.* U.S. Department of Health, Education and Welfare, PHS, 1970.)

(c) The rising cost of land has made it harder to find adequate sites for landfills. If valuable land is used for dumping, or if trash is hauled long distances to less expensive sites, the high cost of discarding trash makes alternative solutions more desirable.

At the present time, many industries and municipalities across the globe are burning trash as fuel, and a great many more large-scale incinerators are being built. In the United States alone, 30 cities and towns are expected to be incinerating garbage by 1980 and using the heat to produce steam for the generation of electricity.

Sewage sludge and animal manure are watery wastes that cannot sustain a flame. However, such organic residues are consumed by certain microorganisms that release gaseous products rich in **methane,** which is an excellent fuel. Some farmers have actually collected methane from cow manure, used the fuel to drive their tractors, and then recycled the residual manure as fertilizer. Although many such small-scale methane generators have been used, few large-scale units are in operation.

15.6 RECYCLING—AN INTRODUCTION

Most refuse contains a wealth of valuable raw materials that can easily be reused or recycled, to produce new products. There are many kinds of recycling paths. Consider, for example, the element phosphorus, which was discussed in Chapter 12. Much of the phosphate ore now being mined is used as a fertilizer and its phosphorus is incorporated into plant tissue, eaten by people, excreted into sewage systems, and concentrated again in sludge. In the United States most of the sewage sludge ends up in the ocean or in sanitary landfills; little is returned to the land as fertilizer. "So what," you may say, "can't we eventually mine the oceans or the old dumps?" The fallacy here, of course, is that we must think of the total environmental consequences of our waste-disposal practices. Mining the ocean may disrupt aquatic food chains. In addition, it is so expensive a process that it may increase the cost of fertilizers prohibitively. The phosphorus would be recycled most efficiently if sewage sludge were used directly as a fertilizer or soil conditioner.

Think about the articles that people normally throw away. What would be the best way to minimize such waste? Obviously, there would be less junk if fewer items were discarded in the first place. That peanut butter jar in your garbage can could be used over and over again if there were a large tub of peanut butter at the grocery store from which you could refill it. The paper towel that you threw away need never have been purchased, for a cloth towel would have worked just as well. Or what about that old automobile that was carted off to the dump? If an automobile

Much energy could be conserved if glass jars were refilled rather than discarded. In this store, cooking oil is dispensed from drums into reusable jars. (Photo by Marion Mackay.)

consumes too much oil and runs inefficiently, the engine can probably be rebuilt and its original performance restored. The body would remain untouched and other bulky components such as the engine block and the transmission could be remachined or readjusted rather than discarded. Similarly, when a refrigerator no longer cools efficiently, the fault generally lies with the compressor, not with the frame or the food compartment, and repair would conserve valuable raw materials.

In general, recycling conserves not only material resources but fuel reserves as well. For example, nearly twenty times as much fuel is needed to produce aluminum from virgin ore as from scrap aluminum, and over twice as much energy is needed to manufacture steel and paper from virgin materials as from scrap.

In many cases, recycling operations also emit less pollution than the original process. Significant quantities of contaminants are released when paper is manufactured from wood pulp or when metal is refined from ore. For example, both pulping plants and smelters discharge various sulfur compounds that foul the air and water (see Chapters 13 and 14). The Environmental Protection Agency recently estimated that recycling all the metals and papers in municipal trash in the United States would prevent the release of over 2000 metric tons of air pollutants and 700 metric tons of water pollutants annually.

But not all items are reparable or reusable. A tire can be recapped only once with safety. Week-old newspapers and spoiled meat are useless to most people. When an item cannot be used in its present condition, it must be destroyed and treated somehow to extract its useful raw materials. For example, used tires can be shredded and converted to raw rubber, old newspapers can be repulped and converted to new paper, and spoiled meat can be rendered and converted to tallow and animal feed. The complex technology available for extraction of useful goods from refuse will be treated in Section 15.7.

At best, recycling processes conserve both energy and materials. However, not every item can be recycled efficiently. For example, imagine that a careless person throws a soft drink bottle onto a roadway in some isolated rural region, and that it falls under the wheels of a trailer truck and is crushed. Crushed glass is a recyclable item, but the collection and concentration of all the crushed glass along rural roadways would consume so much energy as to be grossly inefficient. Therefore, recycling operations, despite their theoretical appeal, lose their effectiveness if the waste is widely dispersed.

One can easily envision an ideal sequence in which durable goods are used for a long time, repaired or patched to prolong their lives still further, and finally broken down into component parts for reuse as raw materials. Some materials would necessarily be drawn out of the cycle, but the size of the drain would be small. The fact is, however, that no

"One evenin' he decided t'go out an' still hunt; try t'kill 'im a bear or somethin'. Sittin' at th' head a'th'swamp an' there's three bear come walkin' out. A small little bear in front, and they's a big he bear in th' center, an' they's a little cub behind this'um. An' he waited 'til this big bear got betwixt him an' a tree t'shoot it wi' his hog rifle so he could save his bullet—go cut it out of th' tree. And he shot this big bear. . . . An' he went home an' took a axe an' cut th' bullet out. He'd take it back an' remold it. Lead was hard t'get, so that's th' way they'd try t'save their bullets." (From Eliot Wigginton (ed.): *The Foxfire Book*. Garden City, N.Y., Anchor Books, Doubleday & Co., 1972. 380 pp.)

"Admit it. Now that they're starting to recycle this stuff, aren't you glad I didn't throw it out?" (From *Saturday Review of Literature*, July 3, 1971. By Joseph Farris.)

such plan has operated effectively in modern society, for complex social and economic factors act as deterrents.

15.7 RECYCLING — TECHNIQUES AND TECHNOLOGY

Much municipal, industrial, and agricultural trash can neither be reused nor repaired, and consequently must be reduced to raw materials suitable for remanufacture. Several techniques are available for this type of recycling.

MELTING

Many materials such as metals, glass, and some plastics can be melted, purified, and recast or remolded.

REVULCANIZING

Rubber is one plastic material that cannot simply be heated and remolded. Raw rubber is gooey and formless and must be reacted with sulfur to bind the individual rubber molecules together in a cohesive form. Used rubber goods can be shredded, broken down chemically, and then rebonded in a process known as **revulcanization**.* Recycled rubber manufactured in this manner lacks the strength and resiliency of material made from virgin stock; therefore, for some applications it is useful only when mixed with more durable fibers.

PULPING AND CONVERTING TO PAPER

Any material containing natural cellulose fiber such as wood, cloth, paper, sugar cane stalks, and marsh reeds can be beaten, pulped, and made into useful fiber. The basic technology behind recycling of paper is as old as the technology of manufacturing paper. The initial step in reclamation of fiber is to mix three parts of waste paper with 97 parts of water in a hydropulping machine. Here the scrap is stirred and beaten vigorously with a device similar to an egg beater until a slurry forms. If paper from municipal sources is being pulped, de-inking chemicals are added to the pulping mixture. The de-inked pulp slurry is screened to remove large objects which might have contaminated the original stock and then rolled through wringers to remove inky water.

***Vulcanization,** named after Vulcan, the Roman god of fire, is the process of heating raw rubber with sulfur or sulfur compounds to make it stronger and more durable.

Wastepaper cycle. (Courtesy of Container Corporation.)

Small impurities are removed in a centrifugal separator and the fibers are then converted into paper by conventional procedures.

As the depletion of forest lands becomes severe, fibers from agricultural waste will undoubtedly become more attractive. When sugar is extracted from cane, the remaining fibrous stalks, or **bagasse,** are well suited to paper production. Bagasse currently contributes 60 million metric tons of solid wastes annually, and could easily be converted into a valuable resource.

COMPOSTING

As we have noted in Chapter 3, the recycling of organic matter by decay organisms produces humus. The controlled, accelerated biodegradation of moist organic matter to a humus-like product that can be used as a fertilizer or a soil conditioner is known as **composting.** This process is a practical method of recycling organic wastes. Almost any plant or animal matter, such as food scraps, old newspaper, straw, sawdust, leaves, or grass clippings, forms an excellent base for a composting operation. To increase the surface area available for decomposition, this stock is first shredded or ground, and packed loosely. The organisms essential to composting need not only cellulose and starch but also nitrogen, phosphorus, potassium, and trace elements. Thus straw, newspapers, and sawdust, which consist mainly of compounds of carbon, hydrogen, and oxygen, do not compost well by themselves but must be mixed with some source of additional chemicals. Manure is an excellent source of nutrients for composting, but chemical fertilizers or materials as diverse as peanut shells or dried blood can substitute for manure.

Aerobic decomposition is generally quicker and more complete than anaerobic action, so a compost pile must be agitated or turned in some way to promote aeration. Additionally, water in appropriate amounts aids the growth of decay organisms.

The simplest such operations are known to backyard gardeners as **compost piles** and to industrial composters as **windrows.** In industrial operations refuse is shredded, and sometimes inorganic matter is removed. Sewage sludge is then added until a favorable chemical composition is obtained, and finally the solids are mixed and piled in long rows. As decay accelerates inside the windrows, heat is generated. In a properly regulated system, temperatures will reach 66°C (150°F), which is hot enough to kill pathogenic bacteria. These windrows are turned periodically by machine for a period of six to seven weeks. The final product is then dried and sold.

Although windrows are cheap and simple to operate, they generate odors and need a great deal of land to accom-

"Compost" is an old form of the word *composed* or *composite.* Later, compost came to mean a composition or combination of anything, as in "to know what malice is . . . what villainy or treachery is, for Satan is but a compost of these" (Jackson, *Creed,* 1640). In cookery, the word became "compote," referring to a mixture of fruit preserved in syrup. In agriculture, compost means a mixture of various ingredients for fertilizing land, and this is the sense in which the word is used here.

A "windrow" is obviously derived from *wind* and *row,* and it means a row of agricultural matter, such as grass or hay, that is set out to be dried by exposure to the wind. In modern industrial agriculture, the *row* is more important than the *wind,* for it is the exposure to operations by machine, rather than by weather, that dictates the arrangement.

modate a large operation. Therefore, several different types of compost reactors have been built. One of these uses a large inclined drum which rotates slowly and pushes the refuse through the system at a controlled rate. If the temperature, airflow, agitation, water, and nutrient levels are carefully controlled, complete composting is possible in five to seven days. During the 1960's and 1970's many commercial composting operations that were established in the United States either went bankrupt or curtailed operations. The primary reason for the failures has been the lack of markets for the compost. Chapter 11 pointed out that farmers have often found inorganic fertilizers cheaper to use than manure. Such circumstances have worked against the sale of compost. However, this situation is not universal. In the Netherlands, for example, a great deal of organic refuse is composted because in that country farmers believe that soil conditioning with humus is valuable in the long run. In North America the demand for humus may increase as the price of inorganic fertilizers rises.

RENDERING

Rendering is the cooking of animal wastes such as fat, bones, feathers, and blood to yield both a fatty product called tallow, which is the raw material for soap, and a non-fatty product that is high in protein and can be used as an ingredient of animal feed. The raw material for a rendering plant comprises wastes from a wide variety of sources—farms, slaughterhouses, retail butcher shops, fish processing plants, poultry plants, and canneries. If there were no rendering plants, these wastes would impose a heavy burden on sewage treatment plants, as well as add pollutants to streams and lakes and nourish disease organisms. At the rendering plant, the waste materials are sterilized and converted to useful products, such as tallow and chicken feed. But the rendering process generates odors. Although these odors can be controlled by methods such as incineration, any control method is subject to occasional interruptions, such as those that might be caused by a power failure.

The community near a rendering plant may feel that the disadvantages outweigh the advantages, but such a judgment is based only on local considerations. As in so many other instances, the cost of environmental improvement is not borne equally by all.

FERMENTATION

Recall from Chapter 11 that yeasts can be cultured on a mixture of sugars derived from agricultural wastes. In this way, such materials as straw, sawdust, orchard prun-

Formation of windrows in composting operations.

The treatment of waste matter was once considered to be one of the "offensive trades," which were characterized by offensive odors. These trades were generally associated with decomposition of animal matter, and therefore centered on dead animals and the products obtained from them (such as glue, fertilizer, and leather) and on the disposal of human wastes and human dead. The Massachusetts Supreme Court cited soap works, glue works, slaughterhouses, rendering plants, and tanneries among these industries. We may now categorize all these as recycling operations

507

Figure 15.6 Destructive distillation. Organic fractions of municipal wastes can be pyrolyzed thermally to yield valuable by-products. (From *Environmental Science and Technology* 7:597, July, 1971. Courtesy of American Chemical Society.)

ings, corncobs, and even hydrolyzed garbage can be converted to food. It is also possible to collect methane from anaerobic composting or yeast culturing.

DESTRUCTIVE DISTILLATION (PYROLYSIS)

Destructive distillation, or pyrolysis, is the process by which a material is decomposed by heating it, in the absence of air, to about 1650°C (3000°F). If a substance like rubber or plastic that is composed of organic molecules is pyrolyzed, the molecules fragment, re-react, and re-fragment many times until an equilibrium is reached. This equilibrium mixture contains a wide variety of different valuable chemical compounds. The pyrolysis products of municipal refuse are shown in Figure 15.6 This technique appears particularly advantageous in that the pyrolysis equipment is essentially a closed system and therefore does not discharge pollutants into the atmosphere.

MUNICIPAL AND INDUSTRIAL SALVAGE

Industrial salvaging comprises a wide variety of operations in which industrial wastes are reused. Some common examples include manufacture of particle board and chipboard from sawmill waste, manufacture of building bricks from fly ash, mine tailings or glass, and manufacture of felt from fur scraps.

Table 15.2 summarizes various recycling routes of some common wastes. All of the processes illustrated in Table 15.2 are useful if the available waste is reasonably

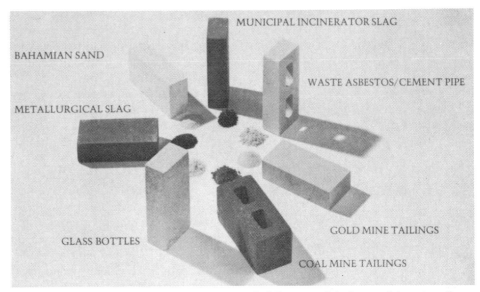

Building materials from solid wastes. (Courtesy of *House and Home Magazine*.)

TABLE 15.2 Various Recycling Routes of Some Common Wastes

WASTE	RECYCLING POSSIBILITIES
Paper	Use the backs of business letters for scrap paper or personal stationery; lend magazines and newspapers to friends, etc. Repulp to reclaim fiber. Compost. Pyrolyze. Incinerate for heat.
Glass	Purchase drinks in deposit bottles and return them; use other bottles as storage bins in the home. Crush and remelt for glass manufacture. Crush and use as aggregate for building material or anti-skid additive for road surfaces.
Tires	Recap usable casings. Use for swings, crash guards, boat bumpers, etc. Grind and revulcanize. Pyrolyze. Grind and use as additive in road construction.
Manure	Compost or spread directly on fields. Ferment to yield methane, use residue as compost. Convert to oil by chemical treatment. Treat chemically and reuse as animal feed.
Food scraps	Save for meals of left-overs. Sterilize and use as hog food. Compost. Use as culture for yeast for food production. Pyrolyze.
Slaughterhouse and butcher shop wastes	Sterilize and use as animal feed. Render. Compost. Pyrolyze.

509

homogeneous, but there is no one-step operation that can process a mixture as varied as municipal waste. For this reason the most important recycling operations carried out in North America today involve the reuse of industrial scrap. The printing department of a large newspaper company generates many tons of waste or unsold paper every week. This paper is clean, compact, and thus ideal for recycling. Similarly, a metal fabricating shop generates large quantities of uncontaminated scrap which can be sold.

The problems in handling mixed municipal refuse have received considerable attention. Some people have proposed enactment of legislation requiring residents and businesses to separate their trash into homogeneous categories, but such a law could hardly be enforced. Law officials would have to examine trash cans and try to ascertain who put the chicken bone in with the beer cans, and that just wouldn't work. But even if people voluntarily segregated their trash, collection problems would arise. The present system of trucking municipal trash is inefficient and costly, and it poses traffic problems in a large city. Imagine the situation that would exist if separate collections were required for wet kitchen garbage, for papers, for cans, for bottles, and for rags. Since it is unthinkable to accomplish such separations at the curb, any successful recycling of municipal refuse would be dependent on some kind of automated garbage-separating plant.

In general, different components can be separated from a given mixture because they differ from one another in their physical properties. Such properties may include size, magnetic susceptibility, mass, resiliency, color, brittleness, and density. Thus, a screen can remove old appliances and tires from municipal trash, and a magnet can remove pieces of iron from nonferrous wastes. Several devices are available to remove heavy materials from light ones. In an air classifier (Fig. 15.7A) mixed refuse is passed over a screen and compressed air is forced up through the system. The force of the air jet is adjusted so that light particles such as lettuce leaves and newspapers will travel up a tube to a conveyor belt while heavy particles such as glass bottles and cans will remain behind. A ballistic separator consists of a rotor which throws objects into a chamber with a constant force. Separation occurs because dense objects meet less air resistance and travel farther than light, fluffy ones (Fig. 15.7B). (Imagine throwing feathers and pennies.) Another type of separator operates by dropping mixed refuse on an inclined conveyor belt (Fig. 15.7C). Heavy and elastic objects bounce and roll downhill against the movement of the belt, while light and sticky ones are carried up. (Imagine dropping bowling balls and banana peels on an "up" escalator.) The cyclone was discussed in Chapter 13 as a method for removing heavy dust particles from an air

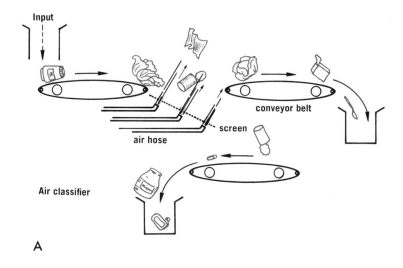

Air classifier

A

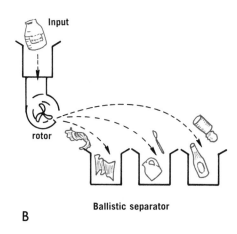

Ballistic separator

B

C   **Inclined conveyer separator**

Figure 15.7 Types of waste-material separators. *A*, Air classifier. *B*, Ballistic separator. *C*, Inclined conveyor separator.

stream by inertial differences. The same principle can be used to separate glass or metal from a water slurry of finely ground refuse. Another device isolates materials by their differential flotation in water. A mixed glass separator developed by the Bureau of Mines is particularly ingenious. The refuse is spread out along a narrow conveyor belt and passes by a series of automatic optical sensors. Each sensor is adjusted to respond to a certain color. When the electric eye sensitive to green, for example, detects a piece of light green glass, it activates a pulse of compressed air which forces the glass off the belt into a waiting bin.

A sequence of different devices must operate in series to effect complete separation. The town of Franklin, Ohio, has built a recycling system that is shown schematically in Figure 15.8. Large items like old swing sets and engine blocks are removed manually, and the remaining refuse is fed into a specially designed hydropulper. The fiber component of the refuse is pulped, and brittle material such as glass and bones is pulverized, but hard solid objects are not affected. Water is added during the pulping operation and

The Franklin, Ohio, recycling center. (Courtesy of Black Clawson Fibreclaim, Inc.)

the fiber slurry is drained off through a screen with a 2 cm mesh size. Anything that has resisted pulping and remains larger than 2 cm in diameter, such as large metal objects or rocks, is removed and washed. The iron is separated magnetically, and the nonferrous metals are separated manually for sale as scrap.

Meanwhile, the fiber slurry is pumped to a liquid cyclone which separates the small, dense inorganic matter that passed through the screening. Glass, bones, small

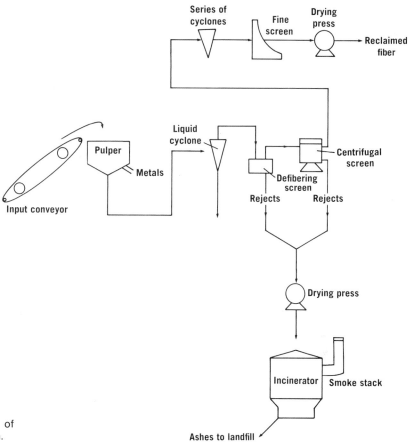

Figure 15.8 Schematic illustration of the Franklin, Ohio, recycling system.

pieces of metal, especially aluminum from cans that break up in the pulper, and miscellaneous dirt, pebbles, and particles of sand are removed in this operation.

The slurry now consists mainly of paper, food wastes, plastics, and textiles. The Franklin plant is designed to extract useful paper fiber from this mixture. This goal is realized in two more pulping and screening operations. The slurry is cleaned again in cyclonic washers and finally long, useful fibers are separated from poor-quality, short ones to be dried and sold. Organic rejects from the various screens and cyclones are considered to be nonrecyclable, but after being dried, they are suitable for use as a fuel.

15.8   THE FUTURE OF RECYCLING

Techniques are available for the recycling of most solid wastes, but the social climate is not favorable for immediate large-scale implementation of these methods (see Table 15.3). In fact, nearly 70 percent of the municipal trash in the United States in 1975 was discarded in the *least* acceptable of the disposal alternatives, that is, in the ocean or in open dumps. Approximately 22 percent was deposited in

TABLE 15.3   Some Factors That Currently Discourage Recycling in the United States

1. Public apathy toward bringing waste to recycling centers.
2. Scarcity of recycling centers.
3. Poor recyclability of many manufactured goods.
4. Tax incentives that favor mining and logging over recycling.
5. Freight rates that favor mining and logging over recycling.
6. Lack of markets for recycled goods owing to
   (a) farmer apathy towards compost, and
   (b) reluctance of many industries to try recycled goods.
7. Poor funding of research on recycling products from trees.
8. Lack of consumer pressure for recycled goods.
9. Zoning laws that discriminate against recycling.

landfills, 8 percent was incinerated (but not all of the heat produced was used industrially), and only 1 percent was recycled.

Recycling has stagnated in our society mainly because the economic externalities of unecological waste practices are generally ignored. For example, in 1976 it cost $2.75 more to process a metric ton of scrap iron than to process a metric ton of iron from the ore (see Table 15.4). Although much less energy is needed to melt scrap iron than to process ore, these savings are offset by the higher labor and transportation costs for the scrap iron. A heterogeneous mixture of scrap must be collected, sorted, and shipped to market; all these operations require much labor. In addition, the railroads charge over twice as much to ship scrap as to ship raw ore. The railroads argue that they are not discriminating against scrap but that higher operating costs lead to higher prices. A single iron mine produces many carloads of ore daily, and an entire train can be routed directly from mine to furnace. But even a large scrap dealer may deliver only one or two carloads of metal per week. These carloads must be added to a larger freight train and then routed and switched many times before they can be delivered. These routing and switching operations are time consuming and account for increased costs. Once scrap arrives at the mill, it must be handled and sorted again—steel is put in one pile, cast iron in another, and so on. This sorting adds yet additional expense. Finally, scrap iron often contains impurities that may weaken new steels or erode furnace walls.

Of course, this economic analysis does not include the cost of discarding used iron. If fuel were taxed in an effort to conserve dwindling supplies, if mills were forced to meet more stringent air pollution standards, or if steel manufacturers were charged with the cost of discarding any unused scrap, iron recycling would become economically more attractive. Alternatively, when the depletion of natural resources becomes more acute, and the costs of virgin materials increase, recycling will become profitable.

In general, European reclamation practice and technology are more advanced than North American simply

TABLE 15.4 Comparison of Production Costs for Recycled Iron and Iron from Virginia Ore*

| COST COMPONENT | COSTS PER METRIC TON | |
	From Ore	From Recycled Scrap
Basic raw material	$31.42	$36.93
Processing costs	13 23	6.61
Scrap handling costs	–	2.76
Furnace repair	–	1.10
Total	$44.65	$47.40

*From *Environmental Science and Technology,* 10(5), May, 1976.

It is more expensive to ship a ton of iron scrap than a ton of iron ore.

because fuels, fiber, and metals are scarcer there. Thus, reclamation affords a more compelling economic incentive.

Industry responds to a variety of pressures, not only economic but also social, political, and legal. A combination of these can be very effective. For example, although the Franklin reclamation plant is slightly less economical than a landfill, the citizens of that city decided that the marginal increase in cost (about 25 cents per family per month) was more than offset by the benefits of the plant. Also, this plant was financed partly with Federal grants to demonstrate the feasibility of recycling, so the choice to enter an "unprofitable" business was also stimulated by governmental policy. The benefits of such a policy are illustrated by the fact that the Franklin operation has served as a model for similar, larger plants in Miami (Florida), Tokyo (Japan), and elsewhere.

Other examples of this sort are abundant. The General Services Administration of the United States Government, the City of New York, the Bank of America, Coca-Cola, and Canada Dry have all ordered large quantities of recycled paper for stationery or annual reports, although at present there is no economic advantage to the use of recycled paper. Finally, it was mentioned earlier that polyvinyl chloride (PVC) containers are potentially troublesome because hydrochloric acid and vinyl chloride are generated when they are incinerated. Public pressure has impelled some food and cosmetic packagers to suspend the use of PVC containers in favor of bottles made of less damaging plastics.

Another way to accelerate recycling rates is to pass laws that encourage reuse. For example, the states of Vermont, Oregon, and Maine have posed a mandatory tax on all nonreturnable bottles (see Section 15.9). However, the deposit required may be too low. Considering the total environmental cost of a discarded bottle, including the energy and raw materials of manufacture as well as the litter and waste disposal problem, a bottle's true cost in 1975 was close to 25 cents. Perhaps a law requiring a 25-cent deposit on glass bottles would be an effective inducement for returning them.

The world cannot afford to wait to recycle a material until a resource is on the brink of exhaustion. Recall from Section 15.6 that complete recycling will never be realized, for some matter will be dissipated beyond the reaches of any reasonable recovery system. Therefore, some supply of raw raterials will always be needed, and our natural resources must be conserved now to insure adequate ore, fertilizer, and fiber for future generations.

15.9 CASE HISTORY I: *Recycling of Beverage Containers*

A century ago, milk, beer, cooking oil, and other liquids were shipped to "general stores" where a retail customer's bottle or jar could be filled from a large drum or carboy. Many granular items such as coffee and nuts that are now available in glass or metal containers were scooped from kegs into cloth sacks which could be used over and over again. In time, glass packaging became increasingly popular, especially for foods. Later the soft drink, dairy, and beer industries added a new approach to glass packaging—the deposit bottle. These generally well-constructed bottles cost more to manufacture than the two- or five-cent deposit values assigned to them. The bottling companies relied

In 1972, the energy used to manufacture nonreturnable bottles in the United States was 62 billion kilowatt hours *more* than that which would have been needed if returnables were used. That difference represents enough energy to supply almost ten million Americans with electric power for a year.

on the fact that the deposit provided sufficient incentive to the customer to ensure the bottle's return. This system worked so well at first that a deposit bottle averaged about 30 round trips. Those that were broken, lost, neglected, or used as flower vases and therefore never returned

Bottling plant capable of filling 1000 bottles per minute.

were, of course, replaced with new bottles. But gradually, consumers became so indifferent to the deposit that the average number of round trips declined to four. Moreover, the cost of washing bottles increased, so bottling companies initiated a shift to no-deposit, no-return containers.

This shift to throwaway glass has led to a tripling of energy consumption in the bottling industry. Today, 1.7 kilowatt hours of electricity, enough to light a 100-watt bulb for 17 hours, is needed to manufacture and transport *one* 12-ounce throwaway bottle.* Glass from a no-deposit, no-return bottle can be recycled, but it must be transported to the recycling plant, crushed, sorted according to color, remelted,

---

*Bruce Hanon: *System Energy and Recycling, A Study of the Beverage Industry.* Center for Advanced Computation Document #23, University of Illinois Press, 1973.

and reprocessed. Overall, more energy is needed to recycle glass from a no-deposit bottle than is needed to manufacture one from virgin materials. It is therefore obvious that raising the deposit for glass bottles and selling only returnable ones would be environmentally more sound than the present utilization of throwaways.

In the 1940's some beer and soft drink manufacturers shifted partially to the use of steel cans as containers. Cans are lighter, easier to ship, and less susceptible to breakage than glass. But steel cans are hard to open, so in the late 1950's aluminum cans appeared on the market, and soon thereafter "flip-tops" were added. Aluminum cans require even more energy to manufacture than glass bottles do. But because they are cheaper to handle *for the beverage bottlers and shippers,* they have been marketed extensively during the past decade. Since aluminum can be remelted efficiently, considerable quantities of energy could be saved if aluminum cans were recycled. However, despite active recycling campaigns, only about 20 percent of the aluminum cans sold in the United States are returned.

Many environmental groups have mounted advertising campaigns to educate people about the problems of beverage containers, and numerous recycling centers have been established throughout North America. These campaigns,

Energy is conserved when aluminum is recycled.

dependent mainly on voluntary cooperation, have met with only limited success. In 1972 the state of Oregon initiated a more comprehensive program to outlaw throwaway containers. The Oregon bottle bill banned flip-top cans and placed a mandatory 2 to 5 cent deposit on all beverage containers. Within the first year, the total quantity of roadside litter was reduced by 23 percent, and the energy conserved amounted to enough fuel to heat 11,000 housing units for an entire year. Public enthusiasm later declined somewhat, so that in 1974 and 1975 more returnable bottles were discarded, but considerable savings were still realized. Bottle bills have been enacted in Vermont and Maine, but similar laws have been defeated by voters in Colorado, Michigan, and Massachusetts.

15.10 CASE HISTORY II: *Junk Automobiles*

The scrap automobile presents a formidable problem. Traditionally, salvage dealers have stored inoperative cars in large lots (Fig. 15.9) and have sold individual parts. Salvage of this sort is an ecologically valuable operation, for functional automobile components are reused to replace worn-out components of operating vehicles.* However, because junkyards are ugly, many localities are attempting to restrain their operation. In some towns, the owner of the yard must pay a tax on all vehicles in his lot; in others, high and expensive fences must be built to hide the auto hulks from public view.

After they are stripped, cars are usually compacted into bales and shipped, often by railroad, to the steel mill. At the steel mill, some of the iron in the compacted cars is recovered. Recovery of a relatively pure metal, such as bare copper wire, is technically simple. It requires only remelting, and recasting into ingots. Recovery of metal is much more involved when a complex object like an automobile is to be scrapped. Large nonferrous components such as seats, tires, radiators, or batteries can be easily removed to leave behind a hulk that is purer in iron than the original whole vehicle. But it is expensive to remove small or firmly fixed objects such as copper wire, padded dash covers, and glass from a junk car. As a result, it had become customary to compact partially stripped vehicles into bales, and to ship them in this condition to steel mills. Because metallurgists customarily extract iron from ore that contains 70 percent rock, sand, and other nonmetallic matter, the technology is available to remove glass, plastics, asbestos, and related nonmetallic impurities from melted auto bodies. On the

---

*Recall the "automobile cycle" from Chapter 2.

Figure 15.9 Auto wrecking yard enclosed by chain-link fence in Denver, Colorado. (Courtesy of Luria Brothers and Co., Inc., a division of the Ogden Corporation.)

other hand, many metals are chemically similar to each other and respond in the same way to processing procedures. As a result, the separation of small amounts of copper from large quantities of iron is not feasible with conventional equipment and techniques. Moreover, copper is an undesirable component of most steels, for copper-iron alloys are not strong. In the past, autos were reused in steel manufacture despite the copper content. Recently, however, copper contamination has become more serious and the value of baled automobiles has decreased. To understand why, imagine that the steel used for manufacture of a car in 1940 contained no copper. When that car was scrapped and melted down, the steels contained, let us say, 0.1 percent copper. Therefore, the body and engine block of the next automobiles were already contaminated, so when they were melted along with the electric wire used in them, the copper content of the second batch of scrap steel increased. Thus, over the course of time, steels have suffered increasing contamination. In addition, the many electrical devices in modern cars (radios, push-button windows, multiple tail lights, etc.) increase the total amount of copper per car. Therefore, the value of automobile scrap has decreased in recent years.

As a result of all these and many other factors affecting the economics of salvaging automobiles (see Table 15.5), the cost of moving an old car to a junkyard is frequently higher than it is worth. This situation has resulted in an increasing rate of abandonment of cars in the United States—over 2 million cars were abandoned in 1975 alone.

But efforts are under way to increase the efficiency of salvage operation. One technique is to shred an automobile into myriads of small pieces in a giant grinder (Fig. 15.10), and then to separate the iron shreds from the strands of copper magnetically. Shredders are expensive, however, and must therefore be located in urban areas where a large supply of scrap is available to return the investment. Junk cars in rural areas cannot be shipped economically to the shredder plants. Alternatively, metallurgists have devised a process to extract copper from automobiles by immersing the whole vehicle in a molten salt bath. Contaminated waste salt from other industrial operations can be used, thus solving two solid waste problems simultaneously. A third solution relies on the fact that aluminum, a good conductor of electricity, separates easily from molten steel. Therefore, if aluminum wire replaced copper wire in cars, the problem would be alleviated. Of course, various social solutions are conceivable, such as levying a disposal tax on new cars, and then using that income to strip old cars.

TABLE 15.5 Some Factors That Control Automobile Recycling*

1. Price paid by scrap processor to wreckers for stripped auto bodies.
2. Cost of transportation of a junk car to a wrecker or scrap processor.
3. Cost of transportation of scrap from processor to consumer.
4. Magnitude of scrap market within a competitive area.
5. Current level of scrap prices, particularly for No. 1 heavy melting steel, shredded scrap, electric furnace grades, cupola grades, No. 2 heavy melting steel, No. 2 bundles, bundled No. 2 steel, automotive slab, motor block, lead, and copper scrap.
6. Lack of consumer confidence in automotive scrap specifications and quality control.
7. Lack of small amounts of money at critical points in the disposal stream.
8. Price paid to owner for automobile at time it is no longer useful as a vehicle.
9. Cost of land for storage of junk cars.
10. Lack of financial cost to owner in leaving a junk car standing on his property, since an auto license is no longer required.
11. Copper content of automotive scrap.
12. Zoning ordinances and their enforcement.
13. Automobile inspection laws.
14. Automobile damage in minor accidents.
15. Price of used or rebuilt parts for maintaining a used car.
16. Cost of producing hot metal from iron ore.
17. Cost of pig iron fed to cupola furnaces in foundries.
18. Local governmental facilities for collecting and disposing of abandoned automobiles.
19. Cost of removing radiator, electric motors, wiring, and other copper-bearing parts, and preparing them for sale as scrap.
20. Cost of burning or otherwise stripping and disposing of upholstery, tires, and other non-metallic materials.
21. Cost of baling automotive scrap.
22. Cost of shredding automotive scrap.
23. Type and size of ironmaking or steelmaking process.
24. Ratio of blast furnace hot metal capacity to steelmaking capacity in an integrated operation.
25. Lack of commercial metallurgical method for removing copper and other unwanted metals from molten iron.

*Abridged from *Automobile Disposal—A National Problem*, U.S. Government Printing Office, 1967, p. 839.

Figure 15.10 Vehicle hulks being conveyed into a scrap shredder, and the resulting scrap. (Courtesy of Luria Brothers and Co., Inc., a division of the Ogden Corporation.)

TAKE-HOME EXPERIMENTS

1. **Solid waste disposal.** Sort through your waste and garbage containers at home. List the items that (a) need not have been purchased in the first place, (b) could have been reused, and (c) could be easily recycled. Weigh your garbage daily for a week. Then start a program to reduce the solid waste in your household and repeat the weighings. How much difference has your program made? How have your personal habits been changed? Locate the recycling centers in your area. Compile data on (a) the time required to prepare cans, bottles, and paper for recycling and to bring them to the center, and (b) the cost of transporting them.

2. **Recycled paper.** In this experiment you will manufacture recycled paper from old newspaper. Cut a square of newspaper approximately 30 cm on a side and shred it into small pieces. Fill a large bowl ¼ full

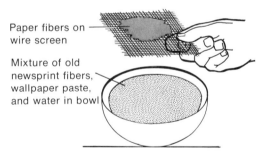

Recycling: Making New Paper from Old

of water, add the shredded paper, and let the mixture soak for an hour or two. Beat the mixture vigorously with an electric mixer or hand-operated egg beater until the paper breaks up into fibers and the mixture appears creamy and homogeneous. Next, dissolve two heaping tablespoons of starch or wallpaper paste in ½ liter of warm water, add this solution to the creamy slurry, and stir. Take a piece of fine window screen and dip it into the solution. Lift the screen out horizontally, as shown in the sketch. If you have done this correctly there will be a fine layer of paper fibers on the screen. Stir up the contents of the bowl and redip the screen carefully as many times as is needed to build up a layer about $1/4$ cm thick.

This layer must now be pressed and dried. To do this, place some cloth towels on a tabletop and lay the screen over the towels so that the fiber layer is facing upward. Cover the fibers with a piece of thin plastic (a plastic bag is fine). Now squeeze down evenly on the plastic, using a block of wood for a press. Most of the water should be squeezed from between the fibers, through the screen, and on to the cloth towel. Set the screen out to dry for a day or two, peel off the newly fabricated recycled paper, and write a letter to a friend!

PROBLEMS

1. **Disposal of manufactured products.** Which of the following manufactured products will undergo biological recycling, industrial recycling, or accumulation as a solid waste? (a) a paste made from casein (a milk protein); (b) a polyethylene squeeze bottle used as a container for mustard; (c) a copper drainpipe from a wrecked house that is sold to a dealer in scrap metals; (d) a woolen sweater; (e) the gold filling in an extracted tooth; (f) the steel in an automobile body that is returned as scrap to the mill; (g) a "no-deposit" soda bottle; (h) the coffee grounds you used to make this morning's coffee.

2. **Mine wastes.** Explain how mine wastes differ from municipal, industrial, or agricultural wastes.

3. **Types of wastes.** Give an example of a biodegradable, a combustible, a toxic, an odorous, and an inert solid waste.

4. **Garbage collection and disposal.** Consider the following three commercially available devices: (a) the home garbage disposal unit, (b) the home garbage compacter, and (c) plastic garbage pail liners. Discuss the relative merits of each with respect to the litter problem, the collection problem, and the eventual disposal problem.

5. **Land disposal.** Explain the difference between an open dump and a sanitary landfill.

6. **Land disposal.** Compare economically and ecologically the use of low and high land as waste disposal sites.

7. **Sanitary landfills.** Glass forms a stable base for landfill areas because it doesn't decompose, settle, or release pollutants into groundwater. Since currently most trash accumulates in landfills do you feel that the beneficial aspects of glass in a landfill would provide a convincing argument in favor of non-deposit bottles?

8. **Incineration.** Discuss the advantages and disadvantages of incineration as a method of waste disposal.

9. **Incineration.** Plastics, made from coal and oil, have a high heat content. If garbage incinerators were commonly used in the United States, and the resulting heat of combustion were used industrially, would you feel that plastic packaging would be an advantageous way to use fossil fuels twice? Defend your answer.

10. **Recycling.** Sand and bauxite, which are the raw materials for glass and aluminum, respectively, are plentiful in the Earth's crust. Since we are in no danger of depleting these resources in the near future, why should we concern ourselves with recycling glass bottles and aluminum cans?

11. **Recycling.** As mentioned in the text, broken or obsolete items can be repaired,

broken down for the extraction of materials, or discarded. Which route is most conservative of raw materials and energy for each of the following items? (a) a 1948-model passenger car that doesn't run; (b) a 1972-model passenger car that doesn't run; (c) an ocean liner grounded on a sandbar and broken in two; (d) an ocean liner sunk in the central ocean; (e) last year's telephone directory; (f) an automobile battery that won't produce current because the owner of the car left the lights on all day; (g) an empty ink cartridge from a fountain pen?

12. **Planned obsolescence.** An old-timer complains that years ago a man could store canned milk in the creek for three years before the can would rust through, but now a can will only last one year in the creek. Would you agree with the old man that cans should be made to be more durable? What about automobiles? Explain any differences.

13. **Recycling.** During World War II, when rubber was scarce in the United States, someone suggested that industry should develop a process to scrape rubber off curves on roadways because that is where most tire wear occurs and then re-form the reclaimed material into new tires. What do you think of this suggestion? Explain.

14. **Industrial salvage.** Why is industrial salvage more economically attractive than salvage from municipal refuse?

15. **Technology of recycling.** Describe the following processes: Revulcanization, repulping, composting, and rendering. State what materials can be treated by each of these processes, explain how they are treated, and describe what products are produced.

16. **Paper recycling.** At the present time, recycling of paper is a marginally profitable business, and many repulping mills have gone bankrupt in recent years. List some of the economic externalities associated with the production of paper from raw materials. Who bears these costs? How could the burden be shifted to encourage paper recycling?

17. **Recycling municipal trash.** A family lives in a sparsely populated canyon in the

northern Rocky Mountains. Their household trash is disposed of in the following manner: Papers are used to start the morning fire in the potbelly stove; food wastes are either fed to livestock or composted. Ashes are incorporated into the compost mixture; metal cans are cleaned, cut open, and used to line storage bins to make them rodent-resistant; glass bottles are saved to store food; miscellaneous refuse is hauled to a sanitary landfill. Comment on this system. Can you think of situations where this system would be undesirable? Do you think that it is likely that many people will adopt this system?

18. **Industrial salvage.** Manufacture of felt from fur scraps is a marginally profitable enterprise. One factory operates as follows: Fur scraps are first decomposed by heating them in a vat with dilute sulfuric acid. The useful fibers are extracted, and the remaining liquid, which consists of water, sulfuric acid, and decomposed animal skin, is dumped untreated into a river. The cost of purifying the effluent would drive this particular business bankrupt. Discuss the overall environmental impact of this fur-scrap recycling center.

19. **Recycling and the Second Law.** Referring to the discussion on page 502 of the difficulty in recovering phosphate fertilizer from ocean sediments, Professor Peter Frank of the Department of Biology of the University of Oregon, who reviewed this material, wrote, "The Second Law of Thermodynamics comes in with a vengeance. . . ." Explain what he meant.

20. **Recycling municipal trash.** The organic wastes that cannot be converted to paper fiber are incinerated at the Franklin, Ohio, reclamation center. Can you think of an alternate use for this scrap? Compare the desirability of your disposal method with incineration from an environmental standpoint.

21. **Recycling municipal trash.** Design a municipal refuse reclamation center which uses different equipment than the Franklin, Ohio, plant. Draw a flow sheet of your center and explain each operation.

22. **Recycling.** Environmental organizations have been active in establishing collection centers for old newspapers, cans, and so

on. Can you think of other activities which these groups might engage in which would produce increased recycling?

23. **Recycling.** List the most efficient recycling technique and the resultant products for each of the following: (a) steer manure; (b) old clothes; (c) scrap lumber; (d) aluminum foil used to wrap your lunch; (e) a broken piece of pottery; (f) old bottle caps; (g) stale beer; (h) old eggshells; (i) a burnt-out power saw; (j) worn-out furniture; (k) old garden tools; (l) tin cans; (m) disposable diapers. How does your choice of technique depend on your location?

24. **Bottle recycling.** In 1976 voters in Colorado were asked to decide whether or not to levy a mandatory tax on nonreturnable cans and bottles. The beverage industry strongly opposed the law. In one brochure published by a major beer manufacturer it was stated,

Claim: Amendment #8 [the proposed bottle law] would conserve energy and resources.

Fact: Any savings in coal consumption result-

ing from the law would be offset by an increase in the consumption of gasoline, natural gas and water.

Returnable bottles are heavier than non-returnable cans. Manufacturing their heavy, durable carrying cases would require an increase in energy consumption.

Bottles would also require twice as much space as cans. Trucks would have to make at least twice as many trips to haul refillable containers, to say nothing of the extra trips to pick up empties, resulting in increased gasoline consumption.

Washing re-usable bottles requires five times more water than cans. Heating the water for sterilization and removing the detergents that are used means increased energy consumption.

Examine this statement critically. Have all the facts been presented fully and accurately, or do you feel that the brochure states the case incompletely? Is it reasonable or misleading? Defend your position.

25. **Glass bottles.** Discuss some of the factors which led to the use of no-deposit, no-return bottles.

26. **Automobile salvage.** Outline some economic and technical problems of the automobile salvage business.

BIBLIOGRAPHY

A compilation of valuable articles on solid wastes is:
Solid Wastes. (Environmental Science and Technology Reprint Book.) Washington, D.C., American Chemical Society, 1971.

Most of the popular books on the environment that are cited in the bibliography of Chapter 3 include some discussion of solid wastes. For more technical information, refer to the following:
Andrew W. Breidenbach: *Composting of Municipal Solid Wastes in the United States.* Environmental Protection Agency Publication, Washington, D.C., U.S. Government Printing Office, Stock number 5502–0033, 1971.
Richard B. Engdahl: *Solid Waste Processing.* Public Health Service Publication No. 1856. Washington, D.C., U.S. Government Printing Office, 1969.
Bruce Hanon: *System Energy and Recycling, A Study of the Beverage Industry.* Center for Advanced Computation Document #23, University of Illinois Press, 1973.
N. Y. Kirov: *Solid Waste Treatment and Disposal.* Ann Arbor, Michigan, Ann Arbor Science Publishers, 1972.
Fred C. Price, Steven Ross, and Robert L. Davidson, eds.: *McGraw-Hill's 1972 Report on Business and the Environment.* New York, McGraw-Hill Publications, 1972.
United States Bureau of Mines: *Automobile Disposal—A National Problem.* Washington, D.C., U.S. Government Printing Office.

16

NOISE

16.1 HEARING

Noise, like odor, is a form of environmental stress that produces a direct sensation. But noise is not a substance, and it does not accumulate; as soon as its echo dies, it is gone. Its effects on the body, however, are not always so easily erased.

The human ear is sensitive to sounds over an extremely wide range of intensities. To appreciate this capability, we must understand that the production of sound requires *power* and that sound intensity can be related to the power of the source that produces it. It takes a *powerful* person to pull the bell rope to ring a heavy bell, and it requires *electrical power* to make a loudspeaker produce sound. As described in Chapter 8, electric power is measured in watts. The power of a small electric light bulb, such as the one inside your refrigerator, is typically about 15 watts. The sound power of a symphony orchestra, playing loudly, is no more than this! To put it another way, the same power that lights up the refrigerator bulb could generate the sound intensity of a symphony orchestra.* And to refer again to the human ear, we may say that the sound produced by 15 watts of power is very loud. Of course, we can hear much louder sounds, such as those from a jet plane or a rocket engine, whose sound power can reach millions of watts. And even more remarkably, we can hear sounds of so little power that they are constantly produced by the routine transfers of energy all around us—the rustle of a leaf, the patter of raindrops on soft earth, a whisper of human speech. The loudest sounds we can hear, at the edge of pain, are billions of times more powerful than the softest whispers, at the edge of silence.

But the human ear is capable of much more than merely sensing sounds over such a wide range of intensities. It is also sensitive to the individual pure tones that make up complex sounds; it transmits to the brain the complex information contained in speech; it serves as a direction and range finder; and, almost incidentally, it helps

---

*Remember, we are counting only the power that produces sound. Of course heat is also produced by friction, muscle action, etc., but we are excluding the power needed to generate this heat.

people to balance on two feet. It should hardly come as a surprise, then, to learn that the human ear is a complex organ, constructed of delicate parts that are readily subject to damage, and that such damage also affects other parts of the body.

A diagram of the human ear is shown in Figure 16.1. The outer ear serves the important function of funneling sound waves into the ear canal. At the inside end of the canal is the **eardrum.** Attached to the inside of the eardrum is a network of three tiny bones (hammer, anvil, and stirrup), which lie in the middle ear and form a mechanical bridge to the inner ear. When a sound wave enters the ear, the eardrum vibrates; these vibrations cause the three bones to vibrate synchronously and thus to transmit the energy of the sound wave to the window of the inner ear.

The inner ear is a complex structure that has two broad functions: (1) the conversion of sound-induced mechanical vibrations into nerve impulses, and (2) maintenance of a sense of balance and spatial orientation. In this section we are concerned with only the first of these. The part of the inner ear that is responsible for the conversion of sound-induced mechanical vibrations into nerve impulses is called the **cochlea.** This structure, set deep in the skull, has a shape similar to that of a tiny snail. When the energy of incident sound is transmitted to the inner ear as described above, the fluid inside the cochlea is also made to vibrate. These vibrations excite a second membrane which in turn excites the thousands of tiny hairs that form part of the **organ of Corti** (located in the cochlea), and, in ways that are not fully understood, the vibrations are converted into nerve impulses which travel via the auditory nerve into the appropriate part of the brain. These impulses are then interpreted as sound by the individual.

Now, obviously, "hearing" a sound is a complex process with many steps, and each step provides a place for something to go wrong. When any step in the sequence fails, hearing acuity may be diminished or even lost entirely. For

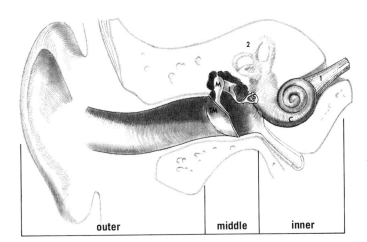

Figure 16.1 TM: tympanic membrane (ear drum); M: malleus (hammer); I: incus (anvil); S: stapes (stirrup); C: cochlea; 1. auditory nerve to brain. 2. semicircular canals.

525

example, an external canal full of wax, a perforated eardrum, a middle ear filled with pus such as happens with the common ear infections, or a diseased organ of Corti or auditory nerve—all these may produce varying degrees of hearing loss.

16.2 SOUND AND NOISE

That which we hear is called **sound.** Unwanted sound is called **noise.** Let us deal with each in turn.

SOUND

Snowy woods, a silent environment.

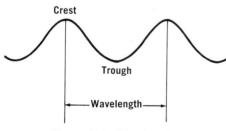

Figure 16.2 Wavelength.

To understand how sound is produced, just consider its opposite—silence. Think of the quietest occasions of your life. Perhaps standing in the woods after a snow on a windless winter day was one such instance. Certainly, you would not think of an occasion on which you were moving, whether walking, running, or riding, because motion itself produces sound. In fact, one relationship between motion and sound is suggested by the use of the word *still*, which means both motionless and quiet. We may infer, then, that sound is associated with motion. Motion, we know, is a form of energy. It follows that sound must be produced by some source that transmits its *energy of motion* to us by some means that our ears can detect.

How can such transmission take place? We note that when the bell is struck, it vibrates. When the vibration stops, so does the sound. If the bell is struck in a vacuum, it will also vibrate, but no sound can be detected. From this experiment it can be deduced that the motion of the bell somehow moves the air, which then moves some receiving device in the ear. This transfer of sound energy through air occurs in the form of a wave. We usually visualize wave motion as water waves, especially ocean waves striking a beach, or ripples in a pond or swimming pool. Water that is still ("quiet" water) has a smooth, level surface. Water waves are disturbances; they alter the normal level so as to make it higher in some places and lower in others. The highest places are called **crests;** the lowest, **troughs.** The distance between successive disturbances of the same type, such as between neighboring crests, is called the **wavelength** (Fig. 16.2). The rate at which a disturbance moves is the **speed of the wave.** The number of disturbances that pass a given point per unit time is the **frequency.** The relationship among these three attributes is given by the equation:

$$\text{speed} = \text{wavelength} \times \text{frequency}$$

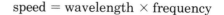

Small water waves (ripples).

All these characteristics of the water wave also apply to the sound wave, except that the nature of the disturbance is different. Instead of manifesting itself as crests and troughs, as in disturbances of the water level, the sound wave is a succession of compressions and expansions that disturb the normal density of the medium (such as air) in which they are propagated. This type of wave is called an **elastic wave** and can be illustrated by the action of a coiled spring, as shown in Figure 16.3. Imagine that a bump on a rotating wheel hits the end of the spring twice per second. The resulting compressions, and the expansions which follow them, travel along the spring at a speed that depends on the properties of the spring (not on the rate of rotation of the wheel). Let us say that this rate is one meter per second. The frequency must be two beats per second, because that is established by the speed of rotation of the wheel. Therefore the wavelength is computed as follows:

$$\text{wavelength} = \frac{\text{speed}}{\text{frequency}} = \frac{1 \text{ m per sec}}{2 \text{ per sec}} = \tfrac{1}{2} \text{ m}$$

Air is a springy substance; a squeezed balloon snaps back when released. Therefore, elastic waves can be propagated in air. This propagation is sound when the frequencies of the waves lie within the range that the ear can detect. Work must be done to beat out the successive compressions; thus, sound is a form of energy, and the rate at which it is transmitted (energy per unit time) is the sound power we discussed in the preceding section. Figure 16.4 is a schematic representation of sound waves in air,

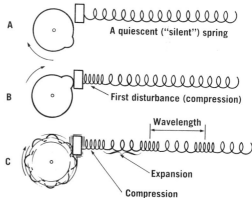

Figure 16.3 Elastic waves. *A*, Wheel starts to rotate twice per second. *B*, First impact. *C*, Continuous production of waves. Frequency = 2 per second. Wave speed = 1 meter per second.

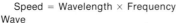

Speed = Wavelength × Frequency

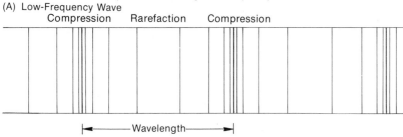

(A) Low-Frequency Wave

Compression   Rarefaction   Compression

|◄————Wavelength————►|

(B) High-Frequency Wave

|◄—Wavelength—►|

Figure 16.4   Sound waves.

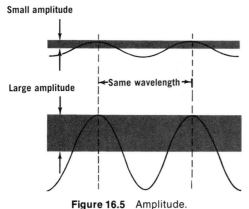

Small amplitude

Large amplitude |◄—Same wavelength—►|

Figure 16.5   Amplitude.

showing the relationship between frequency and wavelength.

The speed of sound in air under normal conditions on Earth is 334 meters/sec (1096 ft/sec, or 747 mi/hr). Any object, such as an airplane, that travels slower than sound is said to be **subsonic;** one that is faster is **supersonic.**

The pitch of any given sound is determined by the frequency of the waves that produce it. The energy of a given sound, however, is not determined by the frequency, wavelength, or wave speed. Thus, if you slap the surface of a pond gently once per second with a spoon, you will make waves at a frequency of one beat per second. If you slap the water hard once per second with a paddle, you will still make waves at a frequency of one beat per second, and they will not travel any faster, but they will be bigger waves because you have been working harder. The disturbances, or the heights of the crests and depths of the troughs, will be greater. The magnitude of the disturbance is called the **amplitude** of the wave (Fig. 16.5). The power of a wave thus depends on the rate at which work must be done to create the disturbance.

To dramatize the difference between sound power and sound frequency, imagine a four-engine plane with only one engine running. The frequency of the sound depends on the type of engine and its operation; the speed of the sound is a property of the air, not of the engine. Now we start the other three engines, so that they operate just like the first.

The frequency of the sound has not changed, nor has the speed of the wave. But four engines are four times as powerful as one engine. The sound power is therefore four times as great.

Sound power is related to loudness, but the two are not the same. We shall discuss loudness in more detail in the next section.

NOISE

Noise can be defined, simply, as unwanted sound. This concept is straightforward enough, but it does not teach us how to predict which sounds will be disliked. After all, a given sound may be music to one person but noise to another, pleasant when soft but noise when loud, acceptable for a short time but noise when prolonged, intriguing when rhythmic but noise when randomly repeated, or reasonable when you make it but noise when someone else makes it. Of all the attributes that distinguish between wanted and unwanted sound, the one that we generally consider the most significant is loudness. There is ample evidence that exposure to loud sounds is harmful in various ways, and, in any event, loudness tends to be annoying; therefore, the louder a sound is, the more likely it is to be considered noise.

There are yet other subtleties to complicate matters. People often associate noise with power, and in that context noise becomes desired. For example, some homeowners choose noisier lawn mowers and vacuum cleaners over quieter ones in the mistaken idea that the noisier ones work better or faster. And some drivers of "hot rods" or of trucks remove the exhaust mufflers from their vehicles to help "soup them up." In these instances it is true that such removal reduces the resistance imposed on the exhaust gases and thereby may increase the efficiency of the engine, but the benefit is small (two to three percent) and the cost is high (impaired hearing). Some of these attitudes date back to the 1940's, however, and have been undergoing a reversal in favor of quietness during the heightened environmental awareness of recent years.

More difficult to modify, perhaps, is the association of noise with social recognition. A loud noise connotes authority, even though what it utters may be nonsense. Such is the case, all too often, with loud speech, a loud motorcycle, loud music, or roaring toys, especially toy guns and firecrackers.

16.3 LOUDNESS AND THE DECIBEL SCALE

We have noted that, other things being equal, the louder a sound, the more likely it is to be considered noisy. And with good reason—loud sounds tend to interfere with our activities and as we shall see, they can be physically harmful. But loudness is not energy, or pressure, or frequency, or anything else that can be measured with a physical instrument. Loudness is a sensation, and if you want to know how loud a sound is, you must get the answer from the person who hears it. However, it is possible to obtain useful information from instruments by measuring some physical property of sound that is *related* to the human perception of loudness. Air must vibrate against the eardrum before sound can be transmitted to the brain. Therefore, it is reasonable to use a device that responds to *sound pressure* (see box at left) and to relate its response to perceived loudness.

The instrument that measures sound is called a decibel meter, and a scale of such values is called a decibel (abbreviated dB) scale. Now, recall that the human ear is sensitive to sounds over an extremely wide range of intensities, so that if our scale were directly proportional to sound intensity, it would suffer the inconvenience of ranging from very small to very large numbers. Furthermore, the scale would be awkward, for it would not start at zero,

---

DEFINITIONS AND UNITS RELATED TO SOUND

Acoustic power, or *sound power,* is the sound energy per unit time radiated by a source. It is measured in watts. *Sound intensity* is the sound power that passes through a unit area (watts/meter^2). A point source of sound that radiates, say, 10 watts becomes the center of an ever-expanding field of sound. The intensity of the sound diminishes as the radius of the field increases, though the power of the source remains unchanged. The "intensity of a barely audible sound" is usually taken to be one-trillionth of a watt per square meter, or 10^{-12} watt/m^2. *Sound pressure* is the force per unit area exerted by sound of a given intensity. *Loudness* is the listener's subjective judgment of the intensity of the sound as it is received.

Decibel meter. Photo courtesy of Gen-Rad, Inc. (formerly General Radio Company), Concord, Mass.

but at some small number that represents the softest audible sound. To avoid these complications, the decibel scale (Table 16.1) is set up as follows:

(a) The scale starts at zero dB, which represents the softest sound that is audible to the human ear.
(b) Each *tenfold increase* in sound intensity is represented by an *additional* 10 dB. Thus, a 10-dB sound is 10 times as intense as the faintest audible sound. (That still isn't very much.) The sound level in a quiet library is about 1000 times as intense as the faintest audible sound. Therefore the sound level in the library is $10 + 10 + 10$ or 30 dB. To summarize:

Sound	Sound level in decibels
Faintest audible sound (threshold of hearing)	0
A leaf rustling, at 10 times the intensity of the threshold of hearing.	10
Sound level in a broadcasting studio, at 10 times the level of the rustling leaf. (This is also 100 times the level of the faintest audible sound.)	20
Sound level in a quiet library, at 10 times the level of the broadcasting studio. (We are now at 1000 times the threshold of hearing.)	30

TABLE 16.1 Sound Levels and Human Responses

SOUND INTENSITY FACTOR	SOUND LEVEL, dB	SOUND SOURCES	EFFECTS		
			Perceived Loudness	Damage to Hearing	Community Reaction to Outdoor Noise
1,000,000,000,000,000,000	180	• Rocket engine			
100,000,000,000,000,000	170				
10,000,000,000,000,000	160				
1,000,000,000,000,000	150	• Jet plane at takeoff	Painful	Traumatic injury	
100,000,000,000,000	140			Injurious range; irreversible damage	
10,000,000,000,000	130	• Maximum recorded rock music			
1,000,000,000,000	120	• Thunderclap			
100,000,000,000	110	• Textile loom • Auto horn, 1 meter away • Riveter • Jet fly-over at 300 meters	Uncomfortably loud	Danger zone; progressive loss of hearing	
10,000,000,000	100	• Newspaper press			
1,000,000,000	90	• Motorcycle, 8 meters away • Food blender			Vigorous action
100,000,000	80	• Diesel truck, 80 km/hr, 15 m away • Garbage disposal	Very loud	Damage begins after long exposure	
10,000,000	70	• Vacuum cleaner			Threats
1,000,000	60	• Ordinary conversation • Air conditioning unit, 6 meters away	Moderately loud		Widespread complaints
100,000	50	• Light traffic noise, 30 meters away			Occasional complaints
10,000	40	• Average living room • Bedroom	Quiet		No action
1000	30	• Library • Soft whisper			
100	20	• Broadcasting studio	Very quiet		
10	10	• Rustling leaf	Barely audible		
1	0	• Threshold of hearing			

532

Motorcycle noise.

Traffic noise.

Jack hammer noise.

Grinding noise.

(c) Decibel levels are not directly additive. If one rustling leaf is 10 times the intensity of the faintest audible sound, two rustling leaves are 20 times as intense, not 100 times as intense, and therefore *not 20 decibels louder.* It would take 10 rustling leaves to be 10 times as intense, and only then would the decibel level go up to 20 dB.

(d) But we haven't answered the question, "If one rustling leaf is 10 dB, what is the decibel level of two rustling leaves?"

Figure 16.6 is a numerical chart (called a nomograph) that will answer this question for you, if you follow the

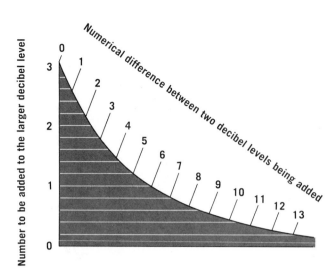

Figure 16.6 Nomograph for adding decibel levels.

procedure outlined in the box. The solution is that if the sound of one rustling leaf is 10 dB, the sound of two rustling leaves is 13 dB. Now let us try another combination of decibel levels: We will take our rustling leaf (10) dB) into the quiet library (30 dB). How much difference does the leaf make? If you now protest that this is a silly question with an obvious answer, you are right, and the meaning of the decibel scale is becoming real to you. The little leaf could scarcely make any difference because, after all, even quiet readers shuffle around a bit, turn the pages now and then, or click the latch on a briefcase, so how could they hear one more leaf, even if it is from a tree instead of from a book? Your common sense is telling you that 30 dB (the library) + 10 dB (the leaf) *cannot* equal 40 dB. In fact, if the leaf makes practically no difference, then 30 dB + 10 dB is only barely more than 30 dB! This example, too, is shown in the boxed section. If you wish to

---

EXAMPLES

Procedure	Two rustling leaves	A rustling leaf in a library
Step 1. Write down the two decibel levels to be combined.	One leaf = 10 dB The other leaf = 10 dB	The library = 30 dB The leaf = 10 dB
Step 2. Subtract one from the other.	10 − 10 = 0 dB	30 − 10 = 20 dB
Step 3. Find this difference on the curved scale and then move horizontally to the left to find the corresponding number on the vertical scale.	Zero on the curved scale corresponds to 3 on the vertical scale.	20 dB doesn't even appear on the curved scale of Figure 16.6, which shows no values above 13. If we imagine the extension of the curved scale farther to the right, the corresponding point on the vertical scale would be practically zero.
Step 4. Add this number to the larger decibel level from step 1, or to either level if they are the same, and you have the answer.	10 + 3 = 13 dB	30 dB + "practically zero" = "practically" 30 dB or, in plain language, the leaf doesn't make much difference.

534

calculate decibel levels from equations, refer to Appendix C-16.

We have been discussing decibel levels rather than loudness, but we must remember that loudness is a sensation perceived by people, not a number on the dial of an instrument. Decibel meters respond to sound pressure, and so does the human ear, but the human ear also responds to the tonal qualities of sound, and particularly to its frequency, so it is important for us to learn about these very interesting attributes.

16.4 TONE AND THE DECIBEL SCALE

Think of a wheel that is mounted vertically, such as the wheel of an automobile or a bicycle. Let us attach a leaking pen to the rim of the wheel by means of a freely rotating pivot, so that the pen always hangs straight down. The pen will drip and make an ink spot on a piece of paper below it (Fig. 16.7A). Now we rotate the wheel; the pen rotates with it, but always remains vertical. The dripping ink will therefore oscillate back and forth and will describe a straight line on the paper (Fig. 16.7B). Let us now slide the paper at constant speed at a right angle to the plane of the wheel. As the wheel continues to rotate, the pen continues to move back and forth, but it no longer leaves a straight line on the paper. Instead, we have a wavy line that looks very much like the silhouette of water waves, or like Figure 16.7C. This shape is said to be **sinusoidal.** What has all this to do with sound? It is this: If we replace our leaky pen

Figure 16.7 Generating a sine curve. *A,* Wheel and paper stationary. *B,* Wheel rotating. *C,* Wheel rotating and paper moving.

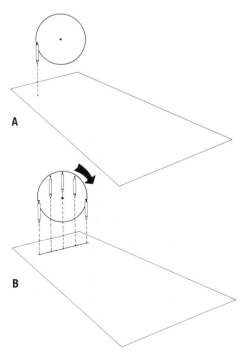

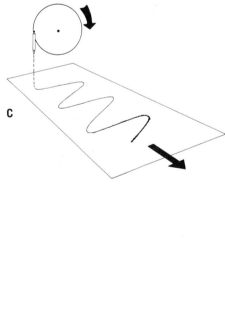

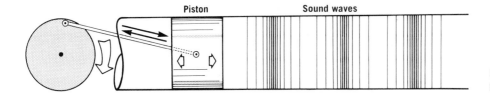

Figure 16.8 Generating a pure tone.

with a rod that drives a piston that beats against a column of air, we will create a sound wave in which the compressions and rarefactions of the air will alternate sinusoidally (that is, like a sine curve).* Such a sound is said to be a "pure tone" (Fig. 16.8).

Of course, not many natural sounds are pure tones. Any sound that can be represented by a wave form which repeats itself periodically, however, can be imagined to be made up of the sum of a series of sine waves.† The frequency of a tone, then, is the number of cycles that pass a given point per second; this unit used to be called, simply enough, "cycles per second," or cps. It is now called Hertz, Hz, after the German physicist Heinrich Rudolph Hertz. We also use the kilohertz, kHz. The relationships are:

$$Hz = cps$$
$$kHz = 1000 \ Hz$$

What of sounds that are neither pure nor periodic, sounds that do not seem to contain any internal repetitions, such as the rustle of leaves in a tree, or the gurgling of water through a drain, or the hiss of steam from a pipe, or a rumble of thunder? Can we compare them with each other in terms of frequency? The answer is yes—a hiss of steam is obviously higher pitched than a rumble of thunder, because the pressure variations in the hiss, even though they are not rhythmic, occur much more rapidly. These differences are shown in Figure 16.9.

The sensitivity of the human ear depends on the *frequency* of the sound to which it is exposed. Very high or very low notes are not heard as well as those at intermediate frequencies. Figure 16.10 shows this relationship. Note that the ear is most sensitive to pure tones at about 4000 Hz. The important information that this curve provides about the ordinary decibel scale is that it does not match the human ear very well; it is fairly good in the 1000- to 4000-Hz range, but at much higher or lower frequencies the scale would predict a loudness that the ear does not hear. We can correct for this mismatch by subtracting the

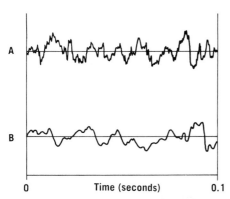

Figure 16.9 Wave forms of a hiss and a rumble: *A*, Hiss—higher frequency; *B*, rumble—lower frequency.

*See Appendix C-16 for the trigonometric description of a sine curve.
†The general statement of this theorem, first made by the French mathematician J. B. J. Fourier in 1822, is, "Every finite and continuous periodic motion can be analyzed as a simple series of sine waves of suitable phases and amplitudes."

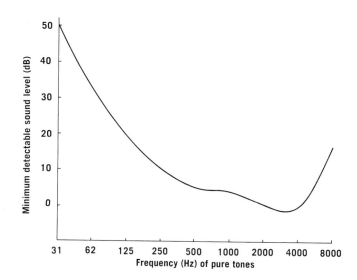

Figure 16.10 Minimum detectable sounds for human listeners.

required number of decibels from the actual sound level at the various frequencies where such corrections are necessary. The most widely used scale of this sort, called the decibel-A, or dBA scale, is shown in Figure 16.11. Decibel meters can be constructed to filter out sound at different frequencies in accordance with the dBA scale.

Human beings, however, are not decibel meters; when they are exposed to a given sound and are asked, "How loud is it?" they introduce a certain subjectivity into their answers. To reconcile this difference between man and machine, a **subjective loudness scale** has been developed. This scale is based on a comparison of any given sound with that of a pure tone having a frequency of 1000 Hz. Let us say that we wish to rate the subjective loudness of some particular sound—for example, that of a motorcycle at a distance of 8 meters. A given person is offered a selection of pure 1000-Hz sounds of different loudnesses, and is asked to pick the one that sounds just as loud as the motorcycle. Let us say that he chooses a 90-decibel sound as the best

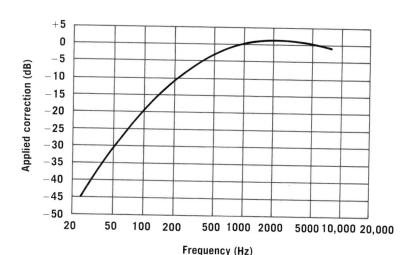

Figure 16.11 The dBA scale.

537

match to his ears. If this choice represents the typical, or the average response of many people, then the motorcycle at 8 meters is said to have a "loudness level" of 90 **phons,** even though its own decibel rating in terms of sound power might be some other number, such as 88 decibels. The **phon** is thus a unit of subjective loudness, which is matched to a decibel scale for a pure 1000-Hz sound.

We must, of course, consider the subjective description of loudness. If you told someone who was unfamiliar with loudness scales that a motorcycle had a loudness level of 90 phons, or produced a loudness of 88 decibels, it would mean nothing to him. It would be more informative, if less precise, to say that a motorcycle sounds "very loud." Therefore it is convenient to append a scale of purely verbal description, such as "quiet," "very loud," etc., alongside a decibel scale of loudness. Table 16.1 shows the perceived loudness descriptions of various sound levels. The upper level is about 120 decibels or a little more; such sounds effectively saturate our sense of hearing, and greater sound powers are not perceived as being louder. Of course, they may be more harmful, but we are discussing only perceptions. Note the extreme range of the scale: a 120-decibel sound is 10^{12}, or one trillion times more powerful than a zero-decibel sound.

16.5 THE EFFECTS OF NOISE

Noise can interfere with our communication, diminish our hearing, and affect our health and our behavior.

INTERFERENCE WITH COMMUNICATION

We have defined noise as unwanted sound. Let us think for a moment about the sounds that we do want: live or recorded speech or music, danger warnings such as the cry of a baby in a distant room or the rattle of a rattlesnake, pleasant natural sounds such as the chirp of a bird or the rustle of leaves in a gentle breeze. We want to hear these sounds at the right level—not too loud or too quiet—and without interference. Noise keeps us from what we want to hear, with the result that we do not hear it so well, or at all, or that the sound we want to receive must be unpleasantly loud for us to get the message.

It is true that modern technology enables us to compensate for its noisiness. We can turn up the volume of our radio or television sets if the traffic outside gets too noisy. Actors, musicians, and public speakers can likewise amplify their sounds. Even if such electronic aids fail to transmit the message clearly, the chances are that we may be able to read about it in the next day's newspaper. Such options were not always available. Before the advent of electricity there were no volume controls to turn up, and no highspeed presses to produce a flood of newspapers and other printed matter; in fact, there was very little artificial lighting at night to read by. Instead, people used their ears to gain information. The tolling of the church bells was not solely a musical recital; it provided information, announcing festivals, births, deaths, and the danger of attack. The town crier was the verbal news reporter. Medieval manuscripts were written with a minimum of punctuation and separation between words; such visual aids were deemed unimportant because the text was intended to be read aloud so that others might hear, not to be merely scanned speedily with the eye while the scanner maintained both external and internal silence.

If you think that the quality of communication afforded by loud sounds superimposed over a noisy background is equivalent to that of soft sounds rising above silence, venture out some quiet winter day into a snowy woodland and hold a muted conversation there, or sing, or listen to the little natural sounds that noisier environments extinguish. Then take a ride on a subway with your girl or boy friend and yell sweet nothings at her or him.

LOSS OF HEARING

An occasional noise interferes with desired sounds, but we recover when quiet is restored. However, if the exposure to loud noise is protracted, some hearing loss becomes permanent. The general level of city noise, for example, is high enough to deafen us gradually as we grow older. In the absence of such noise, hearing ability need not deteriorate with advancing age. Thus, inhabitants of quiet societies, such as tribesmen in southeastern Sudan, hear as well in their seventies as New Yorkers do in their twenties.

It is important to understand that most instances of loss of hearing that result from environmental noise are not traumatic; in fact, the victim is often unaware of the ef-

fect. Let us picture a worker who completes his first day in a noisy factory. Of course, he recognizes the noisiness, and may even feel the effect as a "ringing in the ears." He will have suffered a temporary hearing loss that is localized in the frequency range around 4000 Hz, as shown in Figure 16.12. This means that he will not hear moderately high frequencies so well, but low-frequency and very high-frequency sources will be unaffected. As he walks out of the factory, then, most sounds will seem softer. His car will seem to be better insulated, because he will not hear the rattles and squeaks so well. He will judge people's voices to be just as loud as usual, but they will seem to be speaking through a blanket. He will also feel rather tired.

By morning he will be rested, the ringing in his ears will have stopped, and his hearing will be partly but not completely restored. The factory will therefore not seem quite so noisy as it was on the first day. As the months go by he will become more and more accustomed to his condition, but his condition will be getting worse and worse. Can he recover if he is removed from his noisy environment? That will depend on how noisy it has been and how long he has been exposed. In many cases, his chances for almost complete recovery will be pretty good for about a year or so. However, if the exposure continues, hearing loss becomes irreversible, and eventually he will become deaf. Look at Figure 16.12*B* and *C* to see a typical downward progression caused by prolonged exposure to industrial noise.

In general, noise levels of about 80 decibels or higher can produce permanent hearing loss, although, of course, the effect is faster for louder noises, and it is somewhat dependent on the frequency. At a 2000-Hz frequency, for example, it is estimated that exposure to 95-decibel noise (about as loud as a power lawn mower) will depress one's

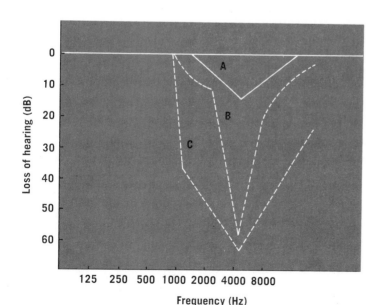

Figure 16.12 Patterns of hearing loss from exposure to industrial noise. *A*, Temporary loss of hearing. *B*, After 20 years. *C*, After 35 years.

540

hearing ability by about 15 decibels in 10 years. Occupational noise, such as that produced by bulldozers, jackhammers, diesel trucks, and aircraft, is deafening many millions of workers. The fact that women in technologically developed societies hear better than men is undoubtedly related to the fact that they are less exposed to occupational noise; in undeveloped, quiet societies, women and men hear equally well. This difference between the sexes in developed societies may diminish as more effort is exerted to make industry quiet, as more women work in industry, and as more people are exposed to noisy household appliances or to radio or television speakers turned up high so that they may be heard over the vacuum cleaner. Battle noises, such as those made by tanks, helicopters, jets, and artillery, are so deafening that many American soldiers who undergo combat training suffer enough hearing loss so that they can no longer meet the requirements for enlistment into combat units.

Is some damage to hearing inevitable? Perhaps, if we are to live in a developed society at the present state of technology; otherwise we had better choose to live elsewhere, and in that case, of course, we may be paying other kinds of prices. But then how much damage might we consider "acceptable"? That question is not so easy; it requires a careful balancing of judgments and values. One frequently cited criterion is that the noise should not produce any permanent damage to hearing that exceeds the values shown in Figure 16.12A after a 10-year exposure. This means that after 10 years in the factory, it is "acceptable" for the worker's hearing to be impaired permanently to the same degree that the hearing of the worker described earlier was impaired temporarily after one day in the noisy factory. Reread the description of his experience to understand what this represents. Do people actually agree that such impairment is acceptable? Some do; others do not. Some can afford to choose quiet environments in which to work; for others, the demands of livelihood preclude such luxuries. In any event, if we agree to accept such damage (to be truthful, the usual situation is that some people agree that *other* people should accept the damage), some specific noise limits must be established. These are noise levels to which people may be exposed without suffering what we have called unacceptable damage. Perhaps a better way to say it is that these are noise levels *which must not be exceeded*. Table 16.2 presents a set of these values.

What about extremely loud noises? Recent concern over exposure of people to rock music stems from the fact that such music is often indeed very loud. Sound levels of 125 decibels have been recorded in some discothèques. Such noise is at the edge of pain and is unquestionably deafening. Noise levels as high as 135 dB should never be experienced, even for a brief period, because the effects can be instantaneously damaging. Such an acoustic trauma

TABLE 16.2 Maximum Acceptable Noise Exposures

DURATION (per day)	LIMIT (dBA)
1.5 minutes or less	120
3 minutes	110
7 minutes	103
15 minutes	97
30 minutes	93
1 hour	90
2 hours	87
4 hours	85
8 hours	85

might occur, for example, as the result of an explosion. If the noise level exceeds about 150 or 160 dB, the eardrum might be ruptured beyond repair, and the ear bones displaced or broken. In addition, the action of the inner ear may be so violent that the hair cells are destroyed.

OTHER EFFECTS ON HEALTH AND BEHAVIOR

As we have already discussed in many contexts, a living organism, such as a human being, is a very complicated system, and the effects of a stress or a disturbance follow intricate pathways that may be very difficult to elucidate. Having read this book thus far, you should be skeptical if you are told that a disturbance great enough to deafen you will have no other effects. Indeed, many investigators believe that loss of hearing is not the most serious consequence of excess noise. The first effects are anxiety and stress reactions or, in extreme cases, fright. These reactions produce body changes such as increased rate of heart beat, constriction of blood vessels, digestive spasms, and dilation of the pupils of the eyes. The long-term effects of such overstimulation are difficult to assess, but we do know that in animals it damages the heart, brain, and liver and produces emotional disturbances. The emotional effects on people are, of course, also difficult to measure. We do know that work efficiency goes down when noise goes up.

INFRASOUND AND ULTRASOUND

The frequency range of normal human hearing is 50 Hz to 15,000 Hz (more or less). Frequencies below 50 Hz are called **infrasound.** Although we cannot hear these vibrations, we can often feel them, because some parts of the human body, including internal organs, do resonate in this range. As a result, prolonged exposure to powerful infrasonic vibration can cause physical damage. The danger is insidious because the long infrasonic wavelengths (up to about 15 meters) are not readily absorbed in air and so can travel for several kilometers from a source and still retain much of their intensity.

Ultrasound occurs in frequencies above 15,000 Hz. Ultrasonic devices are widely used in industry for sealing packages, welding, cutting, drilling, and cleaning. They are also used in medical diagnosis. Unfortunately, we know very little about the effects of ultrasound on the human body.

16.6 NOISE CONTROL

Noise is transmitted from a source to a receiver. To control noise, therefore, we can reduce the source, interrupt the path of transmission, or protect the receiver.

REDUCING THE SOURCE

The most obvious source reduction is simply the reduction of the sound power. Don't beat the drum so hard, or ring the bell so loud, or run so many trucks or motorcycles, or mow the lawn with a power mower so often. There are obvious limitations to this type of solution; for example, if we run fewer trucks, we will have less food and other essentials delivered to us.

Even if we do not reduce the sound power, we may be able to reduce the noise production by changing the source in some way. Our purpose in pushing a squeaky baby carriage is to move the baby, not to make noise. Therefore, we can oil the wheels to reduce the squeaking. Thus, machinery should be designed so that parts do not needlessly hit or rub against each other (see Fig. 16.13).

It might be possible to modify technological approaches so as to accomplish given objectives more quietly. Rotary saws instead of jackhammers could be used to break up street pavement. Ultrasonic pile drivers could replace the noisier steam-powered impact-type pile drivers.

We could also change our procedures. If a city sidewalk must be broken up by jackhammers, it would be better not to start early in the morning, when many people are asleep, but later in the day, when many have left for work. Aircraft takeoffs could be preferentially routed over less densely inhabited areas.

Control of noise is a complex and sophisticated technology, and it is most effective when it is applied to the origi-

Figure 16.13 Hydraulically operated shear has less impact noise than mechanically driven shear. (Courtesy of Pacific Industrial Manufacturing Company.)

Insulated noise test chamber. (Courtesy of Lockheed California Company.)

nal design of the potentially offending source. All too often a device or a machine or an entire industrial facility is designed with a view only to maximize its capacity to carry out its assigned function. If it turns out to be excessively noisy, an acoustical engineer may be called in to "soundproof" it. Under such circumstances, the engineer may be forced to accommodate to features of construction that should never have been accepted in the first place. Therefore, much of his effort may necessarily be applied, not to the source, but to the path between sound and receiver.

INTERRUPTING THE PATH

We have learned that sound travels through air by compressions and expansions. It also travels through other elastic media, including solids such as wood. Such solids vibrate in response to sound and therefore do not effectively interrupt its transmission, as many residents of apartment houses will readily attest. However, we could use various materials that vibrate very inefficiently, such as wool or lead, and absorb the sound energy, converting it to heat. (Very little heat is involved; the sound power of a symphony orchestra will warm up a room about as much as a 15-watt electric heater.) Sound-absorbing media have been developed extensively; they are called **acoustical materials.** We could also build interruption of the sound waves mechanically into more kinds of machinery; devices that function in this way are called **mufflers** (see Figs. 16.14, 16.15, and 16.16). Finally, we may be able to deflect the sound path away from the receiver, as by mechanically directing jet exhaust noise upward instead of down. Such deflection is, in effect, an interruption between source and receiver.

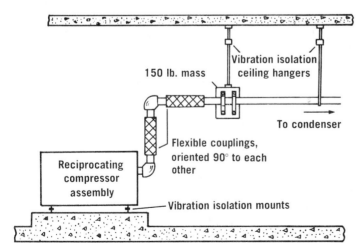

Figure 16.14 Mounting of compressor and piping to isolate vibration noise. (From L. L. Beranek, ed.: *Noise Reduction.* Copyright © 1960 by McGraw-Hill Book Company. Used with permission of McGraw-Hill Book Company.)

Figure 16.15 Sound absorbers suspended close to noise sources. (Courtesy of Elof Hannson, Inc., & Sonosorber Corporation.)

Figure 16.16 Vibration isolation of printing presses required to reduce noise in office on floor below. (From L. L. Beranek, ed.: *Noise Reduction.* Copyright © 1960 by McGraw-Hill Book Company. Used with permission of McGraw-Hill Book Company.)

Man wearing acoustical earmuffs while using chainsaw.

PROTECTING THE RECEIVER

The final line of defense is strictly personal. We protect ourselves instinctively when we hold our hands over our ears. Alternatively, we can use ear plugs or muffs as shown in Figure 16.17. (Stuffing in a bit of cotton does very little good.) A combination of ear plugs and muff can reduce noise by 40 or 50 decibels, which could make a jet plane sound no louder than a vacuum cleaner. Such protection could prevent the deafness caused by combat training, and should also be worn for recreational shooting. Degrees of protection are shown in Fig. 16.18.

We can also protect ourselves from a noise source by going away from it. In a factory, such reduction of exposure may take the form of rotating assignments so that different workers take their turns at the noisy jobs.

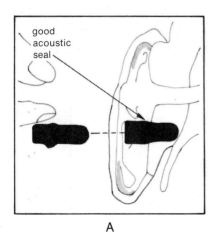

A

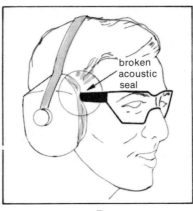

B

Figure 16.17 Protecting the receiver. *A,* Ear plugs must be inserted to a depth to ensure an acoustical seal. *B,* Placing ear muffs over eyeglass frames can break a proper accoustical seal.

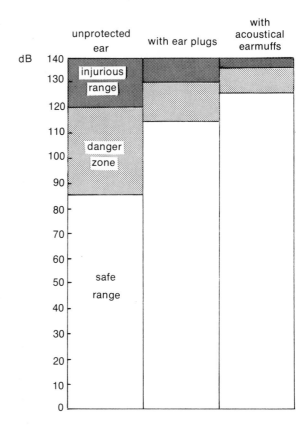

Figure 16.18 Using ear protectors reduces the risk of injury to hearing. The figure shows the danger zone at 1000 Hz.

16.7 LEGISLATION AND PUBLIC POLICY CONCERNING NOISE CONTROL

Noise is hardly a new phenomenon, and the annoyance it produces has been recognized since ancient times. Julius Caesar banned chariots from Rome because their wheels clattered on the stone streets. English Common Law recognized that freedom from noise is essential to the full enjoyment of a private dwelling place. Literature is replete with references to the effects of noise, especially from the loudest source, which is battle. For example, the medical writings of C. H. Parry (1825) describe an incident in which Admiral Lord Rodney was deafened for 14 days following the firing of 80 broadsides from his ship, H.M.S. *Formidable,* in the year 1782.

Despite these early beginnings, federal noise legislation in the United States is a relatively recent development. Until the 1960's, noise control was chiefly handled by state and local governments. However, recognition that community ordinances could not cope adequately with aviation noise led in 1968 to an amendment to the Federal Aviation Act, the first federal legislation of its kind dealing specifically with this problem. The measure gave the Federal Aviation Administration authority to prescribe standards for measuring and controlling civilian aircraft noise including sonic booms.

547

Subsequently, the U. S. Department of Labor in 1969 established limits for occupational noise exposure in certain companies under government contract. The Occupational Safety and Health Act of 1970 broadened this approach with a comprehensive on-the-job safety program for workers in the private sector and required each federal agency to establish an occupational safety and health program for employees. In 1971 the Department of Labor set an occupational noise standard, and an Executive Order later required federal occupational safety and health programs to be consistent with this standard. Other executive orders required federal agencies to conform to federal noise emission standards for various products and to state and local standards for environmental noise control, and to regulate noise from off-road vehicles on public lands. These and other governmental actions led to widespread recognition of noise as an environmental problem and to enactment of the Noise Control Act of 1972 (Public Law 92-574). This landmark legislation set forth a new policy governing all federal agencies with respect to a national goal of quiet. It declared that the policy of the United States was "to promote an environment for all Americans free from noise that jeopardizes their health or welfare." Under this law, the U. S. Environmental Protection Agency (EPA) coordinates all federal programs relating to noise research and noise control.

EPA has the authority to prescribe and amend standards limiting noise generation for products identified as major sources of noise. These include construction and transportation equipment, motors, engines, electrical and electronic equipment. Under this authority, the EPA in 1974 published final regulations establishing noise limits for trucks and buses over 10,000 pounds (4536 kg) gross vehicle weight engaged in interstate commerce, and for portable air compressors. In 1974 the Agency also set noise limits for new medium and heavy trucks, new diesel and diesel-electric locomotives, and railroad cars. EPA may issue regulations for products in other categories if necessary to protect public health or welfare.

The Agency also is authorized to require the labeling of domestic or imported consumer products as to their noise-generating characteristics or their effectiveness in reducing noise.

EPA conducts and finances research on the effects of noise (including sonic booms) on humans, animals, wildlife, and property. The Agency also conducts research to determine acceptable levels of noise, and develops improved methods for measuring, monitoring, and controlling this environmental problem.

Other federal agencies, such as the Departments of Transportation, Labor, Commerce, Defense, Housing and Urban Development, and the Interior, the Federal Aviation

Administration, and the Federal Highway Administration, also conduct research, devise abatement procedures, and set noise limits relating to their activities. All such programs are coordinated through the EPA.

How do such programs work? Let us illustrate with some examples. Suppose you own a truck or bus whose gross weight exceeds 10,000 lb (4536 kg), and you wish to engage in interstate commerce. The law says that your vehicle must not exceed the following maximum permissible exterior noise levels:

- 88 dBA at 50 ft (15 m) under stationary runup
- 86 dBA at 50 ft for speeds under 35 mi/hr (56 km/hr)
- 90 dBA at 50 ft for speeds over 35 mi/hr

In addition, the EPA offers information as to where noise problems may originate, describes specific procedures for locating them, and provides guidelines on the selection of equipment such as mufflers and tires.

However, even such a universally approved environmental objective as noise control may occasionally engender knotty problems of public policy. For example, some years ago environmentalists in the Department of Housing and Urban Development (HUD) declared that certain New York City slums were too noisy to be suitable for new government-subsidized housing. New York officials replied that the government position amounted to saying "that the poor people can live in their old houses in rundown neighborhoods, but not in new houses," and that "to be told that you have to be worried about noise levels of brand new, safe, sanitary housing in Harlem, when hundreds of thousands live in slums, is patently ridiculous."

This type of controversy pits the striving of the poor for decent improvement of living conditions in their own communities against government regulations that may be based on different assumptions about where people like to live and how much money they can spend. For example, the HUD standard specifies a maximum of 65 decibels for 8 of each 24 hours. Local officials stated that the cost of properly insulating government-subsidized housing to meet this federal standard would price the projects out of the federal program and bar them from receiving federal funds, a condition popularly known as catch-22. What is implied here is not merely that such governmental regulations are inconsistent, but that they are fundamentally exclusionist and prejudiced against those who are poorer, less privileged, and less powerful than the average citizen. This accusation, like those against population stabilization programs, is not easy to answer, for although strict application of noise regulations will make it more costly to build new housing for the poor, relaxation of the standards will make the new housing less comfortable for people to live in.

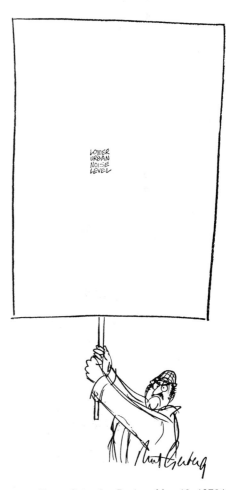

LOWER URBAN NOISE LEVEL

(From *Saturday Review*, May 13, 1972.)

The SST is a passenger aircraft that travels faster than sound and at much higher altitudes than subsonic airplanes. Higher speed requires more power, and more power makes more noise. Near airports, the noise problem is associated with the SST's approach and the rapid climb shortly after takeoff (Fig. 16.19), although the speeds at these times are subsonic. The engines on an SST must be small in diameter to provide optimal streamlining, and the noise from jet exhaust increases very rapidly (for a given engine thrust) as jet diameter is reduced and its speed increased.

When the SST reaches supersonic speed in flight, another effect, the **sonic boom,** occurs, and this, too, creates problems.

If there is one factor that has been common to all of the environmental controversies surrounding various SST programs, it is technical confusion. Just what is a sonic boom? Is it something that occurs when the aircraft "breaks the sound barrier," like cracking through some kind of wall? If an SST on takeoff were as loud as, say, 10 ordinary jets, does that mean it is 10 times as loud as one jet? Or 10 times as annoying to people?

We can answer one of these questions directly from what we have learned about the decibel scale. Recall that a tenfold increase in sound intensity is represented by an increase of

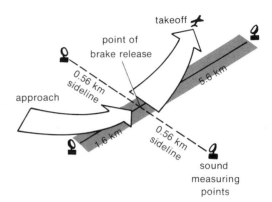

Figure 16.19 Points at which airplane approach, takeoff, and sideline noises are measured.

10 dB, and that sound intensity is produced by sound power. Obviously, the sound power of 10 jets is ten times as great as that of one jet. Therefore 10 jets are 10 dB louder than one jet. And if the sideline noise of an SST sounds like that of 10 ordinary jet planes, the SST is 10 dB louder than one ordinary jet plane. However, this answer does not tell you how much louder the SST sounds to you (remember, you are not a decibel meter), nor how much more annoyance it would cause you; nor does it tell what you would do about it. Experiments show that a sound 10 decibels louder than another is judged

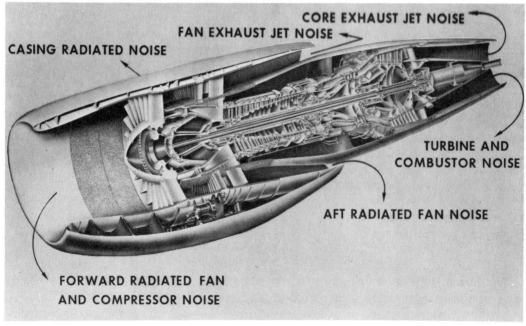

Jet engine.

to be "about twice as loud," but it sounds nevertheless just like 10 of the weaker sounds all sounding simultaneously. Experiments on airport noise show that a single aircraft 10 decibels louder than another produces about the same *annoyance* as 10 separate flights of the quieter craft spread throughout the day. This relationship is used by the government and by airport operators in planning land use around airports.

Now let us consider the phenomenon most closely associated with the SST—the sonic boom. To visualize this effect, think of a speed boat moving rapidly in the water. Its speed is greater than that of the waves it creates, and it

therefore leaves its waves behind it. Moreover, the wave energy is being continuously reinforced by the forward movement of the boat. The result is a high-energy wave, called a **wake,** that trails the boat in the shape of a V and that slaps hard against other vessels or against the shoreline. The sonic boom is a high-energy air wave of the same type. The tip of the wake moves forward with the airplane, while the sound itself moves out from the wake at the usual speed. The faster the airplane, the more slender is the wake.

To understand the geometry of the wake, study Figure 16.20. A stationary object (Fig. 16.20A) remains in the center of the circular

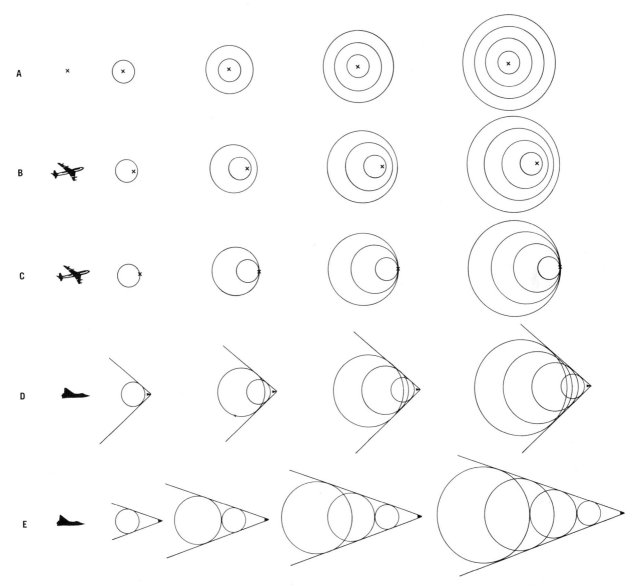

Figure 16.20 Wave patterns of subsonic and supersonic speeds: *A*, stationary; *B*, subsonic; *C*, sonic, Mach 1; *D*, supersonic, Mach 1.5; *E*, supersonic, Mach 3. From left to right, the waves are shown at equal time intervals as they expand.

waves it generates. The waves from a moving object will crowd each other in the direction of the object's motion (Fig. 16.20B). The object is, in effect, chasing its own waves. Recall that sound travels in air at sea level at a speed of 334 m/sec (1096 ft/sec, 747 mi/hr); any speed less than this is said to be **subsonic**. When the speed of the object equals the speed of the wave, the object will not see any waves before it; it will just be keeping up with them (Fig. 16.20C). In air, such a speed is said to be **sonic**; and speeds greater than this are **supersonic**. Such speeds are usually measured in Mach numbers:*

$$\text{Mach number} = \frac{\text{speed of object}}{\text{speed of sound}}$$

If an object is traveling at the speed of sound, then the numerator and the denominator of the equation are the same, and the Mach number equals 1. Mach 2 is twice the speed of sound, Mach 3 is three times the speed of sound, and so forth.

Figure 16.20C and D shows the wave patterns at supersonic speeds. Note that the object is always *ahead* of its waves. A passenger in an SST would therefore not hear the sound of its motion. Instead, the waves will crowd in on each other, and the effect will be a significant elevation of pressure at the advancing boundary of the overlapping wave fronts.

Of course, an airplane travels within its medium, and not, like a boat, on the surface of its medium. Therefore the outlines shown in Figure 16.20C and D are two-dimensional projections of what are really conical shapes. Furthermore, an airplane is more than a point in space, and therefore a whole series of such cones will be generated. It is sufficient to consider only the nose and the tail of the plane, and represent the entire space between the forward and rear cones as the volume of the disturbance, as shown in Figure 16.21. The tail of the airplane leaves a partial vacuum behind it as it rushes along, so that the overall shape of the sonic boom is like the letter N: a rise above normal pressure, then a fall below, and finally a recovery, as shown in Figure 16.21C.

To be struck unexpectedly by a sonic boom can be quite unnerving. It sounds like a loud, close thunderclap, which can seem quite eerie when it comes from a cloudless sky. The duration of the

---

*After Ernst Mach, 1838–1916, a physicist who made important discoveries about sound.

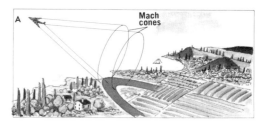

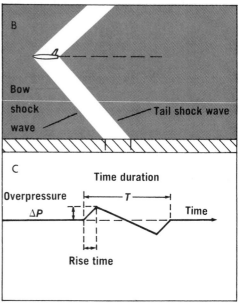

Figure 16.21 Sonic boom. *A*, Double Mach cones and their intersection with the ground surface. *B*, Profile of boom on ground surface. *C*, N-shaped pressure pattern.

boom is only about $1/10$ to $1/2$ second, and the two distinct pressure changes may be heard separately, or they may seem to merge into one if the time is very short. Depending on the power it generates, the sonic boom can rattle windows or shatter them or even destroy buildings. It is important to avoid the misconception that the sonic boom occurs only when the aircraft "breaks the sound barrier," that is, passes from subsonic to supersonic speed. On the contrary, the sonic boom is continuous and, like the wake of a speedboat, trails the aircraft all during the time that its speed is supersonic. Furthermore, the power of the sonic boom increases as the supersonic speed of the aircraft increases.

During the late 1960's an American prototype of a commercial SST was being developed by the Boeing Aircraft Corporation under financial support by the U. S. government. This project provided a great many jobs for the people living in the Seattle, Washington, area (Boeing's home) and thus was a significant asset to the economy of the region. However, as

The Concorde SST.

it became evident that the SST aircraft could give rise to environmental problems, significant opposition to the project was mounted. A conflict arose between environmentalists and industrial and labor groups. One fundamental question that arose was, "How annoying would noise from the SST actually be?"

It has been pointed out that annoyance is a subjective experience. Thus one would expect that a survey of annoyance caused by jet noise near Seatac airport (Washington) would yield different results from a similar survey at J. F. Kennedy (New York) or at Stapleton (Denver). The final decision on the Boeing SST was negative, and the Congress stopped its funding in 1971.

On the other hand, a decision as to whether or not to allow an SST of foreign manufacture to use a country's domestic air space or airfields does not involve the same questions about domestic jobs. In the case of the British-French Concorde SST, an important fact was that the Concorde already existed, and, in the words of one emotional author, for the United States to deny its "right to fly ... would be an incredible rejection of a monumental effort made by two of our best international friends" and would interrupt progress toward development of a quieter supersonic craft. (A curious argument, this latter one.)

One might be tempted here again to an even-handed presentation of alternatives, and to point out that banning SST's would reduce noise as well as stratospheric air pollution (see page 413), while accepting SST's would promote human progress and technical development.

But the arguments in favor of relaxing noise standards to admit the SST are not to be compared with a similar relaxation of standards to allow, for example, construction of more low-cost housing. After all, the SST can save only a few hours of travel time for only a tiny fraction of the population. In the light of the many problems that beset people around the globe, it is difficult to imagine that human betterment is being held back because subsonic jets are too slow. The argument about technical progress also needs close examination. If the purpose of building an SST is to learn how to develop better SST's, the argument is circular. If the purpose is to gain some other beneficial "spin-off," it is difficult to understand how such efforts can be more effective than either fundamental research, which can yield entirely unforeseen advances, or applied research specifically directed toward recognized and urgent human needs.

"At least they have agreed to show only silent movies." (From *Industrial Research*, Nov., 1976.)

TAKE-HOME EXPERIMENTS

1. **Loudness.** Carry this book around with you for a few days so you can refer to Table 16.1 (page 532). Make a diary of various sounds you hear, recording your information in a form such as that given in Problem 6. Estimate the decibel levels as well as you can from the columns in the table that show the sound levels of various sources and their perceived loudness. Try to include the following sources: (a) a television, or radio at normal listening volume in your room or apartment; (b) the central study area in the school library; (c) your environmental science classroom; (d) the street outside your classroom; (e) the background noise in your room at night; (f) the school cafeteria; (g) a local factory or construction job site. Which sounds are annoying? Offer suggestions for reducing the perceived loudness in each of the instances where the sound is annoying.

2. **Sound.** Select a convenient constant source of sound, such as a ringing alarm clock, and try to reduce the loudness you hear from it by the following means: (a) stuff some cotton loosely into your ears; (b) hold your hands over your ears; (c) use both the cotton and your hands; (d) submerse your head in the bathtub (face up) until your ears are underwater; (e) if you can borrow a pair of earmuffs of the kind used in factories or at airports, try them. *(Safety Note:* Don't try to stuff any small hard objects in your ears; the results could be harmful.)

Inaudible
(zero loudness)

Loudness to
naked ear

Draw a straight line (of any length) in your notebook, labeling one end "inaudible" and the other end "loudness to naked ear." Mark the positions of each of the sound-reducing methods on the line at a point that corresponds to the loudness you heard. For example, if you think that one of the methods reduced the loudness by half, mark its position halfway along the line. If you think it reduced it by only 25 percent, mark it at 1/4 of the length away from the "naked ear" end. Discuss the reasons for your findings.

3. **Sound power.** Hang a strip of tissue paper in front of the loudspeaker of a sound system (see sketch). Shut off the sound, close the windows, and turn off any fan or air conditioner in the room. The paper should hang motionless. Now switch the sound on and turn up the volume. Describe the effect on the paper and explain your observations.

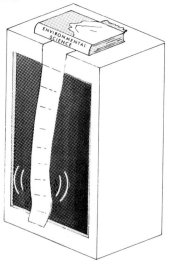

PROBLEMS

1. **Vocabulary.** Define elastic wave; wavelength; frequency; amplitude.

2. **Pure tone.** What is a pure tone? Which of the following sounds would you consider to be the purest: the croak of a frog; the sound of an orchestra; the sound of a tuning fork?

3. **Noise.** What is noise? Do you think it would be feasible to develop an instrument that would indicate how noisy a given sound is? Defend your answer.

4. **Noise.** Which of the following phenomena would you be willing to classify as noise? Defend your answers. (a) Your neighbor's

dog barking while you are listening to music; (b) your own dog barking to warn you of a smell of smoke.

5. **Loss of hearing.** Explain why curve C of Figure 16.12 is said to characterize deafness, even though it shows no loss of hearing below 500 Hz.

6. **Noise control.** A man carries a decibel meter with him for a day and records the following readings in his diary:

7:00 A.M.	Baby crying.	84 dB
7:30	Dishwasher in kitchen.	70
7:45	Garbage truck, 150 feet away.	90
8:00	Traffic noise while waiting for bus.	81
8:45	Arrived at entrance to office. Noise of jack-hammer on sidewalk.	106
9:00–12:00 Noon	Average sound in office.	45
12:00–1:00 P.M.	Noise in restaurant— dishes, etc.	45
5:00–5:30	Rode home on subway (windows open).	90– 111
6:00	Mowed lawn with power mower.	93

Offer suggestions for reducing the perceived loudness of each of these various noises.

7. **Definitions.** Define or explain: Mach number; sonic boom.

8. **"Inevitable" damage to hearing.** On page 541 it was stated that some damage to hearing may be "inevitable" to people living in a developed society. Do you agree? If so, defend your answer. If not, describe some ways that you could live which would prevent damage to hearing over a lifetime.

9. **dBA scale.** If you turned Figure 16.10 up-side down and held it in front of a mirror, would it look something like Figure 16.11? Explain the relationship.

10. **Noise control.** Figure 16.22A shows a simplified sketch of a section of interior house wall made up of two sheets of wall-board attached to verticle wooden supports, called studs. A plan of the same wall section (looking down) is shown in Figure 16.22A′. Three other plans are also shown: B, sheets of lead (a good sound ab-sorber) are placed between the wallboards and the studs; C, sheets of lead are placed

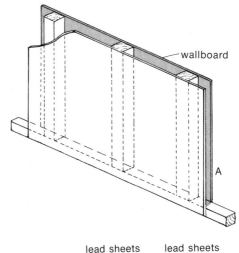

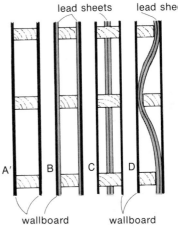

Figure 16.22 Sections of interior house walls. A and A′, No insulation. B, C, and D, Various arrangements of lead insulation.

between the studs; and D, alternate studs touch each wallboard, and sheets of lead are snaked in between. Rate these struc-tures according to their ability to serve as sound insulators. Defend your ratings.

11. **Legislation.** Identify the legislation that establishes national policy in the United States for promoting a quiet environment. What is the legislative approach to achiev-ing this objective?

12. **Public policy.** List some federal agencies in the United States that are active in problems of noise abatement. Which agency coordinates all these activities?

13. **Public policy.** If a governmental agency is concerned with noise, what kinds of activi-ties can it carry out?

14. **Public policy.** All of the following policies are undesirable:
    (a) Displacing people from their communities as a means of thinning out population to compensate for improper sound insulation in new housing.
    (b) Delaying the construction of new housing because proper sound insulation would be too expensive.
    (c) Failing to provide a quiet environment in new housing.

    Suppose, as a public official, you felt that you were forced to choose one of these three policies. How would you make your decision? Do you think you could avoid making any of these choices? If so, what other policy would you pursue?

    The following questions require arithmetic or algebraic reasoning or computation. You may also have to refer to Appendix C-16.

15. **Pure tone.** When we say that the shape of a "pure tone" is sinusoidal, which of the following statements do we imply? (a) The molecules move along in a wavy motion like water waves. (b) A graph of air density vs. distance will be a sine curve. (c) A graph of the degree of increase or decrease of air pressure above or below atmospheric pressure vs. distance will be a sine curve. (d) A graph of pressure vs. time at any one point will be a sine curve. (e) The molecules move back and forth like the inking of the paper in Figure 16.7B. (f) Since the air does not ripple, there is no real sine function, but rather we are using a figure of speech to compare a "pure" tone with the pure symmetry of a rotating wheel.

16. **Combinations of sine waves.** Given two sine curves of different frequencies, as shown in Figure 16.23: (a) Construct the combination of these curves by adding them, using the numerical values on the vertical scale. (b) If one of the curves were shifted slightly to the right or left, would the sum of the two curves still show a periodic variation?

17. **Decibels.** A person hears a cry in the woods that is 1000 times the intensity of the faintest audible sound. What is the sound level in decibels?

18. **Decibel scale.** Table 16.1 lists sound intensity factors as well as sound levels (dB). How are these two lists related to each other? Can you express the relationship in equation form?

19. **Sound intensity.** The sound intensity of a motorcycle at a distance of 8 meters is 90 decibels. How many times more intense than the faintest audible sound is this motorcycle sound?

20. **Loudness.** As stated in Section 16.4, the scale of subjective loudness is based on a pure tone of 1000 Hz. Calculate the wave length of this tone if the speed of the sound is 334 m/sec.

21. **Decibels.** A four engine jet plane stands on the runway. The sound of each engine is 100 dB. What is the sound level when one engine is running? Two? Three? All four?

22. **Decibel scales.** Can there be a negative decibel level? A dog can hear sounds inaudible to a human. Suppose a dog could just hear a sound whose intensity is 10^{-13} watt/m^2; what is the decibel level to which he would be sensitive?

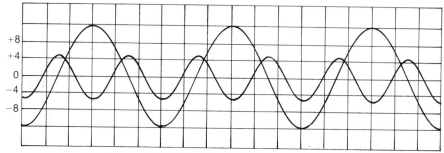

Figure 16.23 Combination of two sine waves.

23. **Sound pressure level.** As you have already learned, a sound wave produces regions of high pressure in the air through which it travels, and the sound intensity is proportional to the square of the pressure, p^2, or:

$$\text{Sound intensity} \propto p^2$$

Derive a formula for sound intensity in terms of sound-pressure level, SPL, expressed in decibels.

24. **Mach numbers.** A rocket is moving through the lower atmosphere, where the speed of sound is 334 m per second, at Mach 1.5. What is the speed of the rocket?

BIBLIOGRAPHY

A delightfully written, non-mathematical, yet authoritative paperback book that covers the entire field very well is:
Rupert Taylor: *Noise.* Baltimore, Penguin Books, 1970. 268 pp.

Two books that emphasize the effects of noise are:
Karl D. Kryter: *The Effects of Noise on Man.* New York, Academic Press, 1970. 632 pp.
William Burns: *Noise and Man.* 2nd Ed. Philadelphia, J. B. Lippincott, 1973. 459 pp.

For a basic text on noise control, refer to:
Leo L. Beranek: *Noise Reduction.* New York, McGraw-Hill, 1960. 753 pp.

A description of methods of measurement, published by a manufacturer of noise instruments, that also includes a catalog section, is:

Arnold P. G. Peterson and Ervin E. Gross, Jr.: *Handbook of Noise Measurement,* 7th ed. Concord, Mass., General Radio, 1972. 322 pp.

Four popular books that take up the environmental aspects of noise are:
Theodore Berland: *The Fight For Quiet.* New York, Prentice-Hall, 1970. 370 pp.
Robert Alex Baron: *The Tyranny of Noise.* New York, St. Martin's Press, 1970. 294 pp.
Henry Still: *In Quest of Quiet.* Harrisburg, Pa., Stackpole Books, 1970. 220 pp.
Clifford R. Bragdon: *Noise Pollution: The Unquiet Crisis.* Philadelphia, University of Pennsylvania Press, 1971. 280 pp.

EPILOGUE

ADAPTATION TO ENVIRONMENTAL CHANGE

As you have seen by now, the environmental sciences cover a very broad range of topics. The questions that environmental scientists ask involve the workings of many complex systems, and knowledge must be drawn from a host of different disciplines. Biology, chemistry, physics, medicine, mathematics, psychology, economics, sociology, the law—the list goes on and sounds almost like a compendium of all human knowledge. As you have also seen, however, our knowledge in many areas is so abysmally incomplete, or the questions we are facing so enormously complicated, that the conclusions often sound like a timid "I don't know," or "Some authorities think . . . while others argue that. . . ." And so on.

The reason that the environmental sciences have come to exist at all, of course, is that human activity is changing the face of our planet in unprecedented ways and, perhaps most important, *with unprecedented speed.* As you know from Chapters 3, 4, and 5, change is no stranger to the Earth, even in the absence of human activity. The cycles of the seasons themselves, the faces of an ocean shore after a brutal storm, the near destruction of a forest after a naturally occurring forest fire—all these kinds of environmental alterations have been going on throughout geologic time and will continue to occur as long as the Earth exists. The life forms that have evolved on the planet have done so in parallel with these naturally occurring changes, and in general, environmental disruptions from such natural "disasters," as people like to call them, do not cause significant long-term perturbations of the biosphere.

However, when we speak of pollution by persistent pesticides or radioactive wastes, or depletion of the atmosphere's ozone layer by fluorocarbons, we are dealing with environmental changes that the biosphere has not experienced before. These changes are potentially disruptive partly because the Earth's living forms do not have an unlimited amount of time to adapt to them. The list of species endangered or already extinct as a direct result of human activity (Chapter 5) bears eloquent testimony to the fact that many organisms have found these challenges insurmountable.

If the study of ecology has one lesson, it is that the relationship between organisms and the environment is a reciprocal one. Fallen leaves, needles, and twigs maintain the composition of forest soil, which in turn promotes the growth of the trees themselves. The holes alligators dig in the Everglades regulate the yearly water flow and thus maintain the integrity of the swamp. Moreover, the whole biosphere maintains proper atmospheric conditions for itself. Such pictures imply homeostatic balance, and balance engenders stability. As already stated, however, the Earth does not provide an unchanging environment. Many changes, such as those noted above, occur over the short term; others, such as climatic variation, have a much broader time scale. If, for example, we could be transported back only about 50,000 years, the climate and terrain of much of North America would be unrecognizable.

We also know that some life forms have survived through many environmental changes; others have succumbed. The process of accommodation to change is called **adaptation.**

Certain forms of life have survived for millennia because they are well adapted to their biological and geological environment. There are basically two adaptive mechanisms. One of these is **genetic variation**. In every species the genes available (usually called the **gene pool**) are not constant in time but undergo continuous variation. As the environment changes, new genetic variants may be favored over previously well-adapted ones. In addition to this genetic mechanism of adaptation, most animals can also learn from their experience and alter their behavior accordingly. Thus, for example, a dog or cat learns to respond to the whims of its master. Compared with other species, however, the human being has developed to an extraordinary degree this mechanism of adaptation by learning. Of all animals, people possess by far the greatest ability to reason in new situations and the largest body of accumulated knowledge that is passed from generation to generation. The human ability to accumulate and transmit knowledge allows us to deal with environmental challenges in ways that have no real parallel in any other species.

Since an understanding of the potential of living organisms for adaptation is crucial for the student who wishes to understand the total impact of environmental changes, this book concludes with a brief consideration of some of

GENETIC ADAPTATION

As described in Chapter 5, eohippus, an early ancestor of the modern horse, was well adapted to life in the swamplike forest that it called home. As conditions changed, eohippus "evolved" gradually into forms that resembled increasingly the horse familiar to all of us. As you probably know from previous courses in biology, the physical appearance of an animal or plant is largely determined by its genetic composition; identical twins look so strikingly similar because they have inherited exactly the same set of genes from their parents. Not only the gross appearance of an organism, however, but also many of its biochemical and physiological characteristics are determined by its genetic constitution.

But don't go away with the notion that genes are the only determinants of observable character traits. If, for example, one were to separate identical twins at birth and rear them separately in two different environments—for example, with an upper-middle-class Christian family in Grosse Pointe, Michigan, and a family of Hindu beggars living in the streets of Varanasi, India, under conditions of semistarvation—you would very likely find pronounced differences at, say, age 21 in body height and weight, skin color, intelligence, and outlook on life. In short, the environment plays a critical role in the expression of much genetic information.

Genes are located on long rodlike structures called **chromosomes,** situated in the nucleus of each cell. If the gene is changed in some way, or more accurately, if the chemical composition of the gene is changed, one or more observable character traits may be altered as a result. How do genes change? There are two general mechanisms for genetic variation:

(a) Genetic recombination. In the course of sexual reproduction, existing chromosomes are shuffled into new combinations of old traits. This process may change the structure of individual genes or may simply provide new mixtures of old genes.

(b) Mutation. A chemical change occurs in the biochemical structure of an individual gene.

Why do mutations occur? We probably do not know all the forces involved in the appearance of mutations inside living cells, but undoubtedly two of the most important causes are radiation and certain chemicals. Both of these may initiate a complex series of reactions leading to a change in gene structure. Whether a mutation results in some kind of *observable* change in the individual animal,

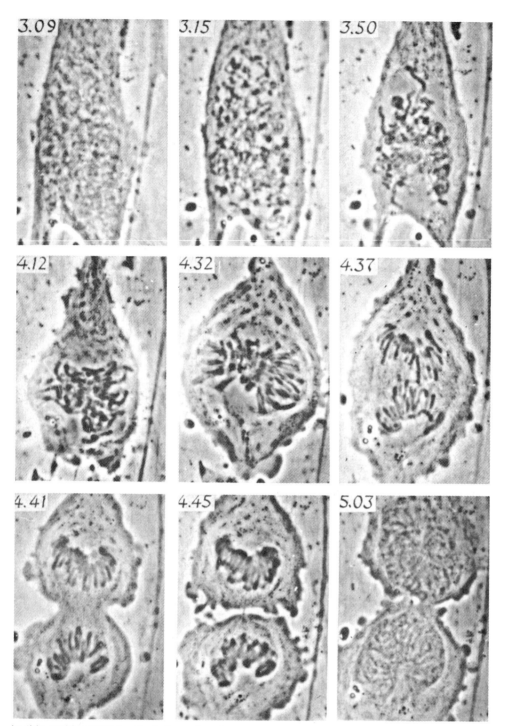

In this rapid sequence of photographs of a dividing cell, the chromosomal material condenses and becomes visible under the microscope and, at the time of cell division, is apportioned equally between the two daughter cells. (From Bloom and Fawcett: *A Textbook of Histology,* 8th ed. Philadelphia, W. B. Saunders Co., 1962.)

however, depends on the particular nature of the mutation; further discussion of these issues may be found in any good text on genetics. It is most important to realize that those mutations of particular concern are the ones that occur in the sex cells, that is, the cells giving rise to the spermatozoa in the male and the eggs cells of the ovary in the female. A mutation in the sex cells of an organism may be passed on during reproduction to subsequent generations. A mutation may be either detrimental or advantageous. Since there are always more ways for a well-integrated system to go wrong than right, however, detrimental mutations are more likely than advantageous ones. The advantageous mutations account for genetic adaptation. The concepts of fitness and natural selection, discussed in Chapter 5, are central to an understanding of how advantageous characteristics are retained by a population.

We should emphasize at this point two important features of the concept of fitness. In the first place, it makes no sense in most cases to talk about an abstract concept of fitness. Characteristics that are advantageous in the jungle could be disastrous in the desert. One must almost always speak of fitness with reference to a particular environment. (We add the "almost" to the preceding sentence because some mutations, such as an animal without reproductive organs, would have a fitness of zero, independent of environmental variations. But most mutations that result in viable progeny do not involve such extreme deviations as this.) Second, the term fitness refers solely to reproductive advantage. For example, the largest and strongest lion in Africa may have won every territorial battle he ever fought, but his fitness will be zero if for any reason he is incapable of reproduction.

An example of natural selection will serve to tie some of these ideas together. In the mid-nineteenth century in England, rare black variants of the common (white) pepper moth were noted for the first time. By the end of the nineteenth century, however, this once-rare color variant had become the predominating variety in most of the industrial sections of Britain. Such a phenomenon has since been noted in many other species of moths in Europe and North America as well. The most likely explanation for these striking observations was that the dark color protects the black moths from their natural predators (birds) in environments darkened by pollution in heavily industrial areas. If so, one might predict that any individual moth with genes for black coloration will preferentially survive and thus have a reproductive advantage over more lightly colored individuals. Hence these color genes will be enriched in subsequent generations.

Though this sounds like a reasonable explanation, it needs experimental verification (that is, a demonstration that color could be protective in different environments). This verification has been provided by experiments in which equal numbers of black and white moths were released together, both in woods whose trees were darkened by pollution and had no normal lichen on their bark and in normal woods far from polluted areas. It turned out that there was great preferential survival of the light variants in the normal woods and of darkened variants in

Moths having industrial melanism. (From Helena Curtis: *Biology.* New York, Worth, 1968. Courtesy of Dr. H. B. D. Kettlewell.)

the polluted setting. This shows that the color of the moth may serve a protective function and strongly suggests a role for color in natural selection, at least for moths.*

But the story of **industrial melanism** (Greek *melas=* black), as this phenomenon has been called, is even more interesting. At the beginning of this century, genetic analysis of the dark variants showed that the dark color was the result of a single dominant gene, which presumably arose from the original "normal" gene that gives moths their light coloration. Fifty years later, however, when the analyses were repeated, the results showed two interesting findings: (a) Dark variants now are darker than they were at the beginning of the century, and (b) the darkness is no longer due solely to a single gene, but rather to a complex of genes working in concert. In other words, we may reasonably suppose that dark color is of such selective advantage in certain environments that additional genes were selected to intensify this character. This means that more moths with high melanin content survive to breeding age. As well as providing a beautiful example of the power of natural selection, this example also shows the frequently complex relationship between gene action and observable character traits. Frequently an observable apparently simple character such as the color of a moth or indeed human skin color is due to the interaction of many genes located at different places on the chromosomes, or even on different chromosomes. Therefore, one can hardly speak of "a gene" for skin color.

---

*It is also interesting to note that in regions where pollution control measures have cleaned the air, the proportion of black variants has diminished in favor of the more lightly colored ones.

ADAPTATION BY LEARNING

Thus far, we have concentrated on the genetic means of adaptation, which have played such a large role in the evolution of life. In addition to this most basic method of adaptation, many animals have evolved mechanisms for learning from their own prior experience. For example, in a part of Africa a fence was erected around a game park inhabited by many animals, including lions. Within one season the lions adapted their hunting habits to conform with and even to utilize the fence. One of the lions would chase its prey toward the fence. On approaching the barricade, the prey would, of course, veer to one side or the other, whereupon it was seized by other lions that had stationed themselves there as part of the trap. Development of this scheme required reasoning, memory, and communication among the hunters and was a rapid response to environmental change.

Further examples of animal learning are provided by observations made by biologists of a troupe of macaque monkeys living on Koshima Island, Japan. In order to supplement the diet of these animals, the biologists had been scattering grains of wheat on the beach. A three year old member of the troupe named Imo, dissatisfied with the tedium of picking the grains out from the sand one by one, invented the astonishing technique of scattering handfuls of sand and wheat together onto the water surface. The sand, of course, sank while the wheat floated at the top and could readily be skimmed off to be eaten. After a short period this method spread throughout the troupe. It was learned most readily by juvenile members from 2 to 4 years old, who then taught it to their mothers and siblings. The

mothers then taught it to the infants. Because of the new intensive exposure to water that all these activities provided, many members of the troupe took to playing in the water, and a few even learned to swim!

Of all the creatures ever to have lived, however, humans have the most prodigious powers of reason, memory, and communication. They alone seem to be able to accumulate large and diverse quantities of knowledge and pass increasingly complex thoughts on to future generations. For example, early humans learned, like the lion, to herd game toward existing traps such as cliffs or tar pits. But then they went a step further by constructing traps in places where there were none. From traps they invented stockades and learned to retain and domesticate animals. Finally, from domesticated wild animals, they went on to breeding genetic variants of animals that suited their needs better.

The ability to amass knowledge in a cumulative way for future generations is dependent on the capacity for symbolic expression of thought, that is, language. As we already mentioned, *Homo sapiens* is not the only species with the ability to communicate; group hunters, like the lion, must somehow "talk" to each other. Social animals, like bees, are also able to communicate. When a bee locates a source of food, it returns to the hive and, by a special dance on the combs of the hive, imparts this knowledge to the other members of the colony. Apparently, the speed of the dance reflects the distance of the source of food from the hive, and the angle that the dancing animal makes with the vertical indicates the direction of the source from the hive. Birds do not sing for the benefit of poets or lovers; their songs serve very important functions as danger warnings or mating calls.

What most nonhuman languages have in common, as far as we know, is that the individual animals are born not only with the ability to learn to "speak" the language but also with the specific vocabulary of the language already wired into the nervous system. After returning to the hive, the successful bee can tell its hive mates about the distance or location of a specific food source from the hive. It tells them nothing, for example, about the general principles of foraging for food. (It need not, of course, for these "principles" are wired in as well.) Recently, however, our insights into the capacity of animals for language has been immeasurably broadened by discoveries concerning the language capacity of the porpoises and apes. Several studies have focused specifically on the chimpanzee and gorilla. Because these animals do not seem to be able to vocalize sounds as humans do, the investigators working with them decided to teach them either sign language (as deaf people are taught) or the manipulation of controls on a computer. By these means several apes have been taught a few hundred "words," which they can both understand and use. The words involve objects (nouns) or actions (verbs) and can have emotional as well as nominative content. Perhaps most striking, however, has been the finding that apes use their vocabulary in innovative ways. When given a stale piece of cake, for example, one referred to it as a "cookie-rock," thus combining two previously learned signs to describe a new situation. While these observations indicate that the higher apes are capable of manipulating symbolic language to some degree, they do not tell us much about natural modes of communication between these creatures.

Human language, of course, is characterized by an incalculably large number of sensible word combinations. Although there are clearly definable patterns to human speech, it has none of the fixed qualities exhibited by lower animals such as bees. Second, human language, as well as allowing communication about specific bits of information, permits expression of generalities and abstractions.

It is easy to see the enormous selective advantage that the acquisition of language must have conferred on its possessors when it arose in human evolution. A flexible symbolic language allows transmission of acquired experience to others and to subsequent generations. Even if it were possible to conceive of the human intellect without a symbolic language, it is hard to see what kind of progress could be made if the Pythagorean theorem or the principles of visual perspective in art or the equally tempered scale in music had to be rediscovered by each generation. Language has made these tasks unnecessary. From the time of the emergence of *Homo sapiens* until now, a period of at most a few tens of thousands of years, which is very short on the evolutionary time scale, people have learned to exist in extreme environments, ranging from the Antarctic shelf to the surface of the Moon, from the tops of the Himalayas to the bottom of the ocean. The mechanism of adaptation in each of these cases has been cultural rather than genetic; that is, people can survive in these situations because of the sum total of what they have been taught by others. The astronaut who orbits the Moon surrounded by many of the comforts of home does so not because he is genetically different from a resident of ancient Athens but because the values, beliefs, and scientific accomplishments of the whole Western cultural tradition have provided him with the means.

It has been said that insects are the most adaptable of animals and that of all animals, they can survive in the widest variety of environments. Certainly, one can find insects in virtually every type of terrain and climate. Remember, however, that with insects one is dealing with nearly three quarters of a million distinct species of animals. Viewed in this way, the true extent of human adaptability becomes clearer. For this single species can live anywhere insects can. To do so, people simply let technology create the proper conditions, and by virtue of appropriate housing, clothing, and food, remain alive and even comfortable.

In the preceding discussion we have treated the two modes of adaptation as if they were somehow equivalent. Nonetheless, there exist important differences between them. In the first place, for humans genetic adaptation is slow compared with cultural adaptation. Since an average human generation time is about 25 years, several centuries or millennia would be necessary for advantageous mutations to be disseminated in the population. Second, the cultural is in many ways a more powerful flexible mode of adaptation than the genetic. To understand why, we must first realize that it is not always possible for a species to adapt genetically to environmental changes. The ability of a species to respond genetically to environmental challenge depends on its pre-existing genetic constitution and on whether it is possible to change, by either a new mutation or genetic recombination, in such a way that adaptive genetic variation will result. It is difficult, for example, to imagine that people could adapt by purely genetic means to all the climatological conditions that cultural methods allow them to master. The anatomical and physiological adjustments that would permit a single species to survive simultaneously in the Antarctic and Africa without clothing or imported food are far too large and varied.

In the history of human development there have occurred environmental changes which, though not so extreme as a drop to polar temperatures, have been radical enough that the human genetic constitution may not have been sufficiently flexible to insure continued survival. Quite obviously, human cultural resources provide an enormous amount of additional flexibility and have allowed extension of the human domain over the entire planet. As noted above, however, the total lifetime of *Homo sapiens* as a species has been fairly short on the evolutionary time scale, and in this short lifetime, cultural achievements have also included certain weapons that threaten life's continued existence. In this capacity to destroy themselves utterly, people are also unique among animals.

THE INTERACTION BETWEEN CULTURAL AND GENETIC MECHANISMS

Up to now we have been speaking of cultural and genetic adaptive mechanisms as quite unrelated entities. It has probably occurred to you that there must be a very close relationship between the two. First of all, the very existence of culture in human society is a reflection of the genetic foundation on which the anatomy and physiology of the human nervous system are based. In one sense, the brain is only another human organ, and like the heart, kidneys, and gonads, its development in fetal life and early childhood is in part determined by the organism's genes. We are indeed very far from even a remote grasp of how human genes, together with environmental influences, direct the formation of an intact nervous system. As stated above, however, no matter what the details of the mechanism, the selection for such a nervous system must have been very powerful because of the enormous benefits that accrue to animals possessing one.

Although genes can be said to influence culture, the reverse is also true, and we are only beginning to appreciate some of the subtler features of such a relationship. Unfortunately, the very term "natural selection," when translated as "survival of the fittest," conjures up images of human beings locked in deadly struggles with saber-tooth tigers and other now-extinct horrors. Perhaps at some time in the past people did face important challenges of this sort, and under these circumstances one might reasonably imagine that the largest, strongest men, who could run fastest and fight best, might survive preferentially. Although this is an oversimplification, one can at least identify some of the selective forces operating in this kind of life.

How has modern society changed this pattern? We must now be scrupulous in observing the distinction between "fitness" as ordinarily used, and fitness in the Darwinian sense. The latter refers only to reproductive advantage, and, defined in this way, one's height, body build, and physical strength may be of secondary importance. In fact, the types of heritable characteristics that endow their possessors with reproductive advantage in modern societies cannot easily be determined. Nevertheless, a few statements can be made.

First of all, the aggregation of people in cities has probably influenced certain patterns of disease resistance, which are most likely genetically inherited. European Jews, for example, have been a predominantly urban people for several centuries. Tuberculosis is an infectious disease that has been predominantly urban in its distribution. Present evidence indicates that Jews as a group now enjoy significantly greater resistance to serious infection by the tubercle bacillus than do members of other groups, such as blacks, who have not been exposed to the organism for so many generations. Presumably, generations of exposure to tuberculosis have allowed those to survive preferentially who tend not to be killed or seriously disabled by the disease prior to their reproductive period.

Selective pressures are also modified by many of the scientific and technical advances of modern society, of which medicine is one of the most obvious. Consider diabetes, a common disease that is clearly inherited. Before the discovery and availability of insulin, many children who became ill with diabetes died before the reproductive period. Others might survive through young adulthood, but because of the severe manifestations of the illness, they were at a reproductive disadvantage with respect to the unaffected population. Insulin has changed this situation markedly, and while diabetes continues to be a severe illness for many of those affected, diabetics incontestably survive better than formerly. This means, of course, that they also tend to have more children than before, and hence we can expect the gene(s) for diabetes to increase in frequency as time goes on, at least in populations where reasonable medical care is available. Thus in several centuries a greater fraction of such a population will be either overt diabetics or carriers of the disease than is true today.

Another interesting example is the influence of dietary intake on the development of coronary heart disease. Although we do not yet know the specific biochemical steps that lead to the development of arteriosclerosis, research suggests that the dietary intake of certain kinds of fats may be important; those with a high intake of such substances seem to have an increased risk of developing heart disease from about age 45 onward. But, you may ask, if natural selection is so powerful, why has human metabolism not evolved so that such high intakes of these substances are not harmful? The answer to this question is somewhat speculative. In the first place, since coronary heart disease causes illness and death largely in the post-reproductive period, selection for factors that reduce its occurrence might not be very powerful. Nevertheless, even if selective pressures are not very powerful, they will produce observable effects if allowed to operate for a long enough time. A probably more important reason has to do with the fact that the general condition of most people was and continues to be characterized by scarcity of food rather than overabundance. Only in a few wealthy areas of the world does the overfeeding problem exist at all. Viewed in this way, it is not surprising that physiological mechanisms have not yet evolved that can deal adequately with the glut of food in the affluent North American diet.

All the previous examples have dealt with one disease or another. When we consider what effect modern culture has on inherited patterns having nothing to do with disease, the answers become even more speculative and sometimes even difficult to define. Intelligence, for example, is a human quality generally considered desirable, and one might naively think that a technologically oriented society such as ours would select for intelligence if in fact such a quality is heritable. But again, we must be careful. What we are asking can be better formulated in the question: Are intelligent people at a reproductive advantage in this society? If we define intelligence (in an admittedly circular manner) as that quality which allows its possessors to do well on intelligence tests, the answer is probably no. The economically disadvantaged have less access to formal education than the well-to-do, and this deprivation undoubtedly depresses their scores on so-called intelligence tests. Since these segments of American society generally have the largest families, they can be considered to be at a reproductive advantage over those with higher test scores and fewer children.

The fact that we cannot easily identify the kinds of selection that are taking place in human societies today does not mean that natural selection has ceased to exist as a force in determining the human future. Natural selection will always exist whenever genetic variation is present in a population and whenever individuals fail to reproduce equally because of genetically determined factors. In such situations those with a reproductive advantage will make a larger genetic contribution to future generations than those at a disadvantage. It is upon this single factor that the environment may exert its most telling influences on the future of mankind.

In conclusion, what can be said? The two types of adaptation do not permit universal flexibility. The types of environmental changes that we have been speaking about in this book are very wide-ranging and are operating on a compressed time scale never seen before on this planet. Moreover, the possibilities for adaptation using the two mechanisms are not unlimited; nor is the genetic mode of adapting always possible. Extensive cultural adaptation may be effective in many situations for people and some of the "higher" animals, but it is not clear how the subtle kinds of environmental changes we have been dealing with, such as persistent pesticides or depletion of the ozone layer, would be subject to cultural adaptations by any species except possibly *Homo sapiens*.

It is precisely because the potential for adaptation is limited that preservation of an intact environment is one of the most pressing issues facing humanity. It does not have the immediate emotional impact, say, of avoiding nuclear war or mass starvation, but in the long run the consequences of environmental destruction are no less grave. We have tried to show in this book that the various issues of concern in preserving the environment are not simple. In addition, the reasons for environmental deterioration are the result not so much of human malevolence as of ignorance or powerful economic pressures. Frequently, the path leading to environmental conservation appears to prejudice our short-term economic interest in matters such as the Alaska pipeline or the use of broad-spectrum insecticides. At other times, practices are holdovers from another era (for example, whaling) and are no longer economically, ecologically, or morally justifiable. In still another situation the shortage and/or high cost of oil and gas seems to be leading us either back to other fossil fuel sources, the use of which will complicate the problem of air pollution control, or forward to radioactive sources, which present serious environmental problems of their own.

The prognosis for the future appears very poor indeed if every issue involving environmental conservation degenerates into a heated and unproductive name-calling contest between the "environmental lobby" and the "industrial complex," who allegedly care for nothing except clean air and high profits, respectively. It would be better for all people to come to regard the environment as a personal resource, the intactness of which is essential for this and all subsequent generations.

BIBLIOGRAPHY

Several excellent works dealing with various aspects of evolution are:

Theodosius Dobzhansky: *Genetics and the Evolutionary Process.* New York, Columbia University Press, 1970. 505 pp.

—————: *Mankind Evolving.* New Haven, Yale University Press, 1962. 381 pp.

Ernst Mayr: *Populations, Species, and Evolution.* Cambridge, Mass., Harvard University Press, 1970. 453 pp.

George Gaylord Simpson: *The Meaning of Evolution.* New York, Bantam Books, 1971. 333 pp.

A summary of the research on industrial melanism is given in:

H. B. D. Kettlewell: "The Phenomenon of Industrial Melanism in the Lepidoptera." *Ann. Rev. Entomol.,* **6:**245, 1961.

A recent, comprehensive account of the societal relationships within various animal species is:

E. O. Wilson: *Sociobiology.* Cambridge, Mass., Harvard University Press, 1976. 697 pp.

APPENDIX

A THE METRIC SYSTEM (INTERNATIONAL SYSTEM OF UNITS)

The metric system (Système International d'Unités, abbreviated SI) is internationally recognized and is now used in nearly all the nations in the world. Terminology in environmental science has often used a mixture of metric and other units. More recently, however, the shift has been toward exclusive use of the metric system. This book is concerned with four of the fundamental metric units. These are:

Quantity	*Unit*	*Abbreviation*
length	meter	m
mass	kilogram	kg
time	second	sec
temperature	degree Celsius	°C

Larger or smaller units in the metric system are expressed by the following prefixes:

Multiple or Fraction	*Prefix*	*Symbol*
1000	kilo	k
1/100	centi	c
1/1000	milli	m
1/1,000,000	micro	μ

LENGTH

The **meter** was once defined in terms of the length of a standard bar; it is now defined in terms of wavelengths of light. A meter is about 1.1 yards.

One **centimeter,** cm, is 1/100 of a meter, or about 0.4 inch.

The unit commonly used to express sizes of dust particles is the **micrometer,** μm, formerly called the **micron.** There are one million micrometers in a meter, or 25,400 micrometers per inch.

AREA

The metric unit is the **hectare** (abbreviated ha), which equals 10,000 m², or about 2.47 acres.

MASS

The **kilogram,** kg, is the mass of a piece of platinum-iridium metal called the Prototype Kilogram Number 1, kept at the International Bureau of Weights and Measures, in France. It is equal to about 2.2 pounds.

One gram, g, is 1/1000 kg.

One metric ton is 1000 kg = 2204.6 lb.

VOLUME

Volume is not a fundamental quantity; it is derived from length.

One **cubic centimeter,** cm³, is the volume of a cube whose edge is 1 cm.

One **liter** = 1000 cm³.

TEMPERATURE

If two bodies, A and B, are in contact, and if there is a spontaneous transfer of heat from A to B, then A is said to be *hotter* or at a higher temperature than B. Thus, the greater the tendency for heat to flow away from a body, the higher its temperature is.

The Celsius (formerly called Centigrade) temperature scale is defined by several fixed points. The most commonly used of these are the freezing point of water, 0°C, and the boiling point of water, 100°C.

The Fahrenheit scale, commonly used in medicine and engineering in England and the United States, designates the freezing point of water as 32°F and the boiling point of water as 212°F.

ENERGY

The metric unit of energy is the **joule.**

The unit commonly used to express heat energy, or the energies involved in chemical changes, is the **calorie,** cal. One calorie is 4.184 joules. The energy of one calorie is sufficient to warm one gram of water 1°C.

The **kilocalorie,** kcal, is 1000 calories. This unit is also designated Calorie (capital C), especially when it is used to express food energies for nutrition.

Some other units of energy are:

1 erg = one ten-millionth of a joule.

1 watt-hour (the energy released when one watt of power is delivered for one hour) = 860 cal.

1 kilowatt-hour (kW-hr) = 1000 W-hr = 860 kcal.

1 Btu (British thermal unit) is the heat required to raise the temperature of one pound of water 1°F = 252 cal.

POWER

Power is energy per unit time. The metric unit is the **watt**.

One watt is equal to one joule per second.

One kilowatt = 1000 watts.

One horsepower = 745 watts.

B CHEMICAL SYMBOLS, FORMULAS, AND EQUATIONS

Atoms or elements are denoted by symbols of one or two letters, like H, U, W, Ba, and Zn.

Compounds or molecules are represented by formulas that consist of symbols and subscripts, sometimes with parentheses. The subscript denotes the number of atoms of the element represented by the symbol to which it is attached. Thus H_2SO_4 is a formula that represents a molecule of sulfuric acid, or the substance sulfuric acid. The molecule consists of two atoms of hydrogen, one atom of sulfur, and four atoms of oxygen. The substance consists of matter that is an aggregate of such molecules. The formula for oxygen gas is O_2; this tells us that the molecules consist of two atoms each.

Chemical transformations are represented by chemical equations, which tell us the molecules or substances that react and the ones that are produced, and the molecular ratios of these reactions. The equation for the burning of methane in oxygen to produce carbon dioxide and water is:

$$CH_4 + 2O_2 \rightarrow CO_2 + 2H_2O$$

Each coefficient applies to the entire formula that follows it. Thus $2H_2O$ means $2(H_2O)$. This gives the following molecular ratios: reacting materials, two molecules of oxygen to one of methane; products, two molecules of water to one of carbon dioxide. The above equation is balanced because the same number and kinds of atoms, one of carbon and four each of hydrogen and oxygen, appear on each side of the arrow.

The atoms in a molecule are held together by chemical

bonds. Chemical bonds can be characterized by their length, the angles they make with other bonds, and their strength (that is, how much energy would be needed to break them apart).

$$H—O \atop \diagup \atop H$$

The formula for water shows that the molecule contains two H–O bonds. The length of each bond is about 1/10,000 of a micrometer, and the angle between them is 105°. It would require about 12 kcal to break all of the bonds in a gram of water. These bonds are strong, as chemical bonds go.

In general, substances whose molecules have strong chemical bonds are stable, because it is energetically unprofitable to break strong bonds apart and rearrange the atoms to form other, weaker bonds. Therefore, stable substances may be regarded as chemically self-satisfied; they have little energy to offer, and are said to be energy-poor. Thus, water, with its strong H–O bonds, is not a fuel or a food. The bonds between carbon and oxygen in carbon dioxide, CO_2, are also strong (about 1.5 times as strong as the H–O bonds of water), and CO_2 is therefore also an energy-poor substance.

In contrast, the C–H bonds in methane, CH_4, are weaker than the H–O bonds of water. It is energetically profitable to break these bonds and produce the more stable ones in H_2O and CO_2. Methane is therefore an energy-rich substance and can be burned to heat houses and drive engines.

C SUPPLEMENTS TO CHAPTERS

This Appendix contains supplementary material for some of the chapters. The number headings refer to the chapters in which the material appears.

C-3 *THE ECOLOGY OF NATURAL SYSTEMS*

Acidity and the pH Scale

Hydrogen ions (shown as H^+ in the table) render water acidic. The original meaning of acid is "sour," referring to the taste of substances such as vinegar, lemon juice, unripe apples, and old milk. It has long been observed that all sour or acidic substances have some properties in common, notably their ability to corrode (rust, or oxidize) metals. When the attack on a metal by an acidic solution is vigorous, hydrogen gas (H_2) is evolved in the form of visible bubbles.

Acidic solutions also conduct electricity, with evolution of hydrogen gas at the negative electrode (cathode). These circumstances imply that acid solutions are characterized by the presence of positive ions bearing hydrogen. A hydrogen ion, or proton, designated H^+, cannot exist as an independent entity in water because it is strongly attracted (in fact, chemically bonded) to the oxygen atom of the water molecule. The resulting hydrated proton is formulated as $H(H_2O)^+$, or H_3O^+. The simpler designation H^+ may therefore be regarded as an abbreviation.

Some slight transfer of protons occurs even in pure water.

$$H_2O + H_2O \leftrightarrows H_3O^+ + OH^-$$

Hydroxyl ions, OH^-, can neutralize H_3O^+ ions by reacting with them to produce water, as indicated by the arrow pointing left. The concentration of H_3O^+ and of OH^- in pure water at 25°C is 1.0×10^{-7} moles/liter.* This solution is said to be *neutral*, because the concentrations of the two ions are equal. When the hydrogen ion concentration is greater than 1.0×10^{-7} moles/liter at 25°C, the solution is *acidic*. Hydrogen ion concentrations are usually expressed logarithmically as pH values, where pH $= -\log_{10}$ (hydrogen ion concentration).

Recall that the logarithm of a number "to the base 10" is simply the *number of times 10 is multiplied by itself* to give the number. As we can quickly see, when the number is a multiple of 10, its log is simply the number of zeros it contains. When the number is 1 divided by a multiple of 10, its log is *minus* the number of zeros in the denominator. Therefore, when the number expresses hydrogen ion contration, the pH is *plus* the number of zeros in the denominator:

Concentration of H^+ (moles/liter)	Calculation of pH	
1/10	$\log 1/10 = \log 10^{-1} = -1.$ pH $= 1$	more acidic ↑
1/10,000,000	$\log 1/10,000,000 = \log 10^{-7} = -7.$ pH $= 7$	neutral ↓
1/1,000,000,000	$\log 1/1,000,000,000 = \log 10^{-9} = -9.$ pH $= 9$	more basic

Any pH less than 7 connotes acidity. The lower the pH of a body of water, the more prone it is to be corrosive and, thereby, to become polluted with metallic compounds. Acidic waters coursing through lead pipes will therefore become more toxic than pure water. And grapefruit juice standing in an iron cup develops a terrible taste.

---

*One mole of hydrogen atoms is the amount that weighs one gram.

C-6 THE GROWTH AND CONTROL OF HUMAN POPULATIONS

To compute doubling time from rate of growth, think of an analogy with compound interest at the bank. If a rate of growth (interest rate) is applied once a year to a population of size P_o (capital in the bank), the population (capital) at the end of one year is:

$$P_1 = P_o + P_o r = P_o (1 + r)$$

More generally, if population growth is compounded n times each year, then the population in year t is

$$P_t = P_o(1 + r/n)^{nt}$$

It is reasonable to suppose that populations grow continuously or that $n \to \infty$. From elementary calculus,

$$\lim_{n \to \infty} (1 + r/n)^{nt} = e^{rt}$$

The doubling time, t_d, is then the solution of the equation:

$$2P_o = P_o e^{rt_d}$$

Or, taking logarithms on both sides of the equation:

$$t_d = 0.693/r.$$

C-12 CONTROL OF PESTS AND WEEDS

Chemical Formulas of Pesticides

Organochlorides

Dichlorodiphenyltrichloroethane (DDT)

Dichlorodiphenyldichloroethane (DDD)

Aldrin

Lindane (containing 99% γ-isomer)

Organophosphates

Parathion

Malathion

A Carbamate

Sevin®
(1-naphthyl-*N*-methylcarbamate)

$$O \parallel$$

O—CNHCH₃

Chemical Formulas of Polychlorinated Biphenyls (PCB's)

Biphenyl itself is a hydrocarbon whose molecules consist of two benzene rings joined to each other:

A polychlorinated biphenyl is a compound in which several of the hydrogen atoms of biphenyl are substituted by chlorine atoms. There are various possibilities, one of which is:

C-13 AIR POLLUTION

Chemical formulas of chemical oxidants

Ozone, O_3

Peroxyacetyl nitrate, $C_2H_3O_5N$, an oxidant component of "smog"

579

Structural formulas of benzene and related hydrocarbons.

Benzene Toluene Phenanthrene

3, 4-Benzopyrene Anthracene Chrysene

C-16 NOISE

The Decibel Scale

Physicists have created a unit that defines a tenfold increase in sound intensity, and named it a **Bel,** after Alexander Graham Bell. If the sound of a garbage disposal unit is 10 times as intense as that of a vacuum cleaner, it is one Bel more intense. A rocket whose sound is a million, or 10^6, times as intense as the vacuum cleaner is therefore 6 Bels more intense. This definition leads to the following relationship:

Difference in intensity between two sounds, X and Y, expressed in Bels $= \log_{10} \left(\dfrac{\text{sound intensity of X}}{\text{sound intensity of Y}} \right)$

It happens that the Bel is a rather large unit, so that it is convenient to divide it into tenths, or decibels, dB, as follows:

$$1 \text{ Bel} = 10 \text{ decibels}$$

Then,

$$\text{Intensity in decibels} = \text{intensity in Bels} \times \frac{10 \text{ decibels}}{\text{Bel}}$$

Therefore,

$$\begin{matrix} \text{Difference in intensity} \\ \text{between two sounds,} \\ \text{X and Y, expressed in} \\ \text{decibels, dB} \end{matrix} = 10 \ \log_{10} \left(\frac{\text{sound intensity of X}}{\text{sound intensity of Y}} \right)$$

Finally, it is very convenient to start our scale somewhere that we can designate zero, and the most convenient point is at the softest sound level which is audible to the human ear. We will call this level zero decibels. We can now write our equation for the decibel scale as follows:

$$\text{Intensity in decibels of any given sound} = 10 \times \log_{10} \left(\frac{\text{intensity of the given sound}}{\text{intensity of a barely audible sound}} \right)$$

Example 1: The sound of a vacuum cleaner in a room has 10 million times the intensity of the faintest audible sound. What is the intensity of the sound in decibels?
Answer:

$$\left(\frac{\text{sound intensity of vacuum cleaner}}{\text{faintest audible sound}} \right) = 10{,}000{,}000$$

$$\log 10{,}000{,}000 = 7$$

$$\begin{aligned} \text{Sound intensity} &= 10 \times \log 10{,}000{,}000 \\ &= 10 \times 7 \\ &= 70 \text{ decibels} \end{aligned}$$

Example 2: What is the intensity in decibels of the faintest audible sound?
Answer: This had better turn out to be zero, which is what we promised the log scale would provide:

$$\left(\frac{\text{intensity of the faintest audible sound}}{\text{intensity of the faintest audible sound}} \right) = 1, \text{ and } \log 1 = 0$$

Thus,

$$\text{Sound intensity} = 10 \times \log 1 = 10 \times 0 = 0 \text{ decibels}$$

Example 3: Referring back to page 534, if the sound level in a library is 30 dB, and the sound level of a rustling leaf is 10 dB, what will be the sound level in the library if we bring in the rustling leaf?
Answer: The difference between the two decibel levels is $30 - 10 = 20$ dB. Substituting this in the equation, we have

$$20 = 10 \ \log \left(\frac{\text{sound intensity of library}}{\text{sound intensity of leaf}} \right)$$

581

$$\log \left(\frac{\text{sound intensity of library}}{\text{sound intensity of leaf}} \right) = \frac{20}{10} = 2$$

$$\frac{\text{sound intensity of library}}{\text{sound intensity of leaf}} = \text{antilog } 2 = 10^2 = 100$$

Now, if x/y = 100, then x + y = 101 y. Therefore,

$$\frac{\text{sound intensity of library + leaf}}{\text{sound intensity of library}} = \frac{101}{100} = 1.01$$

Difference in dB levels between library + leaf and library alone $= 10 \log 1.01 = 0.04$

Therefore the decibel level of the library with the rustling leaf is 30 dB + 0.04 dB = 30.04 dB. So, you see, the leaf hardly makes any difference.

The Sine Curve

Recall that in a right triangle the sine of an angle is the ratio of the length of the opposite side to that of the hypotenuse. Therefore, the height, h, of any point on the rim of the circle from the horizontal is proportional to the sine of the angle that it makes with the horizontal, as shown below:

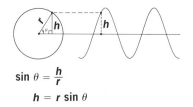

$$\sin \theta = \frac{h}{r}$$
$$h = r \sin \theta$$

TABLE OF RELATIVE ATOMIC WEIGHTS (1969)*

ELEMENT	SYMBOL	ATOMIC NUMBER	ATOMIC WEIGHT	ELEMENT	SYMBOL	ATOMIC NUMBER	ATOMIC WEIGHT
Actinium	Ac	89		Mercury	Hg	80	200.59
Aluminum	Al	13	26.9815	Molybdenum	Mo	42	95.94
Americium	Am	95		Neodymium	Nd	60	144.24
Antimony	Sb	51	121.75	Neon	Ne	10	20.179
Argon	Ar	18	39.948	Neptunium	Np	93	237.0482
Arsenic	As	33	74.9216	Nickel	Ni	28	58.71
Astatine	At	85		Niobium	Nb	41	92.9064
Barium	Ba	56	137.34	Nitrogen	N	7	14.0067
Berkelium	Bk	97		Nobelium	No	102	
Beryllium	Be	4	9.01218	Osmium	Os	76	190.2
Bismuth	Bi	83	208.9806	Oxygen	O	8	15.9994
Boron	B	5	10.81	Palladium	Pd	46	106.4
Bromine	Br	35	79.904	Phosphorus	P	15	30.9738
Cadmium	Cd	48	112.40	Platinum	Pt	78	195.09
Calcium	Ca	20	40.08	Plutonium	Pu	94	
Californium	Cf	98		Polonium	Po	84	
Carbon	C	6	12.011	Potassium	K	19	39.102
Cerium	Ce	58	140.12	Praseodymium	Pr	59	140.9077
Cesium	Cs	55	132.9055	Promethium	Pm	61	
Chlorine	Cl	17	35.453	Protoactinium	Pa	91	231.0359
Chromium	Cr	24	51.996	Radium	Ra	88	226.0254
Cobalt	Co	27	58.9332	Radon	Rn	86	
Copper	Cu	29	63.546	Rhenium	Re	75	186.2
Curium	Cm	96		Rhodium	Rh	45	102.9055
Dysprosium	Dy	66	162.50	Rubidium	Rb	37	85.4678
Einsteinium	Es	99		Ruthenium	Ru	44	101.07
Erbium	Er	68	167.26	Samarium	Sm	62	150.4
Europium	Eu	63	151.96	Scandium	Sc	21	44.9559
Fermium	Fm	100		Selenium	Se	34	78.96
Fluorine	F	9	18.9984	Silicon	Si	14	28.086
Francium	Fr	87		Silver	Ag	47	107.868
Gadolinium	Gd	64	157.25	Sodium	Na	11	22.9898
Gallium	Ga	31	69.72	Strontium	Sr	38	87.62
Germanium	Ge	32	72.59	Sulfur	S	16	32.06
Gold	Au	79	196.9665	Tantalum	Ta	73	180.9479
Hafnium	Hf	72	178.49	Technetium	Tc	43	98.9062
Helium	He	2	4.00260	Tellurium	Te	52	127.60
Holmium	Ho	67	164.9303	Terbium	Tb	65	158.9254
Hydrogen	H	1	1.0080	Thallium	Tl	81	204.37
Indium	In	49	114.82	Thorium	Th	90	232.0381
Iodine	I	53	126.9045	Thulium	Tm	69	168.9342
Iridium	Ir	77	192.22	Tin	Sn	50	118.69
Iron	Fe	26	55.847	Titanium	Ti	22	47.90
Krypton	Kr	36	83.80	Tungsten	W	74	183.85
Lanthanum	La	57	138.9055	Uranium	U	92	238.029
Lawrencium	Lr	103		Vanadium	V	23	50.9414
Lead	Pb	82	207.2	Xenon	Xe	54	131.30
Lithium	Li	3	6.941	Ytterbium	Yb	70	173.04
Lutetium	Lu	71	174.97	Yttrium	Y	39	88.9059
Magnesium	Mg	12	24.305	Zinc	Zn	30	65.37
Manganese	Mn	25	54.9380	Zirconium	Zr	40	91.22
Mendelevium	Md	101					

*Based on the assigned relative atomic mass of $^{12}C = 12$.

GLOSSARY

abortion, induced – The artificial termination of pregnancy.

activated sludge – See under *Sludge*.

adaptation – The process of accommodation to change; the process by which the characteristics of an organism become suited to the environment in which the organism lives.

adsorption – The process by which molecules from a liquid or gaseous phase become concentrated on the surface of a solid.

aerobiosis (*adj.*, aerobic) – Bacterial decomposition in the presence of air.

aerosol – A substance consisting of small particles, typically having diameters that range from 1/100 micrometer to 1 micrometer. See *Fume; Smoke*.

air pollution – The deterioration of the quality of air that results from the addition of impurities.

air quality standards – Specifications for the concentrations of air pollutants that are considered to be tolerable.

air resource management – The strategy of using the atmosphere as a repository for pollutants, to the extent that the air is judged to be able to accommodate them without exceeding previously established standards of quality.

albedo – A measure of the reflectivity of a surface, measured as the ratio of light reflected to light received. A mirror or bright snowy surface has a high albedo, whereas a rough flat road surface has a low albedo.

altruism – Devotion to the interests of others.

amensalism – A two-species interaction in which the growth of one species is inhibited, while the growth of the second is unaffected.

amplitude (of a wave) – The magnitude or height of a wave, measured as the distance between the zero point of the wave to the point of maximum displacement. The amplitude is one half of the vertical distance between the crest and the trough.

anaerobiosis (*adj.*, anaerobic) – The biological utilization of nutrients in the absence of air.

aquaculture – The science and practice of raising fish in artificially controlled ponds or pools.

arithmetic growth – In population studies, growth characterized by the addition of a constant number of individuals during a unit interval of time. For instance, if there are x individuals in year 0 and $x + a$ in year 1, arithmetic growth implies $x + na$ in year n.

atmosphere – The predominantly gaseous envelope that surrounds the Earth.

atom – The fundamental unit of the element.

atomic nucleus – The small positive central portion of the atom that contains its protons and neutrons.

atomic number – The number of protons in an atomic nucleus.

auricle – The external, visible portion of the ear that funnels sound waves into the ear canal.

autotroph – An organism that obtains its energy from the sun, as opposed to a *heterotroph,* which is an organism that obtains energy from the tissue of other organisms. Most plants are autotrophs.

background radiation – The level of radiation on Earth from natural sources.

baleen – A set of elastic, horny plates that form a sievelike region in the mouths of certain whales. Plankton-eating whales have no true teeth, only baleen.

Bel – Ten decibels.

benthic organism – A plant or animal that lives at or near the bottom of a lake, river, stream, or ocean.

biochemical oxygen demand (B.O.D.) – A measure of pollution of water by organic nutrients that recognizes the *rate* at which the nutrient matter uses up oxygen, as well as the total quantity that can be consumed.

biodegradable – Refers to substances that can readily be decomposed by living organisms.

biomass – The total weight of all the living organisms in a given system.

biome – A group of ecosystems characterized by similar vegetation and climate, and which are collectively recognizable as a single large community unit. Examples include the arctic tundra, the North American Prairie, and the tropical rain forest.

biosphere – That part of the Earth and its atmosphere which can support life.

biotic potential – The maximum rate of population

growth of a species that would result if all females bred as often as possible, and all individuals survived past their reproductive age.

birth rate – The number of individuals born during some time period, usually a year, divided by an appropriate population. For example, the crude birth rate in human populations is the number of live children born during a given year divided by the midyear population of that year.

black lung disease – A series of debilitating and often fatal diseases that affect the lungs of miners who work in underground coal mines.

bloom – A rapid and often unpredictable growth of a single species in an ecosystem.

branching chain reaction – A chain reaction in which each step produces more than one succeeding step.

breakwater – See *Groin*.

breeder reactor – A nuclear reactor that produces more fissionable material than it consumes.

calorie – A unit of energy used to express quantities of heat. When calorie is spelled with a small c, it refers to the quantity of heat required to heat 1 gram of water 1°C. (This definition is not precise, because the quantity depends slightly on the particular temperature range chosen.) When Calorie is spelled with a capital C, it means 1000 small calories, or one kilocalorie, the quantity of heat required to heat 1000 grams (1 kilogram) of water 1°C. Food energies for nutrition are always expressed in Calories. The exact conversions are:

1 cal = 4.184 joules
1 kcal = 4184 joules

carnivore – An animal that eats the flesh of other animals.

carrying capacity – The maximum number of individuals of a given species that can be supported by a particular environment.

census – A count of a population.

chain reaction – A reaction that proceeds in a series of steps, each step being made possible by the preceding one. See also *Branching chain reaction*.

chelating agent – A molecule that can offer two or more different chemical bonding sites to hold a metal ion in a clawlike linkage. The bonds between chelating agent and metal ion can be broken and reestablished reversibly.

chemical bond – A linkage that holds atoms together to form molecules.

chemical change – A transfer that results from making or breaking of chemical bonds.

chemosterilant – A chemical capable of sterilizing a living organism.

chromosome – Rod-shaped structures within the nucleus of the cell, on which are located the genes that determine the hereditary characteristics of the organism.

climate – The composite pattern of weather conditions that can be expected in a given region. Climate refers to yearly cycles of temperature, wind, rainfall, etc., and not to daily variations.

climax system – A natural system that represents the end, or apex, of an ecological succession.

cochlea – A snail-shaped structure in the inner ear, which contains a fluid that vibrates in response to sound waves.

colloid – Material composed of minute particles, generally within a size range too small for gravitational settling but large enough to scatter light.

commensalism – A relationship in which one species benefits from an unaffected host.

common mode failure – An accident in which a single event causes multiple failures, and thereby knocks out redundant systems. See *Redundancy*.

competition – An interaction in which two or more organisms try to gain control of a limited resource.

composting – The controlled, accelerated biodegradation of moist organic matter to form a humuslike product that can be used as a fertilizer or soil conditioner.

continental drift – The theory stating the continent-sized masses of the Earth's crust are slowly moving relative to one another.

contraception – Prevention of conception.

control rod – A neutron-absorbing medium that controls the reaction rate in a nuclear reactor.

cooling tower – A large towerlike structure used to cool water from an electric generating station or any other industrial facility.

core (of the Earth) – The central portion of the Earth, believed to be composed mainly of iron and nickel. See also *Crust, Mantle*.

critical level – In general, a critical condition relates to a point at which some property changes very abruptly in response to a small change in some other property of the system. In ecology, a population is said to be reduced to its critical level when its numbers are so few that it is in acute danger of extinction. In nuclear science, a critical condition is said to exist when a chain reaction continues at a steady rate, neither accelerating nor slowing down.

critical mass (in a nuclear reaction) – The quantity of fissionable material just sufficient to maintain a nuclear chain reaction.

crude rate – A vital rate with the entire population of some area as the denominator.

crust (of the Earth) – The solid outer layer of the Earth; the portion on which we live. See also *Mantle; Core*.

cyclone (for air pollution control) – An air cleaning device that removes dust particles by throwing them out of an air stream in a cyclonic motion.

death rate – The number of individuals dying during some time period, usually a year, divided by an appropriate population. For example, the crude death rate in human populations is the number of deaths during a given year divided by the midyear population of that year.

decibel (dB) – A unit of sound intensity equal to $1/10$ of a Bel. The decibel scale is a logarithmic scale used in measuring sound intensities relative to the intensity of the faintest audible sound.

demographic transition – The pattern of change in vital rates typical of a developing society. The process can be outlined briefly as follows. Birth and death rates in preindustrial society are typically very high; consequently, population growth is very slow. Introduction, or development, of modern medicine causes a decline in death rates and hence a rapid increase in population growth. Finally, birth rates fall, and the population grows slowly once more.

demography – That branch of sociology or anthropology which deals with the statistical characteristics of human populations, with reference to total size, density, number of deaths, births, migrations, marriages, prevalence of disease, and so forth.

deoxyribonucleic acid – See *DNA*.

desert – A climax system in which rainfall is less than 25 centimeters per year. These systems are barren and support relatively little plant or animal life.

desiccate – To deprive thoroughly of moisture.

detergent – A cleaning agent that acts by serving to bind water molecules to molecules of grease or other soiling substances.

detritus – Traditionally, the accumulation of small particles of rock worn away by weathering. In ecology, the word has recently been used to describe all the nonliving organic matter in an ecosystem.

detritus feeders – Organisms that consume organic detritus.

DNA – A substance consisting of large molecules which determines the synthesis of proteins and accounts for the continuity of species.

doubling time – The time a population takes to double in size, or the time it would take to double if its annual growth rate were to remain constant.

dust – An airborne substance that consists of solid particles typically having diameters greater than about one micrometer.

eardrum – A membrane in the ear canal that vibrates when sound waves enter the ear. Also called the *Tympanic membrane*.

ecological niche – The description of the unique functions and habitats of an organism in an ecosystem.

ecology – The study of the interrelationships among plants and animals and the interactions between living organisms and their physical environment.

economic externality – That portion of the cost of a product which is not accounted for by the manufacturer but is borne by some other sector of society. An example is the cost of environmental degradation that results from a manufacturing operation.

ecosystem – A group of plants and animals occurring together plus that part of the physical environment with which they interact. An ecosystem is defined to be nearly self-contained, so that the matter which flows into and out of it is small compared to the quantities which are internally recycled in a continuous exchange of the essentials of life.

ecosystem homeostasis – The control mechanisms within an ecosystem that act to maintain constancy by opposing external stresses.

ecotone – A transitional area between two different types of ecosystems, containing characteristics of both, yet having a unique character of its own.

electron – The fundamental atomic unit of negative electricity.

electrostatic precipitator – A device that electrically charges particulate air pollutants so that they drift to an electrically grounded wall from which they can be removed easily.

element – A substance all of whose atoms have the same atomic number.

energy – The capacity to perform work or to transfer heat. See also specific types of energy.

Environmental Protection Agency (EPA) – A branch of the United States government devoted to protection of the environment.

environmental resistance – The sum of various pressures, such as predation, competition, adverse weather, etc., which collectively inhibit the potential growth of every species.

epilimnion – Upper waters of a lake.

estuary – A partially enclosed shallow body of water with access to the open sea and usually a supply of fresh water from the land. Estuaries are less salty than the open ocean but are affected by tides and, to a lesser extent, by wave action of the sea.

euphotic zone – The surface volume of water in the ocean or a deep lake which receives sufficient light to support photosynthesis.

eutrophication – The enrichment of a body of water with nutrients, with the consequent deterioration of its quality for human purposes.

expectation of life – The number of years an infant can be expected to live under a specified schedule of death rates.

exponential growth – See *Geometric growth*.

extrapolation – The prediction of points on a graph outside the range of observation.

fermentation – An anaerobic process by which certain microorganisms consume sugars, starches, or cellulose to produce various organic by-products, particularly alcohols. In this manner, some low-quality organic wastes can be converted to useful fuels or animal feeds.

fire climax – A condition in which the continuance of a given ecosystem is maintained by fire.

First Law of Thermodynamics – See *Thermodynamics*.

fission (of atomic nuclei) – The splitting of atomic nuclei into approximately equal fragments.

fitness – The quality or characteristic of having reproductive advantage.

food chain – An idealized pattern of flow of energy in a natural ecosystem. In the classical food chain, plants are eaten only by primary consumers, primary consumers are eaten only by secondary consumers, secondary consumers only by tertiary consumers, and so forth.

food web – The actual pattern of food consumption in a natural ecosystem. A given organism may obtain nourishment from many different trophic levels and thus give rise to a complex, interwoven series of energy transfers.

Freon – A trade name of the Dupont Company that refers to the class of chlorofluorocarbons. The compounds that may be implicated in stratospheric pollution are Freon-11 ($CFCl_3$), and Freon-12 (CCl_2F_2).

frequency – The number of wave disturbances (can be measured as the number of crests) that pass a given point in a specific amount of time. Frequency is usually expressed in cycles/sec, or hertz. 1 Hz = 1 cycle/sec.

fume – An aerosol that is usually produced by combination of smaller particles.

fusion (of atomic nuclei) – The combination of nuclei of light elements (particularly hydrogen) to form heavier nuclei.

Gaia – The ancient Greek goddess of the Earth. This word has recently been used to describe the biosphere and to emphasize the interdependence of the Earth's ecosystems by likening the entire biosphere to a single living organism.

gas – A state of matter that consists of molecules that are moving independently of each other in random patterns.

gaseous diffusion – The movement of a gas in space by the random motions of its molecules. Lighter gases diffuse faster than heavier ones.

gene pool – The aggregate of all genes in an interbreeding community.

generator – A device that converts mechanical power to electrical power. Generators must be powered by some external source of energy. In most commercial applications, steam or flowing water drives a turbine, and the generator is connected to the spinning shaft of the turbine.

genetic variation – The process by which the structure of genes, and hence the heritable characteristics of an organism, change in time.

geometric growth – In population studies, growth such that in each unit of time, the population increases by a constant proportion. Also called *exponential growth*.

geothermal energy – Energy derived from the heat of the Earth's interior.

Green Revolution – The realization of increased crop yields in many areas due to the development of new high-yielding strains of wheat, rice, and other grains in the 1960's.

greenhouse effect – The effect produced by certain gases, such as carbon dioxide or water vapor, that causes a warming of the Earth's atmosphere by absorption of infrared radiation. The term is an inappropriate analogy to greenhouses, which were once thought to keep themselves warm by admitting sunlight but retaining infrared radiation. However, it has been shown that most of the heat retention in greenhouses results from the conservation of warm air. The atmosphere, of course, is open, so the mechanism is not the same.

groin – A stone or concrete structure built perpendicular to a beach to interrupt coastal currents and trap sand in a local area. Groins impede the normal flow of sand along a shore.

gross national product (GNP) – The total value of all goods and services produced by the economy in a given year.

half-life (of a radioactive substance) – The time required for half of a sample of radioactive matter to decompose.

heat – A form of energy. Every object contains heat energy in an amount that depends on its mass, its temperature, and the specific heat of the materials of which it consists.

heat engine – A mechanical device that converts heat to work.

hectare – A metric measure of surface area. One hectare is equal to 10,000 square meters or 2.47 acres.

herbicide – A chemical used to control unwanted plants.

hertz (Hz) – A unit that measures the frequency of a wave form. When one crest of a wave passes a given point every second, that wave is said to have a frequency of one hertz.

$$1 \text{ hertz} = 1 \text{ cycle/sec}$$

heterotroph – An organism that obtains its energy by consuming the tissue of other organisms.

home range – The area in which an animal generally travels and gathers its food.

homeostasis – See *Ecosystem homeostasis*.

hormone inhibitors – A class of compounds that block the action of juvenile hormones. These compounds can be used as insecticides, for if sprayed properly, they will disrupt the life cycles of specific insect larvae and kill them.

humus – The complex mixture of decayed organic matter that is an integral part of healthy soil.

hydrocarbon – A compound of hydrogen and carbon.

hydroelectric power – Power derived from the energy of falling water.

hydrological cycle (water cycle) – The cycling of water, in all its forms, on the Earth.

hypolimnion – The lower levels of water in a lake or pond which remain at a constant temperature during the summer months.

inbreeding – The mating of closely related individuals. Inbreeding generally weakens a population.

industrial melanism – The shift in color from light to dark of moths inhabiting areas in which the surfaces of trees and other objects have been darkened by industrial pollution.

infrasound – Elastic waves of frequency below the minimum for human hearing (about 50 Hz).

inversion – A meteorological condition in which the lower layers of air are cooler than those at higher altitudes. This cool air remains relatively stagnant and causes a concentration of air pollutants and unhealthy conditions in congested urban regions.

ion – An electrically charged atom or group of atoms.

isotopes – Atoms of the same element that have different mass numbers.

joule – A fundamental unit of energy. 4.184 joules = 1 calorie.

juvenile hormone – A chemical naturally secreted by an insect while it is a larva. When the flow of juvenile hormone stops, the insect metamorphoses to become an adult. These compounds can be used as insecticides because if sprayed at critical times they will interrupt the natural metamorphoses and eventually kill specific insect pests.

krill – Small, shrimplike crustacea that grow in large numbers in the cold waters of the southern oceans. Krill represent a primary food supply for many species of whales.

lava – The material produced when magma pours onto the surface of the Earth rapidly through fissures in the crust. A site where lava appears is called a volcano.

law of limiting factors – A law which states that, under steady state conditions, the growth of a population is limited by that essential material which is least available.

leaching – The extraction by water of the soluble components of a mass of material. In soil chemistry, leaching refers to the loss of surface nutrients by their percolation downward below the root zone.

legal standing – Having the necessary requirements to bring a case to court.

legume – Any plant of the family *Leguminosae,* such as peas, beans, or alfalfa. Bacteria living on the roots of legumes change atmospheric nitrogen, N_2, to nitrogen-containing salts which can be readily assimilated by most plants.

mach number – A unit of measure used to express speeds that are of the order of magnitude of the speed of sound.

$$\text{mach number} = \frac{\text{speed of object}}{\text{speed of sound}}$$

macroeconomic effect – Economic force on the entire society, such as a change in total production, general price level, employment, or economic growth.

magma – A fluid material lying in the upper layers of the Earth's mantle, consisting of melted rock mixed with various gases such as steam and hydrogen sulfide.

magnetohydrodynamic generator – A type of electrical generator that operates by passing ions through a magnetic field. MHD systems are more efficient than conventional mechanical generators because there are fewer moving parts and hence fewer frictional losses.

mantle – The solid but partly semiplastic portion of the Earth that surrounds the central core and lies under the crustal surface.

mass number – The sum of the number of protons and neutrons in an atomic nucleus.

meteorology – The science of the Earth's atmosphere.

metric system – See *Système International d'Unités.*

MHD generator – See *Magnetohydrodynamic generator.*

microclimate – The local climate of a very small area, such as the underside of a rock, the inside of a rotten log, or the center of a pile of organic debris.

microeconomic effect – Economic effect on an individual family, firm, or industry.

mineral reserves – The estimated supply of ore in the ground.

mist – An airborne substance that consists of liquid droplets typically having diameters greater than about 1 micrometer.

moderator – A medium used in a nuclear reactor to slow down neutrons.

molecule – The fundamental particle that characterizes a compound. It consists of a group of atoms held together by chemical bonds.

mutualism – An interaction beneficial and necessary to both interacting species.

natural selection – A series of events occurring in natural ecosystems which eliminates some members of a population and spares those individuals endowed with certain characteristics that are favorable for reproduction.

natural succession – The sequence of changes through which an ecosystem passes during the course of time.

neutralism – The inconsequential case of very little interaction between two species.

neutron – A fundamental particle of the atom that is electrically neutral.

niche – See *Ecological niche.*

nuclear reactor – A device that utilizes nuclear reactions to produce useful energy.

nucleus – See *Atomic Nucleus.*

omnivore – An organism that eats both plant and animal tissue. Common omnivores include bears, pigs, rats, chickens, and people.

open dump – A site where solid waste is deposited on a land surface with little or no treatment.

ore – A rock mixture that contains enough valuable minerals to be mined profitably with currently available technology.

organochlorides – A class of organic chemicals that contain chlorine bonded within the molecule. Some organochlorides, such as DDT, are effective pesticides. They are generally broad-spectrum, and long-lived in the environment.

organophosphates – A class of organic compounds that contain phosphorus and oxygen bonded within the molecule. Some organophosphates, used as pesticides, are broad-spectrum and extremely poisonous, although they are not long-lasting in the environment.

osmosis – The process by which water passes through a membrane that is impermeable to dissolved matter. The water goes from the less concentrated to the more concentrated solution. See also *Reverse osmosis.*

ossicles – The collection of three tiny bones inside the ear that transmit sound energy from the eardrum to the inner ear.

oxidant – An oxidizing agent. Oxidants in polluted air typically contain O—O chemical linkages.

oxidation – The addition of oxygen to a substance. More generally, oxidation is a loss of electrons.

"oxygenate" – A compound that contains carbon, hydrogen, and oxygen. The term is used more commonly in air-pollution parlance than in pure chemistry.

ozone – Triatomic oxygen, O_3.

parasitism – A special case of predation in which the predator is much smaller than the victim and obtains its nourishment by consuming the tissue or food supply of a larger living organism.

particulate – Consisting of particles.

PCB – See *Polychlorinated biphenyl.*

perpetual motion machine – A machine that will run forever and perform work without the use of an external energy supply. Such a machine is impossible to build.

pheromone – A substance secreted to the environment by an individual and received by a second individual of the same species, in which a specific behavioral reaction is released. Such reactions include alarm, sex attraction, and trailing. When pheromones are discharged artificially into the environment, they lead to inappropriate behavior in the organisms that respond to them.

phon – A unit of subjective loudness that is matched to a decibel scale.

photosynthesis – The process by which chlorophyll-bearing plants use energy from the sun to convert carbon dioxide and water to sugars.

phytoplankton – Any microscopic, or nearly microscopic, free-floating autotrophic plant in a body of water. There are a great many different species which exist in a community of phytoplankton; these plants occur in large numbers and account for most of the primary production in deep bodies of water.

plankton – Any small, free-floating organism living in a body of water. See *Phytoplankton* and *Zooplankton.*

plasma – A gas at such a high temperature or pressure that the electrons have been stripped from their atoms, resulting in a mixture of nuclei surrounded by rapidly moving electrons.

Pleistocene Age – The geologic age encompassing the time span from one million to ten thousand years ago. The ice ages, the emergence of man, and the extinction of many species of mammals all occurred during this time.

pollution – The impairment of the quality of some portion of the environment by the addition of harmful impurities.

pollution tax – A tax on a polluter that is determined by the quantity of pollutants emitted. Also called *residual charge.*

polychlorinated biphenyl (PCB) – A class of organochloride chemicals that are structurally related to DDT. They were used widely in the plastics and electrical industries until they were found to be potent environmental poisons.

population bloom – See *Bloom.*

population density – The size of a population divided by the area in which the members live.

population distribution – The composition of a population categorized by several variables, often age and sex.

population ecology – The branch of ecology dealing with the size, growth, and distribution of populations of organisms.

power – The amount of energy delivered in a given time interval.

prairie – An extensive area of fairly level, predominantly treeless land. Prairies are characterized by an abundance of various types of grasses.

predation – An interaction in which some individuals eat others.

predator – An animal that attacks, kills, and eats other animals. More broadly, an organism that eats other organisms.

pressure – Force per unit area.

price elasticity – The strong influence of price on the utilization of a product or service.

price inelasticity – The insensitivity to price of the utilization of a product or service.

primary consumer – An animal that eats plants.

primary treatment (of sewage) – The first stage of removal of impurities from water, generally by simple physical methods such as screening and settling.

protocooperation – A relationship between two individuals that is favorable to both of them but is not essential for the survival of either.

proton – A fundamental particle of the atom that bears a unit positive charge.

putrefaction – The anaerobic decomposition of proteins.

pyrolysis – The process by which a material is decomposed by heating it in the absence of air, sometimes yielding valuable products.

radical (chemical) – A very reactive type of molecule characterized by one or more unpaired electrons.

radioactivity – The emission of radiation by atomic nuclei.

radioisotope – A radioactive isotope.

rate of natural increase – The difference between the crude birth and crude death rates. Also called the crude reproductive rate.

recycling – The process whereby waste materials are reused for the manufacture of new materials and goods.

reduction – The removal of oxygen from a substance. More generally, reduction is a gain of electrons.

redundancy – Superabundance. In the context of safety systems, redundancy refers to the provision of a series of devices that duplicate each other's functions and that are programmed to go into operation in sequence if a preceding device in the series fails.

rendering – The cooking of animal wastes such as fat, bones, feathers, and blood to yield tallow (used in the manufacture of soap) and high-protein animal feed.

replacement level – The level of the total fertility rate which, if continued unchanged for at least a generation, would result in an eventual population growth of zero.

residual charge – See *Pollution tax.*

reverse osmosis – The application of pressure to direct the flow of water through a membrane from the more concentrated to the less concentrated phase, which is the opposite of the normal osmotic flow.

revulcanization – A process whereby scrap rubber is heated in the presence of sulfur compounds to manufacture recycled rubber products.

rotenone — A pesticide extracted from the roots of certain East Indian plants.

sanitary landfill — A site where solid waste is deposited on a land surface, compacted, and covered with dirt to reduce odors, and prevent disease and fire.

saprophyte — An organism, usually a plant such as a mold or fungus, that consumes the tissue of dead plants or animals.

savanna — A type of tropical or subtropical prairie that is subject to seasonal patterns of rainfall. Savannas are common in central Africa.

sea breeze — A local wind caused by uneven heating of land and ocean surfaces.

Second Law of Thermodynamics — See *Thermodynamics.*

secondary consumer — A predator that eats an animal that eats plants.

secondary treatment (of sewage) — The removal of impurities from water by the digestive action of various small organisms in the presence of air or oxygen.

SI system — See *Système International d'Unités.*

sigmoid curve — A mathematical function that is roughly S-shaped, characterized by an initially slow rate of increase, followed by rapid increase, and followed again by a slow, near zero rate of increase.

sine wave — A smoothly oscillating symmetrical wave form that is produced by a pure tone. Mathematically it is defined as a wave described by the following equation: $y = sine\ x$.

sinusoidal — Having the character of a sine wave.

sludge — Wet residues removed from polluted water. When the sludge is laden with microorganisms that promote rapid decomposition, it is said to be *activated.*

smog — Smoky fog. The word is used loosely to describe visible air pollution.

smog precursor — An organic gaseous air pollutant that can undergo chemical reaction in the presence of sunlight to produce smog.

smoke — An aerosol that is usually produced by combustion or decomposition processes.

solar cell — A semiconductor device that converts sunlight directly into electrical energy.

solar collector — A device designed to concentrate solar energy for a useful purpose.

solar energy — Energy derived from the sun.

solid (crystalline) — A rigid state of matter in which the atoms or molecules are arranged in an orderly pattern.

sonic boom — The sharp disturbance of air pressure caused by the reinforcing waves that trail an object moving at supersonic speed.

sonic speed — The speed of sound.

species — A group of organisms that interbreed with other members of the group, but not with individuals outside the group.

steady state — A condition of a system in which the inflow of materials or energy equals the outflow.

sterilization — A procedure which renders a person incapable of fathering or conceiving a child.

strip mining — Any mining operation that operates by removing the surface layers of soil and rock, thereby exposing the deposits of ore to be removed.

subsonic speed — Less than the speed of sound.

supersonic speed — Greater than the speed of sound.

synergism — A condition in which a whole effect is greater than the sum of its parts.

system — An assemblage of objects united by some form of regular interaction or interdependence; an organic or organized whole.

Système International d'Unités (SI) — Commonly called the metric system. A system of measurement used in all scientific circles and by lay people in most nations of the world. The standard units in the SI system are: length — meter; mass — kilogram; time — second; electric current — ampere; temperature — degree Celsius; luminous intensity — candela; and amount of substance — mole.

taiga — The northern forest of coniferous trees that lies just south of the arctic tundra.

technological fix — An approach to solving a social problem, such as environmental degradation, by technical or engineering methods.

temperature — A measure of the warmth or coldness of an object with reference to some standard. Temperature should not be confused with heat. Heat is the energy that a hotter body transfers to a colder one.

tertiary, or "advanced," treatment (of sewage) — Any of a variety of special methods of water purification, such as adsorption or reverse osmosis, which are more effective than simple physical or biological processes for special pollutants.

thermal pollution — A change in the quality of an environment (usually an aquatic environment) caused by raising its temperature.

thermocline — Middle waters of a lake, where temperature and oxygen content fall off rapidly with depth.

thermodynamics — The science concerned with heat and work and the relationships between them.

First Law of Thermodynamics — Energy cannot be created or destroyed.

Second Law of Thermodynamics — It is impossible to derive mechanical work from any portion of matter by cooling it below the temperature of the coldest surrounding object.

thermonuclear reaction — A nuclear reaction, specifically fusion, initiated by a very high temperature.

tidal energy — Energy derived from the movement of the tides.

total dependency ratio — The ratio of the elderly plus the young to the total number of working-age people.

total fertility rate — The total number of infants a woman can be expected to bear during the course of her life if birth rates remain constant for at least one generation.

transpiration — The controlled evaporation of water vapor from the surface of leaf tissues.

trophic levels — Levels of nourishment. A plant that obtains its energy directly from the sun occupies the first trophic level and is called an autotroph. An organism that consumes the tissue of an autotroph occupies the second trophic level, and an organism which eats the organism that had eaten autotrophs occupies the third trophic level.

tundra – Arctic or mountainous areas that are too cold to support trees and are characterized by low mosses and grasses.

tympanic membrane – See *Eardrum*.

ultrasound – Elastic waves of frequency above the maximum for human hearing (about 15,000 Hz).

urbanization – A demographic process characterized by movement of people from rural to urban settlements, from small towns to large cities, and from large cities to their suburbs.

viable – Capable of living.

vital event – In demography, a birth, a death, a marriage, a termination of marriage, or a migration.

vital rate – The number of vital events occurring in a population during a specified period of time divided by the size of the population.

volcano – A fissure in the Earth's crust through which lava, steam, and other substances are expelled.

water cycle – See *Hydrological cycle*.

water pollution – The deterioration of the quality of water that results from the addition of impurities.

wave – A periodic disturbance in some medium. A wave carries energy from one point to another but there is no net movement of materials. Electromagnetic waves are qualitatively different from all other waves in that they can be propagated in a vacuum.

wavelength – The distance between successive disturbances of the same type in a wave, such as between neighboring crests.

work – The energy expended when something is forced to move.

zooplankton – Microscopic or nearly microscopic free-floating aquatic animals that feed on other forms of plankton. Some zooplankton are larvae of larger animals, while others remain as zooplankton during their entire life cycle.

INDEX

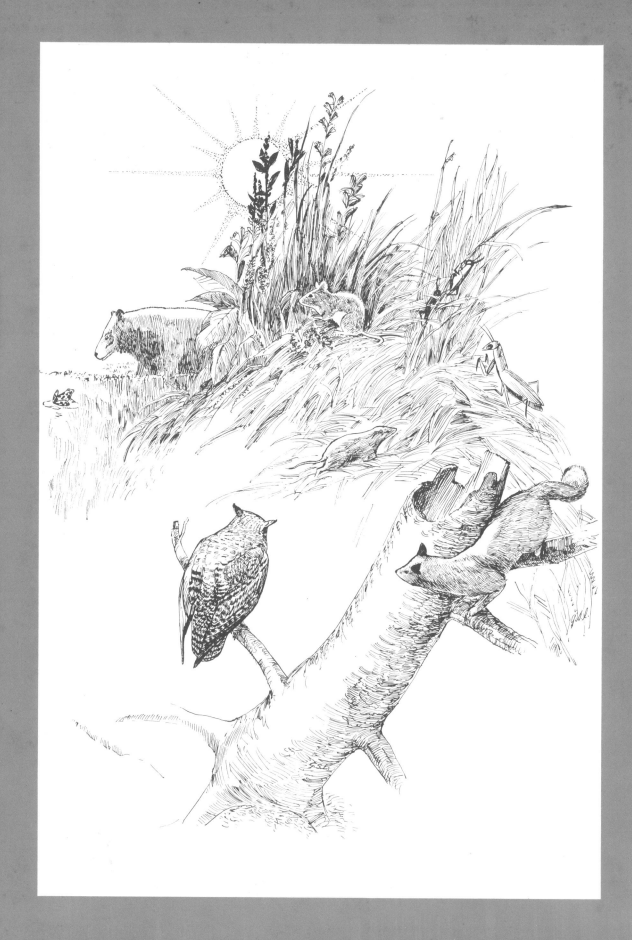